IET CONTROL, ROBOTICS AND SENSORS SERIES 135

Cognitive Sensing Technologies and Applications

The IET International Book Series on Sensors

IET International Book Series on Sensors—Call for Authors

The use of sensors has increased dramatically in all industries. They are fundamental in a wide range of applications from communication to monitoring, remote operation, process control, precision and safety, and robotics and automation. These developments have brought new challenges such as demands for robustness and reliability in networks, security in the communications interface, and close management of energy consumption. This book series covers the research and applications of sensor technologies in the fields of ICTs, security, tracking, detection, monitoring, control and automation, robotics, machine learning, smart technologies, production and manufacturing, photonics, environment, energy, and transport.

Proposals for coherently integrated international multiauthored edited or coauthored handbooks and research monographs will be considered for this book series. Each proposal will be reviewed by the IET Book Series Editorial Board members with additional external reviews from independent reviewers. Please e-mail your book proposal to: vmoliere@theiet.org or author_support@theiet.org.

Cognitive Sensing Technologies and Applications

Edited by
G.R. Sinha, Bidyadhar Subudhi, Chih-Peng Fan and Humaira Nisar

The Institution of Engineering and Technology

Published by The Institution of Engineering and Technology, London, United Kingdom

The Institution of Engineering and Technology is registered as a Charity in England & Wales (no. 211014) and Scotland (no. SC038698).

First published 2023

The Institution of Engineering and Technology
Futures Place
Kings Way, Stevenage
Hertfordshire SG1 2UA, United Kingdom

www.theiet.org

British Library Cataloguing in Publication Data
A catalogue record for this product is available from the British Library

ISBN 978-1-83953-689-2 (hardback)
ISBN 978-1-83953-690-8 (PDF)

Typeset in India by MPS Ltd

Cover image: Waitforlight / Moment via Getty Images

Contents

About the editors

G.R. Sinha is a professor at the International Institute of Information Technology Bangalore (IIITB) India. He has been visiting professor in Taiwan, Italy, and Sri Lanka. His research interests include cognitive science, computer vision, biometrics, medical image processing and ICT tools. He has published 280 research papers, book chapters, and books. He owns two Australian patents and one German patent in his area of research. He is an associate editor of several journals. He has been ACM Distinguished Speaker, IEEE Distinguished Lecturer and CSI Distinguished Speaker. He is a senior member of the IEEE, Fellow of the Institute of Engineers India and Fellow of the IETE (India). He is also a member of several national professional bodies such as ISTE, CSI, ISCA, and IEI. He received his PhD degree in electronics & telecommunication engineering from Chhattisgarh Swami Vivekanand Technical University (CSVTU) Bhilai, India.

Bidyadhar Subudhi is a professor at the School of Electrical Sciences and Dean (R&D) of the Indian Institute of Technology (IIT), Goa, India. His research interests include machine learning and adaptive systems, biomedical engineering, marine robotics and microgrid systems. He has published his research in over 150 reputed international journals and 70 conference papers and has edited 3 books. He has been appointed as a distinguished speaker by ACM (2020–2023). He was elected as fellow of the Indian National Academy of Engineering for his distinguished contribution in Electrical Engineering. He serves as a technical committee member of the IEEE Intelligent Control Group, and an associate editor for *IEEE Transactions on Sustainable Energy* and *IEEE Access*, and he is on the editorial board of the IEEE Technology Conference. He is a senior member of the IEEE, a fellow of the IET (UK), the Institution of Engineers (India) and the Institution of Electronics & Telecommunication Engineers (India). He received his PhD degree in Control System Engineering from the University of Sheffield, UK.

Chih-Peng Fan is a full professor in the Department of Electrical Engineering at the National Chung Hsing University, Taiwan. His research interests include deep-learning for digital image processing and pattern recognition, digital video coding and processing, digital baseband transceiver design, VLSI design for digital signal processing, and fast prototype of DSP systems with FPGA-based and embedded SOC platforms. He has over 100 publications, including technical journals, technical reports, book chapters, and conference papers. He is an IEEE CTSoc Representative at the IEEE Systems Council's AdCom (2020–2021); member of

IEEE CTSoc Awards Committee, secretary of the IEEE CTSoc Sensors and Actuators Technical Committee. He is also a member of several journal editorial boards. He is a supervisor of the Taiwan Consumer Electronics Society, a member of the IEEE, the IEICE, and the Taiwan IC Design Society (TICD). He received his PhD degree in Electrical Engineering from the National Cheng Kung University, Taiwan, ROC.

Humaira Nisar is a full professor and head of the Master of Engineering (Electronic Systems) Programme in the Department of Electronic Engineering at the Universiti Tunku Abdul Rahman, Malaysia. Her research interests include signal and image processing, image analysis for wastewater treatment, biomedical imaging, neuro-signal processing and analysis, brain computer interface, and neurofeedback. She has published over 150 international journals, conference papers and book chapters. She has also edited three books. She is an academic editor for IEEE Access and PLOS. She is a senior member of IEEE and a member of IEEE Signal Processing Society, Malaysia Section. She has also served as an auditor for IEEE EMBS Society, Malaysia Section. She is serving as professional technologist under the Malaysia Board of Technologists, professional engineer under the Pakistan Engineering Council and Graduate Member of the Board of Engineers Malaysia. She holds a PhD degree in Information and Mechatronics from Gwangju Institute of Science and Technology, South Korea.

Preface

Scientific community and research fraternity, and even common people using advanced technologies, modern gadgets and accessories, now talk about ChatGPT, Metaverse, edge computing, Web3.0, Industry 4.0, and so on. The success of all these modern and merging applications of technologies and artificial intelligence (AI) lies in most important component called as sensors. The sensing technologies are becoming smarter and smarter with the advent of new communication and computational technologies. The abilities of sensors are therefore getting extended to beyond all the boundaries. This is why researchers have been implementing metaverse and digital avatar concepts successfully. The essence of data interpretation intelligently depends on AI/machine learning (ML) techniques and also the sensing mechanisms. Talking about digital avatars, exact replicas or virtual representations of persons or objects could be created that shall be useful in marketing, sales, advertisement, and of course most interactive experience of customers. Avatars and their realistic thoughts are incomplete without sensors and smart sensing methods because sensing only brings all details from the customer or an enterprise through avatars and ultimately benefits the end-users with interactive and most exciting experience with immense satisfaction about the products. Interactive experience in any field comes from using augmented reality (AR) and virtual reality (VR) and the reality, rather extended reality is experiences in more impressive manner due to brilliant roles of sensors employed to capture the data and use of suitable AI methods.

Digital twins are other examples where we can create digital version of an object, a system, or even a human being. The monitoring of a system, improvement of a system and better operation of the system can be achieved. On a remote system, we can assess the performance of either a system or an object of the system where twin or virtual representation of a system is present. The system or even the components and processes could be converted as digital twins and using the 3D representations of those twins, handling the errors or faults with any processes or any component of the system can be efficiently done. So, in any application related to digital twins or digital avatars, Metaverse, Web 3.0, or future Internet, the cognitive sensing plays an important role. The smart sensing gives interactive and immersive experience to the users. The users can go to the web pages and get immersive experience intelligently as per their customized needs. Also, users can get the service from the web smartly and this can only happen if the sensing technology used is smart and cognitive. The abilities of sensing technologies shall be of course cognitive, having all cognitive abilities in order to better interpret the data, better classify the data and provide appropriate decision making. Driverless car is another area where a number of sensors due to their intelligence and cognitive abilities guide the vehicles or cars without assistance of any drivers. So, the role of cognitive sensing is very useful in all modern applications of visualization, automation, monitoring, and navigation.

Cognitive sensing finds numerous applications in a variety of fields such as computer vision, natural language processing (NLP), AI-enabled gadgets, driverless cars, smart environment monitoring, smart agriculture, smart city, and healthcare. Robotics and remote sensing are also exciting applications of cognitive sensing where weather related information can be brought more appropriately with the use of remotely sensed data and the area of robotics brings the transformation in the field of industries, healthcare, and smart manufacturing. The robots are used in different parts of the manufacturing to obtain more efficient production and the quality assessment. In assistive technologies and brain computer interface (BCI), the robotics and similar devices play significant role, whose cognitive abilities are extended day by day due to the smart sensing methods and modern sensors. So, the future of cognitive sensing opens lot of exciting avenues through interactive experience in almost all domains of science, engineering, mathematics, and technologies for various applications in years to come.

Due to most emerging applications of cognitive sensing, limitless capabilities of sensing in unlimited number of most exciting and modern applications, this becomes imminent to understand the theory, fundamentals, concepts and essential reviews on cognitive sensing methods, their working principles, implementation strategies, role of AI and ML and their applications in several areas such as agriculture, transport, healthcare, navigation, manufacturing, self-driving car, automobile, automation, control, assistive devices, medical Internet of things, industrial Internet of Things and so on. A need of a literature that can highlight on all the fundamental aspects, theoretical backgrounds and their potential use in the number of application areas of cognitive sensing technologies is felt very much and to meet this need, this book on cognitive sensing technology and applications presents an overview, necessary theory, conceptual background, possible implementation strategy, critical review, advantages and disadvantages, and their potential applications.

The book has been edited after compiling 18 different chapters from authors from various geographical areas working the areas of cognitive sensing. The chapters focus on various aspects of cognitive sensing and use of the sensing technologies for various applications. The chapter-wise description of all the chapters is briefly given here.

Chapter 1 presents an introduction to the book and highlights on almost all aspects of cognitive sensing covered in the book with introductory note. This covers cognitive sensors, sparse sensing, cognitive IoTs, wireless network, AI and ML, for several applications such as smart city, smart environment monitoring, rehabilitation, robotics, and smart agriculture. The benefits of potential use of smart and cognitive sensing are also included with some important drawback and limitations in using the sensors and their cognitive abilities. In Chapter 2, authors have presented about hardware architectures, some sparse signal recovery approaches. Sparse sensing is an emerging area of cognitive sensing where a number of signal processing methods are implemented using suitable architecture and therefore the compressive sensing and signal processing is built on the hardware. This is presented using compressive sensing, and the sparse signal recovery is explained with a suitable mathematical fundamentals and backgrounds using a number of approaches. The compressive sensing has been explained that uses analog hardware architecture for gradient-based reconstruction, and approach has been presented in very detail with. The proposed architecture scheme, oscillator outputs for the hardware that how the different parts respond to the signal processing of compressive sensing are discussed. This chapter forms a very important basis for hardware needed for implementing cognitive sensing and its simulation. The architecture is explained with the help of algorithm as well, which is used using suitable simulation for the architecture. The chapter looks research centric but in fact the simulation results substantiate the fundamentals which are presented in the chapter and therefore some number of research-based results and their discussions are present to establish the simulation-based validation of the hardware, explained in the chapter on a smart sensing. Chapter 3 discusses performance evaluation of cognitive sensor framework for IoT applications in healthcare and environment monitoring because the wireless sensor network which is implemented in order to deploy the sensors for agriculture and environment monitoring applications or any other applications, the performance assessment is very important. So the importance, utility, historical perspective, evolution of cognitive sensors is presented specially with a focus on weather monitoring, environment monitoring and how exactly the performance evaluation framework is developed is explained in this chapter. Wireless sensor data plays an important role in aircraft, satellite radar, and the remote sensing applications and also in environment monitoring. For that, the performance of sensor using a number of performance metrics is presented in this chapter that can monitor different attributes either of environment, air quality, radiation, soil quality depending on the application covered. Some discussion on performance framework of cognitive sensing in healthcare is also covered.

Chapter 4 discusses about cognitive sensors and technologies for rehabilitation and therapeutic applications. In Chapter 5, ensemble machine learning based intelligent system for human activity recognition is explained that analyzes the sensory data and interprets intelligently with the help of ensemble ML technique for human activity recognition. The area of human activity recognition is significant in several applications of surveillance and security-related applications. The role of cognitive sensing is smart understanding of sensory data. This chapter also has presented some results in order to indicate how the cognitive system sensor response to the sensory data and with the use of ensemble ML interprets the data intelligently which helps in classifying the data for human activity recognition. Chapter 6 presents an approach of handling challenges in sensing biomedical data and for which some approaches of de-noising are presented in addition to different ways of acquiring brain signals using sensing techniques. In Chapter 7, cognitive workload assessment is discussed with different ways of sensing the brain EEG signals using different types of brain signal acquisition and sensing device. Chapter 8 presents role of sensing and its application in detection of Parkinson's disease by applying non-linear signal decomposition and machine learning techniques. Chapter 9 introduces acquiring a gait kinematics using sensors and IoTs and then analysis of gait kinematics using machine learning techniques. A number of wearable sensors are used to capture kinematic gaits and their kinematic analysis. The kinematic gait analysis has been presented that includes direct and inverse kinematics under laboratory systems and mathematical modeling and machine learning under wearable systems. The necessary review, search methodology and comparative analysis between laboratory system and variable system has been presented with examples.

Chapter 10 discusses about cognitive sensors and cognitive IoTs used for industrial and biomedical signal processing applications. Chapter 11 presents intelligent automation using IoTs and ML methods. The IoTs that employ AI and smart sensors has been used for the automation purpose and the fundamental background, historical concept, classification and ML models used, physical design of IoTs, IoT protocols, different layers in protocol, and IoTs technologies for smart sensing used for automation tasks has been discussed. In the chapter, emphasis on smart street lighting, smart parking, smart road surveillance, structural health monitoring, emergency response, energy response, environment, and fire detection-related applications have been given. Chapter 12 presents recent trends in cognitive science replications for smart manufacturing and control. This chapter gives an overview of instrumentation and control system used in manufacturing and control, utilizing the sensory data and different types of industrial and temperature sensors. Other sensors such as force sensors, flow sensors, pressure sensors, and strain sensors have been also discussed, along with how the sensory data is captured and interpreted intelligently with the help of ML models in the process of control applications. Cognitive sensing requirement for IoT sensors and their design considerations is also discussed in addition to energy economy and harvesting required for sensor operation. The chapter also highlights potential benefits of IoTs based on manufacturing and control applications of cognitive sensors used for smart manufacturing applications. Chapter 13 presents a study of cognitive sensors used in robotics, UAVs, and drones. Chapter 14 suggests sensory data fusion and processing the smart agriculture application which involves crop quality assessment, crop damage assessment, and a smart planning used in the smart agriculture area. The different types of sensors which are used in a smart agriculture in the assessment of crop quality using different IoTs for disease detection based on non-imaging spectroscopy and analyzing the hyperspectral data has been discussed. In smart agriculture, detection of disease in a different crop is very important. The data fusion technique has been explained that operates on different types of sensory data which are used in the area of agriculture, and based on that it suggests the remedial measures, corrective steps in order to have the sustainable agriculture.

Chapter 15 highlights the value of incremental class learning as a significant advancement in the field of agricultural plants that can pave the way for smart agriculture. Chapter 16 presents disastrous susceptibility analysis in remote sensing application using sensory data with the special highlight on landslides, floods and how the sensing helps in this application, explained. The case study related to landslides in Malaysia has been explained with the data collection approach and preprocessing used in a spatial database, and also susceptibility analysis using artificial neural network. The evaluation and assessment are explained in the analysis of floods in Ganga River in India as well. The work is on susceptibility analysis and therefore related to preprocessing and susceptibility analysis, the appropriate mathematical modeling, and to some extent some results are also included so that the susceptibility analysis is justified.

In Chapter 17, various forms of spatial networks that are practiced in the real-world scenarios elucidate on the road transportation network structures and their properties. Several measures used for quantifying the transportation network performance such as connectivity and coverage using graph theory are deliberated. Finally, future research direction and application is discussed in Chapter 18.

The chapters of the book are structured and presented in such a way that anyone without much knowledge of cognitive sensing technologies can have understanding of the content and applications presented there. The book shall be useful for readers of multiple disciplines studying at various levels of undergraduate, Master and doctoral studies, in addition to research scholars and practicing scientists working in the field of smart sensing techniques.

Chapter 1

Introduction to cognitive sensing technologies and applications

G.R. Sinha[1], *Bidyadhar Subudhi*[2], *Chih-Peng Fan*[3] *and Humaira Nisar*[4]

Abstract

This chapter introduces cognitive sensing technologies and related terminologies with their adequate introduction, significance and applications. Cognitive sensing is incomplete without smart sensors and Artificial Intelligence (AI) and thus the role of AI is immense in demonstrating the capabilities of modern sensors in numerous applications such as driverless cars and humanoid robots. Sensing technologies and intelligent data processing make the applications more accurate and meaningful, in other words, cognitive sensing technology has a significant role to play in making automated and smart systems; realizable and functional. The overview of cognitive sensing covers cognitive sensors; cognitive Internet of Things (CoIT); issues and challenges in cognitive sensing and IoTs; and emerging applications of cognitive sensing technologies.

Keywords: Cognitive sensing; Internet of Things (IoT); Intelligent sensing; Smart sensing; Sensors; Machine learning (ML); Artificial Intelligence (AI)

1.1 Cognitive sensing and sensors

With the advent of enhanced communication capability, computing capabilities, the scope of sensors and the sensing technologies have been tremendously growing for the last few years. In the current decade, smart sensors and intelligent computing technology of cognitive sensing is instrumental in the transformation and advancement in the manufacturing and healthcare sector. Applications such as robotics and automation, smart manufacturing, smart environment monitoring

[1]International Institute of Information Technology Bangalore (IIITB), India
[2]Department of Electrical Engineering, Indian Institute of Technology Goa, India
[3]Department of Electrical Engineering, National Chung Hsing University, Taiwan
[4]Department of Electronic Engineering, Universiti Tunku Abdul Rahman, Malaysia

systems, smart agriculture, smart medicine, smart navigation, and driverless car are becoming a reality. In all these sectors, communication technologies, sensor technologies, computing capabilities, as well as Artificial Intelligence (AI) and machine learning (ML) techniques are playing a significant role in the development of Internet of Things (IoT) for deploying the sensors in numerous applications. Sensing technology increases, enhances supply chains, IoT communication, and data analytics which are further dependent on the communication technologies.

The role of cloud-based AI solutions has helped in spreading the impact of sensors and their usage in many applications. However, in such applications, the data dimensionality, latency time, and dependence on cloud are some of the major challenges in the implementation. The researchers are focusing on the small data, rather than connecting all the sensors and using AI and ML at different edge devices. Sensors are deployed to acquire the real-time data and at different edge devices. Hence, smaller microcontrollers or similar devices are utilized to interpret the data captured by the sensor rather than depending on the cloud every time thus enhancing the sensing capability. For example, in healthcare, IoT can transform the healthcare sector by reaching out to the public for mass screening. In this way, the health-related data for several patients can be analyzed and diagnosed by a physician or a radiologist anywhere in the world. This chapter thus presents an introduction of cognitive sensors and cognitive sensing technologies for smart sensing.

The role of AI in cognitive sensing is very important due to several factors such as application, data quality, data type, and sensors used. There are many applications of cognitive sensing [1–5], for example autonomous vehicles [6] and humanoid robots [7]. Cognitive technical systems have strong cognition abilities such as learning, reasoning, thinking, perception, planning, organizing, and decision making. The AI-governed cognitive sensing system is not only the smart but it also simplifies processes, and decreases the computation load resulting in real-time automated, smartly managed, smartly organized processes and tasks.

The areas such as neuroscience [8], neurocognitive psychology, cognitive science, robotics, and cognitive engineering are deeply impacted by cognitive sensing. Smart sensor networks and robotics in healthcare utilize the perception and cognition abilities and assist in automated systems with excellent decision-making capabilities, such systems practice motor control, have controlled perception, knowledge representation, learning capabilities, reasoning and planning and are actually the expert systems and these expert systems are widely used in a smart infrastructure, smart manufacturing, smart agriculture, and smart health care where the cognitive abilities are of immense importance. The cognitive sensing technologies need to have the role of AI in exploiting the cognitive abilities of human brain so that the execution of the task can be optimally performed with the minimum error and the maximum throughput. The abilities of reasoning and planning in any smart system are enhanced with the help of cognitive sensing. So, the cognitive technical systems are being developed in realizing extremely high-performance driverless car and autonomous robotic systems for various modern applications having most appropriate, most accurate prediction and learning capabilities to adapt to the real-time needs for the real-time applications.

Hence, the cognitive process in a cognitive network can perceive the current network conditions, and plan, decide, and act on those conditions. Hence, network can learn and adapt and use these adaptations to make decisions in the future, all while considering the end-to-end goals.

1.1.1 Cognitive nodes and system

In implementing the cognitive sensing technologies, it is important to place the different sensors at different nodes, as these cognitive nodes can understand the cognitive data and can interpret the sensory data, and hence can represent knowledge [9] and has the ability for reasoning and learning. The node may contain some type of microcontroller [10] or similar computing device with ML method deployed at the node in order to provide the reasoning or to exhibit the reasoning ability just by using the current knowledge database. The learning is used for long-term planning and always involves feedback from several observations. So, the node can be considered intelligent and can act as a cognitive node, it is affected by several factors like knowledge base as it takes input from the past actions.

1.2 Cognitive IoTs (CoIT)

IoT, also known as IoTs, is a powerful technology and IoT devices are composed of a combination of AI and the sensor technologies. IoT can serve in a number of sectors such as healthcare. Many tools and devices were seen during the COVID-19 pandemic in the health care, such as robotic systems or AI controlled devices were deployed for the patient care during the pandemic time. The ability to capture the data intelligently and its smart interpretation makes IoTs extremely powerful for health care and smart manufacturing, etc. Under extreme and difficult conditions, robotic-based IoTs or AI-controlled robots can be deployed to monitor the growth or certain parameters in the production in the industries.

There is a very interesting area known as brain–computer interface (BCI) [11,12] that uses the concept of cognitive science in which the brain signals are captured with the help of sensors and using appropriate biosignal processing, brain signals are interpreted and translated into suitable commands which are used to operate either a wheelchair or robots, robotic arms [13] and also can be used in assistive technologies, either in the area of healthcare or similar applications. The brain signals which are captured with the help of different types of sensors are not only understood and translated into commands but also used to generate an appropriate feedback signal that can assist in a number of rehabilitation schemes through which the persons with brain disabilities and disorders [12] are diagnosed. The feedback signal is used to correct or reverse the abnormality present in the brain of a person suffering with some disability and disorder such as Alzheimer's disease, Parkinson's disease, brain stroke, and paralysis. The IoTs are playing a significant role in these areas that we can see nowadays; a portable brain capture device, portable medical IoTs (MIoT) that are portable and transportable, can be used to record the biomedical signals or other medical parameters in real time. An appropriate monitoring, analysis, and diagnosis can be provided.

CoIT is another emerging area where cognitive sensing finds its relevance. Cognitive IoT was actually coined first time in 1999 [14–17]. As the IoT can think, learn, and understand similarly, CoIT has the ability of automatic selection of most meaningful objects in the given application where IoTs are deployed. The basic difference between cognitive IoTs and cognitive radio network (CRN) [2,16,18,19] is that the CRN is actually a paradigm to effectively utilize radio spectrum very similar to CoIT which can choose most useful data from multiple cameras or IoTs. In CRN, there are important layers in which the network management is done effectively. These layers cover physical world, cyber world, and social world. At the top, service evaluation layer is present, after that the decision-making layer helps in selecting the network and in spectrum planning and implementing that is directly connected with service evaluation layer. This is related with the evaluation of performance and provision of services like everything as a service, quality of service (QoS), quality of Internet, and efficiency. Sometimes, this decision making, service provisioning, and evaluation of performance that cover service evaluation layer and decision-making layer is turned into social network. The quality of service given to the human beings is taken care of in these two layers. After decision making layer going further down, there is a layer called as data semantic knowledge layer. Sensing control layer is related to cyber world. At the bottom, there is a physical world. We include several features related to sensing perception, action, data analytics, semantic derivation, knowledge representation, association, and learning by understanding. All these features are covered in the data semantic knowledge layer and sensing control layer which contribute to the cyber world and are located at the bottom of cognitive IoTs physical environment which includes objects, things, or the resources. Hence, in the framework of cognitive IoT, the massive data analysis can take place which can deal with massive data. The noisy, heterogeneous, highly dimensional and the nonlinear data can be analyzed very efficiently with the help of a CRN. Combining CRN and IoT is known as cognitive IoTs, hence it processes high-dimensional data. Parallel and distributed processing can help in dealing with the high dimension. There are some metrics such as joint matrix and matrix recovery which are suggested as metrics for dealing with the challenges in the cognitive IoTs. Knowledge discovery is a very important factor in CoIT that includes semantic derivation and knowledge discovery. Semantic derivation includes context, standardization, and ontology whereas knowledge representation includes association clustering analysis and outlier analysis. Hence, the CoIT is an intelligent network combined with IoT so that the task and the applications that are performed over a network or over a smart and intelligent network can be performed intelligently with the maximum quality that impacts the society and the world. Hence, in this way, the implementation of CoIT is carried out at the large scale. The cost and computational efficiency is reduced, and energy efficiency or storage efficiency are improved.

The cognitive sensing method also has a few issues and challenges such as interoperability and heterogeneous networks for which the different standards are required. Major issues are listed below:

- Interoperability and standards [16]
- Heterogenous networks and data [20]

- Machine to machine standardization is required with the help of specific methods such as self-organizing networks [21]
- Open framework for cognitive IoTs [7,15,22]
- Fault tolerance [23]
- Adaptable interoperability
- Distributed energy efficient data processing is needed
- Security and routing-related challenges [24]
- Cognitive computing and resource management [25]
- Smart connectivity-related issues
- Data perception and scalability [26]

While implementing CoITs and its deployment for certain applications, above issues need serious attention for efficient and seamless data sensing and analysis.

1.2.1 Sparse sensing

In the implementation of CRN, the spectrum sensing is a very important factor. The number of sensing techniques and technologies involves high processing time, hardware cost, and computational complexity. The selection of suitable algorithm makes the sensing in the spectrum application and cognitive radio application as a compressive sensing. So the performance of sensing technique is improved by using a sparse recovery algorithm and the sparse representation is measured and then using suitable algorithms such as Bayesian compressive sensing greedy algorithm, convex and relaxation algorithm that are used to compress the sensing data to get improved analysis in terms of computational complexity, the hardware cost, and processing time. The sparse representation of any sensed data can be very easily computed in terms of wavelet transform and discrete cosine transform. Coefficients and these transforms can be converted into sparse matrices and using the suitable sparse recovery algorithm, we can reduce the complexity of the algorithm. There are several algorithms in the literature for sparse sensing that reduces the sensing technique load resulting in compressive sensing. This is an interesting application of sensing where the sense sensory data is distributed, and the analysis and the data captured is compressed so that the computational efficiency is improved. There are many models where sparse sensing is used, such as universal software-defined radio peripheral and applications where the sparse recovery is used and the sensing technique [5,27–29] is made as the compressive sensing technique.

1.3 Challenges in smart sensing

A number of challenging factors [20,30–32] are there in the area of implementation of a smart sensor for healthcare applications. For example, lack of standardization and interoperability. There are several tools that are available and also a number of data formats with which we need to work. This poses a big challenge as using the sensors is intermittent and sometimes as there is no network connection. We are talking about 5G capability, computing capability, and communication capability and when we deploy the smart sensors over a network using edge computing and other devices, if the network connection is limited, then that really poses the challenge in the effective implementation.

Another factor is the heterogeneity and fragmentation of data in digital healthcare systems. It depends on the number of computation models, data availability and optimization techniques. The data type must be homogeneous in nature, then it requires a different type of analysis and modeling methods. Also, the validity of sensor data, the data which is received which is captured or required by the sensors, whether the data is reliable and whether there is a lack of validity. The scalability of users is another important factor. The next bigger challenge is about health awareness and the difficulty involved in acquiring the data because the user's participation is needed and despite motivation and encouragement, there is always an apprehension in providing for the medical data. Clinical challenges and health hazards are also there which are imposed by nano sensors and biosensors which are to be deployed for the healthcare application. Privacy, confidentiality, and the regulatory frameworks are issues which do not allow to work on several factors leading to the medical-based research deploying the sensors and IoTs.

Some important factors related to cognitive sensing technologies are to be considered. Most importantly, the hardware that includes communication and signal processing hardware, interfaces, and storage are the major challenges where the cognitive sensing can face difficulties if not implemented and interpreted intelligently. The robustness and reliability of the hardware is another factor. For example, if any component or any part of the hardware fails to respond in the designed way then the whole system fails, and reliability will be at stake. Lastly, there are several components that are working in tandem where signal processing is trying to integrate, so signal integrity and the analysis of unwanted signal are an important challenge in the area of cognitive sensing.

One more important factor that is seen as a challenge in the cognitive wireless sensor networks is the security. If the network is not secure, then the intrusion is possible and the whole purpose of using the cognitive sensing for a real-time application will be at stake. The security must be considered as a very serious factor and suitable protocols and the techniques need to be incorporated into the system to handle different attacks and intrusions related to communication attacks because there may be different types of attacks that can pose challenges in the security like communication-related attacks, power consumption-related attacks, or policy-related attacks or cryptographic attacks. These are to be taken very seriously and suitable approaches and methods that need to be used especially in medical IoT when deploying the cognitive sensing technology, cognitive sensing method, the security and privacy are very important factors. All the challenges that are there in cognitive sensing need to be addressed. There are some more factors that are very much important in the areas of medical IoTs where the cognitive sensing or spectrum sensing is used related to the interference by the number of signals and the number of other nearby sources.

1.4 Applications of cognitive sensing technologies

Due to the emergence of smart sensors, intelligent computing, and ML techniques, the sensing technology has been improving tremendously and therefore the application of sensing and interpretation of sensed data is found in numerous modern and exciting applications such as health care [26,33–36], agriculture, precision agriculture [37–40], smart manufacturing [32], driverless car, robotics in manufacturing, environment monitoring, smart city [35], smart education, remote sensing applications [1,8,41], and so on. The abilities of the intelligent computing and AI, and the advancement in the sensor technology play a very important role in the efficacy of the applications which are extremely useful for the society obviously with the consideration of ethical and data privacy issues. In this section, a few very important and emerging applications are discussed.

1.4.1 Healthcare

There is a transition from healthcare 4.0 to 5.0 [26] due to the emergence of smart sensors and the rollout of ML and AI techniques. The optimization methods are exploited to realize the sensors at miniature level that helps in designing the modern compact sensors which can be easily deployed at edges, using edge computing methods. In healthcare, the IoTs are brilliant products utilizing the capabilities of sensor technologies, miniature techniques and, of course, the importance of AI and ML methods have tremendous capabilities to make use of smart sensors for telemedicine and telehealth care. The patient care be incorporated using these smart sensors, capturing the medical data and providing the diagnosis and analysis even for the faraway places.

There are different types of sensors that are used in healthcare depending on what health-related parameters we intend to observe. For example, to observe heart rate, the variation in heart rate or ECG signal can be acquired with the help of skin electrodes. Similarly, to observe breathing rate and the respiratory variation, we generally observe the respiration with the help of a piezoelectric sensor. The cardiovascular system or any hypertension assessment blood tracer is recorded with the help of a pressure sensor and muscle activity is recorded in the form of electromyography with the help of skin electrodes. To understand and study brain disorder and disability, we record EEG and other signals with the help of biosensors that are deployed over the head with the help of cap-like structured electrodes. These biosensors and nano biosensors are the areas of ongoing research for bringing these nano sensors in the form of IoT in the monitoring of glucose, cholesterol, and sodium levels in the body to assess various body parameters and health parameters. This is due to the fact that the power of sensor technology and IoT is being referred as Internet of Medical Things (IoMT) [19] allowing miniature sensors and intelligent data processing itself. Figure 1.1 shows how exactly cognitive sensing works in healthcare sector. Various types of sensors are used to capture biomedical and medical modalities and subjected to suitable AI/ML methods for either of automating the tasks, treatment planning, diagnosis, reporting, drug research, and also in assistive clinical decision making.

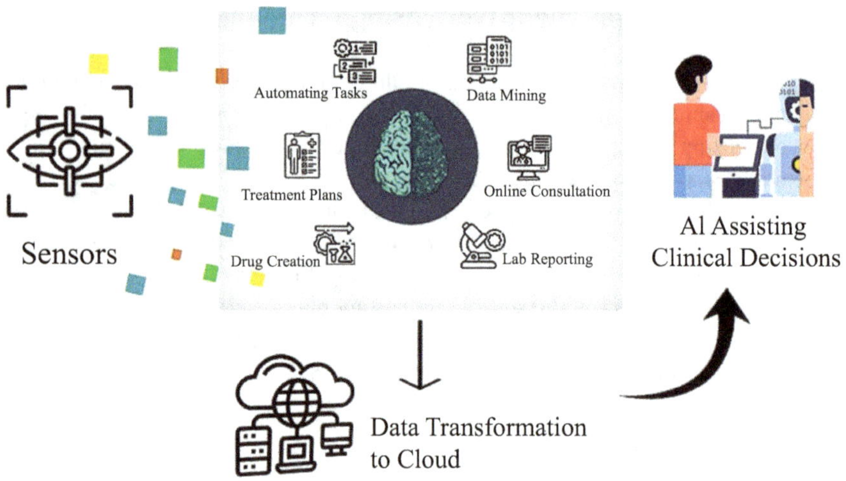

Figure 1.1 Cognitive sensing in healthcare

The role of smart sensors in healthcare will involve several major disciplines and subject areas such as big data, robotics, cloud computing, computing capability, assistive technology, drone technology, and assistive care management. As we are currently in the industry 4.0 age and the healthcare sector is moving very fast towards health care 5.0, the performance of healthcare will be great in providing the health care service to a large number of masses with better outreach and improved diagnosis system. The cognitive sensing will allow the sensory data that can be captured from any part of the world and is interpreted intelligently that can be analyzed by a physician or a radiologist or a doctor sitting anywhere across the world. The diagnosis, medical imaging and its application will have a very good impact on the cognitive sensing method.

1.4.2 Smart grid and environment monitoring

Sustainable environment monitoring uses sensors to detect the contaminants and the pollutants. In addition to it, it is important to check whether smart greenhouse environmental control, smart greenhouse monitoring, wireless sensor monitoring, and smart farming is implemented successfully. The main types of sensors that are used in environment monitoring are trace metal sensors available in the form of nano electrode array, miniature chemical flow probe sensor, and radio isotropic sensor. This is used to record the gamma detection. There are biological sensors that are used in addition to these specific sensors. There are general purpose sensors which need to be brought into miniature level using microarray and other technologies and implementing some AI- and ML-based methods with the help of microcontroller-based solution. The most popular sensors that are used in environmental monitoring are moisture sensor which is very important to assess the moisture level. Then comes pressure sensor and transducers, which are used in aviation sector manufacturing, biomedical measurements, automobile sectors where the pressure sensor plays a significant role. The rain sensors are used to assess the presence of rainfall and the planning and monitoring of conservation of water and designing and monitoring of irrigation system. The vibration sensors are very important for acceleration and displacement in the applications where linear velocity and acceleration are assessed in order to have the sustainable environment. Sensor network also needs to be a very smart and wireless sensor network that must have been the architecture connecting all the sensor nodes, capturing the data intelligently and subjecting that data to the computing system where suitable ML methods and the classification method can be applied to classify the sensory data. For water contamination, air pollution, sound pollution, chemical pollution, radiation detection, there are several applications where we need to implement different types of sensors and use ML methods in order to understand and interpret the sensory data and based on the interpretation, it must be able to generate the real-time alert system so that wherever there is an abnormality reported, the remedial measure can be taken, and the environmental parameters can be controlled and the monitoring could be achieved intelligently. The sensors that are used for the assessment of temperature, humidity, air quality, atmospheric parameters, wind speed and direction, rainfall, radiation, gas whether there is any diffusion of gas, water level, pressure, water quality, etc. are used over a wireless sensor network with cognitive ability. Several applications such as air pollution monitoring, forest fire detection, and landslide detection do not only depend on sensors but also on sensing technologies, and the networks that support the sensing technology and sensors, in the implementation of smart environment monitoring system.

The successful implementation of smart grid and its deployment to monitor the power distribution and generation in smart grids are of huge significance where cognitive wireless sensor network plays a very important role through which we can monitor the power generation and the distribution very properly through remote control. So, remote monitoring of electricity transmission, the distribution network can be implemented remotely, the consumer facilities, the satisfaction of consumer, and the consumption of energy smartly can also be assessed. The cognitive sensing can also be deployed for spectrum sensing where the spectrum-based decision and the spectrum-based mobility of the power distribution and the smart grid can also be achieved. CRN [2,16,18,19,42] are used for smart grid application and the cognitive grid networks utilize the smart sensing technique where the sensing along with intelligent interpretation is an inherent part of cognitive sensing that makes the cognitive sensor network very powerful in order to efficiently deploy and implement the smart grid and related applications. In fact, the CRNs are now seen as a promising technology to overcome the spectrum inefficiency and implementing in smart grid applications smartly. The IoTs and other modern tools are discussed to assess the smart grid and its efficacy can be monitored from remote place. Remote monitoring of smart grid [16,21,42] power generation, power distribution, transmission, etc. can be achieved with the help of smart sensing technology. In Figure 1.2, we can see how cognitive sensing plays an important role in smart grid application in effective and optical management and distribution of power.

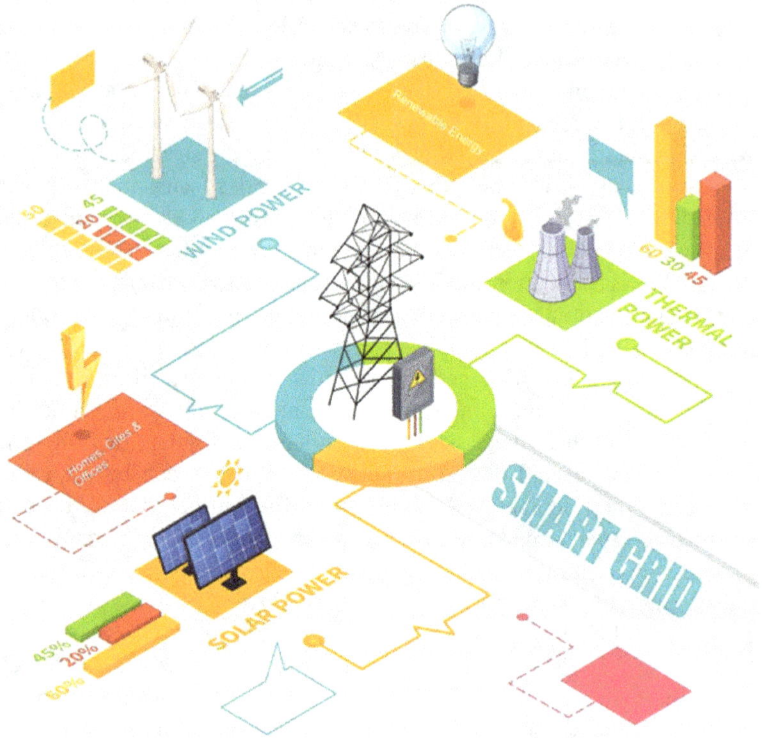

Figure 1.2 Smart grid application of cognitive sensing

1.4.3 Smart city

Smart city, its proper planning, monitoring of smart city parameters, and designing a smart city [35,43–47] with world class smart city specifications include a smart transport system, smart home, smart environment, smart traffic system, smart parking system, smart healthcare, smart education, and smart learning. The number of beneficiary schemes, smart business and smart customer interface are implemented successfully using cognitive sensing that plays a significant role as the data may be from any kind of sensors related to voice, image, video, or any other data. Wherever the sensing is involved, the data will be captured, interpreted intelligently, with the help of suitable ML and AI techniques, the smart city implementation will provide a holistic way of livelihood to the people, as envisaged in sustainable development goals of the United Nations. Other sustainable development goals that include addressing pollution, climate change, quality education, quality health care, and all these have one small component where all the automated and smart systems would require cognitive sensing. This is needed in

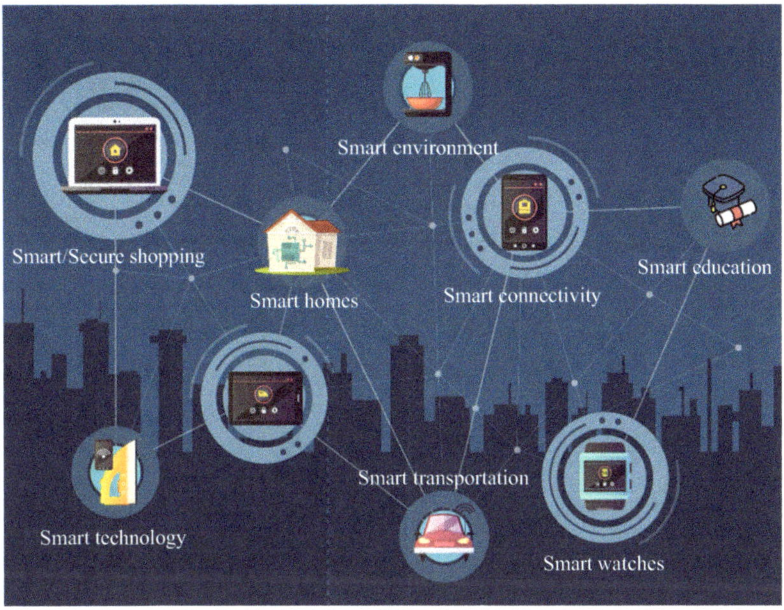

Figure 1.3 Smart city environment

improving the efficacy of the system, be it a smart city or any other smart system. Smart environment monitoring, smart transport, smart parking, and other smart systems are only feasible when we have smart sensing and understanding system embedded with the smart city plan and design. Figure 1.3 depicts a smart city environment where wireless sensor network acts as backbone of data sensing, interpretation, and classification for various smart city usages.

1.4.4 Graphical information system (GIS) and remote sensing

Remote sensing in environmental resource mapping and modeling needs intelligent sensing and there are a number of advances in the area of remote sensing using cognitive sensors and intelligent data analysis. Satellite data captured with the help of smart sensors that may be of the type of airborne remote sensing, spaceborne satellite, remotely sensed data, metrological satellite sensor data, Landsat data, earth observation data, or any similar data. In all these most emerging applications related to remote sensing and GIS [1,8,41], urban mapping and urban planning can be provided with the help of appropriate GIS and the remote sensing applications. Another interesting application of GIS is hydrological applications where satellite data and hydrological resources can be very well monitored and planned, such as mapping of water resources, mapping up hydrological features, flood mapping, and monitoring of a number of lakes in hydrological resources. The hydrological parameters can be estimated so that it can be used for land cover mapping,

Figure 1.4 GIS application

vegetation cover and its estimation, surface cover estimation, heat flux value in different areas, the estimation of soil, water, and the drought monitoring can be implemented with the early warning system. The intelligent sensing is involved in all these applications. A lot of work is going on and there are advances in the space-bound remote sensing, from Sputnik-1 [48] and others. The innovation has been phenomenal and a lot of innovations are taking place and the spatial and temporal data are captured and hyperspectral, multispectral imaging [17] techniques are used to capture the image and the suitable data mining, processing, and intelligent ML techniques are employed to understand the data and deploy that data processing for the several applications related to urban planning monitoring, smart agriculture, navigation, etc. Figure 1.4 shows a GIS system consisting of satellites from which data collection shall take place; imageries; drones; and their analysis and classification.

1.4.5 Smart agriculture

The agriculture sector is also becoming a smart agriculture due to the emergence of wireless sensor networks, IoT, modern sensors, and communication capabilities. There are several sensors that are used in the field of agriculture such as humidity sensor, temperature sensor, soil sensor in order to assess the quality of soil, quality of crop and moisture level, and other factors that are very important in the increased productivity in the area of sustainable agriculture. The Zigbee and GPRS-based technologies [36,49] are available which are low cost and low power technologies that enable the usage of these sensors and the hardware design can easily be achieved. To deploy these Zigbee, GPRS and web-based database servers where we can implement the effective model in order to monitor the different parameters of different types of crops based on the weather and the climate condition, the soil condition and other factors which are responsible in order to get the sustainable growth and progress in the agriculture. Sustainability is a very important factor in agriculture and that is a very important and there the sensing technologies can actually help the farmers in not only improving their productivity but also achieving sustainability in the agriculture and self-reliance on the growth of the agriculture sector.

The smart agriculture sector is also becoming a reality due to the emergence of wireless sensor networks, IoT, modern sensors, and communication capabilities. There are several sensors that are used in the field of agriculture such as humidity sensor, temperature sensor, soil sensor to assess the quality of soil, quality of crop and moisture level, and other factors that are important in the increased productivity in sustainable agriculture. The Zigbee and General Packet Radio Service (GPRS)-based technologies [36,49] are available which are low cost and low power technologies that enable to use these sensors and the hardware design can easily be achieved. To deploy the Zigbee, GPRS and web-based database servers where we can implement the effective model to monitor the different parameters of different types of crops based on the weather and the climate condition, the soil condition and other factors are responsible to get the sustainable growth and progress in the agriculture. Sustainability is a very important factor in agriculture; the sensing technologies can actually help the farmers in not only improving their productivity but also achieving sustainability in the agriculture and self-reliance on the growth of the agriculture sector. Figure 1.5 shows an architecture for smart agriculture and

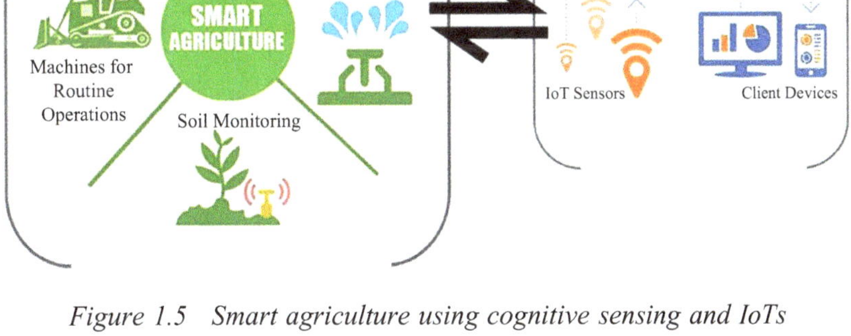

Figure 1.5 Smart agriculture using cognitive sensing and IoTs

it can be seen in the figure that IoTs are in the form of cognitive and smart sensors used for various purposes in smart agriculture sector.

Smart agriculture [37,50,51] uses the signal processing methods and the different technologies such as artificial neural network, fuzzy logic, and deep learning methods for the estimation of soil quality, the crop quality, disease detection in the plants and similar applications where the sensors and edge devices utilizing ML methods are deployed. Based on data understanding intelligently, language translation is needed and the suitable control devices that are used in precision agriculture can be deployed to detect the disease in the images in crop analysis and accordingly the remedial majors can be recommended to smart farming stakeholders. Lot of control devices can be used for such operations where the disease can be controlled, and appropriate monitoring can be achieved. The crop phenotyping is another interesting area that employs the clustering of several hyper spectral signatures. A few ML techniques are used in the smart agriculture area for several crops. For example, in the case of Soyabean, vision-based intelligence is used. The technologies related to genetic algorithm, fuzzy logic, and artificial neural network are used for better classifying the crops and the weeds through the growing. The performance is measured intelligently with the help of hyperspectral data. Suitable data mining techniques can be used. The detection of plant leaf disease can be done with the help of suitable soft computing and image processing techniques in all applications related to fruits as well as detection of fungal and other diseases in fruits and the quality comparison of fruits and plants. The data can also be used to help the farmers to assess the quality of crops and classify the best quality and other qualities in terms of their productivity and marketing purposes.

Several agronomists with the help of AI get a lot of solutions for smart farming to deal with disease and even nutrient deficiencies. Another interesting area in smart agriculture is the agricultural robots that are used at different stages of the agriculture. An interesting application that is also very useful for the farmers is the predictive analysis that uses the study of historical weather patterns, the estimation of rainfall, the pest control, and with the help of drones and cameras. A number of real-time drones [17,52] can be employed to assist the farmers to deal with the data and thus smart agriculture uses a GPS data analytic system, drones, smart sensors, assistive robots, and agricultural robots so that the environmental risk, infection, and disease can be controlled to a significant extent and sustainable agriculture can be achieved where holistic development and growth in the area of the agriculture products can also be obtained.

There are cutting-edge technologies in a geospatial system dealing with geographic information system, global navigation satellite system (GNSS) [29], remote sensing, and other photometric applications that cover a number of utilities such as precision farming, video surveillance camera, radio frequency identification (RFID) [36] tags that are used to evaluate certain interlinked dependencies. In several countries, we have a number of remote sensing satellites with which high-resolution images are being captured and analyzed in real time. In remote sensing, high-resolution imagery with the help of light detection and ranging (LIDAR) [16], and other technologies are used to intelligently analyze the remotely sensed data. In

Global Positioning System (GPS) [49] of aircraft, the orientation and other parameters can be easily studied with the help of smart sensing. Surveying is very important application in the remote sensing where Indian Regional Navigation satellite system from India serves in order to surveying application that uses 3D laser scanning systems to accurately assess the multitudes and various survey points by which the geospatial technology helps in surveying application. The GIS uses several systems like ambient geographic information, volunteered geographic information, and other maps to study different geographies related to earthquake and other conditions, crisis maps which happen across the world. So, to understand and study these kinds of the calamities, several miniaturized computing, radio-based communication network, micro-electromechanical systems (MEMS)-based sensors are available which are deployed over a wireless sensor network and can understand the data intelligently. The number of climate-related calamities and crisis could be averted. The current research in GIS is actually location-based service through which more mobility and the smart environment monitoring are claimed where sensor wave is proposed as a powerful tool as a worldwide wave so that the sensor-based web can be used for evaluating the number of calamities.

1.4.6 Robotics and rehabilitation

Intelligent sensing makes the robotic application complete only when the dynamic movement of robots can be achieved, in applications such as smart manufacturing, smart healthcare, and smart industrial application. The sensing and data processing involved in the robotics along with ML and AI-based approaches makes it intelligent and thus robots are used in numerous applications in the areas of entertainment, games, manufacturing, healthcare, environment monitoring, smart parking, medicines, rehabilitation [7,12,53], patient care, and so on. Even in space missions, the robotics is playing a significant role where a humanoid robot is being sent by different countries in the world. The humanoid robot is a brilliant example of smart sensing that involves a number of cognitive sensors, deep learning capabilities, signal processing, data processing and interpretation. Intelligent understanding by deploying ML techniques becomes essentially important when a robot has to perform a certain task in such applications. Another very interesting and exciting application of robot is to facilitate in the rehabilitation in the health care through which the robots can be deployed to serve the patients. In the adverse environment in difficult manufacturing and automation areas where human approach is extremely difficult, the robots can be sent to difficult locations and the robots must understand the data intelligently. This is only possible when the sensing is cognitive, and the technology is very intelligent. Similarly, in the rehabilitation in healthcare, the sensing from the brain as brain signal sensing and the analysis is very intelligent and brain–computer interface (BCI) can equivalently produce the signal that can be used as feedback signal in the rehabilitation of patients suffering with some brain disorder and disabilities. So, in this application, the role of cognitive sensing is of immense significance. The robotics also provide services in the smart wheelchair, and other assistive devices by which the persons who are having disability in

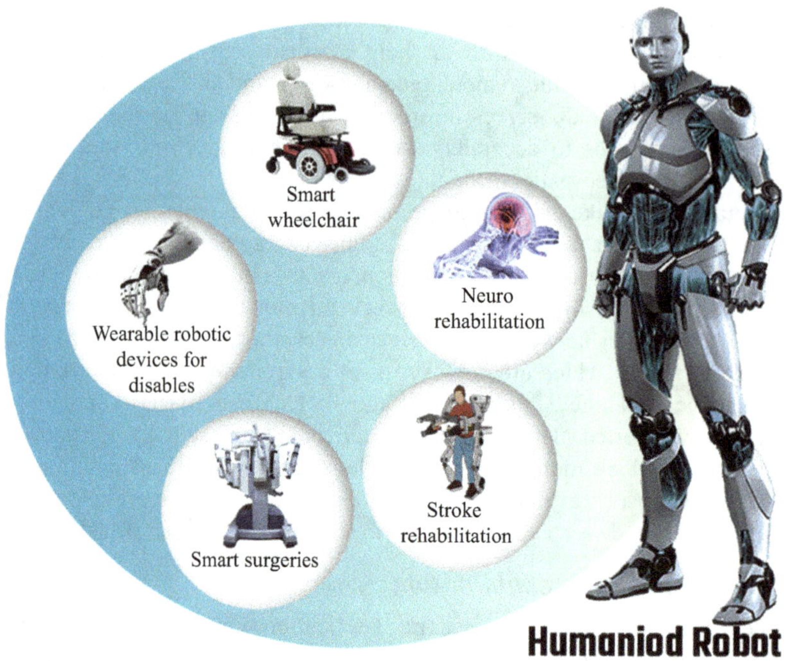

Figure 1.6 Robotics

terms of the mental health will be benefited. In Figure 1.6, we can see various potential applications of robotics and none of the applications is incomplete without using smart and cognitive sensors and the technologies.

Robotics is not only very important for the healthcare and the manufacturing but also in several industrial applications. Industry 4.0 [27] talks about Internet of industrial things (IoIT) [19] and, in industrial Internet of industrial things, the sensing technologies and the ML technologies are deployed in understanding the data and their interpretation to improve the efficacy of different stages of industries. For example, in the steel industry, where the plates and steels are produced and to detect some deformation or defects inside the metals or steels, detection method uses sound or suitable sensors to find exact localization This can be achieved, and the quality of the products can also be improved. In industrial applications, the quality of product and the monitoring of the industrial different stages of industries can be done very well and the real-time monitoring wherever the adverse situation or emergency arises can be obtained. The cognitive sensing can relate to real-time alert and a number of unseen situations like accidents and emergency situations could be averted. This is how the role of AI and ML with smart sensors make the system and processing very intelligent and therefore the cognitive sensing along with robots help to the great extent in the areas of manufacturing, healthcare industries, and several such applications.

1.5 Benefits of cognitive sensing

There are major advantages of cognitive sensing technologies, some of the benefits include the following:

- Automation in the environment for example manufacturing, robotics, and other applications; the automation will become adaptive because of the ability of appropriate decision making, ability to understand data intelligently to the extent of automation; and its timing will also be determined by the cognitive sensing.
- The cognitive sensing will filter out the redundant and useless information and will only provide us with the useful and very important information for the task.
- Better analysis and efficacy due to cognitive sensors that are used in mobile environment. We can implement the sensors using edge computing and powerful cognitive sensing technologies and will deploy the application at different points, therefore the portability, mobility, flexibility, and outreach of data will be much better than that of the work without cognitive sensing technique.
- Improved classification and understanding can be achieved with the help of edge computing at different nodes. The cognitive sensing with the help of ML techniques deployed over edge devices, the better classification with reduced latency time, and redundancy and reduced dependency on cloud architecture will be obtained.
- The cognitive sensing technique will allow fusing the data from an appropriate number of sensors and the amount of data from a particular sensor that has to be allowed can be decided intelligently. The composite sensing can also be obtained that is very important for the manipulation of data from any of the cognitive sensors.

1.6 Assistive technologies and BCI

Assistive technology [11,54–56] covers a number of domains that include environment, communication, hearing, vision, mobility, and self-care. The artificial arteries, virtual reality devices, augmented reality devices, and navigational aids are developed with the help of BCI as smart assistive devices related to environment, smart homes, and cities. For self-care, several solutions are coming up related to health and the emotion monitoring, smart medication, and management. The robots that are used in feeding, smart diapers or mobility and self-care are related with advanced prosthetics and walking aids and advanced wheelchairs. In addition, there are several developments such as non-invasive bone conduction, middle ear implants and so on. In all such emerging assistive technologies, the sensing plays very important role in the understanding of sensed data. The autonomous vehicles, robots, and new materials are being developed using nano-sensors, micro sensors, MEMS technology supporting virtual reality, augmented reality, intelligent manufacturing, IoT, brain computer interface, brain–machine interface, and related exciting applications. These are the actual domains that are of huge importance dealing with the assistive technology and the most important in assistive technology is cognition impairment that impacts the cognitive skills and abilities in persons. Hence, here the wearable technology plays an important role. Assistive robots can play a significant role in the areas such as memory support, time management, and medication dispensing, through which the patients can take advantage. In industries, the assistive technologies can be of help to monitor the environmental parameters specifically that are related to extreme conditions. There are companies that are working in assistive technologies such as Honeywell, Panasonic, and Samsung specifically exploiting cognition. The communication deals with the domains related to hearing, motor abilities, motor tasks, and cognitive tasks.

The BCI is an exciting application of the signal processing where we capture the data from the brains using the different types of BCI systems, namely invasive BCI system, noninvasive BCI system, semi-invasive BCI system through which we understand the brain signals and the numerous tasks are performed that can be controlled by the human brain. This helps in several areas such as human–computer interaction, brain–machine interaction, and the biggest advantages of the smart sensing, analysis, and classification of the brain-related data is in the area of diagnosis of brain disorder and disability, especially for those who are suffering with extreme disabilities. Those who require some assistive technologies or rehabilitation techniques through which an appropriate feedback signal is given to the patients and the disease and disorder, or disability can be improved in the patients. In BCI, the electrodes are used to capture the EEG data or other brain signal data. Hence, better is the quality of electrodes of the information from the brain that can be stored and analyzed. In performing the task by the robots and other assistive technology devices, the meaningful information from the brain plays an important role so the BCI utilizes the cognitive sensing method and the appropriate use of ML and AI algorithms. The appropriate classification of the data captured from the brain is backbone of a lot of applications related to assistive technology healthcare and robotic-based home appliance controls for people who have disabilities as well as old and healthy adults who want to improve their quality of life. Assistive robots may help in operating the home appliances without any assistance and companion robots may provide counseling and companionship. In addition, BCI also helps to develop applications like neurofeedback training that helps to retrain the brain and can decrease the effect of neural, psychological, and behavioral disorder like attention-deficit/hyperactivity disorder (ADHD), addiction, alcoholism, stress anxiety, depression, and executive functions [57,58]. Emotion detection is also achieved through the BCI devices which also play an important role in understanding human psychology and the state of the human mind to solve behavioral and psychological disorders, in addition it is also important for finding best environment for individuals based on their preferences. The emotion detection cannot be achieved without the brain sensing devices along with advanced signal processing algorithms and ML predictors [59].

1.7 Current research and future directions

As we have discussed in the previous sections that cognitive IoT (CoIT), cognitive Internet of vehicles (CIoV) along with Internet are bringing transformations in numerous applications based on science, technology, and engineering. Many concerns related to the advanced technology involve ethical uses of AI, data privacy, and security. Thus, the research in advanced sensors, sensing technology, smart computing, AI, robust approaches for classification and analysis of data, and intelligent data processing are always ongoing. In this section, an attempt is made to understand the potential of future research in the field of smart sensing and its application using cognitive sensors and the most emerging deep learning-based models of ML.

1.7.1 Cognitive Internet of vehicle

The future of cognitive and smart sensing lies in the autonomous driving vehicles on roads where the vehicles could be controlled with the help of Internet. After 5G has rolled out, the tremendous computing and communication power, AI, cloud as well as edge computing provide more mobility and intelligence. The vehicular management is talked about in a new paradigm and is called as cognitive Internet of vehicles [17,60]. It is claimed that a number of major advantages will be associated with the CIoV as through this vehicle to any transportation problems can be solved. It is claimed although the status is still in the research stage that the smart wireless sensor network, CRN and cellular network with the help of suitable AI can make cognitive Internet of vehicles possible that would provide a number of advantages which include the following:

- High-speed mobility will be obtained due to computing power as well as communication power. The complexity and variety of different traffic conditions and road conditions will be there, but since the cognitive sensing is involved, suitable AI and the CRN is involved so the high-speed mobility can be achieved.
- Delay sensitivity in communication can be measured and brought to very small amount. Longer delay means slow computation and thus delay can be reduced so that we can get seamless connectivity.
- Data privacy can be enhanced using intelligent encoding techniques implementing at the edges.
- Efficient utilization of the resource can be made using cognitive intelligence and its perceptive ability.
- Intra-vehicle network and inter-vehicle network will be smart and beyond the network like vehicle to anything, vehicle to any kind of communication can be achieved with the smart and cognitive sensing.
- Reliable decision making that includes things like the vehicular ad hoc network.
- CRN [15,16,18,61] supports at different layers of the implementation like sensing layer, control layer, cognition layer, and communication layer.
- Radio technologies that will be used in the cognitive Internet of vehicles include Wi-Fi technology, wireless LAN technology, and Wi-max technologies that will be providing a good transmission capacity and stable connectivity.

- Seamless communication can be achieved very successfully and efficiently.
- 5G slicing will provide different controls at different layers and the cognitive Internet of vehicles.

1.7.2 Role of ML and robustness

The intelligent sensing in any application is incomplete if emerging ML and AI techniques are not used. Whenever we talk about intelligent sensing, it involves smart sensors along with the AI and ML techniques. The use of ML technique depends on the application and the data. The data involved consists of pictures, videos, real-time scenes, which will be handled by deep learning models that can handle huge. The processing time, the latency, the computational complexity, or the real-time challenges are dealt with deep learning models. The only difficulty with the deep learning models for such applications is the robustness of the analysis. Once we interpret the data using suitable intelligent data processing methods, the performance depends on the analysis, on the quality of data and the methods that were used in preprocessing stages. The appropriate features that were used in the process of feature extraction and suitable ML methods for training also are important in the assessment of robust performance. The robustness is all time challenge in all the applications where computer vision is involved irrespective of the model being used, the performance of a system depends on the appropriate feature set, appropriate use of ML models, appropriate feature extraction methods, and the optimal preprocessing methods. Thus, a lot of research is going on in order to find the robust approaches for preprocessing, feature extraction, training as well as classification because if the preceding stages such as preprocessing, feature extraction are not optimal, then there is no guarantee that the classifiers based on deep learning models are actually producing the appropriate or optimal classified output for several applications. This is a genuine challenge and the research in order to deal the robustness [62] is always ongoing.

1.7.3 Issues in IoTs

There are challenges in IoTs using cognitive sensing technology indicating research directions where a lot of work is being done. However, additional work is needed for IoT deploying at different points at a cloud-based services or edge-based services using the smart sensing and the ML techniques. The factors and components in IoT such as the data connectivity, actions and commands, intelligence, and communication are very important for designing efficient IoTs. All the issues, challenges, and difficulties related with these components need to be addressed. The ecosystem of IoTs must satisfy certain protocols and conditions to provide this service in the different areas for the society and technology.

The factors related to sensors, actuators, hardware, edge devices, cognitive things, intelligent things, and issues in fabrication of sensors, the manufacturing of the elements, the electrodes of the sensors cannot be ignored. High-quality electrodes are needed to capture high-resolution data.

The connectivity-related issues such as networks, protocols, and resources play a significant role in terms of intelligent data exchange. If we use fusion technique between the two sensors or multiple sensors, then data fusion may impact the application of IoT and the data processing involved.

Research efforts are also seen on how to enhance the cognitive functionalities in IoTs for several applications and ongoing research aims to improve the efficacy of IoTs involving more intelligence in the system taking care of ethical and privacy concerns. The functionalities related to analysis, automation, decision making, cognitive ability are all important areas where there are several issues and challenges. Work needs to be done to make suitable decision, and to automate robust approaches so that the intelligent sensing can provide the appropriate commands, actions in the applications related to healthcare, robotics, and so on. The overall ecosystem of IoTs deploying intelligent sensing, taking the quality of service, and ethical issues into consideration, Internet of ethical thing is coined and research is on. With the help of ongoing research and innovation, the very purpose of IoT, utilizing the smart sensing technology, can serve for the improvement of quality of life.

1.8 Conclusions

This chapter has presented an overview, importance, and scope of cognitive sensing technologies with the discussion on possible emerging applications and issues therein.

References

[1] Y. Lim, N. Pongsakornsathien, A. Gardi, *et al.*, "Adaptive human-robot interactions for multiple unmanned aerial vehicles," *Robotics*, vol. 10, no. 1, pp. 1–30, 2021, doi:10.3390/robotics10010012.

[2] P. Kaligineedi, S. Member, and M. Khabbazian, "Cognitive radio cooperative sensing system," *IEEE Transactions on Wireless Communications*, vol. 9, no. 8, pp. 2488–2497, 2010.

[3] B. Hu, "The study on the spectrum sensing technology in wireless cognitive network," in: *6th International Conference on Electronic, Mechanical, Information and Management (EMIM 2016)*, no. Emim, pp. 1880–1885, 2016, doi:10.2991/emim-16.2016.381.

[4] M. A. U. Alam, N. Roy, S. Holmes, A. Gangopadhyay, and E. Galik, "AutoCogniSys: IoT assisted context-aware automatic cognitive health assessment," in: *ACM International Conference Proceeding Series*, pp. 184–195, 2020, doi:10.1145/3448891.3448945.

[5] Y. Arjoune, N. Kaabouch, H. El Ghazi, and A. Tamtaoui, "Compressive sensing: performance comparison of sparse recovery algorithms," in: *2017 IEEE 7th Annual Computing and Communication Workshop and Conference (CCWC)*, pp. 1–7, 2017, doi:10.1109/CCWC.2017.7868430.

[6] M. Beetz, M. Buss, and D. Wollherr, "Cognitive technical systems – What is the role of artificial intelligence?," in: *Lecture Notes in Computer Science*

(including subseries Lecture Notes in Artificial Intelligence and Lecture Notes in Bioinformatics), vol. 4667, LNAI, pp. 19–42, 2007, doi:10.1007/978-3-540-74565-5_3.

[7] A. B. Naranjo-Saucedo, C. Suárez, C. L. Parra Calderón, *et al.*, "Interactive games with robotic and augmented reality technology in cognitive and motor rehabilitation," *Robot. Concepts, Methodol. Tools Appl.*, vol. 3, pp. 1233–1254, 2013, doi:10.4018/978-1-4666-4607-0.ch059.

[8] N. Gillani and T. Arslan, "Intelligent sensing technologies for the diagnosis, monitoring and therapy of alzheimer's disease: a systematic review," *Sensors*, vol. 21, no. 12, p. 4249, 2021, doi:10.3390/s21124249.

[9] C. O. Sereati, A. D. W. Sumari, T. Adiono, and A. S. Ahmad, "Towards cognitive artificial intelligence device: an intelligent processor based on human thinking emulation," *Telkomnika (Telecommunication Comput. Electron. Control.)*, vol. 18, no. 3, pp. 1475–1482, 2020, doi:10.12928/TELKOMNIKA.v18i3.14835.

[10] U. B. Desai and B. N. Jain, "Wireless Sensor Networks: Technology Roadmap," … *Wirel. Sens. Networks IITB* …, p. 121, 2007, http://citeseerx.ist.psu.edu/viewdoc/download?doi=10.1.1.458.8531&rep=rep1&type=pdf

[11] M. Agustina Garcés and L. L. Orosco, "EEG signal processing in brain-computer interface," in: *Smart Wheelchairs and Brain-Computer Interfaces: Mobile Assistive Technologies*, New York, NY: Elsevier, pp. 95–110, 2018, doi:10.1016/B978-0-12-812892-3.00005-4.

[12] J. J. Shih, D. J. Krusienski, and J. R. Wolpaw, "Brain-computer interfaces in medicine," *Mayo Clinic Proceedings*, vol. 87, no. 3, pp. 268–279, 2012, doi:10.1016/j.mayocp.2011.12.008.

[13] H. Nisar, K. H. Way, and Y. Kim Ho, "Brain computer interface: brain computer interface: controlling a robotic arm using facial expressions," *Turkish Journal of Electrical Engineering and Computer Science*, vol. 26, pp. 707–720, 2017.

[14] Y. Miao, G. Wu, C. Liu, M. S. Hossain, and G. Muhammad, "Green cognitive body sensor network: architecture, energy harvesting, and smart clothing-based applications," *IEEE Sensors Journal*, vol. 19, no. 19, pp. 8371–8378, 2019, doi:10.1109/JSEN.2018.2870251.

[15] Q. Wu, G. Ding, Y. Xu, *et al.*, "Cognitive internet of things: a new paradigm beyond connection," *IEEE Internet of Things Journal*, vol. 1, no. 2, pp. 129–143, 2014, doi:10.1109/JIOT.2014.2311513.

[16] M. Ozger, O. Cetinkaya, and O. B. Akan, "Energy harvesting cognitive radio networking for IoT-enabled smart grid," *Mobile Networks and Applications*, vol. 23, no. 4, pp. 956–966, 2018, doi:10.1007/s11036-017-0961-3.

[17] O. Vermesan, M. Eisenhauer, H. Sundmaeker, *et al.*, "Internet of things cognitive transformation technology research trends and applications," in: *Cognitive Hyperconnected Digital Transformation: Internet of Things Intelligence Evolution*, Denmark: River Publishers, pp. 17–95, 2017, doi:10.1201/9781003337584-3.

[18] O. B. Akan, O. B. Karli, and O. Ergul, "Cognitive radio sensor networks," *IEEE Networks*, vol. 23, no. 4, pp. 34–40, 2009, doi:10.1109/MNET.2009.5191144.

[19] A. J. Onumanyi, A. M. Abu-Mahfouz, and G. P. Hancke, "Cognitive radio in low power wide area network for IoT applications: recent approaches, benefits and challenges," *IEEE Transactions on Industrial Informatics*, vol. 16, no. 12, pp. 7489–7498, 2020, doi:10.1109/TII.2019.2956507.

[20] F. Li, K. Y. Lam, X. Li, Z. Sheng, J. Hua, and L. Wang, "Advances and emerging challenges in cognitive Internet-of-Things," *IEEE Transactions on Industrial Informatics*, vol. 16, no. 8, pp. 5489–5496, 2020, doi:10.1109/TII.2019.2953246.

[21] D. Y. Fong, "Wireless sensor networks," in: *Internet of Things Data Analysis Handbook*, New York, NY: Wiley, pp. 197–213, 2017, doi:10.1002/9781119173601.ch12.

[22] P. Jonell, M. Bystedt, D. Kontogiorgos, *et al.*, "Farmi: a framework for recording multi-modal interactions," in: Lr. 2018 – *11th International Conference on Language Resources and Evaluation*, pp. 3969–3974, 2019.

[23] L. B. Bhajantri and P. Baluragi, "A survey on data perception in cognitive internet of things," *Journal of Telecommunications and Information Technology*, vol. 3, pp. 75–86, 2019, doi:10.26636/jtit.2019.131419.

[24] A. Pal and K. Kant, "Smart sensing, communication, and control in perishable food supply chain," *ACM Transactions on Sensor Networks*, vol. 16, no. 1, pp. 1–41, 2020, doi:10.1145/3360726.

[25] K. S. Adu-Manu, C. Tapparello, W. Heinzelman, F. A. Katsriku, and J. D. Abdulai, "Water quality monitoring using wireless sensor networks: current trends and future research directions," *ACM Transactions on Sensor Networks*, vol. 13, no. 1, pp. 1–41, 2017, doi:10.1145/3005719.

[26] E. Mbunge, B. Muchemwa, S. Jiyane, and J. Batani, "Sensors and healthcare 5.0: transformative shift in virtual care through emerging digital health technologies," *Global Health Journal*, vol. 5, no. 4, pp. 169–177, 2021, doi:10.1016/j.glohj.2021.11.008.

[27] M. Taghouti, "Compressed sensing," *Computer Communications Networks from Theory to Practice*, vol. 52, no. 4, pp. 197–215, 2020, doi:10.1016/B978-0-12-820488-7.00023-2.

[28] D. Baron, M. F. Duarte, S. Sarvotham, M. B. Wakin, and R. G. Baraniuk, "An information-theoretic approach to distributed compressed sensing," in: *43rd Annual Allerton Conference on Communication, Control, and Computing* , vol. 2, pp. 814–825, 2005.

[29] Y. Ma, R. Garnett, and J. Schneider, "Active search for sparse signals with region sensing," in: *31st AAAI Conference On Artificial Intelligence AAAI 2017*, pp. 2315–2321, 2017, doi:10.1609/aaai.v31i1.10865.

[30] J. K. Hart and K. Martinez, *Sensor Networks and Geohazards*, 2nd ed. New York, NY: Elsevier Inc., 2022. doi:10.1016/b978-0-12-818234-5.00037-7.

[31] I. Zolotová, P. Papcun, E. Kajáti, M. Miškuf, and J. Mocnej, "Smart and cognitive solutions for Operator 4.0: laboratory H-CPPS case studies,"

Computers and Industrial Engineering, vol. 139, Article 105471, 2020, doi:10.1016/j.cie.2018.10.032.

[32] A. Leone, G. Rescio, P. Siciliano, A. Papetti, A. Brunzini, and M. Germani, "Multi sensors platform for stress monitoring of workers in smart manufacturing context," in: *I2MTC 2020 – International Instrumentation and Measurement Technology Conference Proceedings*, pp. 3–7, 2020, doi:10.1109/I2MTC43012.2020.9129288.

[33] R. Alexandre and O. Postolache, "Wearable and IoT technologies application for physical rehabilitation," in: *2018 International Symposium in Sensing and Instrumentation in IoT Era, ISSI 2018*, no. October, 2018, doi:10.1109/ISSI.2018.8538058.

[34] K. Sweeney, "Motion artifact processing techniques for physiological signals," no. April, 2013.

[35] J. ho Park, M. M. Salim, J. H. Jo, J. C. S. Sicato, S. Rathore, and J. H. Park, "CIoT-Net: a scalable cognitive IoT based smart city network architecture," *Human-centric Computing and Information Sciences*, vol. 9, no. 1, Article number: 29, 2019, doi:10.1186/s13673-019-0190-9.

[36] K. Sohraby, D. Minoli, and T. Znati, *Basic Wireless Sensor Technology*, New York, NY: Wiley, 2007. doi:10.1002/9780470112762.ch3.

[37] D. Paper, "Climate Smart Agriculture and Advisory Services: Approaches and Implications for Future," http://www.manage.gov.in/publications/discussion%20papers/MANAGE-Discussion%20Paper-1.pdf, 2018.

[38] U. Garlando, L. Bar-On, A. Avni, Y. Shacham-Diamand, and D. Demarchi, "Plants and environmental sensors for smart agriculture, an overview," in: *Proceedings of IEEE Sensors*, vol. 2020 October, 2020, doi:10.1109/SENSORS47125.2020.9278748.

[39] A. B. B. Torres, A. R. da Rocha, T. L. Coelho da Silva, J. N. de Souza, and R. S. Gondim, "Multilevel data fusion for the internet of things in smart agriculture," *Computers and Electronics in Agriculture*, vol. 171, no. November 2019, p. 105309, 2020, doi:10.1016/j.compag.2020.105309.

[40] S. L. Ullo and G. R. Sinha, "Advances in smart environment monitoring systems using IoT and sensors," *Sensors (Switzerland)*, vol. 20, no. 11, pp. 1–18, 2020, doi:10.3390/s20113113.

[41] D. M. de Queiroz, A. L. de F. Coelho, D. S. M. Valente, and J. K. Schueller, "Sensors applied to digital agriculture: a review," *Revista Ciencia Agronomica*, vol. 51, no. 5, pp. 1–15, 2020, doi:10.5935/1806-6690.20200086.

[42] A. O. Bicen, O. B. Akan, and V. C. Gungor, "Spectrum-aware and cognitive sensor networks for smart grid applications," *IEEE Communication Magazine*, vol. 50, no. 5, pp. 158–165, 2012, doi:10.1109/MCOM.2012.6194397.

[43] J. Laufs, H. Borrion, and B. Bradford, "Security and the smart city: a systematic review," *Sustainable Cities and Society*, vol. 55, pp. 1–51, 2020, doi:10.1016/j.scs.2020.102023.

[44] C. Study, *Smart City Surveillance* Powered by Honeywell Technologies.

[45] V. Christian and S. Hegade, "Video surveillance options in a smart city securing memory sharing for cloud tenants," 2021Dell Technologies Proven Professional Knowledge Sharing, pp. 1-12, https://education.dellemc.com/content/dam/dell-emc/documents/en-us/2021KS_Christian-Video_Surveillance_options_in_a_Smart_city.pdf.

[46] M. Sookhak, H. Tang, and F. R. Yu, "Security and privacy of smart cities: issues and challenge," in: *Proceedings – 20th IEEE International Conference on High Performance Computing and Communications; 16th International Conference on Smart City; 4th International Conference on Data Science and Systems, HPCC/SmartCity/DSS 2018*, vol. 21, no. 2, pp. 1350–1357, 2019, doi:10.1109/HPCC/SmartCity/DSS.2018.00224.

[47] W. T. Lee, H. Nisar, A. S. Malik and K. H. Yeap, "A brain computer interface for smart home control," in: *2013 IEEE International Symposium on Consumer Electronics (ISCE)*, Hsinchu, Taiwan, pp. 35–36, 2013, doi:10.1109/ISCE.2013.6570240.

[48] B. Valeske, R. Tschuncky, F. Leinenbach, *et al.*, "Cognitive sensor systems for NDE 4.0: technology, AI embedding, validation and qualification," *Technisches Messen*, vol. 89, no. 4, pp. 253–277, 2022, doi:10.1515/teme-2021-0131.

[49] G. V Satyanarayana, "Wireless sensor based remote monitoring system for agriculture using ZigBee and GPS," in: *Conference on Advanced Communication and Control Systems 2013*, vol. 2013, no. Cac2s, pp. 110–114, 2013.

[50] S. L. Ullo and G. R. Sinha, "Advances in smart environment monitoring systems using IoT and sensors," *Sensors*, vol. 20, no. 11, pp. 3113, 2020, doi:10.3390/s20113113.

[51] P. Angin, M. H. Anisi, F. Göksel, C. Gürsoy, and A. Büyükgülcü, "Agrilora: a digital twin framework for smart agriculture," *Journal of Wireless Mobile Networks, Ubiquitous Computing, and Dependable Applications*, vol. 11, no. 4, pp. 77–96, 2020, doi:10.22667/JOWUA.2020.12.31.077.

[52] F. Al-Turjman, *Drones in IoT-Enabled Spaces*, London: CRC Press, 2019. doi:10.1201/9780429294327.

[53] M. C. Pol, G. Ter Riet, M. Van Hartingsveldt, B. Kröse, and B. M. Buurman, "Effectiveness of sensor monitoring in a rehabilitation programme for older patients after hip fracture: a three-arm stepped wedge randomised trial," *Age Ageing*, vol. 48, no. 5, pp. 648–655, 2019, doi:10.1093/ageing/afz074.

[54] J. Friend, *Assistive Technology – Assistive Educational Technology*, 2012. http://assistiveeducationaltechnology.weebly.com/assistive-technology.html.

[55] H. Nisar, K. W. Boon, Y. K. Ho, and T. S. Khang. "Brain-computer interface: feature extraction and classification of motor imagery-based cognitive tasks." in: *2022 IEEE International Conference on Automatic Control and Intelligent Systems (I2CACIS)*, New York, NY: IEEE, pp. 42–47, 2022.

[56] H. Nisar, T. C. Hoe, and R. Nawaz. "Reducing sensors in mental imagery based cognitive task for brain computer interface." in: *2020 14th International Conference on Signal Processing and Communication Systems (ICSPCS)*, New York, NY: IEEE, pp. 1–10, 2020.

[57] R. Nawaz, H. Nisar, V. V. Yap, and C.-Y. Tsai. "The effect of alpha neurofeedback training on cognitive performance in healthy adults." *Mathematics,* vol. 10, no. 7, pp. 1095, 2022.

[58] R. Nawaz, H. Nisar, and Y. V. Voon. "Changes in spectral power and functional connectivity of response-conflict task after neurofeedback training." *IEEE Access*, vol. 8, 139444–139459, 2020.

[59] R. Nawaz, K. H. Cheah, H. Nisar, and V. V. Yap. "Comparison of different feature extraction methods for EEG-based emotion recognition." *Biocybernetics and Biomedical Engineering*, vol. 40, no. 3, 910–926, 2020.

[60] M. Chen, Y. Tian, G. Fortino, J. Zhang, and I. Humar, "Cognitive internet of vehicles," *Computer Communications*, vol. 120, no. October 2017, pp. 58–70, 2018, doi:10.1016/j.comcom.2018.02.006.

[61] A. Mateska, K. Chomu, and L. Gavrilovska, "Cognitive sensor networks implementation in medical environmental surveillance," in: *BODYNETS 2012 – 7th International Conference on Body Area Networks*, 2012, doi:10.4108/icst.bodynets.2012.250008.

[62] N. Koep, A. Behboodi, and R. Mathar, "An introduction to compressed sensing," in: *Applied and Numerical Harmonic Analysis*, New York, NY: Springer, pp. 1–65, 2019, doi:10.1007/978-3-319-73074-5_1.

Chapter 2

Hardware architectures for some sparse signal recovery approaches

Andjela Draganić[1], *Irena Orović*[1,2], *Srdjan Stanković*[1], *Nedjeljko Lekić*[1] *and Marko Beko*[2,3]

Abstract

Over the last couple of decades, sparse signal processing has been constantly the focus of research interests aiming to exploit all benefits of sparseness property. Particularly, the sparsity has been utilized in numerous applications such as compression and coding, spectral analysis, channel estimation, and recently for the development of compressive sensing (CS) approaches and signal reconstruction algorithms. Indeed, intensive development of the approaches for sparse signal recovery in the CS scenario has created a new sensing paradigm and opened up a new perspective of sensing technology design. Moreover, a variety of CS scenarios have been exploited in different real-world applications, not only for sensing purposes but for example in signal/image denoising problems. A number of approaches and algorithms have been designed to cope with under-sampled and random-sampled signal recovery, depending on the signal nature.

Although signal processing algorithms can be tested in software, there is a growing need for hardware realizations. Namely, software implementations of those approaches are generally not suitable enough for real-time applications. Therefore, the necessity for hardware implementations was raised due to the requirement of fast testing, which is hardly accomplished in software. There are various approaches starting from the less complex and much faster analog-circuits-based solutions to the more complex, slower but more precise digital approaches. Depending on the system requirements, analog or digital implementation is used. Generally, for fast and less complex circuits, analog implementation is better, while for slower circuits, a digital implementation is more suitable.

In this chapter, we will provide an overview of the hardware architectures for several approaches based on the gradient descent algorithm. The algorithm belongs

[1]Faculty of Electrical Engineering, University of Montenegro, Montenegro
[2]COPELABS, Universidade Lusófona, Portugal
[3]Instituto de Telecomunicações, Instituto Superior Tecnico, University of Lisbon, Portugal

to the group of convex optimizations and provides successful signal recovery from a relatively small number of acquired samples.

Keywords: Analog hardware; Compressive sensing; Gradient-based reconstruction; Sparse signal reconstruction; Image reconstruction

2.1 Introduction

Compressive sensing (CS) [1–6] is a relatively new approach to signal sensing that opens the possibility for developing fast and energy-efficient sensing systems. The major benefits of the CS approaches are evident in the applications where signals require high temporal or spatial resolution, but exhibit sparsity property when represented in a suitable transformation basis [7–14]. Namely, the acquisition of the sparse signals according to the CS principles could result in a much fewer number of measurements compared to the traditional sampling based on the Shannon–Nyquist sampling theorem, keeping the quality of the recovered signal as in the traditional approach.

The signal is considered sparse in a certain domain if it can be represented with a small number of non-zero values in this domain [1,2,5]. For many real-world and synthetic signals, a suitable basis in which they exhibit sparsity property can be found. Therefore, the CS approach has applicability in a wide range of applications. Much before the CS development, the signal sparsity was exploited in compression and coding algorithms to eliminate the redundancy in the signal having in mind the specific features and the sensitivity of the human sensory system. Those are the lossy compression algorithms such as MPEG or JPEG. The sparsity can also be useful for noise suppression and is applicable in 5G and IoT networks for spectrum improvement and increase energy efficiency [8]. Various approaches in wireless communications, such as spectrum sensing, sparse estimation, sparse detection, and support identification exploit the sparsity property [8,11,13].

Sparsity is one of two preconditions that should be satisfied in order to provide an accurate signal recovery from a reduced and random set of acquired samples. Another precondition is related to the sensing procedure. Namely, the coherence index calculated between measurement matrix and sparsity basis should be as high as possible, to ensure successful recovery from a reduced set of samples [1,2]. If both preconditions are satisfied, the recovery approaches aim to find the sparsest solution in the domain where the signal is considered sparse.

A reduced set of measurements could be the result of an intention to acquire the lowest possible number of samples but still be able to recover the signal. Another scenario of having a reduced set of samples arises when the signal is corrupted by noise. Then, noisy samples can be discarded and then considered missing. The presence of noise and other obstacles that prevent the acquisition of the full set of samples are unintentional ways to produce a reduced set of measurements.

There are many different approaches in the literature devoted to sparse signal reconstruction from an incomplete set of coefficients [15–23]. Generally, the reconstruction approaches can be divided into several groups. The most accurate reconstruction results are provided by using convex optimization algorithms based on ℓ_1-norm minimization, such as Basis Pursuit (BS), Gradient algorithm (GA), and relaxed approaches such as Least Absolute Shrinkage and Selection Operator (LASSO). A convex optimization approach, based on ℓ_1-norm minimization, although providing accurate recovery, is not suitable for real-time applications due to the numerical complexity. The versions with relaxed constraint – convex relaxations, are also numerically pretty demanding. An efficient recovery with reduced numerical complexity can be obtained by applying another group of algorithms, greedy approaches. The representatives of this group are Matching Pursuit (MP) and its modified versions such as the Orthogonal MP, stage wise orthogonal MP (StOMP), and compressive sampling MP (CoSaMP), then Single iteration reconstruction algorithm (SIRA) based on a threshold, as well as its iterative version, Iterative Hard Thresholding (IHT), Iterative Soft Thresholding (IST), etc. [1,2,16–23]. The third group of sparse recovery approaches is the Sparse Bayesian learning (SBL) group [9], widely applied for solving various sparse recovery problems in wireless communications and in audio processing.

The recovery of sparse signals from a reduced set of samples can be challenging both for software and hardware realizations [24–37]. Most approaches require iterative procedures to solve the problem of finding the sparsest solution [24,26,31]. Some of the important requirements for efficient hardware implementation are simple control flow, low memory requirements, well-separable tasks, and suitability for fixed-point arithmetic [32]. However, parallel hardware implementations using analog, digital or combined circuits can enable real-time implementations of the sparse recovery algorithms. Analog implementations are faster and less complex but also less precise compared to digital ones. However, not all approaches are suitable for digital implementations [24,25].

There are a number of hardware architectures proposed for the commonly used algorithms for sparse signal recovery [24–37]. For example, a high-speed OMP realization using Matrix Inversion Bypass (MIB) is proposed in [26]. This modification exploits the results of the previous algorithm iteration, leading to the reduction of the computational complexity. The proposed method is suitable for parallelization, leading to high-speed low-energy consumption hardware implementations.

A low-power fast OMP (LPF-OMP) algorithm, together with the architecture for its hardware realization, is proposed in [27]. Here, the processing time is decreased by decreasing the number of columns to be searched in consecutive steps. Also, the calculation of the pseudoinverse matrix is avoided, which leads to time and storage savings.

A novel architecture of the OMP algorithm, based on incremental Gaussian elimination (IGE), is proposed in [29]. The verification is done in the FPGA while the analog to information converter (AIC) is used to provide compressed measurements. The Gaussian elimination (GE) is used in place of matrix factorization-based pseudo-inversion, to compute the signal estimate.

An architecture for the realization of the iterative method with adaptive thresholding (IMAT) is proposed in [31]. The algorithm uses two multiplication-free transforms, Walsh–Hadamard transform (WHT) and approximate discrete cosine transform (ADCT) in order to compare which transform is more suitable. It is shown that WHT is more efficient in terms of required hardware resources.

The chapter is organized as follows. An introduction to the CS approach and sparse signal recovery is given in Section 2.2. The gradient-based approach for sparse signal recovery is also described in this section – the one-dimensional and two-dimensional versions of the algorithm are provided. Section 2.3 contains a modification of the original one-dimensional gradient-based approach, that is adopted for efficient hardware realization. This part also contains a description of an analog hardware architecture for a one-dimensional gradient-based sparse recovery algorithm. An analog hardware architecture for a two-dimensional approach is described in Section 2.4. Section 2.5 provides hardware architectures for several possible scenarios in image recovery problems. Namely, the described hardware architecture allows under-sampled image recovery but also has a denoising option. It allows image denoising prior reconstruction. Two types of noise are considered: Gaussian and heavy-tailed (impulse) noise. For the image corrupted by impulse noise, the system identifies the corrupted pixels, discards them, and recovers their amplitudes. For the image corrupted by the Gaussian noise, a random sampler circuit randomly selects a certain percentage of pixels to serve as available, prior to the image being sent to the block for reconstruction. Detailed description of the system is provided in Section 2.5.

2.2 CS and sparse signals recovery

Consider an N-length signal $f(n)$ and its transform domain coefficients $F(k)$, $k = 0,\dots N-1$. The signal $f(n)$ and its transformation $F(k)$ are described by the following relations [1,8]:

$$f(n) = \frac{1}{N}\sum_{k=0}^{N-1} F(k)T_k^{-1}(n) \tag{2.1}$$

$$F(k) = \sum_{n=0}^{N-1} f(n)T_k(n) \tag{2.2}$$

where $T_k(n)$ and inverse $T_k^{-1}(n)$ are basis functions, i.e. sparsity basis. If we denote the direct and the inverse transform matrix as $\mathbf{T}_{N\times N}$ and $\mathbf{T}^{-1}{}_{N\times N}$, then relations (2.1) and (2.2) can be written as follows: $\mathbf{f}_{N\times 1} = \mathbf{T}^{-1}{}_{N\times N}\mathbf{F}_{N\times 1}$ and $\mathbf{F}_{N\times 1} = \mathbf{T}_{N\times N}\mathbf{f}_{N\times 1}$. If only D out of N samples are acquired, then the vector of acquired samples, here denoted as $\mathbf{y}_{D\times 1}$, can be described as:

$$\mathbf{y}_{D\times 1} = \mathbf{M}_{D\times N}\mathbf{f}_{N\times 1}, \tag{2.3}$$

where $\mathbf{M}$ denotes the CS matrix, defined as $\mathbf{M} = \mathbf{\Phi}\mathbf{T}$. The matrix $\mathbf{\Phi}$ models the random selection of the signal samples leading to the satisfaction of the second precondition for the CS application.

The system (2.3) is undetermined since the number of available samples D is less than the signal length N (the common case is $D \ll N$). However, we are looking for the sparsest solution of the system (2.3), which is unique in most cases [38].

The first precondition for successful signal recovery from the incomplete set of samples is the signal sparsity in a suitable domain [1,11,38]. The signal can be sparse in the original domain or in a certain transform domain, such as the Fourier domain, discrete cosine transform domain (DCT), wavelet domains, Hermite transform domain (HT), and time–frequency domain [39–43]. When dealing with natural signals, an exact signal sparsity K, is usually unknown a priori. Therefore, the D measurements needed for successful signal recovery should be chosen empirically. Also, there are signals that are not strictly sparse on an appropriate basis, but only approximately sparse, called the compressible signals. Most of their coefficients have values of nearly zero and such coefficients could be disregarded. Therefore, those signals can be considered as sparse, as the error introduced by disregarding small valued coefficients is commonly negligible. Assume that the signal to be estimated is K-sparse, i.e. has K non-zero values in the sparsity domain and N-K non-zero values. A K-sparse vector has K unknown positions and the K corresponding unknown values. Therefore, the number of available samples D should be larger than $2K$ in order to be able to find the solution of the system (2.3) [38].

The second precondition for successful signal recovery is the incoherence between the measurement and the sparsity basic. Higher values of the coherence index allow recovery with fewer measurements. If the coherence index μ for the K-sparse signal $\mathbf{f}$, satisfies the following relation:

$$K < \frac{1}{2}\left(1 + \frac{1}{\mu}\right), \tag{2.4}$$

then the reconstruction is unique. The coherence index μ is defined according to [2]:

$$\mu(\mathbf{\Phi},\mathbf{T}) = \max_{k \neq l} \left| \frac{\langle \mathbf{\Phi}_k, \mathbf{T}_l \rangle}{\|\mathbf{\Phi}_k\|^2 \|\mathbf{T}_l\|^2} \right|, \tag{2.5}$$

where $\mathbf{\Phi}_k$ i $\mathbf{T}_l$ denote row vector and column vector of the matrices $\mathbf{\Phi}$ and $\mathbf{T}$, respectively. The mutual coherence can be defined using the CS matrix $\mathbf{M}$, as the maximal absolute value of the normalized inner product between all columns of the matrix $\mathbf{M}$ [2]:

$$\mu(\mathbf{M}) = \max_{k \neq l, k \geq 1, l \leq M} \left| \frac{\langle \mathbf{M}_k, \mathbf{M}_l \rangle}{\|\mathbf{M}_k\|^2 \|\mathbf{M}_l\|^2} \right|. \tag{2.6}$$

The signal sparsity, i.e. the number of non-zeros in the vector, can be found by minimizing ℓ_0-norm. The ℓ_0-norm is defined according to the relation:

$$\min_{\mathbf{f}} \|\mathbf{f}\|_{\ell 0} \text{ subject to } \mathbf{y} = \mathbf{Mf}. \tag{2.7}$$

However, for the N-length signal whose sparsity is K, ℓ_0-norm minimization requires searching $\binom{N}{K}$ combinations. This is a nondeterministic polynomial – NP hard problem [38] and therefore, other more computationally suitable approaches are used for finding the sparsest solution of the system (2.3). For example, a convex and non-convex relaxations of the problem (2.7) can be used. Commonly used ℓ_1-norm is a convex relaxation of the NP hard ℓ_0-norm. The minimization problem can now be written as:

$$\min_{\mathbf{f}} \|\mathbf{f}\|_{\ell 1} \text{ subject to } \mathbf{y} = \mathbf{Mf}, \tag{2.8}$$

where $\|\mathbf{f}\|_{\ell_1} = \sum_{n=0}^{N-1} |f(n)|$. The sparsest solution can be found using both ℓ_0-norm and ℓ_1-norm minimization. An NP hard problem of the ℓ_0-norm minimization can be solved given enough time, but requires $D \geq 2K$ equations. The convex and easier to solve the ℓ_1-norm minimization requires larger number of equations, $D \geq CK\log(N)$, where C is a constant. The number of equations should be minimal as possible, since less equations requires less samples and consequently, reduce sampling requirements such as power consumption and acquisition time [38].

The problem (2.8) can be solved by using some of the algorithms for sparse signal reconstruction. A number of algorithms are proposed mostly based on the convex optimizations and greedy approaches. Convex optimizations successfully solve the Basis Pursuit problem defined by (2.8), while Basis Pursuit Denoising (BPDN) is applied if we are dealing with the noisy data. The BPDN problem in the Lagrangian form is defined as follows:

$$\min_{\mathbf{f}} \frac{1}{2} \|\mathbf{y} - \mathbf{Mf}\|_{\ell_2}^2 + \beta \|\mathbf{f}\|_{\ell_1}, \quad \beta > 0, \tag{2.9}$$

where β is the regularization parameter. For solving the problem defined by (2.9), primal-dual interior-point method can be used [2]. However, convex optimization methods are unsuitable for hardware realizations due to the requirements for high computing power and high numerical precision [32]. Greedy approaches are computationally less demanding while producing optimal solutions operating in an iterative manner. Some of the algorithms belonging to the group of greedy approaches are the MP and its variations, such as orthogonal MP (OMP), Compressive Sampling MP (CoSaMP), then Iterative hard and soft thresholding (IHT and IST), gradient-based approaches, etc. If the missing samples are considered as variables, then the sparsest solution can be found by iteratively changing the unavailable samples' values while keeping the amplitudes of the available samples unchanged. This is how the gradient-based approach [24,35,40–43] works. This approach will be described in the sequel. The advantage of this approach is that it does not require a signal to be strictly sparse in a certain domain, which makes it suitable for wider applications. Also, it allows recovery with high precision using a relatively small amount of available samples.

The gradient-based algorithm belongs to the group of convex optimization approaches [43]. The algorithm deals with under-sampled one-dimensional and two-dimensional signals and various types of sparsifying basis. Which sparsifying basis will be used depends on the signal nature. First, a one-dimensional approach will be described, together with an analog architecture for its realization.

2.2.1 A gradient-based algorithm for sparse under-sampled signal reconstruction – one-dimensional approach

Let us introduce the gradient-based approach for the under-sampled one-dimensional signal **f** reconstruction [43]. The procedure is as follows. Initially, the signal consists of the available and missing samples. The amplitudes of the missing samples can be set to zero or some other user-defined value. The procedure of finding their exact values can be iteratively performed in the sparsity domain using the steepest (gradient) descent approach. The values of the available samples are changed in each iteration leading to the constant improvement of the concentration in the domain of sparsity. The values are changed for predefined step λ, i.e. for $+\lambda$ and $-\lambda$ and the concentration measure is calculated for both changes, leading to the gradient vector determination. This gradient vector is then used to update the missing samples values [2,19,43].

Assume a set of acquired samples positions $\mathbf{S}_a$ and a set of missing samples positions as $n_i \in \mathbf{S}$. The procedure starts from the signal [8,43]:

$$\mathbf{f}^{(0)}(n) = \begin{cases} \mathbf{f}(n) \text{ for available samples}, & n \in \mathbf{S}_a \\ 0 \quad \text{for missing samples}, & n \in \mathbf{S}_m \end{cases} \tag{2.10}$$

applying the sparsity-measure minimization according to [2,10,43]:

$$\mathbf{f}^{(l+1)}(n_i) = \mathbf{f}^{(l)}(n_i) - \alpha \mathbf{g}(\mu_p), \tag{2.11}$$

where $\mathbf{f}^{(l)}$ denotes the signal in l-th iteration, $\mathbf{f}(n_i)$ denotes the signal at the positions of missing samples, α denotes the iteration step, μ_p is the sparsity measure while $\mathbf{g}$ is the vector that denotes the gradient of the sparsity measure and it is defined as:

$$\mathbf{g}(\mu_p) = \frac{1}{2\lambda N} \frac{\partial \mu_p}{\partial \mathbf{f}(n_i)}\bigg|_{\mathbf{f}(n_i)=\mathbf{f}^{(l)}(n_i)}. \tag{2.12}$$

To ensure the algorithm convergence, it is required for the gradient of the sparsity measure to be a convex function. The sparsity measure μ_p, could be e.g. the ℓ_1 or ℓ_2 norm. It is calculated using the missing samples' positions according to the following relation [43]:

$$\mu_p(\mathbf{F}^{\pm}) = \sum_k \left| \mathbb{T}\left[\mathbf{f}^{(l)}(n) \pm \lambda\delta(n - n_i)\right]\right|^{\frac{1}{p}}$$
$$\lambda \leftarrow \max_{n \in S_a} \left|\mathbf{f}^{(l=0)}(n)\right|. \tag{2.13}$$

Parameter λ is a finite difference step whose optimal value is $\lambda = \alpha/2$ [10,19,43], and $\mathbb{T}$ denotes the transformation operator, $F(k) = \mathbb{T}\{f(n)\}$. Finally, the gradient vector Γ is calculated for each missing sample n_i and has zero value at the available samples' positions:

$$\Gamma^{(l)}(n_i) = \left(\mu_p(\mathbf{F}^+) - \mu_p(\mathbf{F}^-)\right)(2\lambda N)^{-1}. \tag{2.14}$$

The number of iterations of the gradient-based algorithm depends on the stopping criterion defined as [14,19,43]:

$$\varepsilon = \frac{\langle \mathbf{g}^{(l-1)} \mathbf{g}^{(l)} \rangle}{\|\mathbf{g}^{(l-1)}\|_{\ell_2}^2 \|\mathbf{g}^{(l)}\|_{\ell_2}^2}, \quad \varepsilon < \cos(175^\circ). \tag{2.15}$$

If the stopping criterion is satisfied, i.e. when ε becomes larger than $\varepsilon_{ref} \approx 175^\circ$, further concentration improvement for a precision defined by the parameter λ is not possible.

Therefore, the value of the step λ should be decreased to one-third of its previous value [19,43].

2.2.2 *Gradient-based approach for sparse under-sampled image reconstruction*

Similar to the one-dimensional case, the algorithm changes the values of the available signal samples in an iterative manner, improving the concentration in the sparsity domain. An adaptable step $\pm\lambda$ is used with a goal to approach an exact signal value and improve the concentration in the sparsity domain [35].

The two-dimensional gradient-based algorithm can be applied to the image blocks. Here, we will assume 8×8 image block size.

As in the one-dimensional case, the values of the missing pixels are updated by the gradient vector, obtained as a difference between the ℓ_1-norms of the vectors changed for $+\lambda$ and $-\lambda$.

Assume the $N \times M$ image with some of the pixels missing. The algorithm steps can be described as follows:

Step 1: Assume set of the available samples position as $(n_x, n_y) \in \mathbf{S}_a$ are the available pixels. The values of the unavailable pixels are set to zero or to some other arbitrary value. Set $\mathbf{S}_a$ denotes the positions of the available pixels, while the $\mathbf{S}_m$ set contains the positions of the unavailable pixels. Consider $\mathbf{y}$ as 2D signal – image block or the whole image and represent the image with unavailable pixels, defined as [19,35,41,42]:

$$y^{(0)}(n_x, n_y) = \begin{cases} f(n_x, n_y), & \text{for } (n_x, n_y) \in \mathbf{S}_a \\ 0, & \text{for } (n_x, n_y) \in \mathbf{S}_m \end{cases}, \tag{2.16}$$

Step 2: Set the initial value for the step λ:

$$\lambda = \max_{n_x, n_y} \left| y^{(0)}(n_x, n_y) \right|. \tag{2.17}$$

Step 3: Calculate the transformation coefficients $\mathbf{Y}^+$ and $\mathbf{Y}^-$:

$$\begin{aligned} Y^+(n_x, n_y) &\leftarrow \mathbb{T}\left[y^{(l)}(n_x, n_y) + \lambda\right] \\ Y^-(n_x, n_y) &\leftarrow \mathbb{T}\left[y^{(l)}(n_x, n_y) - \lambda\right] \end{aligned} \tag{2.18}$$

where $\mathbb{T}$ is the 2D transform operator (e.g., *2D DCT*).

Step 4: Gradient vector calculation, based on the difference of the ℓ_1 norms of the vectors $\mathbf{Y}^+$ and $\mathbf{Y}^-$:

$$g^{(l)}(n_x, n_y) \leftarrow \frac{1}{NM}\left(\|\mathbf{Y}^+\|_{\ell_1} - \|\mathbf{Y}^-\|_{\ell_1}\right). \tag{2.19}$$

Step 5: Update the missing pixel value:

$$y^{(l+1)}(n_x, n_y) \leftarrow y^{(l)}(n_x, n_y) - g(n_x, n_y) \tag{2.20}$$

The values of the unavailable pixels are changed until the stopping criterion is reached. The stopping criterion is defined as an angle between direction of two successive gradient vectors and can be calculated according to (2.15).

Similar as in the one-dimensional case, when value of the parameter ε becomes lager than $\varepsilon_{ref} \approx 175°$, the step λ needs to be decreased to $\lambda/3$. The procedure is repeated until precision $P > P_{\max}$ is achieved. The P denotes the threshold, defined according to the following relation:

$$P = 10\log_{10}\frac{\sum\sum_{(n_x,n_y)\in\mathbf{S}_m}|y_p(n_x, n_y) - y^{(l)}(n_x, n_y)|^2}{\sum\sum_{(n_x,n_y)\in\mathbf{S}_m}|y^{(l)}(n_x, n_y)|^2}, \tag{2.21}$$

where $y_p(n_x,n_y)=y^{(0)}(n_x,n_y)$ for every new λ value and $P_{\max} \approx 100$ dB [35].

2.3 An analog hardware architecture for gradient-based reconstruction approach

2.3.1 A modified version of the 1D gradient-based approach suitable for hardware realization

The one-dimensional version of the gradient-based algorithm is modified in order to be adapted for an efficient hardware implementation avoiding complex functions and calculations. The modified version of the algorithm is described in the sequel [19,24].

Input:

Step 0: Set $\mathbf{y}^{(0)}(n) \leftarrow \mathbf{y}(n)$

Step 1: $l \leftarrow 0$

Step 2: $\lambda \leftarrow \max|\mathbf{y}^{(l)}(n)|$

Part 1:

Step 3: **Set** $\mathbf{y}_p(n) = \mathbf{y}^{(l)}(n)$,

Part 2:

Step 4: **for** $n_i \leftarrow 0$ to $N-1$ **do**

Step 5: **if** $n_i \in \mathbf{S}_m$

Step 6: $\mu_1(\mathbf{f}^+) = \sum_k \left|\mathbb{T}\left[\mathbf{f}^{(l)}(n) + \lambda\delta(n - n_i)\right]\right|$

Step 7: $\mu_1(\mathbf{f}^-) = \sum_k \left|\mathbb{T}\left[\mathbf{f}^{(l)}(n) - \lambda\delta(n - n_i)\right]\right|$

Step 8: $\mathbf{g}^{(l)}(n_i) \leftarrow \frac{1}{N}(\mu_1(\mathbf{f}^+) - \mu_1(\mathbf{f}^-))$

Step 9: **else** $\mathbf{g}^{(l)}(n_i) \leftarrow 0$
Step 10: **end if**
Step 11: **end for**

Part 3:

Step 12: $\mathbf{y}^{(l+1)} \leftarrow \mathbf{y}^{(l)} - \mathbf{g}^{(l)}$
Step 13: $\varepsilon_l = \frac{\langle \mathbf{g}^{(l-1)} \mathbf{g}^{(l)} \rangle}{\|\mathbf{g}^{(l-1)}\|_{\ell_2}^2 \|\mathbf{g}^{(l)}\|_{\ell_2}^2}$

Part 4:

Step 14: **if** $\varepsilon_l < \cos(175°)$
Step 15: **Increase iteration number** l, $l + 1$
Step 16: **go to 4**
Step 17: **else**
Step 18: $\lambda \leftarrow \lambda/3$
Step 19: $P = \left(\sum_{n \in S_m} \left| \mathbf{y}_p(n) - \mathbf{y}^{(l)}(n) \right|^2 \right) \left(\sum_{n \in S_m} \left| \mathbf{y}^{(l)}(n) \right|^2 \right)^{-1}$
Step 20: **end if**
Step 21: **if** $P > P_r$
Step 22: **Increase iteration number** l, $l + 1$
Step 23: **go to 2**
Step 24: **end if**

Result:

Step 25: **return** $\mathbf{y}^{(l)}(n)$

In order to ease the hardware implementation, the algorithm is separated into several parts that could be performed in parallel. In the proposed analog hardware architecture, the following steps are identified to be implemented [24]:

- According to the Step 0, the signal samples $\mathbf{y}$ are set to the input of the architecture. A digital signal $\mathbf{e}$ is formed, where positions of the available signal samples are set to high voltage (logical 1) and the positions of the unavailable samples are set to low voltage (logical 0);
- According to the Step 2, set the value for the parameter λ;
- The next is implementation of the Part 2, i.e. updating the values of the gradient vector $\mathbf{g}$;
- The values of the unavailable samples and the parameter ε are updated according to the Steps 12 and 13 (Part 3);
- Part 4 checks whether is it necessary to change the parameter λ. If $\varepsilon_i < \cos(175^{\circ})$, the parameter λ is changed to $\lambda = \lambda/3$ and the stopping criterion condition $P > P_r$ is checked. If the stopping criterion is satisfied, the procedure stops and the current values of input signal represent the reconstruction result. If the stopping criterion is not satisfied, the procedure repeats with new value of λ.

  If the condition $\varepsilon_i < \cos(175^{\circ})$ is not satisfied, then the Steps 4–13 are repeated.

When the algorithm reaches minimum of the sparsity measure, the gradient changes direction for 175 degrees. It means that the step λ should be decreased.

2.3.2 *An analog hardware architecture for a modified version of the 1D gradient-based approach*

The proposed analog architecture is shown on a block scheme in Figure 2.1 [24]. The inputs in the architecture are an array **y** that contains the analog signal samples, and an array **e** where positions of the available and unavailable samples are stored. Four blocks that correspond to the parts 1–4 in the Algorithm's description are denoted as PART 1, PART 2, PART 3 and PART 4.

The architecture contains an oscillator circuit, producing the signals *O*1, *O*2, *O*3 and *O*4 that control the performance of the architecture blocks. Signal *START* is used to control the oscillator outputs. The outputs are set to the low voltage level before the start signal activation. Also, at this point, the values of the vector **y** are loaded to the sample and hold amplifiers. When *START* is activated, the oscillator outputs *O*1, *O*2, *O*3, and *O*4 are as in Figure 2.2. Parameters *T*1, *T*2, *T*3, and *T*4 denote the duration of the active states for the signals *O*1, *O*2, *O*3, and *O*4, respectively. Circuit "λ correction" sets the value for the ε. Initial value of this parameter is sent to the output of the "λ correction" circuit when *START* signal is activated, while the signal *CRC* causes the eventual change of λ value [24].

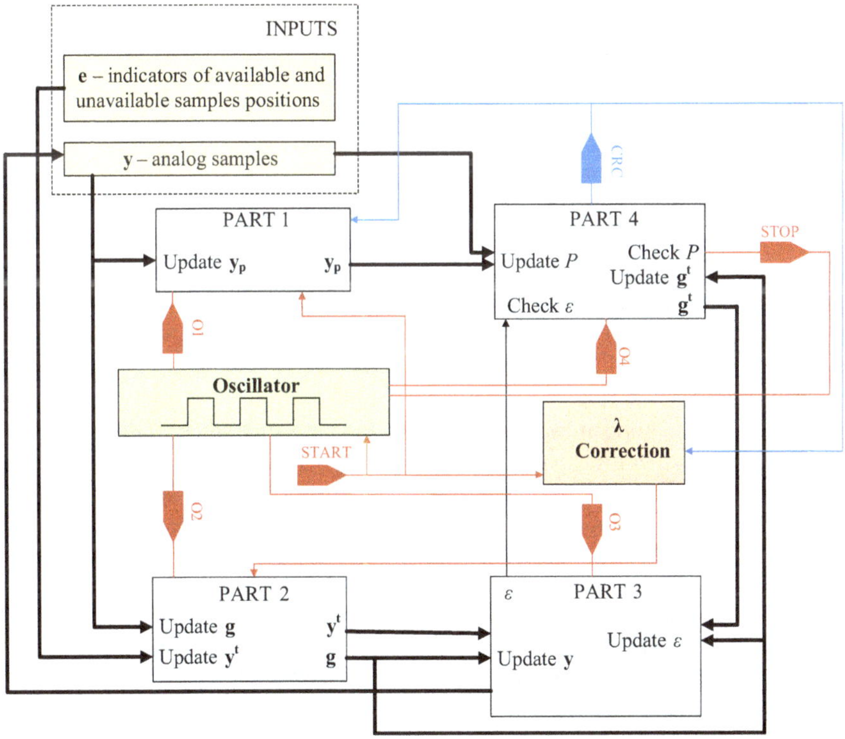

Figure 2.1 The proposed analog architecture scheme for one-dimensional gradient-based signal reconstruction

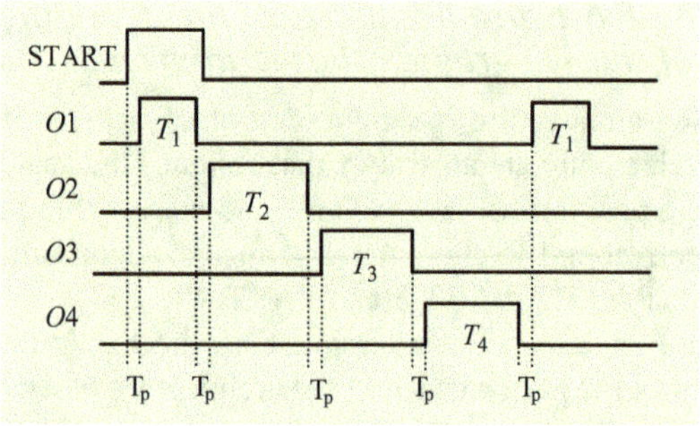

Figure 2.2 Signals O1, O2, O3, and O4 represent the oscillator outputs

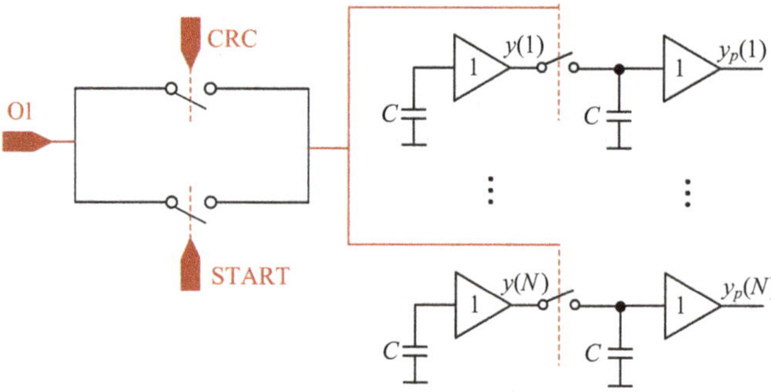

Figure 2.3 The architecture of the PART 1, sample, and hold circuits is controlled with O1, START, and CRC signals

The PART 1 of the architecture temporarily keeps values of $\mathbf{y}$ in the vector $\mathbf{y_p}$. This vector is in fact a reconstructed signal for current value of parameter λ. The values of $\mathbf{y_p}$ vector are updated when the parameter λ changes its value to $\lambda/3$. The $\mathbf{y_p}$ vector is used for value P generation, which is a mean squared error between $\mathbf{y}$ before and $\mathbf{y}$ after applying λ correction.

As it can be seen from Figure 2.1, PART 1 is activated when *START* (or *CRC*) is activated and when *O*1 signal is in the high logical value. When *START* is activated and during the first *T*1 impulse, the initial value of $\mathbf{y}$ is loaded into the vector $\mathbf{y_p}$. Active states of *O*1 and *CRC* signals control further change of vector $\mathbf{y_p}$. The electronic scheme for the PART 1 is shown in Figure 2.3.

Figure 2.4 shows the electric scheme of the "λ correction" circuit. The initial value of the λ is obtained when START is activated. Then the capacitor *C* is loaded to the approximately maximal absolute value of the input signal samples and the further change is controlled by the CRC signal – if CRC is at high logical state, the λ becomes $\lambda/3$ [24].

The block scheme of the PART 2 is shown in Figure 2.5. It is used to update the values of the gradient vector $\mathbf{g}$ (for each missing sample position). This block provides parallel computation gradient samples *g*(*i*). The *O*2 output from the oscillator (when *O*2 is in a high voltage state) activates the PART 2 of the

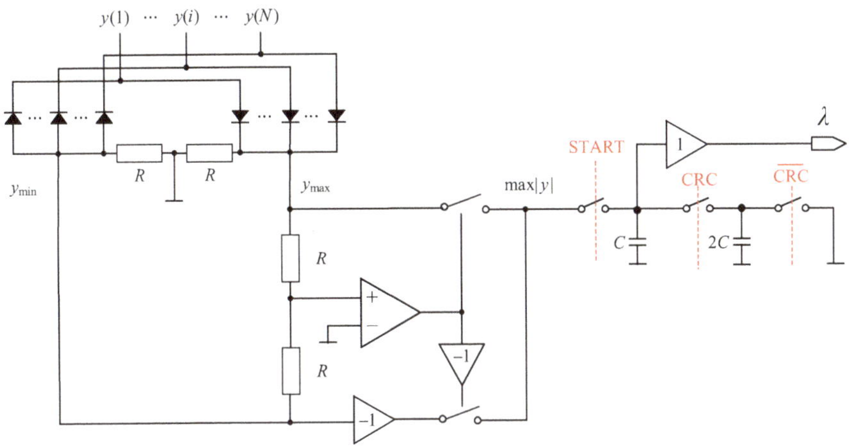

Figure 2.4 The scheme for the λ correction circuit

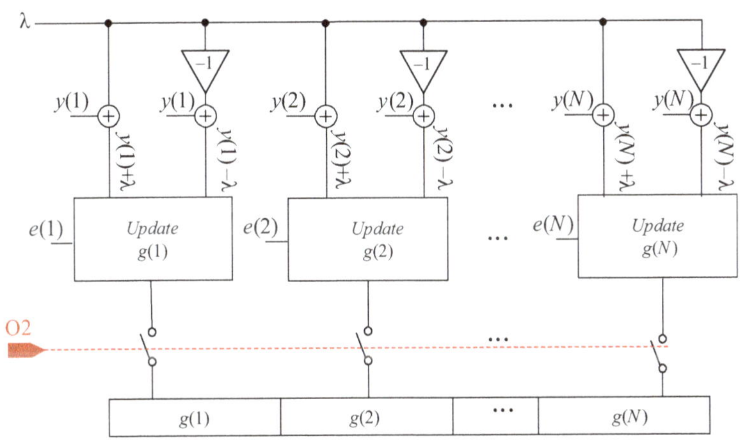

Figure 2.5 PART 2 of the architecture. Gradient vector samples generation and values update.

architecture. Current values of **y** are placed in $\mathbf{y}^{\mathbf{t}}$ (in the algorithm $y^{t} = y^{(i)}$ for the *i*th iteration), i.e. they are loaded into the sample and hold amplifiers $\mathbf{y}^{\mathbf{t}}$.

Gradient samples are calculated in PART 2 of the architecture. Block scheme for the gradient sample generation is shown in Figure 2.6. The coefficient of the transformation matrix **T** is denoted as T_{ij}. The amplifiers are used for representing the coefficients T_{ij}, as it can be seen from Figure 2.6. The amplifiers represented using dashed lines in Figure 2.6, used for calculation of the $y(p)-\lambda$, could be avoided in implementation since the required values could be used from the outputs of the appropriate amplifiers for the $y(p)+\lambda$ calculation (left side of Figure 2.6). The block scheme also contains circuits for absolute value generation and adder

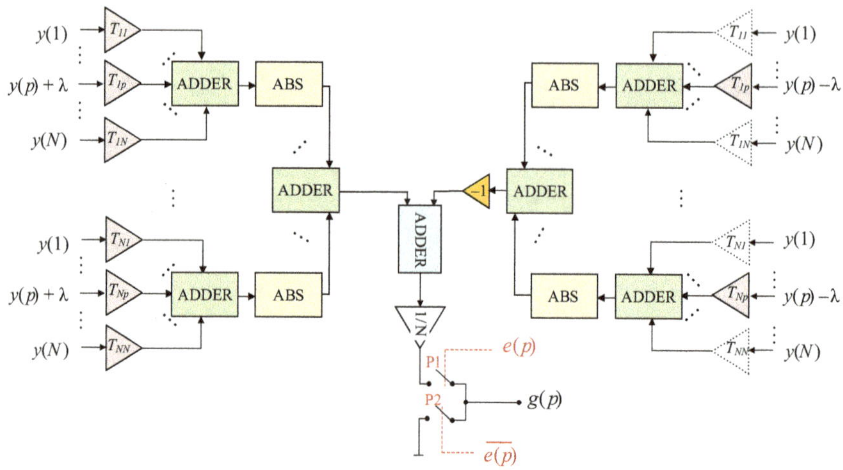

Figure 2.6 Block scheme of the architecture for the generation of the gradient coefficient value g(p), at the position p

circuits. The scheme contains switches $P1$ and $P2$. The switches are controlled by the signal e, i.e. by the indicators of available and unavailable samples, and close the contacts alternately, i.e. [24]:

- In the case of the available sample – $e(p)$ opens $P1$, $\overline{e(p)}$ closes $P2$.
- In the case of the unavailable sample – $e(p)$ closes $P1$, $\overline{e(p)}$ opens $P2$.

Namely, when $P1$ is open, $P2$ is closed and when $P1$ is closed, $P2$ is opened. As the gradient is calculated only for unavailable samples, the $P1$ closes the contact on the positions of unavailable samples and opens the contact at the positions of available samples (the opposite scenario is with $P2$ switch).

Let us provide the complexity of the block for gradient sample generation. This part of the architecture requires $N^2 + N + 1$ amplifiers, $2N$ circuits for absolute value generation, $2N + 2$ adders with N inputs, 1 adder with two inputs, 1 invertor circuit and 2 switches.

The electric scheme of the circuit that performs updating of the $\mathbf{y}$ values is shown in Figure 2.7. It is the PART 3 of the architecture that performs update of the $\mathbf{y}$ signal values from current signal values stored in $\mathbf{y}^t$ (PART 2 of the architecture) and values from the $\mathbf{g}$ ($y^{(l+1)} = y^{(l)} - g^{(l)}$ in the algorithm). This part of the architecture is activated when $O3$ signal is in a high logical state (T3 impulse).

Additionally, PART 3 of the architecture performs update of the ε value. The block scheme of this circuit is shown in Figure 2.8. It consists of analogue voltage adders (denoted as ADDER in the scheme), circuit for square root generation (SQRT), squaring circuit, multiplier circuits (MULT), divider circuit (DIVIDER) and sample and hold amplifiers [24].

The parameter $g^t(i)$ denotes the gradient values from the previous iteration, while the update of $\mathbf{g}^t$ vector is done in the PART 4. The value of ε is led to the input of the circuit for validating the condition for changing the value of λ.

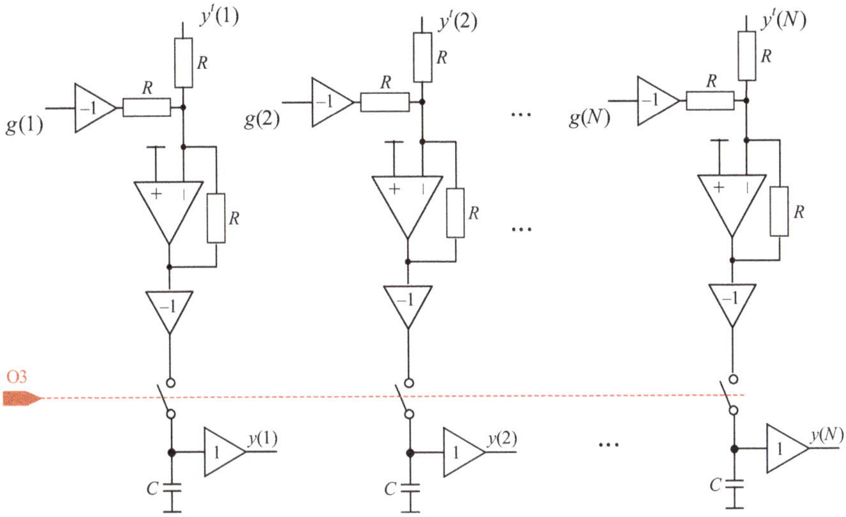

Figure 2.7 Electric scheme of the circuit that performs updating of the y values (PART 3 of the architecture)

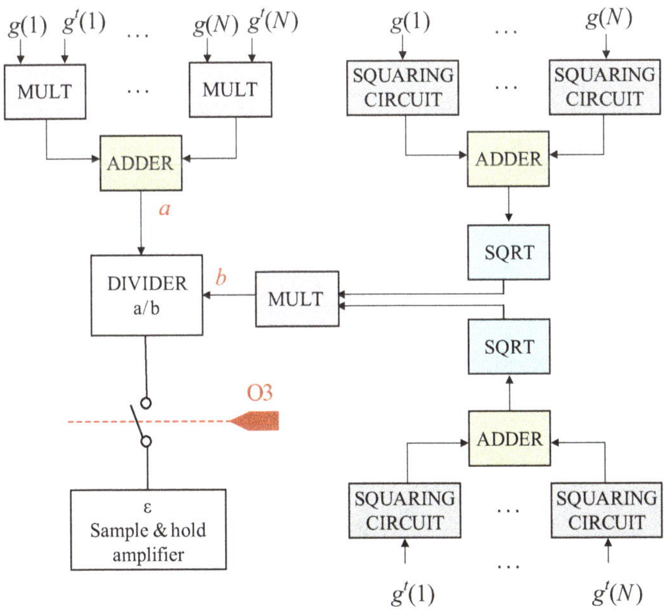

Figure 2.8 Block scheme for a circuits that perform update of the ε value (PART 3 of the architecture)

This part of the architecture for ε value generation has the following complexity: it requires three N-inputs adder circuits, $N + 1$ multiplier, one divider, $2N$ squaring circuits, and two circuit for square root generation [24].

The PART 4 is activated when $T4$ impulse arrives, i.e. when signal $O4$ is in a high logic state. This part has several roles. Specific part within PART4 is used to load current values from the vector **g** into the vector $\mathbf{g^t}$. Another part checks $\varepsilon < \varepsilon_{ref}$ condition and activates the signal CRC (as a trigger to update the value of λ) if the condition is satisfied [24]. This part is shown in Figure 2.9. When signal CRC is in its active state, checking the stopping criterion is allowed.

There is also part for updating the value of the threshold P (Figure 2.10) and part for stopping criterion $P < P_{ref}$ checking (Figure 2.11). The circuit in

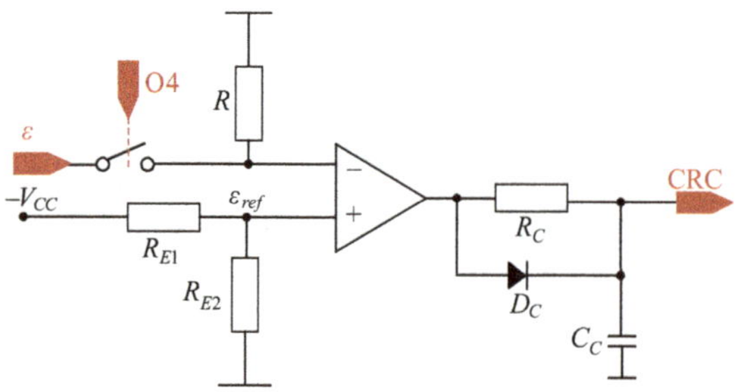

Figure 2.9 Circuit in PART 4 of the architecture used for $\varepsilon < \varepsilon_{ref}$ condition checking. CRC signal is activated if the condition is satisfied.

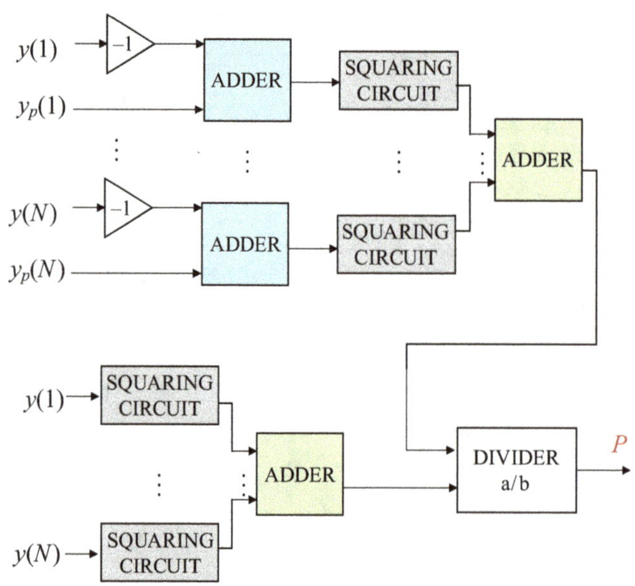

Figure 2.10 Circuit in PART 4 of the architecture used for threshold value P update

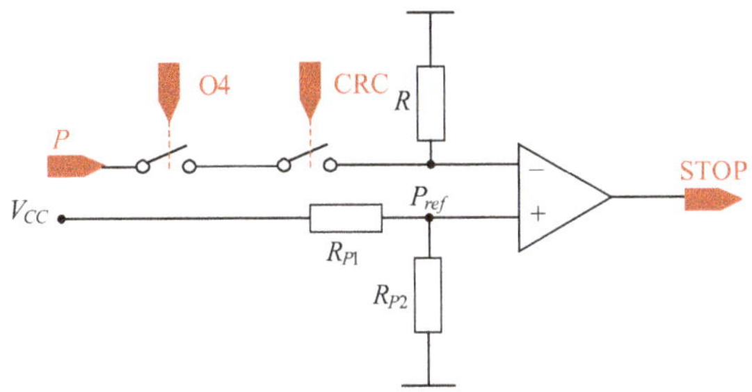

Figure 2.11 Circuit in PART 4 of the architecture used to check the stopping criterion, $P < P_{ref}$

Table 2.1 The number of the most represented circuits required

	Input signal circuits	**PART 1**	**PART 2**	**PART 3**	**PART 4**	**Total**
Amplifiers			$N^3 + N^2 + 3N$	$2N$	N	$N^3 + N^2 + 6N$
Adders			$2N^2 + 4N$	$N + 3$	$N + 2$	$2N^2 + 6N + 5$
Absolute value circuits			$2N^2$			$2N^2$
Sample and hold amplifiers	N	N	$2N$	1	N	$5N + 1$
Switches		$N + 2$	$4N$	$N + 1$	$N + 3$	$7N + 6$
Multipliers/ dividers				$3N + 4$	$2N + 1$	$5N + 5$

Figure 2.10 consists of the $N + 2$ adders, N invertor circuits, $2N$ squaring analog voltage circuits, and 1 division circuit [24].

When stopping criterion is satisfied, the signal *STOP* is activated and it deactivates all oscillator outputs and causes opening of all switches in the hardware. If the stopping criterion is not met, the R_cC_c time constant keeps the high voltage of *CRC* signal for the next $T1$ period. The $T1$ period enables update of sample and hold amplifiers denoted by $\mathbf{y_p}$. Table 2.1 shows the required components per block (only the main and most represented components are listed, such as amplifiers, adders, and multipliers).

2.3.3 An OrCAD simulation

The described architecture is simulated using the PSpice (OrCAD 16.6). Prior discussing the simulation time, let us first list analog devices used for specific circuits realization [24]:

- AD734 analog device is used for the realization of the multipliers and dividers, circuits for squaring and square root generations. This device is a precise four-quadrant high speed analogue multiplier, suitable for the realization of the mentioned circuits.

- LT1394 (comparator, a settling time of 7 ns) [24,44] is used as an analog voltage comparator.
- LT1191 device, an ultra-high-speed operational amplifier, is used to construct amplifiers, inverters, adders and absolute value circuits [24,45].
- AD783 sample and hold circuits are exploited to store the voltage values [24].
- MAX4645 analogue switches [24,46].
- 1N4448 diodes [24,47].

The components that are used in the architecture introduce negligible error. The error affects the number of iterations required for achieving satisfactory reconstruction precision. The maximal introduced error causes just a few more iterations of the algorithm, which does not affect much the reconstruction time. The biggest error is introduced by AD734 device. The typical error is 0.1% of the full scale, or 0.25 in full temperature range from −40 °C to +85 °C [24].

The time T_p (Figure 2.2) is the time required for turning off the analog switches. For the switches used in the described architecture, MAX4645, this T_p time is up to 40 ns. The simulation showed that duration times for the oscillator outputs are $T_1 \approx$ 320 ns, $T_2 \approx$ 740 ns, $T_3 \approx$ 650 ns, and $T_4 \approx$ 610 ns. Therefore, one iteration of the algorithm requires around 2.5 μs, i.e. $T_1 + T_2 + T_3 + T_4 + 4T_p = 2.48$ μs [24]. Depending on the number of iterations, the total calculation time will be obtained. This architecture allows up to 400 iterations within 1 ms [24]. Having in mind that the gradient-based approach for most signals in real applications and for different percentage of available samples provides solution in just a few iterations [19], the number of iterations allowed by this architecture is enough. Therefore, it may be concluded that the described analog architecture provides fast processing and low-complexity realization and it is suitable for real-time applications.

In the sequel, the example of under-sampled signal recovery using the gradient-based algorithm is provided. The performance of the algorithm is tested by using MATLAB® R2015a software. The real QRS complex of the electrocardiogram (ECG) signal is observed [40]. The signal length is 50 samples. The original signal is shown in Figure 2.12, while the signals recovered using 40%,

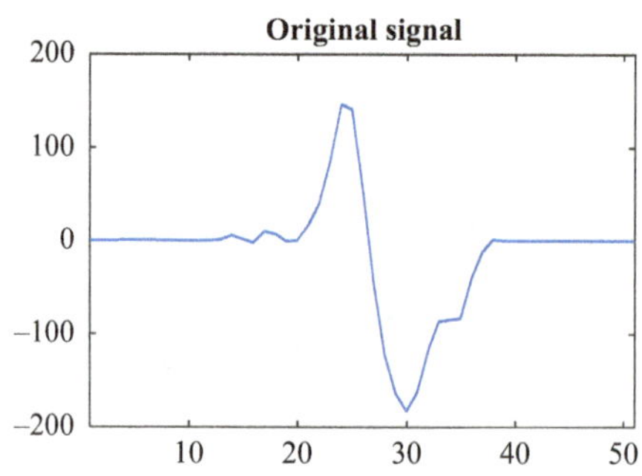

Figure 2.12 The original QRS signal

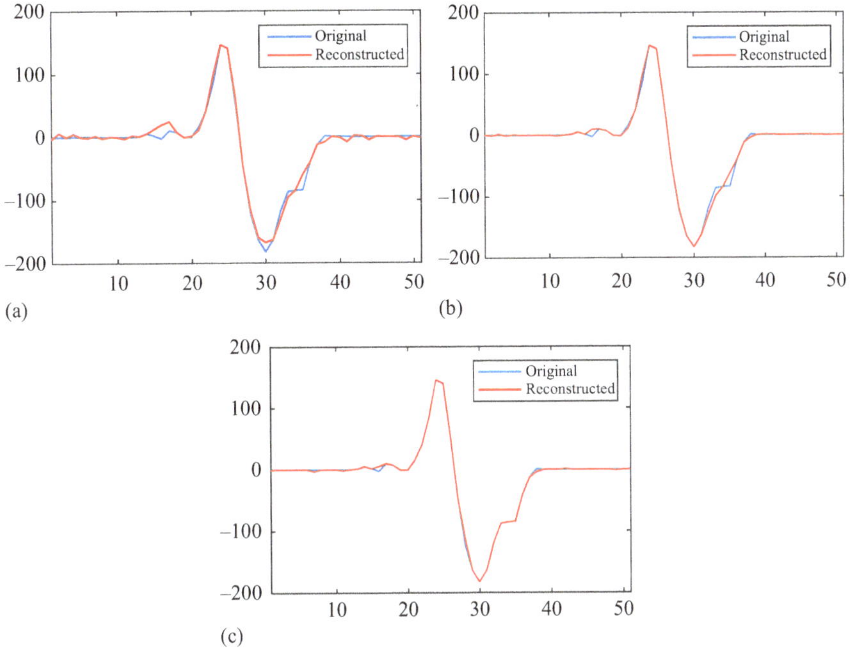

Figure 2.13 The original (blue line) QRS signal and reconstructed (red line) QRS signal using: (a) 40%, (b) 60%, and (c) 80% of the available samples

60%, and 80% of the available samples are shown in Figure 2.13. The reconstruction error in all considered cases is of order 10^{-3}. The number of iterations is set to 50 for all considered percentages of available samples. It is important to emphasize that for a larger percentage of available samples (50% or higher), the reconstruction can be done in around 10 iterations of the gradient-based algorithm.

2.4 An analog architecture for the 2D gradient-based sparse reconstruction algorithm

The analog hardware implementation for the 2D gradient reconstruction algorithm consists of four parts [35]. The described implementation works with 8×8 image blocks. The input of the architecture is matrix $\mathbf{y}$ which is the 8×8 part of the $N \times M$ image and the 8×8 binary matrix $\mathbf{e}$, which indicates the positions of the unavailable pixels that need to be reconstructed (0 indicates the position of available pixels while 1 indicates the position of unavailable pixels).

The main functions of all four parts of the architecture are as follows [35]:

- **Part 1** contains circuit that checks the condition for parameter λ decrease according to the relation: $\varepsilon = \frac{\langle \mathbf{g}^{(l-1)} \mathbf{g}^{(l)} \rangle}{\|\mathbf{g}^{(l-1)}\|_{\ell_2}^2 \|\mathbf{g}^{(l)}\|_{\ell_2}^2}$ and updates step λ if the condition

$\varepsilon_i < \cos(175°)$ is satisfied. This part also contains circuits that provisionally keeps the current values of input block $\mathbf{y}$ in the $\mathbf{y}^{\mathbf{p}}$ and $\mathbf{y}^{\mathbf{t}}$ matrices, and current values of gradient matrix $\mathbf{g}$ in the matrix $\mathbf{g}^{\mathbf{t}}$.

- **Part 2** performs update of the gradient vector $\mathbf{g}$ values according to the relation $\mathbf{g}^{(l)}(n_x, n_y) \leftarrow (\|\mathbf{Y}^+\|_{\ell_1} - \|\mathbf{Y}^-\|_{\ell_1})/(NM)$.
- **Part 3** performs update of the unavailable $\mathbf{y}$ values according to the relation $\mathbf{y}^{(l+1)}(n_x, n_y) \leftarrow \mathbf{y}^{(l)}(n_x, n_y) - \mathbf{g}(n_x, n_y)$ and calculates the value ε.
- **Part 4** checks the stopping criterion and if it is satisfied, the signal *STOP* stops the reconstruction procedure and the current values of the input signal become the reconstruction result. This part also checks the parameter ε update condition and performs update of the parameter P value.

The block scheme is shown in Figure 2.14. The architecture also contains an ℓ_1-output oscillator that controls all of the parts. The oscillator outputs are denoted

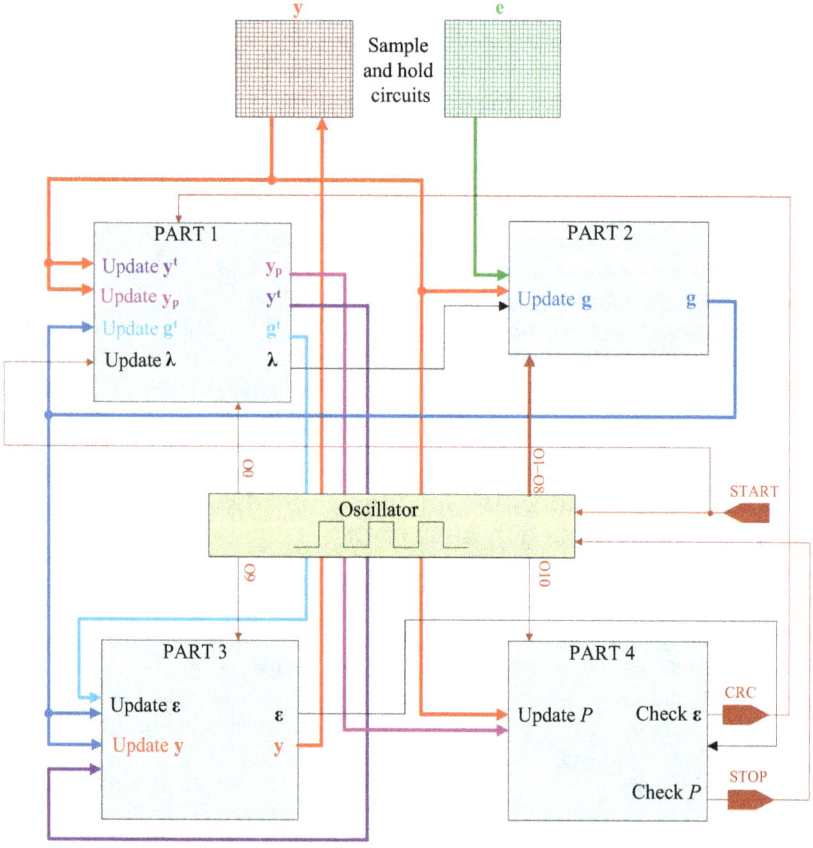

Figure 2.14 The block scheme for 2D gradient-based analog realization

as *O*0–*O*10. After *START* signal activation, signals *O*0–*O*10 change according to the diagram in Figure 2.15. Using sample and hold amplifiers shown in Figure 2.16, the acquired pixels are loaded into the matrix **y** before *START* signal activation. Also, matrix **e** is set at this point. A detailed description of the architecture's parts is provided in the sequel.

Part for the λ value update is the same circuit shown in Figure 2.4, used for the λ correction in one-dimensional implementation of the gradient algorithm. The

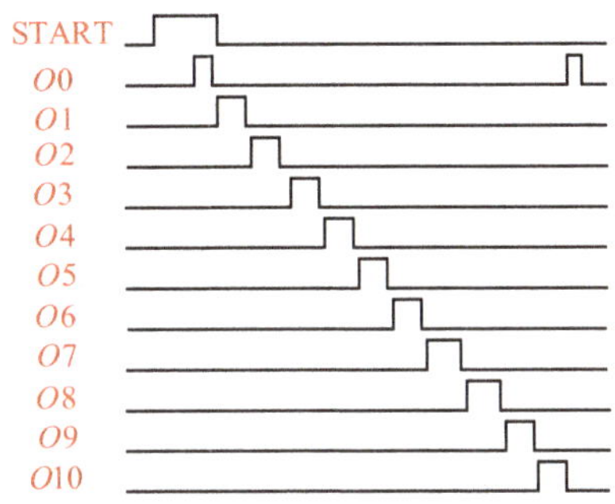

Figure 2.15 Output signals of the oscillator

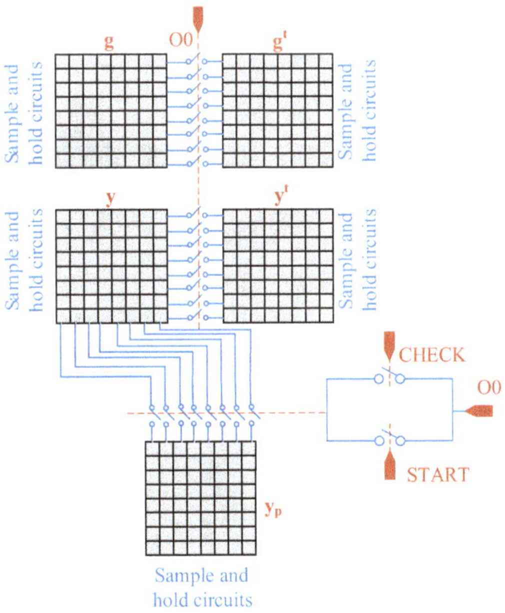

Figure 2.16 Part 1 of the architecture – block scheme of the circuit for placing samples of **y** *matrix into* $\mathbf{y_p}$ *and* $\mathbf{y^t}$ *matrix*

only difference is that the input here is from an 8 × 8 image block, i.e. $y(1,1),\ldots,$ $y(i,j),\ldots y(8,8)$. The circuit function is the same as for the one-dimensional case: the capacitor *C* is loaded to the maximal absolute value of the input signal samples as long as the *START* signal is in its active state. The change of the value λ is dependent on the *CRC* signal – the high voltage level of the *CRC* causes a decrease of the λ value to approximately $\lambda/3$ [35].

The block diagram for part 2, which performs an update of the gradient vector **g**, is shown in Figure 2.17.

This part of the architecture consists of eight circuits that perform parallel updates of eight values of the **g** matrix. In such a way, the complexity of the architecture is eight times decreased compared with the case if all gradient factors are simultaneously updated. The price is paid in processing time, which is eight times increased. However, the processing time is relatively fast and it is still faster compared with the software processing time. The circuit is

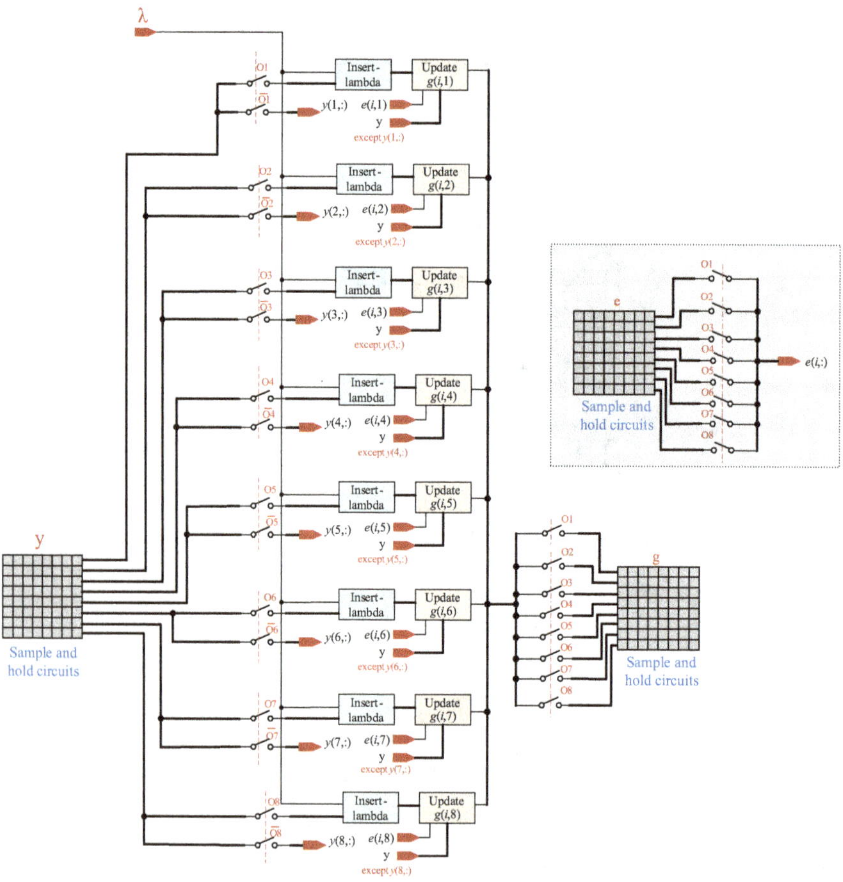

Figure 2.17 Part 2 – block performing update the gradient for every missing sample

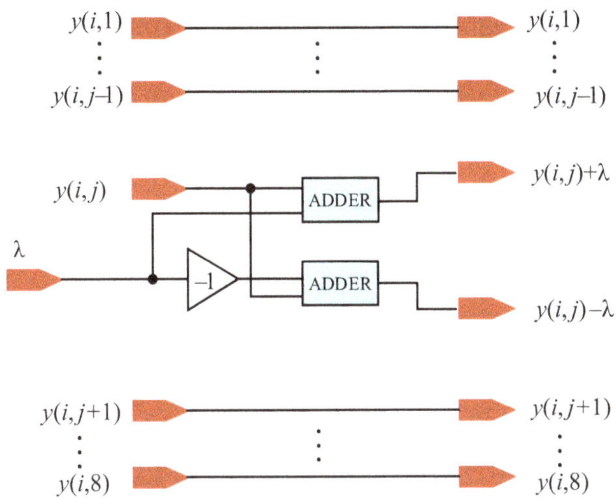

Figure 2.18 Realization of the Insert-lambda circuit. Adding/subtraction of the value λ to/from the sample y(i,j).

controlled with the oscillator outputs $O1$–$O8$. The active state of the corresponding signal $O1$–$O8$ updates the corresponding row of the sample and hold matrix **g**. The architecture contains the *Insert-lambda* part, that adds/subtracts λ to/from the observed input samples. The realization of this part is shown in Figure 2.18 [35].

Block scheme of the part that updates one gradient value is shown in Figure 2.19 [35]. The circuit consists of amplifiers, adders, circuits for absolute value calculation and switches. The coefficients of the transform matrix **T** are constant values, i.e. they are not dependent on the signal. Therefore, it is suitable for the coefficients of the matrix **T** to be implemented using the amplifiers. The gains of the amplifiers correspond to the values of the transform coefficients.

The switches are denoted as *P1* and *P2* and they are controlled by the signals $e(p,j)$.

The switches work in the opposite manner. The available sample at the position (p,j) causes the $e(p,j)$ to open the $P1$ and close the $P2$, in order to keep the available samples unchanged during the reconstruction process. The unavailable sample at the position (p,j) causes the $e(p,j)$ to close $P1$ and to open $P2$.

Part 3 of the architecture performs the update of the ε value and values of the signal **y**. This part is activated with the $O9$ signal from the oscillator ($O9$ has to be at a high voltage level). The updated values of the vector **y** are calculated as the difference between $\mathbf{y}^t$ values and gradient factors **g**. The architecture for updating the **y** values is shown in Figure 2.20, while Figure 2.21 shows the architecture for ε update [35].

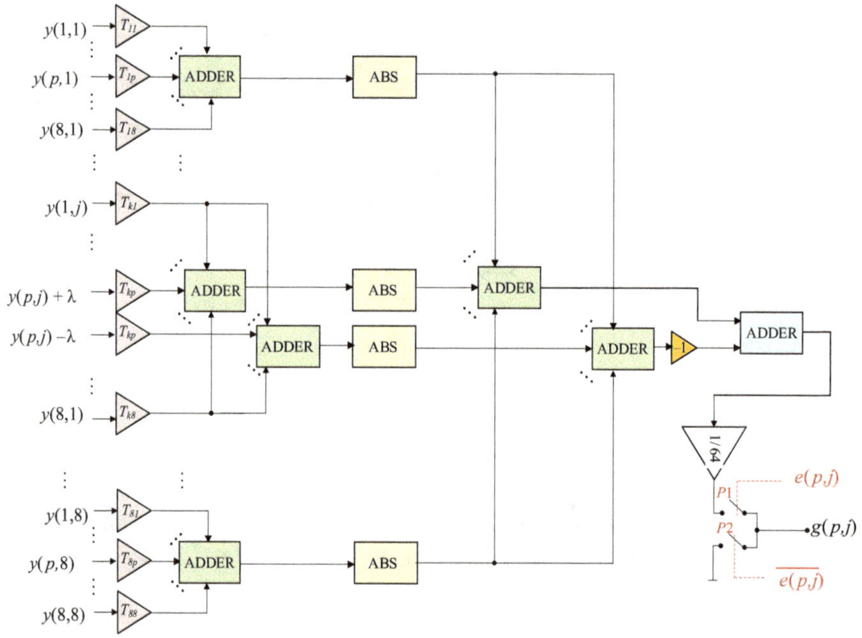

Figure 2.19 A block scheme for updating a concrete value of the matrix **g**

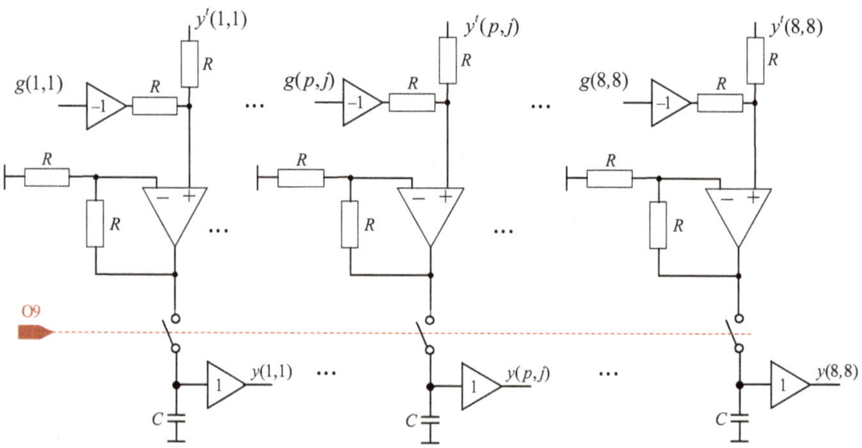

Figure 2.20 Part 3 of the architecture – performs update of the **y** *samples*

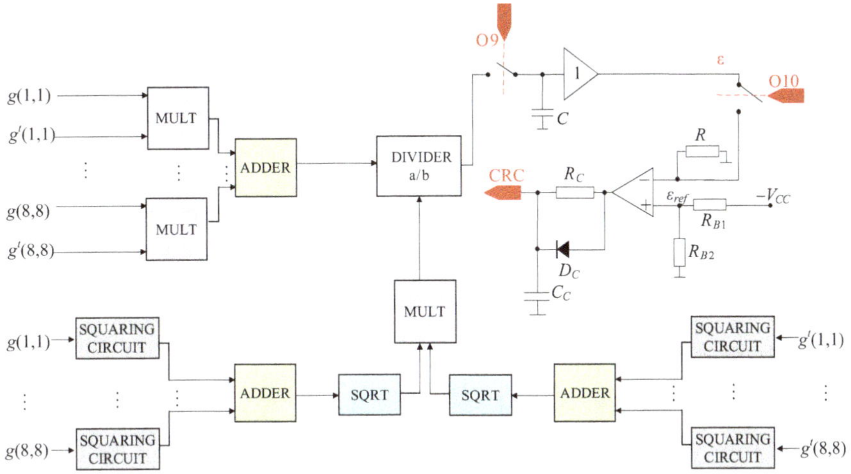

Figure 2.21 Part 3 of the architecture – checking and updating the ε value

An architecture that performs ε value update consists of the analog voltage amplifiers, analog voltage adders, analog voltage division circuits, analog voltage squaring function circuit, and the circuit for square root calculation.

Part 4 of the architecture is controlled with *O*10 oscillator output. It is activated when *O*10 is in the active state. This part performs several operations. It performs checking whether is it necessary to activate CRC signal and consequently update the ε value, updates the value *P* and checks the condition for the stopping criterion. The circuit that checks ε-update and λ-update conditions is shown in Figure 2.21 [35]. Namely, ε-update condition is directly connected with λ-update condition check, that is performed comparing the ε and ε_{ref} values. The signal that initiates the change of λ is *CRC* and it is activated when $\varepsilon > \varepsilon_{ref}$ is valid.

The active state of the *CRC* signal also initiates a stopping the criterion check by comparing P_{ref} with the current value for *P*, at the divider *A*/*B* output (Figure 2.22). Valid condition for stopping criterion causes active state of the signal *STOP*. This state deactivates all oscilloscope outputs and opens *P*1 and *P*2 switches, while the current values in the vector **y** represent the reconstructed samples. If the stopping criterion is not valid, during the next active state of the *O*0, the $R_C C_C$ time constant keeps the *CRC* signal in the high state. The values $\mathbf{y_p}$ are then updated for the new iteration, exploiting new values for λ and $\mathbf{y_p}$.

Part 1 performs temporary storage of the current values of input samples **y** within the matrices $\mathbf{y_p}$ and $\mathbf{y^t}$. The $\mathbf{y_p}$ coefficients are used for the value P_{ref} generation, i.e. for stopping criterion check. The vector $\mathbf{y^t}$ provisionally stores the values used for vector **y** updating.

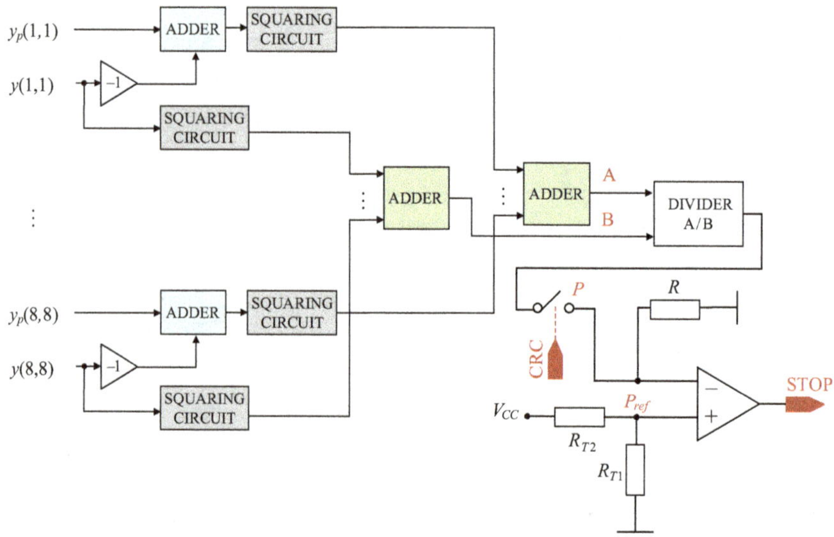

Figure 2.22 Threshold P generation and stopping criterion check (Part 4 of the architecture)

Assuming that we are dealing with 8 × 8 image blocks ($N = 8$), the generation of the specific gradient value $g(p,j)$ requires $2N^2 + 3$ adders (for $N = 8$ it is 131 adder circuits in total), $N^3 + N + 1$ amplifiers (521 amplifier circuits in total), $2N^2$ circuits for absolute value generation, one adder with two inputs, one inverter circuit and two switch circuits [35].

For the value ε generation, the described architecture requires $N^2 + 1$ multiplier, three adders with N^2 inputs, 2 circuits for square root generation, $2N^2$ squaring circuits and 1 divider. It is important to note that functions arccos and log are not implemented. Instead, a direct comparison of the value of the divisor output and the corresponding reference value is performed [35].

A circuit for generating P value consists of N^2 analog voltage adders (adders with 2 inputs), 2 adders with N inputs, N^2 invertor circuits, $2N^2$ circuits for analog voltage squaring function, and one analog voltage division circuits [35].

2.4.1 An OrCAD simulation for the analog 2D gradient-based architecture

The described analog architecture is verified in the PSpice software, OrCAD 16.6 version, using common analog circuits [35]:

- AD734 circuit is used for the realizations of divider, multiplier, circuit for square root generation and squaring circuits. The slew rate is 450 V/μs and the settling time (to 0.1%) is less than 200 ns. This circuit introduces error that is less than 0.1% [35,48].

- LT1394 (comparator) circuit, with a settling time of 7 ns [35,44] is used for the analogue voltage comparison;
- LT1191 operational amplifier is used for implementation of the adders, inverters and amplifier. The LT1191 is ultra-high speed circuit, with settling time (to 0.1%) less than 110 ns and the slew rate 450 V/μs [35,45];
- AD783 sample and hold circuits, with acquisition time (to 0.01%) of 250 ns are used to store the voltage values [35,49];
- 1N4448 for the realization of diodes are used [35,47];
- MAX4645 [35,46] circuits are used as analogue switches.

The biggest error among used analog components is introduced by the AD734 circuit but it is relatively small (less than 0.1%) to significantly affect the accuracy of the realization. The error eventually introduces several more iterations of the gradient-based algorithm, keeping the realization still suitable for real-time implementation. The minimal time required for set on the MAX4645 switches is $T_p = 40$ ns. Minimal times required for oscillator output impulses duration are as follows: for $O0$ impulse approximately $T_{O0} = 480$ ns, for $O1$–$O8$ impulses approximately 940 ns, and minimal times for $O9$ and $O10$ are $T_{O9} = 710$ ns and $T_{O10} = 860$ ns, respectively. Now we can calculate the minimal time required for one iteration of the algorithm: $T_{O0} + 7T_{O1} + T_{O8} + T_{O9} + 10T_p = 9.97\ \mu s \approx 10\ \mu s$ [35].

It is important to mention that the total processing time depends on the number of iterations and the proposed design allows up to 400 iterations within 1 ms [35]. Reconstruction usually requires less than 100 iterations for satisfactory reconstruction accuracy.

Realization in the CMOS technology (e.g. 130 nm, 180 nm, or 340 nm) can lead to a significant reduction of the power consumption, keeping the same or increasing the speed. For example [35]:

- Fast low power analog switch can be realized with power consumption less than 0.05 mW.
- High-speed sample-and-hold amplifier has a power consumption less than 0.5 mW [35].
- High-speed four-quadrant multiplier/divider can be realized with power consumption less than 0.4 mW [35].
- High-speed operational amplifier has a power consumption less than 0.5 mW as well as high-speed analog comparator [35].

Therefore, the power consumption of the proposed architecture can be lower than 1 W. If, for example 100 iterations are performed for image block reconstruction, the energy consumption can be less than 1 mWs. The power consumption for different parts of the architecture is shown in Table 2.2 [35].

The performance of the 2D gradient-based algorithm is demonstrated on the natural image reconstruction. The image is from the MATLAB test image database, and MATLAB version R2015a is used. The grayscale image of 512 × 512 size is observed. The original image is shown in Figure 2.23. The image is separated into

Table 2.2 Power consumption of different parts of the architecture using a new standard of the CMOS technology

	Power consumption
Input signal circuits	<70 mW
Part 1	<180 mW
Part 2	<390 mW
Part 3	<140 mW
Part 4	<110 mW
Oscillator	<60 mW

Figure 2.23 The test image

the 16 × 16 blocks, and the reconstruction is done block-wise. It is assumed several different percentages of missing samples: 20%, 40%, and 70%.

Figure 2.24 shows the reconstruction results. The left column of the subfigures: (a), (b), and (c) displays the images with 20%, 40%, and 70% missing samples, respectively. The reconstructed images are shown in the right column of Figure 2.24. The reconstruction quality is measured by the peak-signal-to-noise-ratio (PSNR), showing satisfactory results. Namely, the PSNR is around 30 dB or higher. In all considered examples, the number of iterations is fixed to 100.

(a)

(b)

(c)

Figure 2.24 *(a) Left – image with 20% of missing samples, right – reconstructed image, PSNR=41.01 dB, (b) left – image with 40% of missing samples, right – reconstructed image, PSNR=35.79 dB, (c) left – image with 70% of missing samples, right – reconstructed image, PSNR=29.19 dB*

2.5 Hardware implementation of combined CS-based image filtering and reconstruction

A hardware design for combined image filtering and the CS reconstruction approach is described in this part [36]. Here, CS is observed from the perspective of signal denoising. Traditional signal denoising techniques use various spectral approaches. For example, in the case of impulse disturbances, the L-estimate approach is applied. It implies discarding the coefficients that are detected as corrupted ones. However, this discarding causes loss of information and consequently lowers the quality of the denoised signal. Therefore, the idea of the approach described in this section is to incorporate the CS approach into the techniques used for denoising and recovering the discarded information, with an ending goal of improving the quality of the final result.

The system allows users to select one of the scenarios: image-filtering-then-reconstruction approach or direct-image-reconstruction. The input of the system can be an image corrupted by noise or an image with unavailable pixels. Both, noisy and missing pixels contain no information about the exact amplitude of the sample and can be subject to the CS recovery. Two types of noise are considered: Gaussian and heavy-tailed (impulse) noise. Therefore, two filtering modules, one for Gaussian and another for impulse noise, are implemented together with the part for the under-sampled signal recovery. Both modules are based on the gradient approach, using DCT as a sparsifying basis. The impulse noise filtering module contains the part performing the L-estimation while the Gaussian noise filtering module contains the random subsampler. The presented approach is a blind image reconstruction procedure and does not depend on the image characteristics.

The system performs three operations:

- reconstruction of the image whose pixels are missing, using a set of available pixels and set of their positions;
- filtering Gaussian noise. Here, it is assumed that the complete set of pixels is available. The image is intentionally randomly subsampled and then reconstructed using a two-dimensional gradient-based approach. As a result, the denoised image is obtained. This procedure can be repeated several times in order to reduce as much noise as possible;
- filtering heavy-tailed (impulse) noise. The samples that are considered noisy are discarded using robust statistics based on the L-estimation approach. The discarded samples are then recovered using two-dimensional gradient-based approach and the denoised image is obtained at the output. In this scenario, it is also assumed that the complete set of pixels is available at the beginning.

For the first function of the described system (image reconstruction), the inputs of the system are the values of the available pixels and their positions. An image with missing pixels could be obtained from a certain CS-designed device. Along with the amplitudes of the available pixels, their positions should be stored. The positions of the available pixels are stored as a logical “1” value while the logical “0” denotes the positions of the unavailable samples. The amplitudes of the unavailable samples could be set to zero or the values could be set to some other arbitrary chosen value (e.g., in the grayscale image, where pixels take values between 0 and 255, the amplitudes of the unavailable pixels could be set e.g. to 128). Having in mind that we are dealing with a two-dimensional signal, two-dimensional address space should be mapped to one-dimensional, in a row-major or column-major order.

The reconstruction approach is the gradient-based procedure. It iteratively calculates a gradient vector, which is used to update the values of the unavailable samples. The gradient is calculated as a difference of the concentration measure (ℓ_1 norm is used) of the 2D discrete cosine transform.

The second function of the system is related to the image denoising in the case of the Gaussian noise. Therefore, it is assumed that the complete image is corrupted. This part of the system architecture has a circuit for random sequence generation and the random sampler circuit, used to randomly select a certain number of pixels to serve as available pixels in the gradient-reconstruction approach. Namely, the image should be randomly subsampled in order to be reconstructed, i.e. in this case – denoised. The random generator is based on the Galois version of the linear feedback shift register. The modification of the linear feedback shift register is proposed. The modified version is able to generate the random numbers sequence whose values may go up to any integer total number of image pixels (or image block pixels). The realization allows the generation of various ranges of random number depending on the total number of pixels that is loaded into the system. Also, the realization is possible using different types of polynomials.

The third function of the system is related to the image denoising in the case when the image is corrupted by the heavy-tailed noise. This part of the architecture uses the architecture that performs the L-estimation. The definition of the L-estimation approach is based on the robust L-estimation theory. It performs the sorting of the weighted pixels sequences and discards (set to zero) the extreme values from the beginning and the end of the sequences. The block for L-estimation consists of the memory unit, the fast sorting circuit using D/A converters, the set of multipliers with predefined exponential functions (representing the Fourier transform basis), the analog comparators, and the α-trimming part of the system. The positions of the discarded samples are stored as zero values, while the positions of available (non-trimmed) values are stored as unit values. A detailed description of the architecture is provided in the sequel, substantiated with the drawings. The block scheme of the architecture is shown in Figure 2.25.

The system consists of four main parts:

- Part for gradient-based reconstruction;
- Part performing the L-estimation;
- Part performing random sampling of the input signal;
- Generator of the random numbers.

The three-state buffer BF1 controls whether the system is active or not. Namely, when logic “0” is loaded to the enable pin *START*, the high-impedance

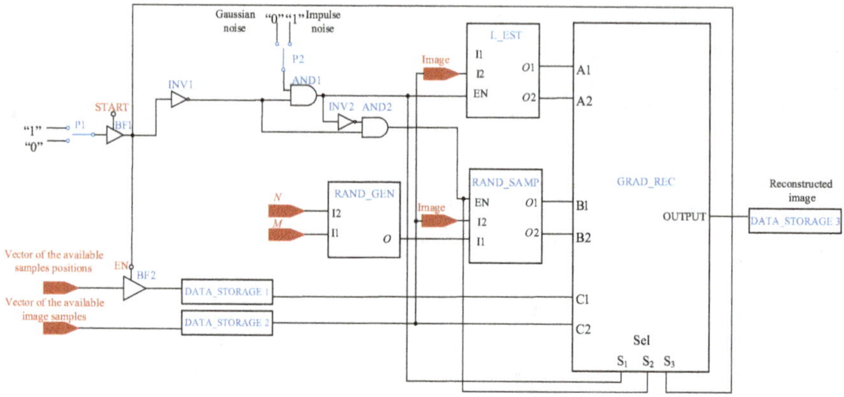

Figure 2.25 CS-based image filtering and reconstruction – block scheme of the architecture

state is at the output. In the case when logic "1" is loaded to the enable pin *START*, the input is passed to the output.

Two switches, denoted as *P*1 and *P*2, control the system. Depending on the position of the *P*1 switch, direct-reconstruction or filtering-then-reconstruction function is performed. The first function is performed when the enable signal *START* at the three-state buffer is logic "1" and the switch *P*1 is in the position "1." In this case, the output enable signal of the second three-state buffer BF2 is "1." The vector of the available samples positions is fed to the DATA STORAGE 1.

When *P*1 is in the position "0," the second function is initiated. The second switch, *P*2, is used to select the type of noise present in the image. Position "1" is for the impulse noise, while position "0" denotes Gaussian noise. Consequently, position "1" of the *P*2 switch requires activation of the circuit performing L-estimation, while position "0" of the *P*2 switch requires activation of the RAND_SAMP and circuit TG.

When *P*1 is in the position "1" and signal START at the tree-state buffer BF1 is logical "1" then the logical "1" is on the output of the BF1 buffer. This value is inverted passing the inverter INV1, causing logical "0" at the outputs of the AND1 and AND2 circuits. Therefore, the enable signals of the L_EST and RAND_SAMP circuits will be logical "0," keeping those circuits inactive. Logical value "1" is sent to the select input S_3 of the circuits performing gradient reconstruction, GRAD_REC circuit. In that case, the S_2 and S_1 are zeros, and the combination on the select inputs is $S_3S_2S_1 = 100$, activating pair of inputs C_1C_2 at the GRAD_REC circuit, leading to the direct reconstruction of the input image.

When *P*1 is in the position "0" and the START signal of the BF1 is logical "1," the logical "0" is fed to the select input S_3. The BF2 buffer is in a state of high impedance. The outputs of the AND1 and AND2 circuits will depend on the other (upper) inputs of those circuits, since the logical "1" is fed to the lower inputs of the AND1 and AND2. Therefore, the select inputs of the GRAD_REC circuits will be

$S_3S_2S_1 = 001$ or $S_3S_2S_1 = 010$, activating A_1A_2 or B_1B_2 pairs of inputs (both pairs are related to the filtering modes).

The switch $P2$ function is to choose the type of noise we are dealing with. The position "1" chooses impulse noise, and causes the output of AND1 to be "1." This logical "1" enables the L_EST circuit and set the GRAD_REC select input S_1 to "1." Therefore, the outputs of the L-estimate circuit will be loaded to the A_1A_2 inputs of the GRAD_REC circuit. The input of the L_EST circuit is an entire image, stored in DATA_STORAGE2. This circuit has two outputs, $O1$ contains a set of image coefficients remaining after the L-estimation, and $O2$ is a vector of their positions. When damaged samples are removed by the L-estimate circuit, the missing pixels are then recovered using the GRAD_REC circuit.

When the switch $P2$ is at the position "0," the Gaussian noise is chosen. The logical "0" at one input of the AND1 causes "0" at its output, i.e. to the enable input of the L_EST circuit. Therefore, the L_EST circuit is not active in this case. However, the upper input of the AND2 is "1" and it causes "1" at the AND2 output. This leads to the activation of the RAND_SAMP circuit. In order to be able to denoise the image corrupted by Gaussian noise using the CS approach, the image should be randomly subsampled. This function is done by the RAND_SAMP circuit. Outputs of this circuit are fed to the GRAD_REC in order to reconstruct the image using an incomplete set of samples and at the same time, reduce the noise. Logical "0" at the AND1 output means logical "0" at the S_1 select input of the GRAD_REC. The output of the AND2 is logical "1," leading this value to the S_2 select input of the GRAD_REC. The combination now is $S_3S_2S_1 = 010$, activating the B_1B_2 pair of inputs. At the first input of the RAND_SAMP circuit, $I1$, the sequence of M integers (randomly selected from the total of N integers) is fed. Those M integers correspond to the positions of the available image pixels, while pixels at the other position are set to zero. The specific circuit, RAND_GEN, generates M out of N randomly permuted integers (N and M are inputs of the RAND_GEN circuit, where $N > M$). The second input of the RAND_SAMP circuit is for the image loaded from DATA_STORAGE 2. The RAND_SAMP has two outputs: $O1$ output contains positions of the randomly chosen image samples, while the output $O2$ contains amplitudes of the randomly chosen samples. The RAND_GEN circuit consists of two parts – the first is the random number generator and the second is the memory unit whose role is to store a sequence of M randomly generated numbers. The block scheme of the RAND_GEN circuit is shown in Figure 2.26.

The random number generator is based on the Galois linear feedback shift register. By performing the XOR operations in parallel, the Galois-based realization is computationally efficient compared to the Fibonacci realization. Each input bit is a linear function of the previous state. The feedback is realized using the XOR circuits, and the feedback function is represented using the feedback polynomial:

$$P(x) = x^{16} + x^{14} + x^{13} + x^{11} + 1. \tag{2.22}$$

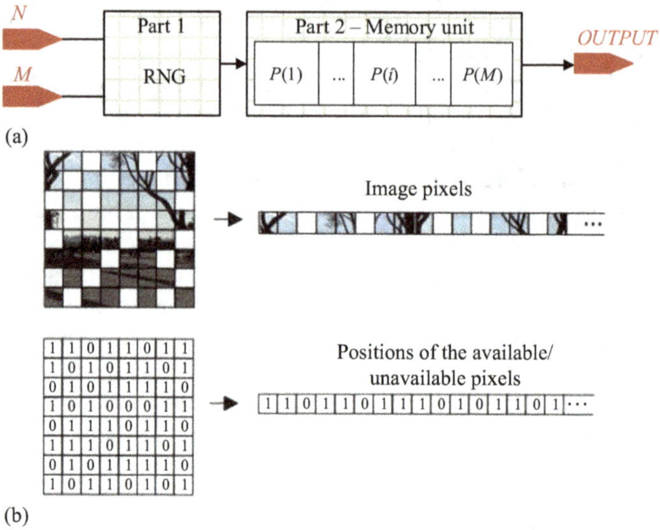

Figure 2.26 *(a) RAND_GEN circuit, consisted of the random number generator and memory unit, (b) upper image – image with missing pixels and how they are stored in memory using row-major order, lower image – missing pixels are considered as zero values in matrix, while available pixels are denoted as value one. The positions are mapped from matrix to vector using row-major order.*

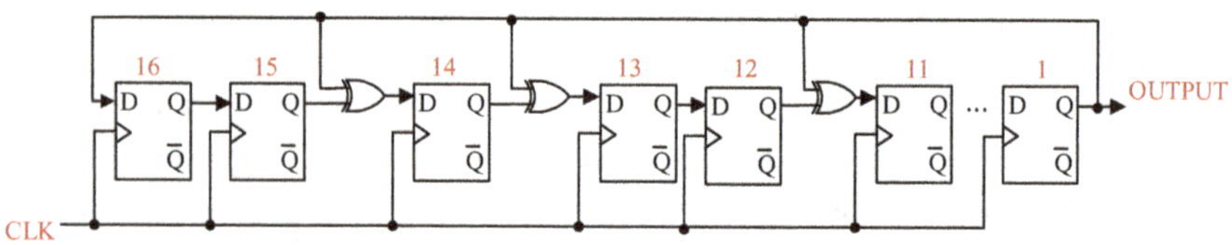

Figure 2.27 *A block diagram for the random number generator realization for the polynomial* $P(x) = x^{16} + x^{14} + x^{13} + x^{11} + 1$

The realization of the 16-bit linear-feedback shift register is done using D-flip flops and it is shown in Figure 2.27. The positions of the XOR circuits are called taps and their number should be even. Here, the taps are on the bits 11, 13, 14, and 16. The taps are first fed to the XOR circuits and then shifted, while the bits that are not taps are shifted to the right.

The OUTPUT could be “0” or “1.” When OUTPUT=“0,” then all bits shift for one position to the right and at the input of the first flip flop (from the left side) will be “0.” When OUTPUT=“1,” the tap bits are changed since the XOR operation is performed with “1” and at the input of the first flip flop will be “1.” The values of the rest bits will be the same as in the previous step. The modified version of the

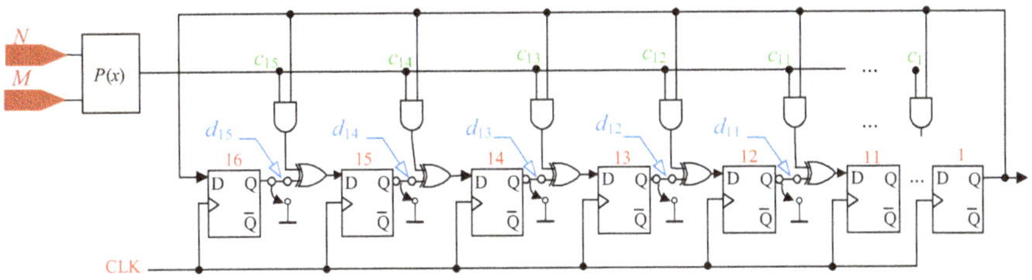

Figure 2.28 An architecture of the modified version of the random number generator

random number generator is shown in Figure 2.28. The maximal output corresponds to the input N of the circuit. Namely, if we are deal with 16-bit number, the N will be $N = 2^{16} = 65{,}536$. The part $P(x)$ shown in Figure 2.28 uses polynomials of different orders. The signals c_1–c_{15} are control signals, and determine the position of tap bits, i.e. determine the XOR circuit which should flip the input bits:

- $c_i =$ "0" signal defining the non-tap;
- $c_i =$ "1" signal defining the tap.

The signals d_1–d_{15} set the first flip-flop in a series and they are set to "0" by default. It means that all switches are in horizontal positions and first flip flop in a series is the flip flop denoted with 16 in Figure 2.28. The value $d_i =$ "1" put the switch into the vertical position, when the logical "0" (ground voltage) is fed to lower input of the XOR circuit. This means that the value of the upper XOR input is fed to its output. Another random number is generated in each clock cycle and the number is stored in memory unit, shown in Figure 2.26(a). The random numbers generation is controlled by the number of CLK cycles. The procedure is finished when the Mth number is generated.

The L-EST circuit is used for denoising in the case when image is corrupted by the impulse noise [50–57]. The circuit calculates the robust L-estimation form of the short-time Fourier transform. The L-estimation is done block by block, where the block size is $B \times B$. The 2D L-estimation procedure can be defined as a composition of two sequential 1D forms. If the image block is defined as $f(x+n, y+n), n \in \left[-\frac{B}{2}, \frac{B}{2}-1\right]$ and one row in the block as $f(x + n, y)$, then we first calculate the 1D form as follows:

$$s = sort\left\{\left(f(x+n,y)e^{-j\frac{2\pi nk_x}{B}}\right)\right\}. \tag{2.23}$$

The sorting operation is done in order to place the corrupted samples to the beginning and at the end of the sorted sequence s (the sorting operation is non-decreasing). Those values will be set to zero (or some arbitrary value) after the α-trimming operation. The L-estimated form of the short-time Fourier transform can be defined as follows:

$$STFT_L(x+n,y) = \sum_{k=-B/2}^{B/2-1} a_k(r_{ksort}(x+n,y) + j \cdot i_{ksort}(x+n,y)), \tag{2.24}$$

where

$$r_k(x+n,y) \in \mathrm{Re}\left[f(x+n,y)e^{-j\frac{2\pi}{B}nk_x}\right]$$

and

$$i_k(x+n,y) \in \mathrm{Im}\left[f(x+n,y)e^{-j\frac{2\pi}{B}nk_x}\right].$$

The weights a_k are designed according to α-trimming filter as follows:

$$a_k = \begin{cases} 1, & \text{for } k \in [(B-2)\alpha, \alpha(2-B)+B-1] \\ 0, & \text{otherwise} \end{cases}. \tag{2.25}$$

The procedure is repeated for the columns of the observed image block. Therefore, the corrupted samples are considered unavailable and the others are considered available. After the applying α-trimming operation, the positions of the α-trimmed samples are considered unavailable and set to “0,” while the positions of the samples that are not trimmed (available positions) are set to “1.” The architecture for the L_EST part is shown in Figure 2.29. The circuit loads the $B \times B$ image block and stores it in the memory. Rows and columns of the block are then loaded using the network made of switches. There are B pairs of switches. The first row with B coefficients is sent to the upper first switch, the second row is sent to the upper second switch, etc. (upper switches are marked with blue color). The switches are controlled by the signal CLK. The first pass closes the upper switches and allows the rows of the image block to pass to the multipliers. There are B sets of multipliers, associated with the one row of the image block, and in each set there are B multipliers. The row are multiplied by $e^{-j\frac{2\pi nk_x}{B}}$, $n \in [-B/2, B/2-1]$, where k_x is the frequency index and $k_x \in [-B/2, B/2 - 1]$. The non-decreasing sorting operation is performed after the multiplication with an exponential factor. For the purpose of fast sorting, an analog sorter is used. The architecture of an analog sorter is shown in Figure 2.30.

The L_TRIM block performs the discarding certain amount of samples placed at the beginning and at the end of the sorted sequence. The positions (from the original vector, before the sorting operation is applied) of the samples that should be kept are stored using the STR circuit. The positions that correspond to the available samples are denoted as “1” while the positions of the unavailable samples are denoted with “0.” The same procedure should be done with the image columns. The lower switches are used for this purpose (switches marked with green color), and they are also controlled by the CLK signal.

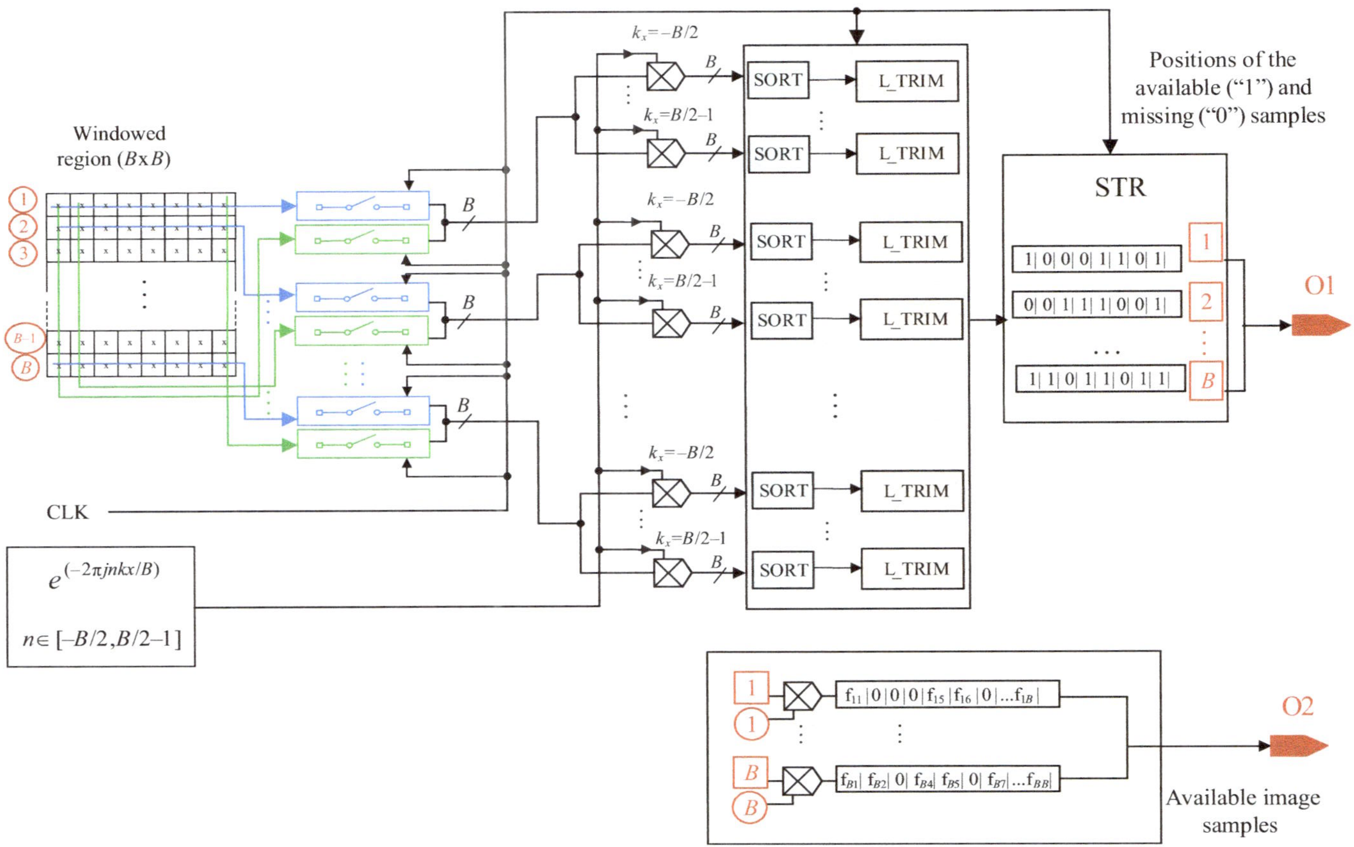

Figure 2.29 An architecture for performing the L-estimation

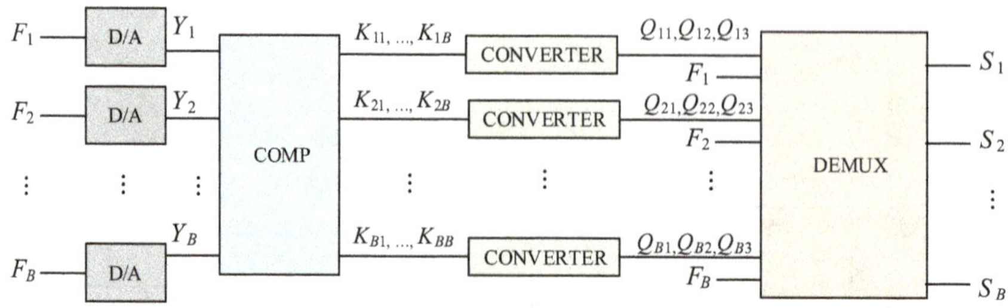

Figure 2.30 Fast sorting circuit used in L-estimation

The first output of the L_EST circuit, $O1$, contains B rows with B values each and contains the positions of the available and unavailable samples ("0" denotes the positions of the available samples, "1" is for the positions of the unavailable samples). The $B \times B$ values from the $O1$ output are multiplied by the original image values ($B \times B$ values from the input of the L_EST circuit). The values f_{ij} denote the available image values, and at the positions of the unavailable samples are zero values.

The architecture of the sorting circuit used in the L-estimation procedure is shown in Figure 2.30. The architecture consists of D/A converters, comparators, binary converters and demultiplexer. First, the inputs $F_1,\ldots,F_B$ are fed to the D/A converters. In order to obtain a non-decreasing sequence, the values $Y_1,\ldots,Y_B$ from the outputs of the D/A converters are fed to the inputs of fast analog sorters (COMP). The outputs of the comparators are binary sequences K_{ij}. The function of the comparators can be described as follows:

$$K_{ij} = \begin{cases} 1, & I_i \geq I_j \\ 0, & \text{otherwise} \end{cases} i = 1,\ldots,B;\ j = 1,\ldots,B;\ i \neq j. \tag{2.26}$$

The number of logical "1" values in the sequences $K_{i1}, \ldots, K_{iB}$ is calculated by the converters which then convert those numbers to the and then convert these numbers to the binary sequences $Q_{i1}Q_{i2}\ldots Q_{Ip}$. The Q sequences will serve as selection signals for the demultiplexer. The demultiplexer determines the position of each input value $F_1, F_2,\ldots, F_B$ within the sorted sequence ($S_1, \ldots, S_L$).

Prior to the reconstruction, all missing pixels are set to the 128 value (midpoint between minimal and maximal amplitude value of the pixels). The procedure is applied to the $B \times B$ block size (8×8 and 16×16 are commonly used block sizes).

The architecture of the GRAD_REC block is presented in Figure 2.31. One of the inputs of the circuits is positions of the available (unavailable) samples, obtained directly from the DATA_STORAGE2, or from the output of the RAND_SAMP circuit, or from the output $O1$ of the L_EST circuit. The input $I1$ is a $B \times B$ matrix, containing values "0" for unavailable and "1" for the available

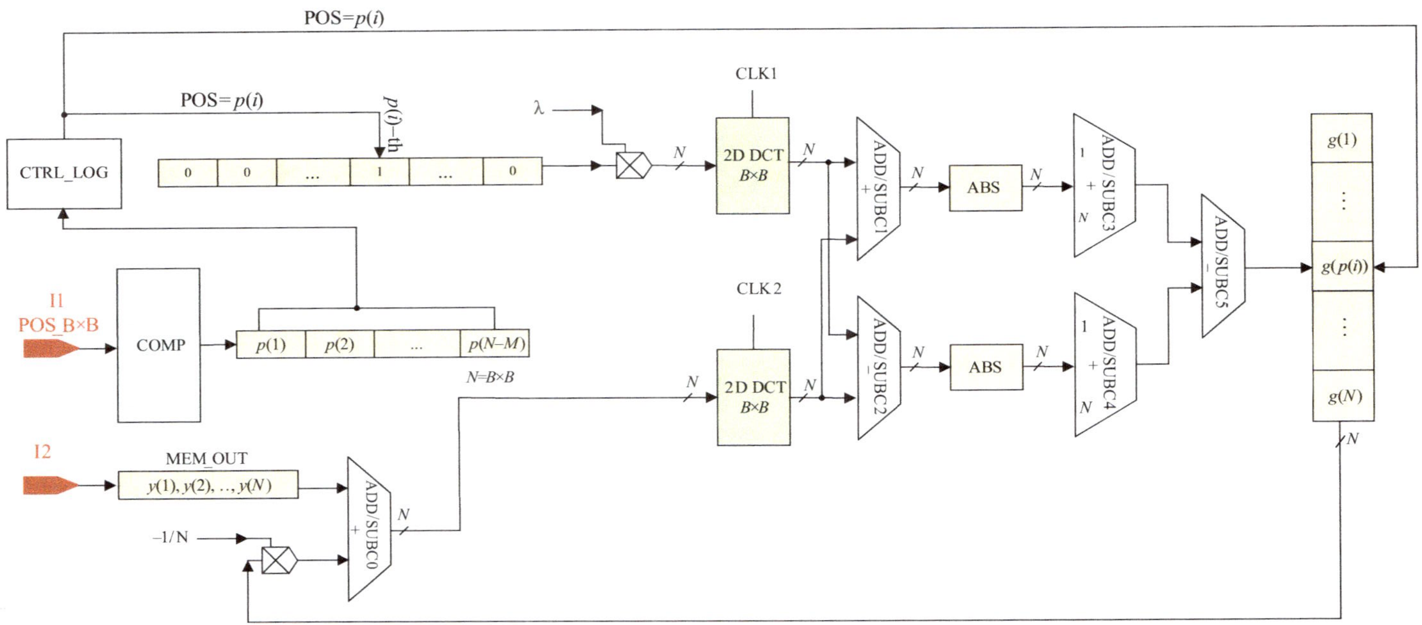

Figure 2.31 The architecture of the GRAD_REC block

pixel. The positions of the unavailable pixels are selected from the input matrix. It is done in the COMP circuit, according to the following pseudocode:

```
r = 0;
        for q = 1:B^2
        if pos_BxB(q)=="0"
             p(r)=q;
             r++;
        end
```

The **p** vector has a length equal to the number of unavailable pixels. The elements of the vector **p** are inputs of the CTRL_LOG circuit. CTRL_LOG circuit outputs a value of the POS index that corresponds to the position of the pixel whose amplitude is currently changing towards its exact value. For POS=$p(r)$, the value "1" is set at the rth position in the vector stored in memory MEM1 (for each r, there is only one element whose value is "1") and the gradient values are assigned to the r-th position in the memory MEM2. The POS index is set according to the following pseudocode:

```
ind = 1;
r = 1;
if r <iter && ind<N-M
        POS = p(ind);
        ind = ind + 1;
        iter = iter-1;
else if
    r<iter && ind = N-M
    ind = 1;
    POS = p(ind);
    ind = ind + 1;
    iter = iter-1;
else
break;
```

The vector from the MEM1 is multiplied by a value λ (this value is stored in the register) and the result is sent to the circuit for 2D DCT calculation. The 2D DCT is calculated for the vector that contains value λ only at the rth position, while the rest of the vector are zeros: $\mathbf{T}(k_1,k_2)$ = DCT2$\{\lambda\delta(p\text{-}p(r))\}$, where δ is the Kronecker function. However, in order to perform the 2D DCT calculation, this vector is rearranged into the $B \times B$ matrix, while at the output of the 2D DCT circuit is again returned to vector $N = B \times B$.

The second input of the GRAD_REC system is denoted as *I2*. Here, the block with missing pixels is fed. The values of the available pixels are the same as originally, while the 128 is set to the positions of the unavailable pixels (where zero values were initially set).

After normalization with the factor $1/N$, the block is sent to the 2D DCT part that calculates 2D DCT transformation over the input image block of the size $B \times B$: $\mathbf{Y}(k_1,k_2) = \text{DCT2}\{y\}$, $y = y(i,j)$, $i = 1,\ldots, B; j = 1,\ldots, B$. The outputs of the 2D DCT circuits are N-length vectors **T** and **Y**. Those vectors are sent to the

Adder/Subtractor circuits, producing N-length vectors:

$$\begin{aligned} &\mathbf{Y}(n) + \mathbf{T}(n),\ n = 1, ..., N \\ &\text{and} \\ &\mathbf{Y}(n) - \mathbf{T}(n),\ n = 1, ..., N \end{aligned}, \tag{2.27}$$

that are further sent to the circuits calculating the absolute value. The DCT transform coefficients are real number so the standard absolute value is calculated. The circuits ADD/SUBC3 and ADD/SUBC4 calculate sum of N entries from the input of those circuits, according to:

$$\begin{aligned} &\sum_{n=1}^{N} |\mathbf{Y}\{n\} + \mathbf{T}\{n\}| \\ &\text{and} \\ &\sum_{n=1}^{N} |\mathbf{Y}\{n\} - \mathbf{T}\{n\}| \end{aligned} \tag{2.28}$$

The outputs from the ADD/SUBC3 and ADD/SUBC4 are subtracted in the ADD/SUBC5. This value is the value of the gradient at the position $p(r)$ and it is stored in MEM2 memory.

One iteration of the gradient-based algorithm is finished when all $p(r)$ positions are updated, $r = 1, \ldots, N-M$. At this point, the POS index is equal to the last element of vector $\mathbf{p}$. The current state of the MEM2 is multiplied by the constant $-1/N$ and the image pixel is updated according to:

$$y(n) = \frac{1}{N}(y(n) - g(n)). \tag{2.29}$$

The procedure is finished when the maximal number of iterations *iter* is reached. This is controlled by the CTRL_LOG circuit. The state in the MEM_OUT memory unit represents the final result, i.e. the recovered image block. Thc same procedure is repeated for each image block. As a result, the entire image is reconstructed.

2.6 Conclusion

CS allows signal acquisition at much lower sampling rates when compared to the traditional sampling procedure required by the Sampling theorem. Consequently, it requires specific signal recovery methods to reconstruct the entire signal from a reduced set of randomly acquired samples. Reduced set of samples can be a result of an intentional and less demanding sensing strategy (in terms of required resources), but can be also a consequence of errors during the acquisition procedure, or a consequence of discarding samples corrupted by certain kinds of noise. By intentionally decreasing the amount of information required for successful signal recovery, the CS approach makes the acquisition procedure faster, lowers the memory requirements for storage of the collected samples, and generally, provides an efficient resource utilization. The possibility of damaged signal recovery is also important in practical applications where noise presence is common.

The literature is covered with a number of approaches for sparse signal recovery. In this chapter, we focused on the gradient-based approach since it can provide satisfactory recovery in a relatively small number of iterations. For example, when considering an image, it can be recovered using around 15% of available samples (depending on the nature of the image, the reconstruction could be possible even with a smaller percentage of the available samples). Therefore, the hardware architectures for the gradient-based algorithm dealing with one and two-dimensional signals are described. The architectures are suitable for real-time performance and provide faster testing compared with the software implementations of the considered approaches. Certain steps of the approaches are parallelized in order to cope efficiently with the problem of signal recovery. Analog architectures are proposed, since they provide faster execution compared to the digital ones, with satisfactory precision. One iteration of the 1D algorithm requires around 2.5 μs, while for the 2D version, the execution time is around 10 μs. It is important to emphasize that the majority of the real-world signals can be accurately recovered within just a few iterations of the gradient-based algorithm. Therefore, the proposed architectures are suitable for real-time implementations. Simulations showed that the maximal error, introduced by the circuits exploited in realizations, is small and can cause just a few more iterations of the algorithm, which is negligible in terms of the execution time.

A specific part of the chapter is devoted to the application of the gradient-based approach in the image denoising problems. Namely, the special architecture for combined image filtering and reconstruction approach is described. The denoising part is designed for the Gaussian and heavy-tailed noise removal. The approach is capable of finding the exact values of the missing or corrupted pixels. The DCT is used as a sparsity basis, but the algorithm works with an arbitrary transformation. Future work may be focused on the additional optimizations of the proposed realizations. For example, power consumption can be additionally reduced by switching off inactive parts of the architectures.

Acknowledgments

The work of M. Beko was funded by Fundação para a Ciência e a Tecnologia under Projects UIDB/04111/2020, ROBUST EXPL/EEI-EEE/0776/2021, and CEECIND/01830/2018.

References

[1] Eldar, Y. C. and Kutyniok, G. *Compressed Sensing: Theory and Applications*. Cambridge: Cambridge University Press, 2012.

[2] Stanković, S., Orović, I., and Sejdić, E. *Multimedia Signals and Systems: Basic and Advance Algorithms for Signal Processing*, 2nd ed. New York, NY: Springer, 2015.

[3] Stanković, S. and Orović, I. 'An approach to 2D signals recovering in compressive sensing context'. *Circuits, Systems & Signal Processing*. 36(4): 1700–1713, 2017.

[4] Brajović, M., Stanković, S., and Orović, I. 'Analysis of noisy coefficients in the discrete Hermite transform domain with application in signal denoising and sparse signal reconstruction.' *Signal Processing*. 150(2018): 145–156, 2018. https://doi.org/10.1016/j.sigpro.2018.04.007.

[5] Candès, E. J. and Wakin, M. 'An introduction to compressive sampling.' *IEEE Signal Processing Magazine*. 25(2): 21–30, 2008.

[6] Orović, I., Papić, V., Ioana, C., Li, X., and Stanković, S. 'Compressive sensing in signal processing: algorithms and transform domain formulations.' *Mathematical Problems in Engineering*. 2016, Article ID 7616393, 16 pages, 2016. https://doi.org/10.1155/2016/7616393

[7] Stanković, S., Orović, I., and Stanković, LJ. 'Polynomial Fourier domain as a domain of signal sparsity'. *Signal Processing*. 30(2017): 243–253.

[8] Han, C., Chen, L., and Wang, W. 'Compressive sensing in wireless powered network: regarding transmission as measurement'. *IEEE Wireless Communications Letters*. 8(6): 1709–1712, 2019. doi:10.1109/LWC.2019.2938517

[9] Zhang, Z., Jung, T. P., Makeig, S., Pi, Z., and Rao, B. D. 'Spatiotemporal sparse Bayesian learning with applications to compressed sensing of multichannel physiological signals.' *IEEE Transactions on Neural Systems and Rehabilitation Engineering*. 22(6): 1186–1197, 2014, doi:10.1109/TNSRE.2014.2319334.

[10] Lakičević Žarić, M., Draganić, A., Orović, I., Beko, M., and Stanković, S. 'Combining gradient-based and thresholding methods for improved signal reconstruction performance'. *Journal of Signal Processing Systems*. 2022, https://doi.org/10.1007/s11265-022-01780-5.

[11] Rao, X. and Lau, V. K. 'Compressive sensing with prior support quality information and application to massive MIMO channel estimation with temporal correlation'. *IEEE Transaction on Signal Processing*. 63(18): 4914–4924, 2015.

[12] Orović, I., Stanković, S., and Beko, M. 'Multi-base compressive sensing procedure with application to ECG signal reconstruction'. *EURASIP Journal on Advances in Signal Processing*. 2021, Article number: 18, 2021.

[13] Qin, Z., Fan, J., Liu, Y., Gao, Y., and Li, G. Y. 'Sparse representation for wireless communications: a compressive sensing approach'. *IEEE Signal Processing Magazine*. 35(3): 40–58, 2018, https://doi.org/10.1109/MSP.2018.2789521.

[14] Vujović, S., Draganić, A., Lakičević, M., *et al.* 'Sparse analyzer tool for biomedical signals'. *Sensors*. 20(9): 2602, 2020, doi:10.3390/s20092602.

[15] Nascimento, J. M. P., Véstias, M. P., and Martín, G. 'Hyperspectral compressive sensing with a system-on-chip FPGA'. *IEEE Journal of Selected Topics in Applied Earth Observations and Remote Sensing*. 13: 3701–3710, 2020, doi:10.1109/JSTARS.2020.2996679.

[16] Blumensath, T. and Davies, M. E. 'Iterative thresholding for sparse approximations.' *Journal of Fourier Analysis and Applications*. 14(5–6): 629–654, 2008.

[17] Tropp, J. A. 'Greed is good: algorithmic results for sparse approximation'. *IEEE Transactions on Information Theory*. 50(10): 2231–2242, 2004.
[18] Candes, E. and Romberg, J. 'L1-Magic: Recovery of Sparse Signals Via Convex Programming'. 2005, https://candes.su.domains/software/l1magic/downloads/l1magic.pdf. pdf (accessed on 19th August 2022).
[19] Stanković, L. J. and Daković, M. 'On a gradient-based algorithm for sparse signal reconstruction in the signal/measurements domain'. *Mathematical Problems in Engineering*, 2016, Article ID 6212674, 11 pages, 2016, doi:10.1155/2016/6212674.
[20] Yang, Y., Sun, J., Li, H., and Xu, Z. 'ADMM-CSNet: a deep learning approach for image compressive sensing'. *IEEE Transactions on Pattern Analysis and Machine Intelligence*. 42(3): 521–538, 2020, doi: 10.1109/TPAMI.2018.2883941.
[21] Figueiredo, M. A., Nowak, R. D., and Wright, S. J. 'Gradient projection for sparse reconstruction: application to compressed sensing and other inverse problems'. *IEEE Journal on Selected Topics in Signal Processing*. 1: 586–597, 2007.
[22] Wang, J., Kwon, S., and Shim, B. 'Generalized orthogonal matching pursuit'. *IEEE Transactions on Signal Processing*. 62(12): 6202–6216, 2012.
[23] Stanković, L. J., Stanković, S., and Amin, M. 'Missing samples analysis in signals for applications to L-estimation and compressive sensing'. *Signal Processing*. 94: 401–408, 2014.
[24] Orović, I., Lekić, N., Beko, M., and Stanković, S. 'An analog hardware solution for compressive sensing reconstruction using gradient-based method'. *EURASIP Journal on Advances in Signal Processing*. 2019, Article number: 61, 2019.
[25] Hahm, M. D., Friedman, E. G., and Titlebaum, E. L. 'A comparison of analog and digital circuit implementations of low power matched filters for use in portable wireless communication terminals'. *IEEE Transactions on Circuits and Systems II: Analog and Digital Signal Processing*. 44(6): 498–506, 1997, doi:10.1109/82.592584.
[26] Huang, G. and Wang, L. 'High-speed signal reconstruction with orthogonal matching pursuit via matrix inversion bypass'. In: *Proceedings of IEEE Workshop Signal Processing System (SiPS)*, 2012, pp. 191–196.
[27] Roy, S., Acharya, D. P., Sahoo, A. K., and 'Fast, O. M. P. Algorithm and its FPGA implementation for compressed sensing-based sparse signal acquisition systems'. *IET Circuits, Devices and Systems*. 15(6): 511–521, 2021, https://doi.org/10.1049/cds2.12047.
[28] Vujović, S., Daković, M., Orović, I., and Stanković, S. 'An architecture for hardware realization of compressive sensing gradient algorithm'. In: *4th Mediterranean Conference on Embedded Computing*, MECO–2015, pp. 189–192, doi: 10.1109/MECO.2015.7181900.
[29] Roy, S., Acharya, D. P., and Sahoo, A. K. 'Incremental Gaussian elimination approach to implement OMP for sparse signal measurement'. *IEEE Transactions on Instrumentation and Measurement*. 69(7): 4067–4075, 2020, doi:10.1109/TIM.2019.2947118.

[30] Draganić, A., Orović, I., Lekić, N., Daković, M., and Stanković, S. 'Architecture for single iteration reconstruction algorithm'. In: *4th Mediterranean Conference on Embedded Computing*, MECO 2015, 2015, pp. 77–80, doi:10.1109/MECO.2015.7181870.

[31] Fardad, M., Sayedi, S. M., and Yazdian, E. 'Hardware implementation of iterative method with adaptive thresholding for random sampling recovery of sparse signals'. *IEEE Transactions on Very Large Scale Integration (VLSI) Systems*. 26(5): 867–877, 2018, doi:10.1109/TVLSI.2018.2791351.

[32] Mächler, P. 'VLSI architectures for compressive sensing and sparse signal recovery'. Doctoral Thesis, 2012.

[33] Huang, G. and Wang, L. 'An FPGA-based architecture for high-speed compressed signal reconstruction'. *ACM Transactions on Embedded Computing Systems*. 16(3): 89-1–89-23, 2017.

[34] Stanislaus, J. L. V. M. and Mohsenin, T. 'Low-complexity FPGA implementation of compressive sensing reconstruction'. In: *International Conference on Computing, Networking and Communications (ICNC)*, 2013, pp. 671–675, doi:10.1109/ICCNC.2013.6504167.

[35] Lekić, N., Lakičević, M., Orović, I., and Stanković, S. 'Adaptive gradient-based analog hardware architecture for 2D under-sampled signals reconstruction'. *Microprocessors and Microsystems*. 62(2018): 72–78, 2018. ISSN 0141-9331, https://doi.org/10.1016/j.micpro.2018.07.009.

[36] Guna, J., Stanković, S., Orović, I., Stanković, L. J., and Daković, M. *Postopek zgoščenega zaznavanja na področju fultriranja in rekonstrukcije slik in naprava za izvedbo postopka. In English: A method of compressive sensing-based image filtering and reconstruction, and a device for carrying out said method.* Univerzitet Crne Gore Elektrotehnicki Fakultet Podgorica, Univerza v Ljubljani. European Patent Application, 2017.

[37] Fardad, M., Sayedi, S. M., and Yazdian, E. 'A low-complexity hardware for deterministic compressive sensing reconstruction'. *IEEE Transactions on Circuits and Systems I: Regular Papers*. 65(10): 3349–3361, 2018, doi:10.1109/TCSI.2018.2803627.

[38] Majumdar, A. '*Compressed Sensing for Engineers*', 1st ed.; London: CRC Press, 2018, https://doi.org/10.1201/9781351261364.

[39] Brajović, M., Orović, I., Daković, M., and Stanković, S. 'Compressive sensing of sparse signals in the Hermite transform basis'. *IEEE Transactions on Aerospace and Electronic Systems*. 54: 950–967, 2018.

[40] Brajović, M., Orović, I., Daković, M., and Stanković, S. 'Gradient-based signal reconstruction algorithm in the Hermite transform domain'. *Electronics Letters*. 52(1): 41–43.

[41] Music, J., Orović, I., Marasović, T., Papić, V., and Stanković, S. 'Gradient compressive sensing for image data reduction in UAV based search and rescue in the wild'. *Mathematical Problems in Engineering*. 2016: 14 pages, Article ID 6827414, 2016, http://dx.doi.org/10.1155/2016/6827414.

[42] Musić, J., Marasović, T., Papić, V., Orović, I., and Stanković, S. 'Performance of compressive sensing image reconstruction for search and rescue'. *IEEE*

Geoscience and Remote Sensing Letters. 13(11): 1739–1743, 2016, doi:10.1109/LGRS.2016.2606767.

[43] Stanković, L. J., Daković, M., and Vujović, S. 'Adaptive variable step algorithm for missing samples recovery in sparse signals'. *IET Signal Processing*. 8: 246–256, 2014.

[44] LINEAR TECHNOLOGY CORPORATION, 7 ns, Low Power, Comparator – LT1394 (datasheet), 16 pages, http://cds.linear.com/docs/en/datasheet/1394f.pdf.

[45] LINEAR TECHNOLOGY CORPORATION, Ultrahigh Speed Operational Amplifier – LT1191 (datasheet), 12 pages, http://cds.linear.com/docs/en/datasheet/1191fa.pdf.

[46] MAXIM, Fast, Low-Voltage, 2.5Ω, SPST, CMOS Analog Switches – MAX4645 (datasheet), 11 pages, https://datasheets.maximintegrated.com/en/ds/MAX4645-MAX4646.pdf.

[47] VISHAY SEMICONDUCTORS, Small Signal Fast Switching Diodes – 1N4448 (datasheet), 4 pages, http://www.vishay.com/docs/81858/ 1n4448.pdf.

[48] ANALOG DEVICES, 10 MHz, Four-Quadrant Multiplier/Divider – AD734 (datasheet), 20 pages, http://www.analog.com/media/en/technicaldocumentation/data-sheets/AD734.pdf.

[49] ANALOG DEVICES, Complete Very High Speed Sample and Hold Amplifier – AD783 (datasheet), 9 pages, http://www.analog.com/media/en/technical-documentation/data-sheets/AD783.pdf.

[50] Stanković, S., Stanković, L. J., and Orović, I. 'Relationship between the robust statistics theory and sparse compressive sensed signals reconstruction'. *IET Signal Processing*. 8: 223–229, 2014.

[51] Stanković, S., Orović, I., Mobasseri, B., and Chabert, M. 'A robust procedure for image watermarking based on the Hermite projection method'. *Automatika – Journal for Control, Measurement, Electronics, Computing and Communications*. 53(4): 335–343, 2012.

[52] Orović, I. and Stanković, S. 'L-statistics based space/spatial-frequency filtering of 2D signals in heavy tailed noise'. *Signal Processing*. 96(Part B): 190–202, 2014.

[53] Stanković, L. J., Stanković, S., Orović, I., and Amin, M. 'Robust time-frequency analysis based on the L-estimation and compressive sensing'. *IEEE Signal Processing Letters*. 20(2013): 499–502, 2013.

[54] Huang, S. and Zhu, J. 'Removal of salt-and-pepper noise based on compressed sensing'. *Electronics Letters*. 46(17): 1198–1199, 2010.

[55] Stanković, S., Orović, I., and Amin, M. 'L-statistics based modification of reconstruction algorithms for compressive sensing in the presence of impulse noise'. *Signal Processing*. 93(11): 2927–2931, 2013.

[56] Stanković, L., Daković, M., and Vujović, S. 'Reconstruction of sparse signals in impulsive disturbance environment'. *Circuits, Systems and Signal Processing*. 35:767–794, 2016.

[57] Orović, I., Lekić, N., and Stanković, S. 'An analogue-digital hardware for L-estimate space-varying image filtering'. *Circuits, Systems and Signal Processing*. 35(2): 409–420, 2016.

Chapter 3

Performance evaluation of cognitive sensor frameworks for IoT applications in healthcare and environment monitoring

Amitabh Mishra[1], Darian S. Lalla[2] and John Compo[3]

Abstract

Wireless Internet of Things (WIoT), based on the concept of ubiquitous computing, attempts to create communication networks by embedding microchips in everyday consumer electronic devices and to set those devices up to share data with each other to provide innovative connected solutions. With the explosion in the number and variety of wireless applications in the last two decades, all of them face a major challenge of availability of frequency spectrum. This challenge can be addressed by resorting to cognitive radio. The cognitive radio scheme uses a radio system that can make an informed decision to dynamically fine-tune its radio operation metrics as it can sense and is mindful of its operational setting. Combining the cognitive radio technology with WIoT can make the application design more robust and universal. In this paper, we discuss the possibilities and challenges involving such cognitive sensors, used for environmental sensing and healthcare applications.

Keywords: Wireless Internet of Things; Cognitive radio; Healthcare; Microclimate; Sensor networks

3.1 Introduction – wireless sensor networks

Wireless Internet of Things (WIoT) [1] utilises the high computing power that comes packed today in extremely small sized, low-priced embedded electronic devices, which are perfect for sensor and interfacing purposes. The easiness of connectivity offered by a strong and diverse networking infrastructure comprising

[1]Department of Computer Science, The University of West Florida, USA
[2]National Weather Service, USA
[3]Environmental Health Division, Florida Department of Health Polk County, USA

of wired, local wireless, cellular, and if required even satellite and other terrestrial communication technologies, this computing power has given birth to several application areas such as industrial control, distributed sensor networks, smart grid, home automation, medical devices, smart travel, and smart agriculture amongst others [2]. While the list indicates disparate applications, they involve electronic devices that can connect to the Internet through wireless access points and locally networked devices, which can extend this connectivity all the way to a cloud.

Wireless sensor networks (WSNs) form an important constituent of WIoT that comprises the usage of low power wireless sensor nodes for sensing of physical quantities such as temperature, pressure, and humidity, and transmission of the sensed data for qualitative and quantitative information over a low radio range. The sensor data is transmitted and routed using wireless links over a network through single or multiple hops with the end goal to reach a sink node or a base station. A connected concept using wireless sensors in a network is that of the Internet of Things (IoT).

IoT is a revolutionary design that enables a pervasive existence of a diversity of entities that can be uniquely identified while being able to communicate such as sensors, radio-frequency identification (RFID) tags, cell phones, and control actuators that can exist in peoples' workplace, home, or when they are moving [3]. Through distinctive addressing structures and communications capacity, these entities would be able to interact with their neighbouring entities to achieve shared aims [4]. The IoT journey has already begun, and it is expected that 10 years ahead of now, such IoT nodes may be embedded in everyday consumer things such as paper documents, food packages, groceries and store items, furniture, and more. Business users will experience the effects of this new paradigm in areas of industrial manufacturing, automation, goods transport, smart travel, agriculture, environmental monitoring, and healthcare. Under this design, IoT seems to play a promising and important role in household living, electronic healthcare, assisted living, public health, and new education styles that are some indicators of probable application scenarios in the decade to come [5].

3.1.1 Cognitive sensors in a WSN

A cognitive WIoT network (CWIoTN) is one in which the sensors have cognitive abilities to change and dynamically adapt their communication parameters, protocols, and channels based on availability and as warranted by performance constraints. A CWIoTN needs to be designed with a key ability of spectrum sensing to find out unused frequency channels and utilise them for either underlay transmissions using low power or for overlay transmissions occupying the unused channel completely. The setup needs to be adaptive because different spectral ranges have separate sets of challenges and coverage area that dictate the distances between sensor nodes and hence the density and number of sensor nodes.

Several of the advantages of a CWIoTN depend on the operating spectrum available for the cognitive radio-based sensor nodes. If the lower range of the frequency spectrum such as the ultra high frequency (UHF) and very high frequency (VHF) is used by sensor network nodes, the nodes would have a higher transmission range and would need lessening of hops to the gateway node and lesser synchronisation errors. Lesser sensor density would be required, and the transmission would penetrate through walls, increasing the reliability in indoor settings. As a trade-off, the sensors would be more expensive due to a design having higher spectrum sensing, dynamic spectrum access and trans-reception capabilities.

3.2 Cognitive sensor networks – importance and utility

A CWIoTN improves the efficiency of spectrum utilisation for improved data traffic and its quality by enabling opportunistic access to the spectrum. Multi-hop WSNs can benefit immensely by going cognitive as they are already constrained for resources. Dynamic spectrum access allows for event-driven communication based on the event characteristics and spectral availability. The technique also helps in reducing the number of collisions and delay in packet transmission due to contention drastically. The delays are reduced partly because of the increase in range and lesser number of hops required for packet routing.

3.2.1 Cognitive sensors – a historical perspective

Cognitive sensors used in a CWIoTN work on the idea that if the radio technology used with the wireless sensor was added with intelligence, it could recognize the services offered by local wireless networks. The sensor radio could exchange communication with the local networks using their favoured protocols, pick up the suitable wireless network for sending the sensor data immediately or receiving control and configuration information from a distant network [6]. Also, it could use the wireless waveforms and frequency pairs that cut down on possible interference with other wireless communication schemes working in the vicinity.

Software-defined radio (SDR) used in cognitive sensors today can credibly accomplish critical tasks that support the user and the network, while conserving the frequency spectrum and curtail its congestion, in addition to helping with assured transmission of critical physiological data.

Digital signal processing (DSP) techniques arose with the requirement to convert analog signal processes to their digital equivalents and the potential of breakthrough techniques such as digital filtering in the industry was recognised quickly [7–12]. Advances in DSP helped the industry in employing new processes that were completely unworkable using the older analog signal processing techniques. New DSP software tools facilitated quick and efficient modeling of intricate DSP algorithms [13,14]. Analog electronic functions realised using huge discrete electronic components were replaced by digital functionality realised at a lower expense in semiconductor design giving smaller sizes with lower power consumption, along with ease of mass production, and a considerable improvement in reliability [15]. These advances in the semiconductor industry made it possible to employ digital signal processing for a better performance on modulation and demodulation of involved radio signals while adding value for the customer, with trustworthiness as well as flexibility of spectral usage.

Networking technologies have evolved fast and have moved from wired local and Internet network to wireless networking used to connect to networks through a wireless access point and other wireless techniques in ad hoc IoT and sensor networks in which surrounding wireless sensor nodes assist as repeaters to collect and forward sensor data packets from other nodes towards a base station used for data collection.

3.2.2 Cognitive sensors – the evolution

Cognitive radios used for CWIoTN are applications that depend on Software-Defined Radios created from digital signal processors fabricated using semi-conductors. The spectral performance metrics of the applications and other smart provisions for the user are affected by complex connections between several radios to accomplish the final performance, while adding capacity and performance benefits for the customer.

The development of cognitive radio for use with new age sensors started in 1987 with the development of a programmable modem by Air Force Rome Labs (AFRL) that attempted to combine several single-purpose radios in one box to prepare a single consumer electronic device that contained an integration of navigation, communications, along with identification electronics.

The essential hardware framework of a modern day SDR defines the carrier frequency, its modulation technique, required encoding for security, and the program to accomplish all of this in software. The hardware resources may include a mix of FPGAs, DSP IC chips, and other required computational resources such as power, processing and memory, to define a broad array of modulation schemes. The application programming interfaces (APIs) are written to include a big library containing details of implementation of protocols and the waveforms to get the radio working, along with vital interfaces to guarantee that the fundamental software organisation provisions for a varied diversity of waveform applications without major modifications or rewriting required, thus ensuring software portability across a variety of hardware platform implementations.

3.3 Use of cognitive sensors in weather monitoring

Keeping an eye on the vital parameters related to the environment is extremely important for ensuring a healthy and hygienic living of the public. Forecasting of weather and warning the public in advance through weather alerts at times of extreme weathers can assist in saving lives and property. Monitoring, recording, and analysing weather, climate and environmental data also help researchers and climatologists in keeping track of weather trends. Weather services from different countries use articles, social media posts, websites, weather information and warning videos, and weather infographics to inform the public about imminent weather hazards. A variety of sensors and communication equipment help in data collection and generation of actionable information. Weather warnings and alerts are communicated to the public through several means such as sirens audible in the local area, local weather radio transmissions and alerts through text messages, social media dissemination on smart devices and cell phones, weather forecast social media handles and dedicated websites.

3.3.1 Monitoring the environment and microclimates

The discipline of environmental science involves multiple sub-disciplines as it combines diverse scientific fields in the research of our planet's climate arrangement, with the influence of human activities on the ecology and the environment, and is dynamic. The field of environmental studies provides a groundwork for this area, empowering environmental experts to investigate natural processes, evaluate exchanges between environmental processes, and discover answers to the weather and environmental challenges.

Wireless sensors available today can be utilised for climate studies, soil investigations, and ecological analysis, along with several other environmental surveys. The sensor network supports monitoring and gathering of environmental data, allowing for more comprehensive and precise appraisals. The system includes integrated data loggers and analysers, weather stations, and software to make them work, which offer environmental researchers reliable instruments for accumulating a broad spectrum of environmental data, making way for improved awareness of environmental settings. Figure 3.1 shows some of the major environment and climate monitoring applications of cognitive sensors in a network.

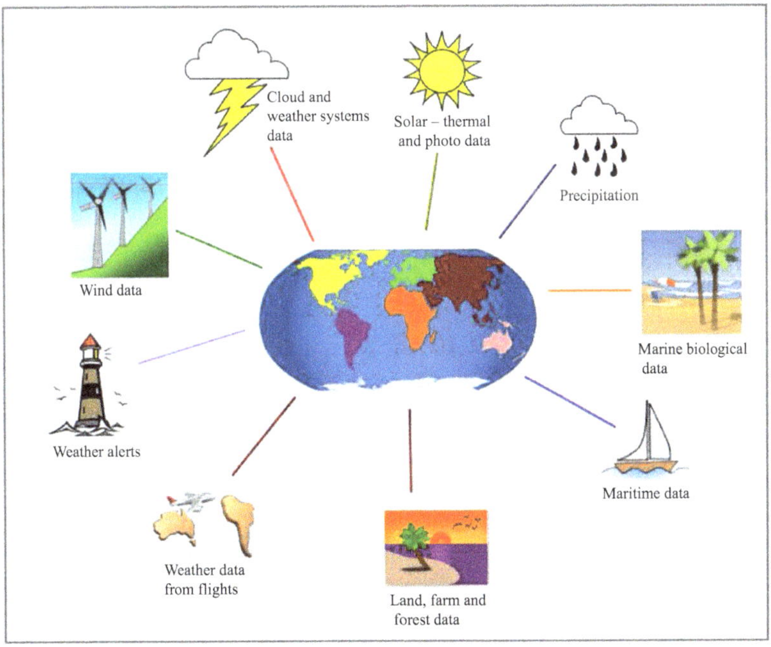

Figure 3.1 Use of cognitive sensors in some critical weather and environment measurements

3.3.2 Monitoring crops and fields

Field monitoring systems today assist farmers and growers save cultivation costs by reducing the amount of water used for irrigation, providing for humidity control, thereby improving the quality of crops, and protecting against extreme climate situations such as excessive heat and frost. Such systems offer measurements on parameters of interest such as ambient temperature, soil moisture, relative humidity, rainfall, leaf wetness, photosynthetically active radiation, solar radiation, wind speed and its direction, and barometric pressure.

WSNs that work on plug-and-play design can be used to create such systems. As no complicated wiring is required, this saves on systems cost while providing better reliability on data gathering. The sensors available today can help create web-enabled systems for monitoring of field conditions. The sensor plug-and-play feature offers ease of use, scalability, and cost effectiveness scaling from small fields to hundreds of acreages of crops. These sensors can be used in open fields or in greenhouses. They help create a 900 MHz mesh that offers reliable signal transmission through foliage and vegetation. The sensors send the data to a cloud-based station with 4G cellular technology incorporating an HMI that allows for real-time visualisation, interactivity, central data collection, system configuration, customisation, and control operations. Such systems can have multiple power options in addition to mains and can work on solar power or battery power too. The HMI also provides for notifications on any alarms that have been configured and set up by the user. Remote access to data and system management is also possible as such systems are cloud-based. The stations and loggers offer cost effectiveness due to easy availability and portability of data with USB or Bluetooth data offload. Integration of the system with map and navigation systems such as Google Maps also improves the utility and meaningfulness of the sensor data. The systems can also run supervisory mathematical models and provide calculated data on parameters such as evapotranspiration. The system can be used to monitor the climate and microclimates over a wide area with rapid, accurate access to sensor data, allowing for possible integration with other data.

3.3.3 Green roof monitoring

Responsible and ecologically aware humans follow practices such as rainwater harvesting for replenishing the rapidly reducing water table. Some also go for green roofs that provide multiple benefits of space utilisation and temperature control inside buildings. Green roof monitoring involves keeping an eye on physical parameters of the roof through sensors such as rain gauges, soil moisture smart sensors, soil temperature smart sensors, air temperature and humidity sensor, solar radiation smart sensor, light sensor, wind speed and direction sensor. WSNs are ideal for measuring the performance implementation of intensive and extensive green roofs. Solar powering of these systems further reduces the carbon footprint of the owner. The web-based sensor systems that allow for remote monitoring of green roofs today present dependable, research-grade operation while supervising green roof conditions. As a part of this, the systems also use sensors that measure stormwater runoff, enhance irrigation planning, and help in maintaining healthy rooftop vegetation. The systems also provide for trend logging, monitoring, and alarm notifications.

3.3.4 Weather observation systems

The collection of weather data around the world by weather observation systems amounts to billions of measurements related to the Earth's atmosphere every day. Studies involving the atmosphere are crucial to every facet of weather forecasting. Mathematical weather models, vital to short and long-term weather prediction, employ observations to initialise conditions consumed by algorithms with high levels of complexity. The relevance and timeliness, quality, and density of observed weather information is vital to guaranteeing the most excellent numerical model output achievable. Weather observations can also offer forecasters real-time data that they could put to good and effective use for realising things like announcing life-saving weather warnings related to safety and evacuation, and for making critical modifications to forecasts vital for aviation.

3.3.5 Weather parameters from upper air

To find out the information about the upper air layers, weather balloons carrying instrument bundles called radiosondes are launched. The sensors in the radiosondes gauge upper-air parameters such as temperature and humidity, barometric pressure, wind direction and velocity. The data are used for mathematical weather modeling and prediction. The findings from these sensors are critical for aviation safety and to detect the possibility for harsh weather.

3.3.6 Surface weather observation

Surface weather observing networks are created with the objective of keeping a consistent and constant, 24/7 watch on the weather conditions for supporting weather forecast and aviation functions, and strengthening the requirements of the climatological, hydrological, aviation and meteorological research groups. The networks use WSNs to sense critical aviation weather parameters within airport runway environments for direct application on site. Surface weather observation networks keep evolving to have better coverage of weather observations for improving the accuracy, usefulness, and timeliness of weather forecasts and warnings.

The audio quality of weather warning messages is of paramount importance. Typical audio-quality forensics system (AQFS) uses parameters from the wireless network and computing equipment to check the quality of audio signals used for transmitting weather warnings. Transmitters used in such broadcasts today have a variety of network connections through which the AQFS can help or automatically perform switchover to another connection technology if one fails. The AQFS can detect multiple scenarios, starting from intermittent audio outages to total loss of audio. It can also identify strange tones, audio anomalies, audio drops and outage due to fading and use this information to make the transmission audio quality better. The analysis performed by an AQFS is presented in Figure 3.2.

Waveform analysis

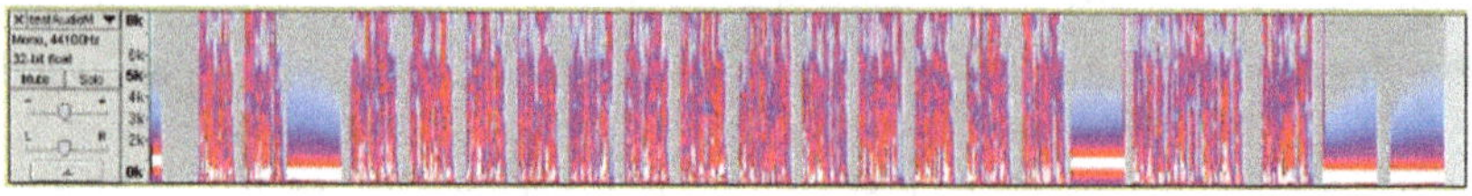

Voice and tone testing

Figure 3.2 Analysis of audio weather warning signals

Weather and climate disasters faced by the world such as drought and storms, wildfires, cyclones, severe floods, and winter storms convoyed with freezing temperatures cause enormous harm to life and property. Recording the details of such catastrophes is important because it aids the appraisal of their influences to the world and verifies the losses [16]. Such chronicling also helps in keeping an eye on weather trends [17].

Agencies from different countries around the world routinely provide the public with daily weather forecasts and severe weather warnings. The agencies also monitor the climate to back economic vitality and help boost the gross domestic product of a country. Such chronicling provides citizens, emergency planners, first responders, first response managers, and key policy/decision makers with dependable evidence that aids the country's economy and helps lower the damages caused by weather associated disorders. The agencies dedicated to weather information service publicise relevant weather forecasts, weather or event watches, audio alerts, and prior warnings to help the public prepare for bad weather in advance. These weather alerts and warnings are communicated to the public through several ways such as locally audible sirens, audio alert transmissions on weather radio channels, text message alerts on cell phones, social media inputs on cell phones, and updates on weather forecast websites. A real-time AQFS system monitors system parameters at the transmission for detecting poor performance of audio announcements and outages. Figure 3.3 shows the design of one such system [18].

3.3.7 Wireless sensor data from anchored buoys

Making weather observations on land is relatively easy compared to doing so across the vast expanse of water bodies across the lakes, seas, and oceans. Weather buoys fitted with multiple wireless sensors are used for gathering observational data from the water bodies for use in weather forecasting. The data are also used for engineering design, scientific study and research plans, emergency response to hazardous waste or chemical spills, as well as in legal proceedings.

Equipped with sensors to measure the common weather parameters such as barometric pressure, wind direction, wind speed, gust parameters, air and sea temperature, these anchored buoys also measure and transmit the direction of wave propagation and wave energy spectra, based on which wave parameters such as

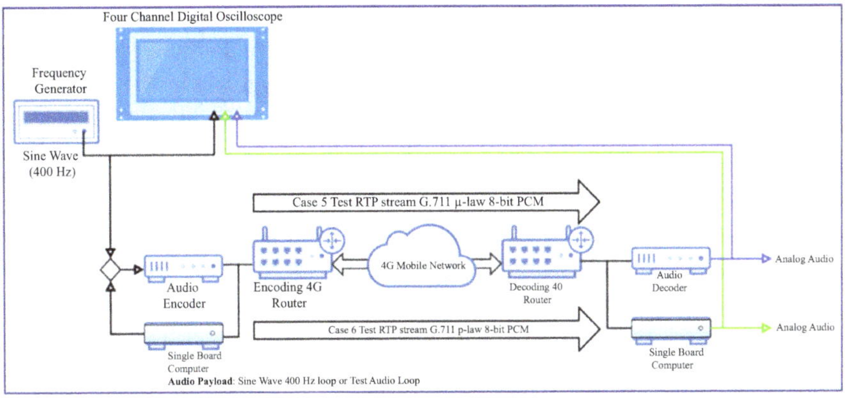

Figure 3.3 Basic design of an AQFS

dominant wave period, average wave period, and significant wave height are deduced. The data is most utilised in operational forecasting, generation of weather warnings, and feeding into atmospheric numerical models.

3.3.8 Sensors on aircraft

Mathematical models for weather prediction get benefitted from the weather and meteorological information received from the sensors that are fitted on commercial transport and cargo aircraft.

Data sensed and reported by these aircraft include air temperature as measured by an immersion thermometer probe measures air temperature. The pitot tube fitted to the airplane's nose contains a barometer that measures static and total air pressure. The pitot tube sensor also helps measure the air speed. The inertial navigation system in the aircraft provides ground speed. Using the two parameters, the wind speed can be computed. All this data furnished by flights over wireless networks when the aircraft are in air helps improve the accuracy and details of forecasts from the prediction mathematical models.

3.3.9 Remote sensing through satellites and radar

Remote sensing involves acquiring information about the parameter of interest without being in contact with the region where the parameter is being measured. Another benefit is that remote sensing can cover large regions from a distance. Wireless sensors fitted in weather satellites can be used to sense the physical parameters associated with electromagnetic energy such as light and heat. Parameters associated with radio waves can be sensed using Doppler radar sensors.

3.4 Cognitive sensors in environmental monitoring

One of the critical issues facing the world today is the fate of the environment. Climate change, global warming, and pollution are just some of the environmental concerns plaguing society. The continuous advancement of technology and infrastructure to support a growing global population has created an increased demand for environmental resources. In the twenty-first century, it has become especially evident that these technological advancements can have unintended and unexpected detrimental effects to the environment, many of which are only noticed or attributed to a specific practice or technology years after the damage has been done. This enhanced understanding of the dynamic relationship between human activity and the environment has led to an increased concern for preserving the environment and protecting it from degradation.

Hence, it is pertinent that systems be developed and implemented to monitor and evaluate the condition and changes in the different components of the environment across several parameters. The collection of this data would aid in the decision-making process for different stakeholders including researchers, businesses, governmental authorities, and the general community. This data would also provide greater insight to the natural cycles and physical, chemical, and biological processes occurring in the environment granting humans a more comprehensive understanding of the environment and its interconnected relationship between its components and systems. Figure 3.4 shows a cognitive sensor network set up for monitoring the important weather and environment parameters.

The use of a CWSN would be a viable and effective tool to facilitate the development of such a system to monitor environmental conditions and changes. A typical environmental cognitive sensing system consists of several environmental measurement nodes wirelessly connected to a central hub. These environmental nodes are distributed across a specific area at designated locations and are designed to measure specific parameters or attributes of specific components of the environment such as air, water, or soil. Measurements are taken by these sensors in tandem and in situ and sent to the central hub for storage and processing. These measurements combined with GIS data, human investigations, and observations to provide a "snapshot" of the current state and quality of environmental conditions at a given point of time.

Initially such a system would function simply to measure, acquire, and store data. Over time however, through the application of machine learning algorithms and human testing, a cognitive model would be designed to "understand" and interpret the raw environmental data and make informed inferences, predictions, and decisions based on the data. As more data is acquired and analysed, the system learns and becomes smarter and grows in capacity and capability in its decision-making. The cognitive cycle for such a system is typically comprised of four main steps:

- Measure – The environmental sensors continuously acquire environmental data and transmit this data wirelessly to a central hub.
- Analyse – The raw environmental data is analysed by using statistical methods and machine learning algorithms to "make sense" of the raw data. At this stage, the system interprets the raw data by extracting meaningful information and performing various calculations to understand what the data conveys about

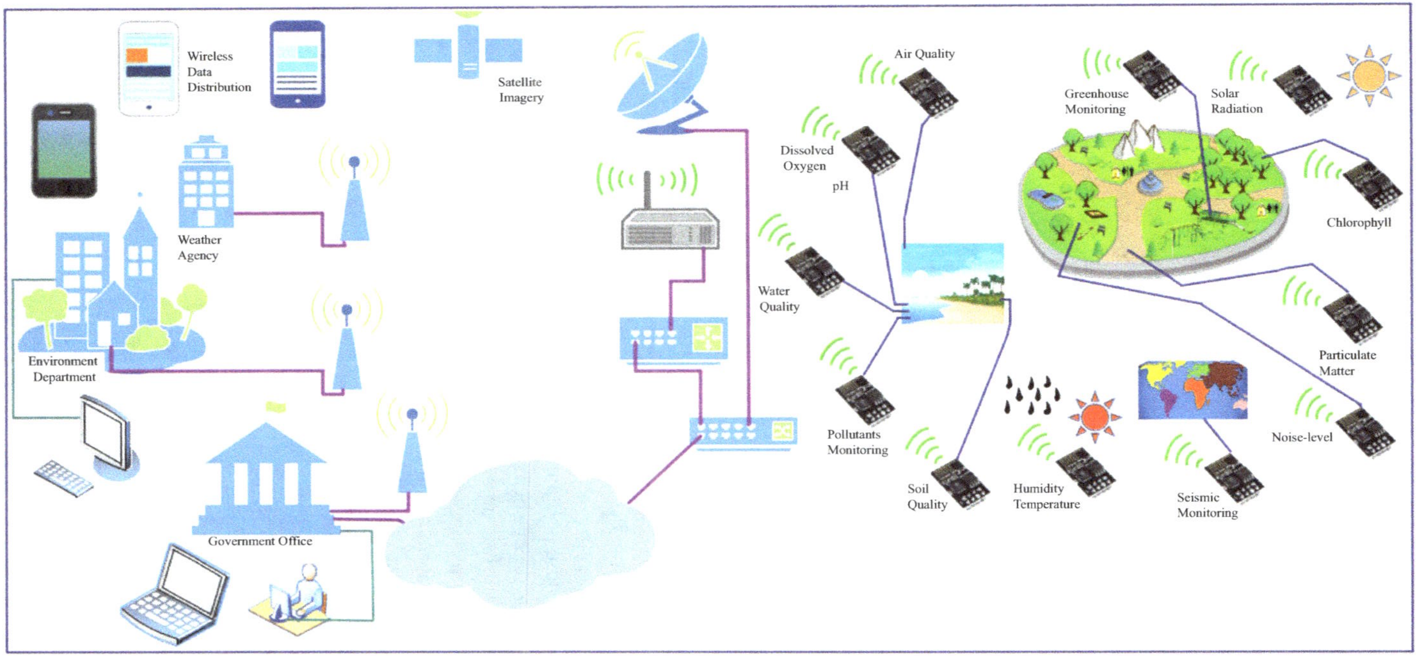

Figure 3.4 A cognitive sensor network setup for weather and environment monitoring

the current state of the environment. At this stage, the system can make predictions or infer what environmental changes are likely to occur or the reason for these changes based on its previously analysed data and supporting observations.

- Decide – After performing its analysis, the system has a clearer understanding of the current environmental conditions, possible future environmental conditions, or reasons for these changes. At this point it will in turn make an informed intelligent decision.
- Respond – The system will perform an action based on its decision parameters in response. For example, a system may transmit a signal to change the orientation of its sensors to better capture environmental data more accurately or provide a specific output to its user to warn the user of an impending issue.

This cycle of measuring, analysing, decision-making, and responding is an ongoing and continuous cycle with each process flowing into each other. In environmental modelling, there are three main components that should be assessed: air, water and soil. For each of these components, there are numerous parameters that can be measured and analysed. The environmental component and its associated parameters to be measured would depend on the purpose of the system, its scope and scale. However, there are some common parameters measured by environmental monitoring systems for determining the quality and health of each of these environmental components.

3.4.1 Air – air quality, noise & weather forecasting

Environmental monitoring of air can be used to determine the general air quality in a specific area or location. For such a system, nodes are in specific designated locations outfitted with sensors that measure the presence and concentration of specific pollutants of interest. These pollutants are commonly referred to as criteria pollutants by the EPA and are a set of chemical compounds or particles that have been proved to have a detrimental effect on air quality and human health. Particulate matter of different sizes, greenhouse gases, ozone, lead, sulphur dioxide, and nitrogen dioxide are some of the most common pollutants that can be analysed by electrochemical sensors to evaluate air quality status of a given area. Noise pollution can have a detrimental impact on human hearing. Audio sensors located in cities can be used to measure the noise levels at different decibel levels by a cognitive system. This system can map out the noise pollution distribution across a city and can work in conjunction with data from health records to determine which activities and areas produce the most noise pollution and their associated health impacts.

Environmental monitoring of atmospheric conditions in the air to determine weather conditions and patterns are a common use of cognitive sensing. In analysing several atmospheric and air parameters, the current weather conditions can be determined. This data can also be modelled to predict the weather conditions in the future. Determining future weather conditions is important for land, sea and aerial navigation, the agricultural industry, public health and safety and urban planning and development.

The main parameters assessed by typical weather monitoring sensors include barometric pressure, ambient temperature, ambient humidity, precipitation, wind velocity, wind direction and solar radiation. Temperature and humidity are measured using a thermometer and a hygrometer. Several commercial IoT sensor resources are available for such measurements. Wind direction and speed are measured using a Wind Vane and an Anemometer. Precipitation gauges are used to measure liquid precipitation. Common technologies used for this are the tipping bucket or weighing equipped with electronic sensors to measure the fill level. Air pressure is measured with a barometer which consists of aneroid wafers which are sealed and have an internal pressure of 29.92 in. of mercury. Any decrease or increase in pressure makes the wafer contract or expand. Solar radiation is measured with a pyranometer.

3.4.2 Cognitive sensors in the monitoring of water quality

All living beings are dependent on water to sustain life. It is a vital component of human life and is fundamental to many recreational, industrial, agricultural, and social activities. It has been observed that many human activities and practices have had significant impacts to the quality and availability of water. Additionally, natural environmental processes can affect water availability and quality. These occurrences in turn affect aquatic biodiversity, land availability, and ecosystem health. Because of this, there is a need for the preservation of water quality and conservation of water sources. Several indicators of water quality, commonly measured by monitoring sensors include the following:

3.4.2.1 Temperature

Water temperature is a measure of how hot or cold the water is an indicator of the total kinetic energy in the water. It is important as water temperature affects the rate at which chemical and biological processes occur in aquatic environments. It influences the amount of dissolved oxygen stored in the water and the solubility of metals in the water. As such, it has a direct bearing on the survivability of specific aquatic fauna and flora. Monitoring sensors equipped with digital thermometers can provide temperature measurements to the central hub.

3.4.2.2 Dissolved oxygen

Dissolved oxygen (DO) involves sensing of quantity of oxygen dissolved in water that is obtainable to aquatic fauna and flora. It is directly connected to the survivability of different aquatic animals. It is especially important in maintaining and supporting fishing activity. In low DO conditions, fish cannot survive and so fish populations dwindle. Low DO conditions are typically indicative of excessive algal bloom which reduce DO via photosynthesis. Monitoring sensors equipped with optical, polarographic, or galvanised DO sensors are typically used to determine DO.

3.4.2.3 pH

pH provides a measure of how acidic or basic a waterbody is. Aquatic species thrive in specific pH ranges and so pH has a direct bearing on the survivability of the aquatic life of a given waterbody. Sudden drastic changes in pH can be an indicator of the introduction of a pollutant or chemical to the waterbody. Monitoring sensors equipped with a pH electrode can measure pH levels and transmit the readings to the central hub.

3.4.2.4 Turbidity and total suspended solids

Turbidity measures the clarity of a waterbody whereas total suspended solids (TSS) measures the concentration of suspended particles in a waterbody. Sudden changes in TSS or turbidity can be signs of erosion, algal bloom, and water pollution. TSS and turbidity are typically measured by a sensor that utilises light to determine how opaque the water is.

3.4.2.5 Chlorophyll a and phycocyanin

Chlorophyll a and phycocyanin are pigmented molecules associated with the proliferation of different types of algae. The presence of these molecules provides an indication of the scale of an algal bloom in a waterbody. Environmental monitoring sensors typically use spectrophotometry sensors to detect the specific wavelengths of light emitted from these molecules to determine the amount of Chlorophyll a and phycocyanin present in a body of water.

Depending on the waterbody being monitored and the purpose of the system, the size, location, and type of water sensor may vary. For example, the LG monitoring buoy is composed of water quality sensors attached to a floating buoy. When placed at different points in a standing body of water, it can monitor the changes in water quality. Another system may use stationary monitors installed along a riverbank to assess the changes in water quality down a watershed.

Satellite imagery from earth orbiting satellites can also be used in remote sensing to determine several parameters based on spectral analysis methods including algal bloom, temperature, and turbidity. Because of the time taken for the satellites to rotate around the earth, there is a long wait time between measurements compared to typical sensors.

Some of these sensors are in fixed stationary locations such as weather stations whereas other sensors may be mounted to weather balloons and deployed into the air to gather data. These systems would communicate with each other and transmit data back to the central hub where it would be analysed to give the holistic status of the current and future weather conditions based on its predictive modelling algorithms. In response to certain conditions, the system may change the orientation and positioning of its sensors to capture data more accurately.

3.4.3 Soil – soil quality & seismic monitoring

Evaluating soil quality by monitoring several parameters is especially useful in studying ecology and in the agricultural industry. Understanding the health of soil can aid in the agricultural industry by driving informed decision making and ensuring best agricultural practices are enacted to allow for maximum crop yield. Parameters such as soil moisture, salinity, and pH are commonly monitored by cognitive environmental systems to assess the changes in soil quality over time.

Seismic sensors are used commonly today to determine the severity and magnitude of earthquake events. Information gathered by these sensors are influencing the materials and practices implemented in the infrastructure of specific areas.

3.4.4 Sensors used in environmental pollution monitoring

For each of these environmental components, cognitive sensing would prove to be a competent and viable tool for monitoring and mitigating environmental pollution. By measuring the changes in air, water and soil quality health parameters with time, the system would be able to detect sudden and drastic changes that appear to be due to unnatural environmental activity. Such phenomenon when studied in conjunction with GIS data, general knowledge of the practices and technologies in the area and human investigations and observations can be attributed to specific pollutants and human activities.

Because this data is being monitored, transmitted, analysed, and responded to by the system in real time, it gives humans the advantage of being able to "catch" a potential disastrous pollution event. At this point, humans can take preventative and corrective actions to mitigate the spread of an environmental pollutant. Humans would also be able to attribute specific pollution occurrences to specific entities, organisations, and industries and in turn implement the appropriate policies, legislations and regulations and decisions to reduce or remove a specific source of pollution.

The sensors used in environmental pollution monitoring would typically be the same sensors used to monitor environmental quality for each of the environmental components. The introduction of a pollutant is likely to enact some quantifiable change to the environmental conditions. If some activity in a specific area is known to produce a specific chemical pollutant into an environment, monitoring equipment can be equipped with specialised sensors designed to measure the concentration of that chemical of interest to determine its introduction and concentration into the environment.

3.5 Cognitive sensors in healthcare monitoring

Physiological parameters of human and animal subjects need to be monitored to assess their health. Such monitoring becomes important for patients, sports persons, deployed soldiers, first responders to hazardous situations and the elderly, who usually have one or more health complications. There has been an increasing need to do so unobtrusively and passively, while also pervasively. This is where wireless body sensor networks and the Internet of Medical Things come in to create a CWIoTN. They involve implantable and wearable body sensors that need to be networked together using wireless for pervasive monitoring of the physiological parameters of the subjects. The concept with one such network is shown in Figure 3.5. It embodies a subject with several sensors connected to their body, while also being linked to a small processor for processing the sensor data, a wireless transmitter to send this data ahead, and a battery to power up the components and modules, creating a "CWIoTN node" that seamlessly integrates with

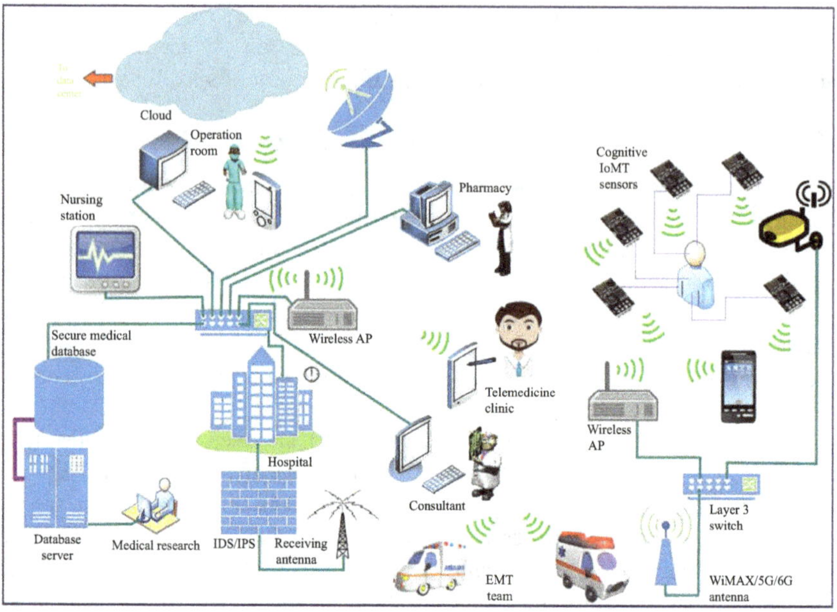

Figure 3.5 A cognitive wireless Internet of Medical Things network

home, workplace, and hospital environments. The data from all the sensors is accumulated and fused before being transmitted over a wireless link to a central base station or server over a wireless LAN link, Bluetooth connection, or cell phone data (GPRS or 5G/4G) network for further monitoring [19]. WBSNs also allow monitoring of the subjects over a prolonged period in comparison with traditional tools that provide a snapshot of the physiological or biochemical parameter of the subject at a point of time. Additionally, no prolonged hospital stays with heavy bills are encountered because the WBSN obviates the hospital or bed confinement for subjects with acute or chronic ailments, while providing comprehensive health monitoring. It then makes all the data available to the physician who can then have a better diagnosis and line of treatment for the subject.

The following sections focus on the applications and design of cognitive sensors and techniques for healthcare.

3.5.1 Diagnostic tools – telestethoscopy

Electronic stethoscopes for listening to heart and breath sounds either use microphones, piezoelectric crystals, or a capacitive diaphragm sensor [20] in the chest piece and transmit those sounds over wireless links to help the caregivers monitor vulnerable patients with cardiovascular disease in their home environments [21].

As a part of comprehensive telecare programs, teleauscultation has also been tried with a videophone-based method for remote cardiopulmonary examination of heart patients, utilising an electronic stethoscope and transmitting the audio signals using an integrated services digital network with very accurate interpretation by the physician at the other end, proving to be very reliable.

3.5.2 Monitoring blood sugar

The dreaded Diabetes mellitus can cause end-organ complication as it is connected to hypertension, renal disease, coronary artery, and peripheral vascular complications amongst others. Data for the United States indicates that every year 24,000 people get affected by diabetes, and close to 56,000 limb loss cases happen due to peripheral vascular disease [22]. Management of diabetes requires regular monitoring of fasting blood glucose based on which medication or insulin administration decisions on frequency and dosage are taken [23]. A recent development is the use of wearable patches for continuous blood glucose monitoring instead of the traditional pin-prick method. Adding wireless capability to these patches could allow for communicating the tissue and blood glucose data for immediate intervention in a closed feedback control loop for delivery of fast-acting insulin through an insulin pump as done in the artificial pancreas project [24,25]. This closed loop arrangement avoids the possibilities of hyper or hypo glycemia and saves the patient from end organ damage.

3.5.3 Surgery

Surgeons performing operative procedures can be helped better if the monitoring of vital parameters uses sensors in a wireless network, thereby allowing them unobstructed work front. The patients can be helped using such wireless-based monitoring during the post-operative recovery phase by not being confined to a hospital ward, and saved from the restrictive, wired monitoring that is also labor intensive, expensive, and prone to error due to human involvement. This automation with the capability to monitor patients pervasively in the hospital or at home through sensors that could be worn or implanted is a win–win for the healthcare provider, as well as the patient in the timely detection of potential postsurgical complications. Imperial College London and Sensixa have developed (e-AR), an ear-worn activity recognition sensor that performs continuous physiology monitoring and uses activity indices to monitor post-operative home recovery, for chronic or episodic events [26]. In another application, after endovascular stenting, an implantable wireless pressure sensor was utilized to monitor pressure in the aneurysm sac to detect and predict the risk of arterial rupture due to high pressure during abdominal aortic aneurysm surgery [27]. Such applications also allow the better-informed caregivers to provide more prompt and effective intervention based on this pervasive monitoring obviating the need for the patient to attend outpatient clinics and allowing for better workflow. This also helps with better capacity utilisation in the hospitals.

3.5.4 Monitoring elderly patients

The increase in average life expectancy in developed and developing countries and the increase in the elderly age group has increased the medical monitoring and care needs for the elderly who are at risk and now are significantly sizeable with the group size consistently increasing [28]. They can be immensely benefited by regular and non-intrusive ubiquitous health monitoring through WBSNs with cognitive sensors. Such monitoring is invaluable during extreme weather times when travel to hospitals is difficult, more so for the elderly persons living all by themselves [29,30]. Moreover, if the ambient sensors are connected to the subject's WBSN, smart decisions can be taken for more rigorous monitoring and early detection of deteriorating health parameters and communication using cognitive radio.

The use of an unobtrusive cognitive sensor network for health monitoring helps with early detection of health problems and lessens the need and risk of hospital admission, connected morbidity or mortality.

3.5.5 Lifestyle, sleep, and wellbeing

The focus today has moved from reactive to proactive healthcare through community healthcare programs. Such programs aim at management and risk reduction due to obesity, diabetes mellitus, and ischaemic heart disease through wearable sensors that provide the patient with direct feedback daily. Numerous wearable devices involving sensors such as multi-axis pedometers conveniently hidden in wristwatches are available today for keeping track of the physical activity for health improvement [31]. Armbands such as "SenseWear" using multiple wireless sensors can keep tracking sleep quality, duration, and disturbances in sleep for reliable, home-based sleep assessment [32]. Measurement of skin temperature, heat flux and galvanic skin response and establishment of relationships between the parameters can be done by respective sensors along with movement during sleep monitored by a multi-axis accelerometer.

3.5.6 Memory glass

Research in wearables is also aiming at providing a context-aware memory aid to the wearer through "memory glasses". This memory glass is a smart glass with an optical, head-mounted display that has an operating system of its own and comes with wireless connectivity for use in augmented reality for applications in the fields of medical, pharmaceutical, and biotechnology. The memory glass finds use in interactive training of medical professionals as it can provide medically accurate and realistic 3D animation, motion graphics, or for creating high end illustrations. It has the potential to provide powerful, related, and positioned learning practises, and to help study the complex interconnections in real world data [33]. Students can use this technology to understand complex subjects better, created through their collaborations with virtual objects, which use inherent data to create life-like learning experiences.

3.5.7 Monitoring the gastro-intestinal tract

The diagnosis of widely prevalent gastro-oesophageal reflux disease (GORD) is provided by a radio-telemetric capsule deployed using an endoscope at the junction of the oesophagus and the stomach in the Bravo pH system [34]. This makes the restrictive and obtrusive practice that traditionally used transnasal catheter-based pH electrodes in the oesophagus for pH monitoring outdated [35,36].

Capsule endoscopy employs a miniature wireless camera placed inside a vitamin-size capsule to click pictures of the patient's digestive tract, after the patient swallows it and the capsule travels through the patient's digestive tract. The pictures are transmitted by an integrate transmitter in the capsule to a recorder worn by the patient. The pictures help doctors see inside the patient's esophagus and small intestine into area that is not easily reachable with traditional endoscopy, and help find the reason behind gastrointestinal bleeding, and diagnose inflammatory bowel diseases, cancers, or polyps.

3.5.8 In vivo monitoring through implants

The Medtronic's "Reveal Insertable Loop Recorder" offers an implantable cardiac monitor for recording the ECG displaying the subject's heart's rate and rhythm during events of mysterious palpitations, dizziness, or fainting [36]. Pressure in an aneurism sac can be read using an implantable pressure sensor during endovascular repair for tracking the status of repair until well after the surgery [27]. A completely implantable wireless surface acoustic wave (SAW) pressure sensor containing quartz wafer resonators can now monitor the blood pressure for heart and pulmonary hypertension patients. This device developed at Imperial College can be powered using wireless charging and does not need batteries.

A neurostimulator linked to subdural leads positioned in the section of the patient's brain that is causing seizures can now monitor electrocorticographic activity and flag abnormalities for helping patients of medically resistant epilepsy. It can even perform therapy by stimulating the area to halt the seizure.

3.5.9 Prosthesis

Knee replacement arthroplasty is a surgical process that is performed to resurface a knee injured due to arthritis, a disease that causes pain and stiffness, mostly in older people. In this surgery, metal and plastic accessories are used to cap the ends of the bones making the knee joint, and the kneecap. Several hundred thousand knee replacement arthroplasties are done worldwide every year, with revisions required to the tune of 8% [37,38]. Smart implant prosthesis developed by Bergmann *et al.* for total knee replacements uses a tibial component in which strain gauges perform force measurements during the gait cycle for analysing the performance of different parts of the implant under varying load conditions.

3.5.10 Design features

The cognitive IoMT sensors used for instrumentation and closed-loop control need to be reliable and one way to increase this reliability is through redundancy and use of multiple sensors in the system. Implantable biosensors now use multiple sensors in an array for improving the accuracy of measurements and thus the system reliability [39].

This strategy of using sensor arrays instead of isolated sensors goes against the important design requirement that the sensor sizes are supposed to be small. This requirement is being met by resorting to nano engineering and MEMS technology for miniaturisation as seen in optical glucose sensors made from fluorescent hydrogel alginate [40] and for closed loop control for drug delivery [41]. The MEMS technology can be used for creating microfabricated non-integrated circuit devices with physical and biomedical sensors at very low cost. Implantable blood pressure sensors using microneedle arrays have been created using MEMS technology [42]. The MEMS technology also offers very efficient and precise sensors at a very low cost for mass production, as seen in case of manufacture of accelerometers for the automobile industry used in vehicle airbag control.

3.6 Challenges for cognitive sensors

The exploding demand from IoT applications in addition to traffic from new smartphone data applications has already started to overwhelm wireless capacity, triggering a spectrum crisis that affects all wireless devices. Lack of availability of new spectrum for wireless data carriers has become a major problem.

The biggest issue that cognitive radio technology faces is that IoT sensors require not only opportunistic spectrum access but need to include heterogeneous radio framework management into their design. Spectrum allocation approaches for wireless applications use fixed designs by default and are not designed to permit a dynamic allocation of frequency bands to different wireless technologies.

Cognitive radios are anticipated to transmit across multiple decades of frequencies by persistently sensing the frequency spectrum of interest and detecting obtainable communication channels. They need to address this in the face of a widening gap between wireless supply and demands owing to multiple new IoT applications popping up in recent times, each needing its own frequency range for communication. The design of such networks needs to address several challenges related to working out usable, wideband signal paths and the synthesis of communication carrier frequencies across the frequency spectrum of interest. The setup needs elaborately designed electronics such as low-noise amplifiers that work across a wide range of frequencies that also have low power dissipation in addition to low noise figure. Designing them for miniature IoT sensors is another level of a challenge.

The design must address global efficiency needs in the spectral domain, as well as in the energy domain. Apart from sensing algorithms, cross-layer signalling and flexible hardware designs, other major obstacles in the use of cognitive sensors for IoT applications include security, intelligence distribution and implementation and addressing delay/protocol overhead.

3.6.1 Challenges in environmental cognitive sensing

One of the challenges in implementing an environmental cognitive sensing system is providing power for each of the measurement nodes. These nodes not only need power to operate to measure their respective parameters but also need to be able to transmit and communicate information between themselves and the central hub. The difficulty of this challenge is dependent on the purpose of the environmental monitoring system, the environmental component and parameters being analysed, the location of the designated site, the environmental conditions of the location, access to infrastructure and scale and scope of the system.

Because of the unpredictable nature of the environment, those sensors left to the elements can also be damaged or destroyed by the natural elements and changing weather conditions. They may need to be replaced, recalibrated, maintained or recovered multiple times depending on their location and type.

Another issue the system may face is in deciding which data types and data sets are meaningful from its initial data acquisition. This problem would be the most prevalent in the initial stages of the system's implementation as the cognitive method and its "thinking ability" is still in its infancy. At this time, the system may not be able to make any accurate or reasonable suggestions based on the gathered data as it has not gathered enough information to make inferences or has not yet learned what the measurements can be indicative of.

3.6.2 Challenges for cognitive sensors in healthcare

One of the major design issues specifically important for designing cognitive sensors for IoT in healthcare involves addressing the excessive heat generation due to switching between wireless technologies. The use of IoT data transmission using radio waves, microwaves and sub-micrometer waves involves radiation which always generates a lot of heat that can be harmful to healthy human cells. With a broader spectrum of frequencies used in cognitive wireless technologies for more efficient and prompt communication of healthcare data from living subjects also comes the risks of additional radiation-caused damage to the subjects. The design of a cognitive healthcare IoT network needs to address this problem.

The intrinsic safety of cognitive radiation for IoT is another factor of concern. As the penetrating power of radio waves is different for different wavelengths, elaborate studies also need to be performed that focus on the safety of living subjects because of the variation in radiation.

As the cognitive sensor technology is expected to work across multiple decades of frequencies, it is also expected to face challenges related to interference from other wireless applications active in the same area of interest. It is important that the routing on critical healthcare data does not suffer due to problems related to interference in their transmission.

Conveying physiological data from living subjects to healthcare providers mandates that wireless sensors do this promptly and within certain maximum limits of latency as mandated by the established Quality of Service (QoS) parameters for such networks.

3.7 Future trends

While the IoT presents opportunities to provide services with great benefit in the form of solutions that empower elderly people to live longer in their own homes, sensors that combine to provide early recognition for fires, hurricanes, and floods, it also poses some great risks that need to be mitigated.

In the past, a system failure would have impacted the users of that system only, with IoT applications, the designers should now consider the users of interconnected systems. There is a great need to build security solutions into CWIoT, because one weak link can expose multiple systems and devices and put many users at risk.

CWIoT systems need to be designed with toughness, value, and business continuity in mind. The CWIoT systems need to be rigorously tested for interconnectedness of the systems before release and deployment. Another major risk involves concerns about the ethical or moral use of IoT solutions, where the possibility for benefit and risk is governed by the application of the solution. IoT solutions need to be finalised with an understanding of the outcomes of how the combined technology could be used and have protections built in against harmful use of this technology.

Acknowledgements

The authors acknowledge the use of Visio drawing software and some free to use images from wpclipart.com for creating the figures.

References

[1] K. J. Singh and D. S. Kapoor, "Create your own Internet of Things: a survey of IoT platforms," *IEEE Consumer Electronics Magazine*, vol. 6, no. 2, pp. 57–68, 2017.

[2] M. Frustaci, P. Pace, and G. Aloi, "Securing the IoT world: issues and perspectives," in *IEEE Conference on Standards for Communications and Networking* (CSCN), Helsinki, Finland, September 2017, pp. 246–251.

[3] L. Atzori, A. Iera, and G. Morabito, "The Internet of Things: a survey," *The International Journal of Computer and Telecommunications Networking Archive*, vol. 54, no. 15, pp. 2787–2805, 2010.

[4] D. Giusto, A. Iera, G. Morabito, and L. Atzori, *The Internet of Things*, New York, NY: Springer, 2010. ISBN: 978-1-4419-1673-0.

[5] G. V. Crosby, T. Ghosh, R. Murimi, and C. A. Chin, "Wireless body area networks for healthcare: a survey," *International Journal of Ad Hoc, Sensor & Ubiquitous Computing*, vol. 3, no. 3, pp. 1–26, 2012.

[6] B. A. Fette, "History and background of cognitive radio technology," in *Cognitive Radio Technology*, Oxford: Newnes, 2006, pp. 1–27, ISBN 9780750679527, https://doi.org/10.1016/B978-075067952-7/50002-7.

[7] T. W. Parks and J. J. McClellan, "Chebyshev approximation for nonrecursive digital filters with linear phase," *IEEE Transactions on Circuit Theory*, vol. 19, pp. 189–194, 1972.

[8] F. Harris, *Multirate Signal Processing for Communication Systems,* Englewood Cliffs, NJ: Prentice Hall, 2004.

[9] L. Rabiner and R. Schaefer, *Digital Processing of Speech Signals,* Englewood Cliffs, NJ: Prentice Hall, 1978.
[10] http://www.mathworks.com/company/aboutus/founders/jacklittle.html
[11] L. R. Rabiner, J. H. McClellan, and T. W. Parks, "FIR digital filter design techniques using weighted Chebyshev approximations," *Proceedings of the IEEE*, vol. 63, pp. 595–610, 1975.
[12] J. Markel and A. Gray, *Linear Prediction of Speech*, Berlin, Heidelburg and New York, NY: Springer Verlag, 1976.
[13] J. Flanagan, *Speech Synthesis and Perception,* New York, NY: Springer Verlag, 1972.
[14] http://www.dspguide.com/filters.htm
[15] A. Oppenheim and R. Schaefer, *Discrete Time Signal Processing,* Englewood Cliffs, NJ: Prentice Hall, 1989.
[16] A. B. Smith and J. L. Matthews, "Quantifying uncertainty and variable sensitivity within the us billion-dollar weather and climate disaster cost estimates," *Natural Hazards*, vol. 77, no. 3, pp. 1829–1851, 2015.
[17] P. Hoeppe, "Trends in weather related disasters – consequences for insurers and society," *Weather Climate Extremes*, vol. 11, pp. 70–79, 2016.
[18] A. Mishra, S. Bagui, and J. Compo, "Design and testing of a real-time audio-quality feedback system for weather broadcasts and a framework for a weather broadcast transmission technology switch," *IEEE Access*, vol. 7, pp. 158782–158797, 2019, doi:10.1109/ACCESS.2019.2950306.
[19] L. BPL, and G. Z. Yang, "Key technical challenges and current Implementations of body sensor networks," in *Proceedings of the Second International Workshop on Wearable and Implantable Body Sensor Networks*, London, UK, 2005, pp. 1–5.
[20] M. E. Tavel, "Cardiac auscultation: a glorious past – and it does have a future!," *Circulation*, vol. 113, no. 9, pp. 1255–1259, 2006.
[21] G. Fragasso, A. Cuko, R. Spoladore, *et al.*, "Validation of remote cardio-pulmonary examination in patients with heart failure with a videophone-based system," *Journal of Cardiac Failure*, vol. 13, no. 4, pp. 281–286, 2007.
[22] A. S. Brown, "Lipid management in patients with diabetes mellitus," *American Journal of Cardiology*, vol. 96, no. 4A, pp. 26–32, 2005.
[23] M. Maldonado, S. D'Amico, M. Otiniano, A. Balasubramanyam, L. Rodriguez, and E. Cuevas, "Predictors of glycaemic control in indigent patients presenting with diabetic ketoacidosis," *Diabetes, Obesity and Metabolism*, vol. 7, no. 3, pp. 282–289, 2005.
[24] S .K. Garg, S. Schwartz, and S. V. Edelman, "Improved glucose excursions using an implantable real-time continuous glucose sensor in adults with type 1 diabetes," *Diabetes Care*, vol. 27, no. 3, pp. 734–738, 2004.
[25] G. M. Steil, A. E. Panteleon, and K. Rebrin, "Closed-loop insulin delivery-the path to physiological glucose control," *Advanced Drug Delivery Reviews*, vol. 56, no. 2, pp. 125–144, 2004.
[26] O. Aziz, L. Atallah, B. Lo, *et al.*, "Ear-worn body sensor network device: an objective tool for functional postoperative home recovery monitoring,"

Journal of the American Medical Informatics Association, vol. 18, no. 2, pp. 156–159, 2011.
[27] M. G. Allen, "Implantable micromachined wireless pressure sensors: approach and clinical demonstration," in *Proceedings of the Second International Workshop on Wearable and Implantable Body Sensor Networks*, London, UK, 2005, pp. 40–43.
[28] R. N. Butler, "Population, aging and health," *British Medical Journal*, vol. 315, no. 7115, pp. 1082–1084, 1997.
[29] W. S. Aronow and C. Ahn, "Elderly nursing home patients with congestive heart failure after myocardial infarction living in New York City have a higher prevalence of mortality in cold weather and warm weather months," *Journals of Gerontology Series A: Biological Sciences and Medical Sciences*, vol. 59, no. 2, pp. 146–147, 2004.
[30] P. J. Koken, W. T. Piver, F. Ye, A. Elixhauser, L. M. Olsen, and C. J. Portier, "Temperature, air pollution, and hospitalization for cardiovascular diseases among elderly people in Denver," *Environmental Health Perspectives*, vol. 111, no. 10, pp. 1312–1317, 2003.
[31] H. W. Kohl, C. L. Craig, E. V. Lambert, *et al.*, "The pandemic of physical inactivity: global action for public health," *Lancet*, vol. 380, no. 9838, pp. 294–305, 2012.
[32] D. M. O'Driscoll, A. R. Turton, J. M. Copland, B. J. Strauss, and G. S. Hamilton, "Energy expenditure in obstructive sleep apnea: validation of a multiple physiological sensor for determination of sleep and wake," *Sleep Breath*, vol. 17, no. 1, pp. 139–146, 2013.
[33] Feng Zhou, H. B. Duh and M. Billinghurst, "Trends in augmented reality tracking, interaction and display: a review of ten years of ISMAR," in *2008 7th IEEE/ACM International Symposium on Mixed and Augmented Reality*, 2008, pp. 193–202, doi:10.1109/ISMAR.2008.4637362.
[34] R. H. Holloway, "Capsule pH monitoring: is wireless more?," *GUT*, vol. 54, no. 12, pp. 1672–1673, 2005.
[35] J. E. Pandolfino, J. E. Richter, T. Ours, J. M. Guardino, J. Chapman, and P. J. Kahrilas, "Ambulatory esophageal pH monitoring using a wireless system," *The American Journal of Gastroenterology*, vol. 98, pp. 740–749, 2003.
[36] T. Vo-Dinh, "Biosensors, nanosensors and biochips: frontiers in environmental and medical diagnostics," in *Proceedings of the First International Symposium on Micro and Nano Technology*, Hawaii, 2004, pp. 1–6.
[37] D. J. Culliford, J. Maskell, D. J. Beard, D. W. Murray, A. J. Price, and N. K. Arden, "Temporal trends in hip and knee replacement in the United Kingdom: 1991 to 2006," *Journal of Bone and Joint Surgery*, vol. 92, no. 1, pp. 130–135, 2010.
[38] D. F. Dalury, D. L. Pomeroy, R. S. Gorab, and M. J. Adams, "Why are total knee arthroplasties being revised?," *The Journal of Arthroplasty*, vol. 28, p. 120, 2013.
[39] W. K. Ward, M. D. Wood, H. M. Casey, M. J. Quinn, and I. F. Federiuk, "An implantable subcutaneous glucose sensor array in ketosis-prone rats: closed loop glycemic control," *Artificial Organs*, vol. 29, no. 2, pp. 131–143, 2005.

[40] J. Q. Brown, R. Srivastava, and M. J. McShane, "Encapsulation of glucose oxidase and an oxygen quenched fluorophore in polyelectrolyte-coated calcium alginate microspheres as optical glucose sensor systems," *Biosensors and Bioelectronics*, vol. 21, no. 1, pp. 212–216, 2005.

[41] R. A. C. Grayson, R. S. Shawgo, Y. Li, and M. J. Cima, "Electronic MEMS for triggered delivery," *Advanced Drug Delivery Reviews*, vol. 56, no. 2, pp. 173–184, 2004.

[42] N. Najafi and A. Ludomirsky, "Initial animal studies of a wireless, batteryless, MEMS implant for cardiovascular applications," *Biomedical Microdevices,* vol. 6, no. 1, pp. 61–65, 2004.

Chapter 4

Cognitive sensors for rehabilitation and therapeutic treatment

Deepeshwar Singh[1], Raghavendra Bhat[2] and Suhas Vinchurkar[3]

Abstract

Cognitive sensors (CS) for healthcare are devices that utilize various sensor technologies to monitor a person's health condition and provide them with appropriate rehabilitation services. These sensors can be worn on the body, embedded in clothing, or integrated into other devices such as smartphones, smartwatches, and tablets. These sensors can monitor various indices of health and diseased conditions. Data related to human health are increasingly integrated with cognitive sensors, which facilitates decision-making and better disease treatment. Further, real-time communication and patient monitoring devices have become a central concept of smart healthcare services. The CS networks consist of a large number of sensors that can collect localized information automatically and intelligently. In addition to information fusion, intelligent sensors, sensing grids, communication protocols, network routing, and complex event processing architectures, CS intersects with other trends in computing. It has been shown that such CS can significantly improve the quality of life and support for the elderly and physically handicapped by designing interactive, adaptable, and intelligent multisensory surveillance systems, which can be equipped with panic buttons, current, bed, and water flow sensors. The use of CS technology, such as sensors, can reduce directional errors and assist in the neural health of patients. Many cognitive sensor systems have been developed for rehabilitation purposes to assist various groups of individuals in overcoming physical or mental health conditions. However, only a few attempts have been made to assist individuals with traumatic brain damage, schizophrenia, or dementia.

Keywords: Cognitive sensor; Health; Cognitive impairment; Rehabilitation

[1]Babasaheb Bhimrao Ambedkar University (Central University), India
[2]Swami Vivekananda Yoga Anusandhana Samsthana (S-VYASA), Bengaluru, India
[3]Electrical Geodesics, Inc., USA

Key message

Smart cognitive sensor offers novel solutions in healthcare and rehabilitation system such as disease prevention, monitoring, early detection of pathologies and minimizing the invasive methods.

4.1 Introduction

Over the last few decades, electronic and communication systems have significantly improved the quality of life and health of humans. This development is focused on improving smart healthcare services by utilizing real-time communication with smart sensing devices in order to provide more comprehensive and efficient healthcare services [1]. Importance has been given to the development of smart and comfortable devices which can aid diagnosis, prevention, and examination of a wide range of physical and mental issues. Smart cognitive sensor-based technologies use cognitive framework which provide patients-centric and high-quality smart healthcare at low costs. Research in this area is still in the infancy stage. However, researchers have been developing various platforms with multiple sensors for assessing the gesture, recognition, tracking, monitoring, etc., to improve the rehabilitation of disabled or older people [2]. Currently, medical electronic devices use smart cognitive sensors which can be effectively implemented for acquiring individual or patient's behavioral or physiological information and outside environmental data that can improve life expectancy [3]. The proliferation of miniature cognitive sensors in devices such as wearable devices, smartphones and tablets, and ambient sensors that are designed for smart homes are playing a substantial part in the remote and continuous monitoring of diseased conditions. In telemedicine, wearable body sensors are very useful and are being used increasingly. These wireless sensors can be placed on the patient's body surface or sometimes implanted inside which acquire critical data that may be related to a health condition, physiology, mental status, or any abnormality inside the body. These remote sensors provide health-related data and assist doctors to plan their treatment. Today, these wireless cognitive sensors are the backbone of intelligent systems that enable these to sense various signals of interest from outside or inside of the body. These smart sensors acquire signals from humans as well as the surroundings thus providing contextual information. Further, the signals are then monitored and processed by embedded processors which provide an automatic response. This feature of the provision of apt feedback and response distinguishes these smart and intelligent technologies from the conventional ones [4]. The healthcare paradigm has recently introduced the use of appropriate sensor devices for patients to enable the remote monitoring of vital signals, the early detection of critical conditions, and the remote control of certain medical treatments. By combining several sensors with software, applications and medical devices can be improved in performance. This method is capable of playing a major role in improving healthcare services in acute as well as chronic situations. In 2011, IEEE

802.15 Task Group 6 approved a draft standard for body area network technology [5,6]. The aim is intended to provide an international standard for low-power, short-range, and extremely reliable wireless communication within the surrounding areas of the human body. This body area network technology is extremely helpful, particularly, in areas with a low number of specialists, for example, remote regions of developing countries. Several remote areas of India and Nepal have already been using this healthcare system. Intercommunication technology permits the acquisition of vital physiological data in and around the body through the placement of wireless medical sensors on the body's surface. In other conditions, these sensors can also be embedded inside the body as implants and monitor any changes in physiology or any abnormalities. Remote monitoring cognitive-based sensors are mostly based on modern communication and information technologies, providing an efficient solution to allow elderly and disabled individuals to live comfortably at home.

The purpose of this chapter is to provide a general overview of the smart cognitive sensor that is used in electronic medical devices as well as its application in everyday life.

4.2 Importance of cognitive sensors in the healthcare system

A rapid development in telemetric or wireless smart cognitive sensors that are interconnected with medical devices allows continuous health monitoring and updates of medical records in real time. Cognitive sensors can be incorporated into wearable medical devices so that they can be used with computer-assisted rehabilitation or early detection of medical conditions. These technologies are helpful in healthcare management seamlessly and provide benefits to:

- Monitoring of health and disease condition in real-time
- Emergencies involving healthcare
- Patient's or individual's medical history and records should be maintained
- Devices that track various parameters related to fitness

Among the important areas of healthcare are the remote monitoring systems, which provide an extensive of the hospital medical system that analyses vital body functions. In addition, smart cognitive systems are providing patients and individuals with comprehensive, reliable, and low-cost healthcare services [7]. With the use of cognitive and smart sensors, patients' states can be monitored in real time and accurately and high-quality healthcare services can be provided promptly and at an affordable price [8]. Sometimes, these sensors can be implanted inside the body, which will collect various health-related physiological measures and aid in the development of reliable medical records. Regardless of their location, these outcomes assist medical practitioners in providing possible solutions and healthcare services more efficiently and effectively. In this process, the health-related data will be transmitted to external devices and processing units to develop the medical report of the patients. Cognitive sensor technologies developed have been utilized to assist individuals with cognitive function and cognitive health in some cases.

4.3 Cognitive sensors and cognitive health

A person's ability to think clearly, learn and remember clearly and execute everyday activities successfully is the result of cognitive health and cognitive functioning. It depends on the health of the brain which brings health and longevity. The brain functions as a command center that facilitates thoughts, memory, movement, and emotion by utilizing complex brain networks. The brain networks face increasingly challenging challenges on a daily basis that threaten their mental and physical health [9]. However, rapid technological advances assist in easing these challenges and emerging as more convenient, efficient, and convenient in a wide range of health conditions. In order to accommodate highly required healthcare services as well as provide low-cost patient-centric healthcare services, cognitively smart healthcare services are in high demand. As a result of the cognitive sensor-based technological development, individuals are able to recognize changes in the environment and human behavior and can be able to reduce hazardous situations and increase efficiency [10]. By introducing deep learning (DL), machine learning (ML), and artificial intelligence (AI) techniques into the smart healthcare system, we can achieve human-like intelligence on a timely basis. However, AI and DL can provide cognitive sensors in health care that can improve decision-making capabilities. These cognitive sensors have recently been used to assist individuals who have difficulties walking, understanding, learning, etc., particularly those who are disabled and elderly. As e-health applications collect medical data and patient information, the cognitive sensor emerges in healthcare environments [11].

There is substantial demand for sensor-based technology in healthcare services, which provides important and critical services to patients and generates revenue. It offers fast, affordable, and low-cost healthcare services. As a result, cognitive sensor-based healthcare and rehabilitation services are being integrated for the purpose of health assessment, which includes utilizing smart wearable devices to measure blood sugar, insulin levels, blood oxygen concentrations, body temperatures, electrocardiograms, electroencephalograms, sleep quality, etc. In these smart sensing devices, data is collected concerning a person's health, it is processed internally, and different features are extracted that reflect the individual's health condition and provide information relevant to that individual.

To quantify the disease condition, it provides an alternative or addition to more expensive and strictly lab-based investigations. These sensors provide physical data in a home-based setting, and may be particularly useful for those with disabilities or older adults. By utilizing cognitive computing and remote health monitoring, the smart healthcare system can analyze and monitor the physical health of patients. The sensors are usually embedded within wearable devices or medical electronic devices that communicate with the base station for the collection of desired data that can then be transmitted to the cloud or to personal devices (such as laptops, smartphones, or data storage devices) and developed into a doctor's report. Any clinical or health conditions are treated by doctors based on the clinical report. A number of advantages are associated with using cognitive or start sensor devices technology, such as facilitating the report quickly and allowing the treatment of patients to go forward without delay. The technology also assists in providing remote monitoring for the patients.

4.4 Smart sensors-based technologies

Currently, smart sensors are used in medical devices and equipment that collect information by means of signals that can be utilized for assessing the health condition of an individual or patient. The sensor-based technologies are useful for health monitoring including the detection, and identification of damage, in addition to mitigating the inference problems that facilitate individual or patient health conditions during emergency situations or even during surgical procedures and operations. Using sensor-based health technologies, health indicators can be automatically analyzed, allowing for accurate detection of any damage and an evaluation of the extent to which this damage is influencing the patient. Through the use of hardware sensors, the sensor-based cognitive system interacts with the external environment, as well as with the operating system that consists of perception and executive processes [11]. In the perceptual process, raw data from hardware sensors is received and processed to produce a perception of the environment. The perception of the environment is also processed by the executive process and passed to the operating system, which determines what action should be taken by the sensors.

The development of such a system, however, requires several processes and hardware to enable the perception, understanding, and execution of the task. As of now, smart cognitive sensors, including physiological smart sensors, are able to record healthcare data as well as assess changes in signals relating to facial expressions, speech, movement, attention, memory, and learning abilities in real time. In order to classify these biosignals for healthy or clinical conditions, biosignals can be transferred to deep learning modules [8]. The assessment will assist in assessing the patient's critical condition and providing emergency assistance. Furthermore, the expensive assessments and healthcare facilities are becoming free or less expensive where patients can monitor very important physiological parameters of their patients in real-time, observe health conditions, and provide preventive measures and feedback [12]. Defense, healthcare organizations, and insurance companies can benefit from smart cognitive sensors since they need real-time, reliable, and accurate diagnostic results that can be monitored efficiently, regardless of whether the patient is in a hospital, clinic, or at home.

Let us understand the concept of sensors, sensor nodes, sensor networks, and cognitive sensing systems.

4.5 Sensors

The concept of a sensor refers to a device that detects changes and occurrences using a physical stimulus and then produces an output signal that can be measured and recorded as a result. It is common for the output signal in this case to be an electrical quantity, but it may also be any measurable signal. Various vital signs are converted into electrical signals by these sensors, which are used in electronic medical and non-medical equipment [13]. This can be used for many purposes, including the implanting of devices, the prevention of illness, assessing the extent of disability, determining the benefits of drugs, and long-term monitoring of health status, which can assist people in better understanding their health problems [12].

In two different ways, sensors provide users with information about their surroundings via electrical signals:

- Sensors that generate an electric signal, such as piezoelectric sensors, are active sensors.
- The ability to modify the electrical signal available to a passive sensor, such as a potentiometer, as a result of environmental stimuli.

Healthcare delivery through electronic and mobile means is progressing and developing more rapidly, and Telehealth systems that are based on the Internet of Things (IoT) are emerging [14]. IoT has developed the interconnection of smart sensor devices to cloud-based computing devices. Cloud computing offers fast processing and large storage capability which provide real-time smart healthcare services. Further, IoT-based technologies when embedded with a cognitive framework provide patients-centric and high-quality smart healthcare at low cost. Earlier it was difficult to have human-like intelligence, but after introducing the concept of AI and DL techniques, human-like intelligence in smart healthcare framework is well accepted. The integration among IoT, AI, and deep learning can provide a seamless, fast, and cognitive-based response which can improve decision-making capability [15,16]. These developed cognitive-based smart sensor devices and technologies facilitated healthcare services to patients with critical cases who need immediate care and fast response to life. Smart healthcare services provide immediate responses to emergency conditions and intelligently detect abnormality or pathogenic conditions [17]. AI-assisted smart healthcare systems interconnect with IoT can recognize the complicated patterns obtained from various clinical procedures and investigations, that will save time and money as shown in Figure 4.1.

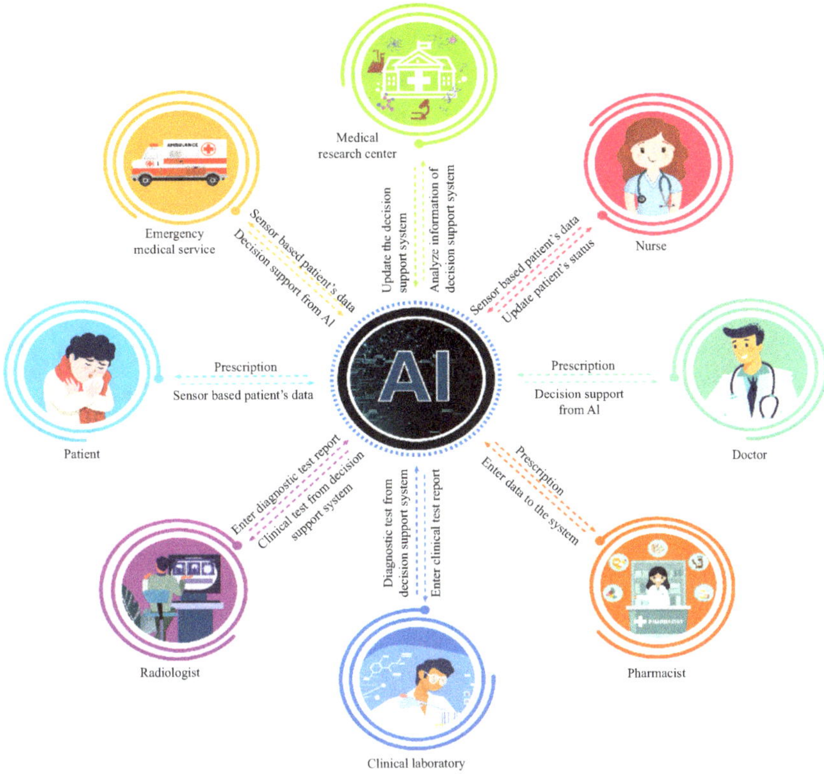

Figure 4.1 In this diagram, AI is shown assisting doctors, patients, nurses, Emergency Medical Services (EMS), pharmacists, radiologists, clinical laboratories, and researchers in the development of smart healthcare systems [18]

4.6 Sensor node

In essence, a sensor node is a component of a sensor network that is capable of collecting sensory data, analyzing it, and communicating with other nodes in the network (as shown in Figure 4.2). In addition to the analog-to-digital converter (ADC), a sensor node also consists of an actuator (optional), a power supply unit (often battery-operated), and other components as well. Sensor nodes consist of sensors, ADCs, microcontrollers, transmitters, actuators, and power supplies.

4.7 Sensor network

A collection of tiny, powered devices connected to a wireless or wired network architecture constitutes a sensor network. These networks monitor various environmental conditions in various settings, including farms, hospitals, and industrial

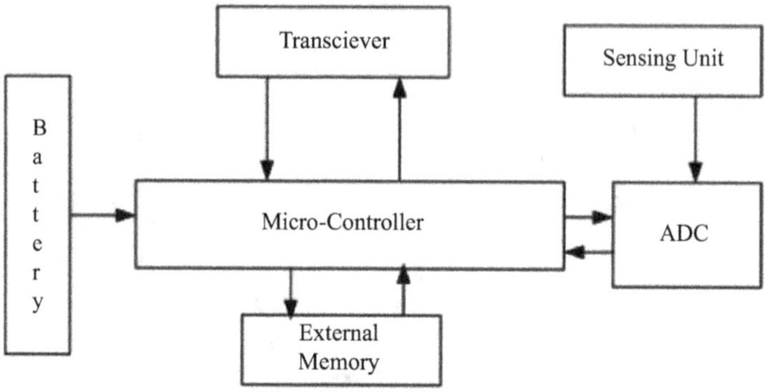

Figure 4.2 Sensors communication with other networks

plants. In order to analyze and use the data generated by the sensor network, it connects to the Internet or computer networks. Sensor network nodes sense and control the environment in support of the system. They enable interaction between persons or computers and the surrounding environment. Typically, sensors are connected to computer networks or the Internet in order to transmit data for use and analysis [19].

4.8 Two categories of sensor networks exist

1. *Wired sensor network*: The sensors in wired sensor networks are connected via ethernet wires.
2. *Wireless sensor network*: Wireless sensor networks (WSNs) link sensors together using technologies like Bluetooth, cell phones, WiFi, or near-field communication (NFC). WSNs now use the industrial, scientific, and medical (ISM) band of transmission, which is unlicensed and free for communication, covering very large areas that are expanding daily. The ISM band is suffering from its narrow transmission band due to the growth in the amount of data generated and the region covered by WSNs, which raises network traffic congestion and the risk of data loss.

4.9 Operation of a sensor network

The way that sensor network's functions is described by the process below (Figure 4.3):

- A sensor node keeps track of the information a sensor collects and sends it to other sensor nodes.

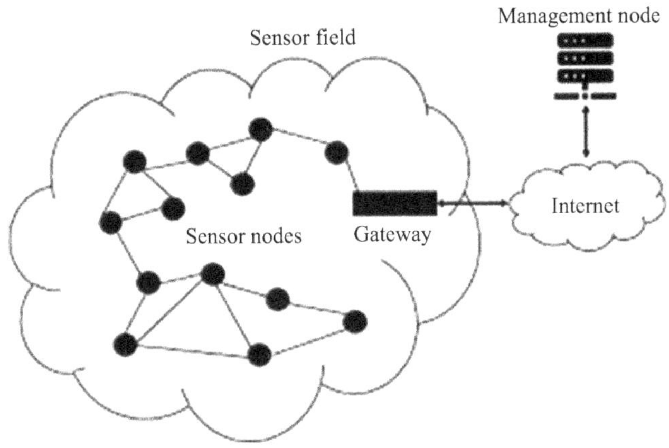

Figure 4.3 Sensor network component

- As data travels through the transmission process and arrives at a gateway node, it may be handled by various nodes.
- The management node, which is controlled by the user and decides what monitoring is necessary, receives the data after it has been delivered to it.

4.10 Cognitive sensing systems

Cognitive sensing systems feature a perceptual processor that transforms raw sensor input into an informative perception of the environment and an executive processor that determines the next sensing action, simulating the perception-action cycle of cognition.

4.11 Cognitive sensors

The word "cognitive" refers to the use of mental processes, including memory, thought, and reasoning [20]. Cognitive sensors are defined as those that make use of certain cognitive processes [21]. In accordance with the principles of self-organized systems, a cognitive sensor is one of the components of a cognitive system. As a result of sensor design, each cognitive sensor can perceive the outside world and construct representations of it within a spectrum of perception [22].

An architecture of cognitive sensors consists of three subsystems: the logical subsystem, the preprocessing subsystem, and the sensory subsystem.

4.12 Sensory subsystem

The sensory subsystem measures a scalar value x, which represents a certain quality of the external environment. Since the cognitive sensor perceives the world as being one-dimensional, we assume that x can select from a set of possibilities.

4.13 Preprocessing subsystem

The preprocessing subsystem performs three primary tasks. The input stream of values originating from the sensory system is first discretized. As a second step, it arranges input values sampled from the stream based on their spatial location. Third, it performs preprocessing on the ordered and sampled collection of values. A logical subsystem can store and process the sequence of preprocessed values.

4.14 Logical subsystem

A characteristic of the logical subsystem is that it is unable to process more than a discrete set of values. Cognitive sensors are based on the ability to compare deals taken by x at various points within the data stream to be examined as part of their working logic.

With the development of cognitive sensors, we are now able to detect the electromagnetic spectrum, to make decisions, to share information, and to regulate the mobility of secondary users. Cognitive sensors can detect the new range and choose the optimum area of it thanks to these new abilities [23].

4.15 Cognitive sensor for rehabilitation and therapeutic treatment

During the last decade, electronics and computing have grown rapidly, leading to advances in the field of healthcare and rehabilitation. Wearables with cognitive sensors have emerged as a promising and potential tool for assessing individual health and performance. In addition to collecting quantitative measurements, these sensors enable the use of these sensors in free-living environments and enable the monitoring of disease effects and recovery. A wearable cognitive sensor enables the collection of big data across different clinical conditions in real-world settings. Beyond the traditional clinical setting, this technology has been widely used to monitor remote patients, to provide mobile health services, and to expand health metrics. Wearable sensors are low-cost, portable, and unobtrusive tools that provide quantitative and continuous information regarding sensory and motor behavior. It provides real-time physical monitoring of disabled people, older, and children who seek continuous support may be due to deteriorating or undeveloped cognitive abilities. As a result of the support provided by smart cognitive sensors and devices, caregivers are able to monitor the physical activity of patients and at the same time improve the patients' functional independence and quality of life [24].

In recent years, a growing body of research has focused on the development of wearable devices with a wide array of applications that allow healthcare experts and clinicians to remotely monitor patients, particularly older adults and patients with chronic conditions (as shown in Figure 4.4). Sensor, communications, and data analysis technologies have been developed over the past few years, which

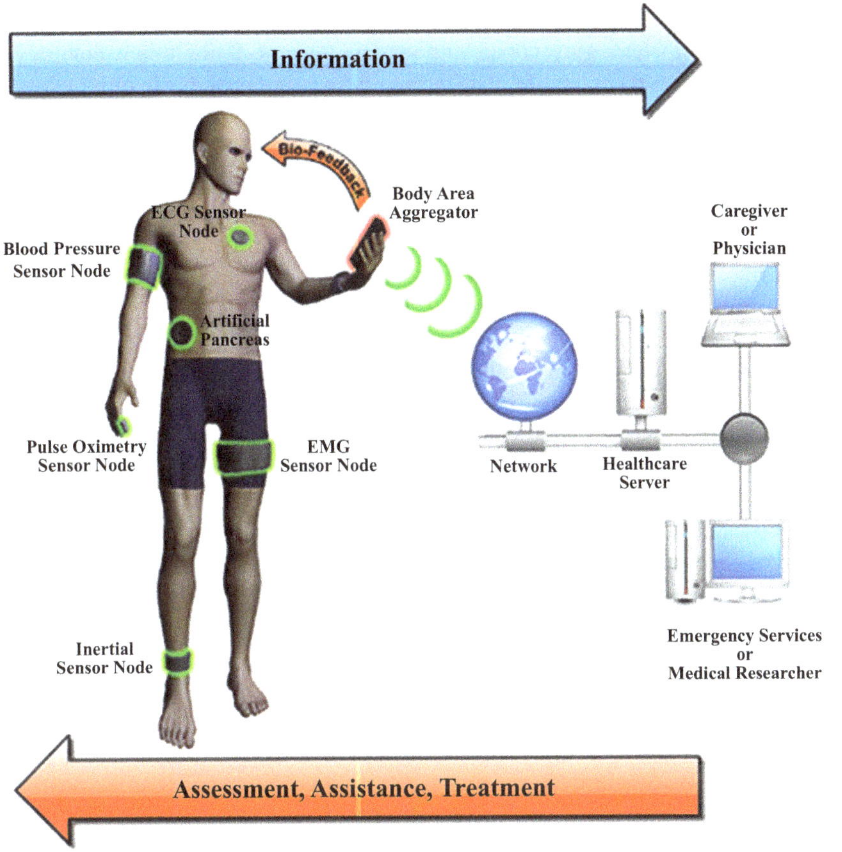

Figure 4.4 *Sensors on a patient or anyone who wants to monitor their physiological condition can communicate from a body area network to the internet for remote analysis. Source: University of Virginia Engineering.*

have enabled researchers and clinicians to implement wearable smart sensor devices.

Currently, cognitive sensors have been used primarily in the rehabilitation of neurological disorders and orthopedic injuries [25]. It is common for wearable sensors to be integrated with a variety of health conditions. These wearable cognitive sensors have been used in therapeutic and clinical settings to assess mobility tests, identify pathological movement, characterize disease stages, manage falls, and recognize activity levels. Furthermore, it has been used to train gait and augment treatment [26]. Due to a wide variety of applications of smart sensor-based devices that have impacted multiple aspects of human life, particularly in healthcare and daily life monitoring, smart sensor-based devices are becoming increasingly popular in recent years. The sensors are primarily used for rehabilitation and treatment purposes. These are comprised of inertial measurement units (IMUs) such as accelerometers, gyroscopes, and magnetometers. Using this technology, it is possible to observe human life in different routines and postures in order to detect changes in location, movement of the body, and rotation of the body in three-dimensional space [27]. Sensors are useful for assessing and recording physical activity and physiological changes. Wearable sensors are intended to reduce issues in the continuous monitoring of physiology, movement, posture, and orientation.

For the purpose of identifying physical activity, four main steps are proposed: the placement of sensors, the process of denoising the signals, classification, and the selection of features from the collected data. For rehabilitation of neurological problems, there are three inertial sensors i.e., gyroscopes, accelerometers, and magnetometers can be mainly placed at different locations of the body. These sensors can collect data from different locations such as the wrist, chest, and thigh. Signals acquired can be filtered with a variety of filtering modules to eliminate any noise resulting from muscle movement, physiological noise, or electrical noise and restore signals to a level that is close to normal. A sensor can assist patients, caregivers, and healthy individuals at various stages of their lives. As a result of cognitive sensor technology, artificial intelligence, and machine learning, seamless cognitive behavior and decision-making capabilities are possible, which can be quite beneficial for critical care units and healthcare facilities.

A group of nine individuals was monitored during walking during verbal instructions and an active feedback system for gait assessment. The motion capture system and force place sensor were used to assess the knee adduction moment during both conditions. During active feedback, the peak knee adduction moment was reduced. This study suggests the potential application of force sensors for knee osteoarthritis. Another smart signal processing approach has been utilized to quantify the parameters of stroke survivors mainly complex movement. The complex movement can be monitored using accelerometer-based step activity in the home and the community assesses real-world performance and guides community-based treatment programs [28]. A goal-oriented physical activity can benefit gait after stroke by increasing the amount of complex movements, the time spent walking, and the number of medium and long walking periods [29].

There are few articles reporting sensor-based rehabilitation of motor functions in patients with functional impairment. The cognitive sensor can be used to analyze the relationship between neurological conditions and gait analysis. In addition to providing a portable and affordable alternative to healthcare practitioners, wearable devices can also provide healthcare professionals with a high level of precision and accuracy when estimating kinematic variables [30–32]. In order to assist medical personnel in assessing the level of problems in patients with other neurological disorders, the device develops a novel contactless method for assessing patients with spinal cord injuries, hemiparetic conditions, and other movement-based problems. Disease diagnosis and rehabilitation can be carried out by monitoring patient gait behavior and their restricted activities using electrostatic gait signals.

4.16 General application of cognitive sensors

There are a few other applications of the cognitive sensor as described below (as shown in Figure 4.5).

4.16.1 Healthcare

Although newer applications include advertising, waste management, forensics, assisted living, and biometrics, intelligent sensors are employed mainly for disease and disorder diagnosis. It allows for real-time patient monitoring in hospitals, which enables the early detection of clinical deterioration [33], promotes large-scale field studies of human behavior and chronic diseases, improves the quality of life for the elderly through intelligent surroundings, and increases the ability of first responders to offer emergency care in significant disasters.

4.16.2 Smart cities

Intelligent sensors capable of monitoring and offering schematic solutions are installed in smart cities. Parking, drainage, traffic, healthcare, and environmental monitoring are critical issues that need to be considered for intelligent monitoring in smart cities.

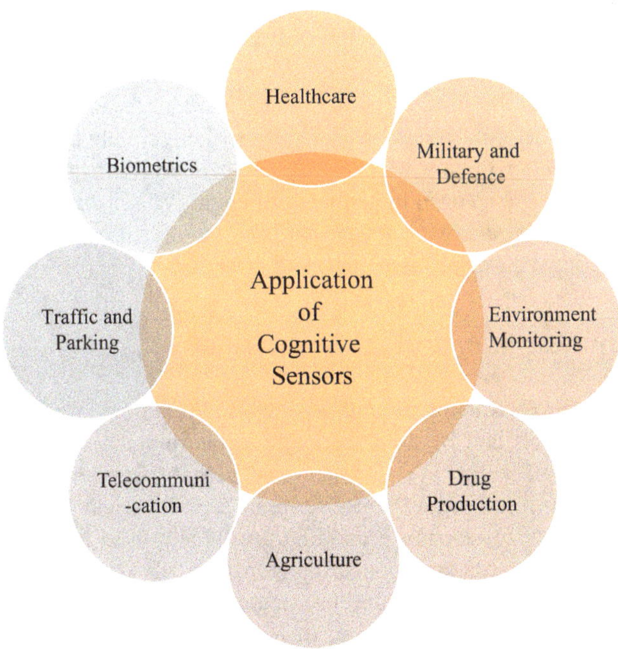

Figure 4.5 Application of cognitive sensors

4.16.3 Home monitoring

Nowadays significance of home monitoring is the most important. An intelligent sensor can help us monitor every activity in our home very effectively. As an illustration, a camera can be used to ensure security, an LDR sensor to control lighting, a temperature and humidity sensor to control the air conditioning, and an air quality sensor to monitor the air quality in the house. With these sensors' help, we can live a safer, more secure and more quality life.

4.16.4 Disaster management

Because of the lack of information about disaster, we lose many people, money and many more things. We may obtain specific advanced knowledge through cognitive sensor networks regarding many natural calamities, such as floods, forest fires, volcanic eruptions, and tsunamis. We can also better manage any disaster with the use of information.

4.16.5 Environment monitoring

Environment monitoring is also a vital application of cognitive sensors. Due to an uncertain environment, many countries face problems such as agriculture, food, crop protection, health, and economics. If we have the proper knowledge of the environment and appropriate data on different environmental parameters, we can manage all the problems very efficiently.

4.16.6 Usefulness in education

The sector of education can also benefit from the usage of cognitive sensors. Biometric sensors can be used to track student attendance, and a camera can be used to maintain order on campus. Sensors also make it simple to convey complex topics. We can monitor the institution's supply chain with the help of sensors; we can create different brainstorming games with cognitive sensors for the student's brain development. At last, we can say that with the use of cognitive sensors, we can make our education more interesting and effective.

4.17 Usefulness in neurological conditions

As one of the methods to explore brain activity and functions, neurophysiological or neuroimaging tools can be used to diagnose neurological conditions such as epilepsy, impairment in function and neuronal activity, stroke, and Alzheimer's disease. In addition to recording real-time brain activities, electroencephalogram and functional magnetic resonance imaging are widely used to detect brain disorders due to their low cost and non-invasive nature. As a result of the real-time smart cognitive sensor, reliable neurophysiological data is produced and processed that can be used to assess performance. In recent years, several researchers have developed smart healthcare frameworks utilizing AI, ML, and DL techniques in order to use the electrophysiological technique for brain-related ailments. As of today, smart sensory embedded technology has been used to diagnose, detect, and observe real-time changes in disease or performance in healthcare services [24]. Using these cognitive sensors, a variety of medical data can be collected, including actions, movements, and emotions, and these signals can be transferred to deep learning, which identifies pathologies and communicates the results back to the cognitive system. Through the cognitive module, the patient's condition can be assessed in order to implement appropriate follow-up procedures. The diagnosis or detection of health conditions includes a screening of ailments, such as depression, brain injuries, stroke, and Alzheimer's disease [34,35].

Wearable devices are also equipped with other sets of sensors which monitor the health of the individual and record different body parameters such as heart rate, blood pressure, breathing, and body temperature. When a significant change occurs in a body parameter, the cognitive sensor sends an alert to the doctor and appropriate person. With the data sent by the sensor, a doctor can easily understand the problem, providing treatment very frequently.

4.18 Limitation of cognitive sensors

Despite the potential advantages of using cognitive sensors to monitor health, a large number of challenges must be overcome before such a system can be implemented on a large scale. It is common for wearable devices to utilize battery technology as well as cultural barriers, such as the stigma associated with the use of medical devices for home-based clinical monitoring.

The cognitive sensor network field is in the process of development with a great deal of open questions. This section illustrates the potential for new research in the field by presenting various difficulties derived from various publications. In healthcare systems, the usefulness of cognitive sensory is limited to only diagnosing or detecting pathological conditions. A cross-check of the accuracy of detection is required. As a result, these sensors can be developed to assist individuals and patients in making decisions, providing home-based treatment for critical conditions, and guiding them in preventing health conditions. Further, these sensory devices should include modules that can assist caregivers and therapists in coping with the challenges of rehabilitation and treatment on a daily basis according to the diagnosis.

4.19 Increased power consumption and sensor lifetime

As a result of the cognitive capability of sensors, there is a high demand for power and more resources, which drives up expenses overall. Networks should be designed so that they use less energy while increasing the usefulness of the model, while using less energy. Nevertheless, several studies are being conducted in this area that utilize techniques such as multichannel cooperative sensing and optimal node selection. Therefore, new suggestions and solutions are required to address this issue.

4.20 Spectrum allocation

Since numerous secondary users may simultaneously request spectrum, a fair spectrum distribution mechanism with a prioritization system is required.

4.21 Interference management

There are many challenges associated with cognitive sensor networks, including the problem of interference caused by cognitive nodes when they are using the spectrum when they are communicating with primary nodes or secondary nodes. As a result of interference, power consumption increases. Therefore, it is imperative that the nodes are designed to minimize interference and maximize spectrum utilization.

4.22 Future scope of cognitive sensors

The notion of healthcare to diagnose patients and provide possible treatment is changing. Electronic devices with a cognitive sensor have the ability to scale and elevate expertise which can lead to improved healthcare knowledge and help organizations assess the individual or patients' data efficiently.

> "*In the healthcare system, the addition of cognitive sensors in medical electronic devices is revolutionizing the system while the use of artificial intelligence and machine learning helps to access patient information and improve treatment delivery.*"

The cognitive sensor should be able to address issues related to the rapidly aging population, elderly care, and neurological conditions. A further requirement is that the development of such devices as well as the underlying information technology be sufficiently flexible to facilitate the rapid development of digital applications and complement applications that can be integrated with existing systems. As a result of the cognitive capabilities provided by electronic devices, the healthcare system is capable of undergoing significant transformation. But in order to do so, it is imperative to identify and add the critical data sources along with the services and processes. As a result, the system will be able to take advantage of the optimized cognitive capabilities. These sensors must be stable, secure, and scalable in order to be used in a healthcare system.

Rehabilitative and medical treatment devices require more than collecting and securing data; a mechanism for accruing benefits must be developed as well. Cognitive sensor-based applications and wearable devices should follow cognitive computing steps and allow us to make any necessary improvements to the healthcare system if necessary.

4.23 Conclusion

The cognitive sensor technology can be applied to a wide range of healthcare, rehabilitation, and clinical treatment areas, including disability care and fitness.

Additionally, it can be used to diagnose and detect any abnormalities occurring at home in a cost-effective and portable manner. This chapter presents an overview of cognitive sensors by demonstrating that cognitive sensors are able to use logic and grasp a situation better than other sensors commonly used in WSNs. A cognitive sensor network utilizes their assets in order to optimize the utilization of the resources already available by allocating spectrum within different paid spectrum bands to optimize the utilization of these resources. A number of applications for smart sensor wearable technology are focused on health and wellness, safety, home rehabilitation, early detection of diseases and assessment of the effectiveness of treatment. The use and difficulties of cognitive sensors have also been discussed. There is however a need for further research into the best sensors for a scenario with high efficiency and low cost. Readers can begin their own research by searching for related applications from this chapter.

References

[1] Alcalá Garrido, H. A., Rivero-Angeles, M. E., Aguirre Anaya, E., Cruz-Perez, F. A., Castellanos-Lopez, S. L., and Hernandez-Valdez, G. (2018). Performance analysis of a wireless sensor network with cognitive radio capabilities in structural health monitoring applications: a discrete model. *International Journal of Distributed Sensor Networks*, 14(5). https://doi.org/10.1177/1550147718774001

[2] Roe, J., Mondschein, A., Neale, C., Barnes, L., Boukhechba, M., and Lopez, S. (2020). The urban built environment, walking and mental health outcomes among older adults: a pilot study. *Frontiers in Public Health*, 8, 575946. https://doi.org/10.3389/fpubh.2020.575946

[3] Gillani, N. and Arslan, T. (2021). Intelligent sensing technologies for the diagnosis, monitoring and therapy of Alzheimer's disease: a systematic review. *Sensors (Basel, Switzerland)*, 21(12), 4249. https://doi.org/10.3390/S21124249

[4] Silverio-Fernández, M., Renukappa, S., and Suresh, S. (2018). What is a smart device? – a conceptualisation within the paradigm of the internet of things. *Visualization in Engineering*, 6(1), 1–10. https://doi.org/10.1186/S40327-018-0063-8/TABLES/5

[5] IEEE 802.15. (2012). *IEEE 802.15 Wireless Next Generation Standing Committee*. https://www.ieee802.org/15/pub/TG6.html

[6] Salehi, S. A., Razzaque, M. A., Tomeo-Reyes, I., and Hussain, N. (2016). IEEE 802.15.6 standard in wireless body area networks from a healthcare point of view. In *Proceedings – Asia-Pacific Conference on Communications, APCC 2016*, pp. 523–528. https://doi.org/10.1109/APCC.2016.7581523

[7] Sodhro, A. H., Sennersten, C., and Ahmad, A. (2022). Towards cognitive authentication for smart healthcare applications. *Sensors*, 22(6), 2101. https://doi.org/10.3390/s22062101

[8] Amin, S. U., Hossain, M. S., Muhammad, G., Alhussein, M., and Rahman, M. A. (2019). Cognitive smart healthcare for pathology detection and monitoring. *IEEE Access*, 7, 10745–10753. https://doi.org/10.1109/ACCESS.2019.2891390

[9] Bassett, D. S. and Bullmore, E. T. (2009). Human brain networks in health and disease. *Current Opinion in Neurology*, 22(4), 340. https://doi.org/10.1097/WCO.0B013E32832D93DD

[10] Belkacem, A. N., Jamil, N., Palmer, J. A., Ouhbi, S., and Chen, C. (2020). Brain computer interfaces for improving the quality of life of older adults and elderly patients. *Frontiers in Neuroscience*, 14, 692. https://doi.org/10.3389/FNINS.2020.00692/BIBTEX

[11] Mateska, A., Chomu, K., and Gavrilovska, L. (2012). Cognitive sensor networks implementation in medical environmental surveillance. In *Proceedings of the 7th International Conference on Body Area Networks*. https://dl.acm.org/doi/10.5555/2442691.2442743

[12] Angelov, G. V., Nikolakov, D. P., Ruskova, I. N., Gieva, E. E., and Spasova, M. L. (2019). Healthcare sensing and monitoring. In *Lecture Notes in Computer Science (Including Subseries Lecture Notes in Artificial Intelligence and Lecture Notes in Bioinformatics), 11369 LNCS*, pp. 226–262. https://doi.org/10.1007/978-3-030-10752-9_10/FIGURES/13

[13] Kantipudi, M. V. V. P., Moses, C. J., Aluvalu, R., and Kumar, S. (2021). Remote patient monitoring using IoT, *Cloud Computing and AI. Intelligent Systems Reference Library*, 209, 51–74. https://doi.org/10.1007/978-981-16-2972-3_3/COVER

[14] Islam, S. M. R., Kwak, D., Kabir, M. H., Hossain, M., and Kwak, K. S. (2015). The internet of things for health care: a comprehensive survey. *IEEE Access*, 3, 678–708. https://doi.org/10.1109/ACCESS.2015.2437951

[15] Parra, D. T. and Guerrero, C. D. (2020). Technological variables for decision-making IoT adoption in small and medium enterprises. *Journal of Information Systems Engineering and Management*, 5(4), Article No: em0124. https://doi.org/10.29333/jisem/8484

[16] Subrahmanya, S. V. G., Shetty, D. K., Patil, V., *et al.* (2022). The role of data science in healthcare advancements: applications, benefits, and future prospects. *Irish Journal of Medical Science,* 191(4), 1473–1483. https://doi.org/10.1007/s11845-021-02730-z

[17] Ali, A. A., Altemimi, A. B., Alhelfi, N., and Ibrahim, S. A. (2020). Application of biosensors for detection of pathogenic food bacteria: a review. *Biosensors*, 10(6), 58. https://doi.org/10.3390/BIOS10060058

[18] Kamruzzaman, M. M. (2020). Architecture of smart health care system using artificial intelligence. In *2020 IEEE International Conference on Multimedia & Expo Workshops (ICMEW)*. https://doi.org/10.1109/ICMEW46912.2020.9106026

[19] Talpur, A., Shaikh, F. K., Baloch, N., Felemban, E., Khelil, A., and Alam, M. M. (2019). Validation of wired and wireless interconnected body sensor networks. *Sensors (Basel, Switzerland)*, 19(17), 3697. https://doi.org/10.3390/S19173697

[20] Cognitive Definition & Meaning – Merriam-Webster. (n.d.). Retrieved October 21, 2022, from https://www.merriam-webster.com/dictionary/cognitive
[21] Sinha, G. R. (2020). Advances in modern sensors. In *Advances in Modern Sensors*. Bristol: IOP Publishing. https://doi.org/10.1088/978-0-7503-2707-7
[22] Serov, A. (2017). Cognitive sensor technology for structural health monitoring. *Procedia Structural Integrity*, 5, 1160–1167. https://doi.org/10.1016/j.prostr.2017.07.027
[23] Haykin, S. (2005). Cognitive radio: brain-empowered wireless communications. *IEEE Journal on Selected Areas in Communications*, 23(2), 201–220. https://doi.org/10.1109/JSAC.2004.839380
[24] Kańtoch, E. (2017). Human activity recognition for physical rehabilitation using wearable sensors fusion and artificial neural networks. In *2017 Computing in Cardiology (CinC)*, 1–4.
[25] Döbrössy, M. D. and Dunnett, S. B. (2001). The influence of environment and experience on neural grafts. *Nature Reviews. Neuroscience*, 2(12), 871–879. https://doi.org/10.1038/35104055
[26] Dowling, A. V., Fisher, D. S., and Andriacchi, T. P. (2010). Gait modification via verbal instruction and an active feedback system to reduce peak knee adduction moment. *Journal of Biomechanical Engineering*, 132(7), 071007. https://doi.org/10.1115/1.4001584
[27] Casilari, E., Álvarez-Marco, M., and García-Lagos, F. (2020). A Study of the use of gyroscope measurements in wearable fall detection systems. *Symmetry*, 12(4), 649. https://doi.org/10.3390/SYM12040649
[28] Danks, K. A., Roos, M. A., McCoy, D., and Reisman, D. S. (2014). A step activity monitoring program improves real world walking activity post stroke. *Disability and Rehabilitation*, 36(26), 2233–2236. https://doi.org/10.3109/09638288.2014.903303
[29] Roos, M. A., Rudolph, K. S., and Reisman, D. S. (2012). The structure of walking activity in people after stroke compared with older adults without disability: a cross-sectional study. *Physical Therapy*, 92(9), 1141–1147. https://doi.org/10.2522/PTJ.20120034
[30] Nutt, J. G., Marsden, C. D., and Thompson, P. D. (1993). Human walking and higher-level gait disorders, particularly in the elderly. *Neurology*, 43(2), 268–279. https://doi.org/10.1212/WNL.43.2.268
[31] Alexander, N. B. (1996). Gait disorders in older adults. *Journal of the American Geriatrics Society*, 44(4), 434–451. https://doi.org/10.1111/J.1532-5415.1996.TB06417.X
[32] Weizman, Y., Tirosh, O., Beh, J., Fuss, F. K., and Pedell, S. (2021). Gait assessment using wearable sensor-based devices in people living with dementia: a systematic review. *International Journal of Environmental Research and Public Health*, 18(23), 12735. https://doi.org/10.3390/IJERPH182312735
[33] Gao, T., Pesto, C., Selavo, L., *et al.* (2008). Wireless medical sensor networks in emergency response: implementation and pilot results. In *2008*

IEEE International Conference on Technologies for Homeland Security, HST'08, pp. 187–192. https://doi.org/10.1109/THS.2008.4534447

[34] Kamnitsas, K., Ledig, C., Newcombe, V. F. J., *et al.* (2017). Efficient multi-scale 3D CNN with fully connected CRF for accurate brain lesion segmentation. *Medical Image Analysis*, 36, 61–78. https://doi.org/10.1016/J.MEDIA.2016.10.004

[35] Shoeibi, A., Khodatars, M., Ghassemi, N., *et al.* (2021). Epileptic seizures detection using deep learning techniques: a review. *International Journal of Environmental Research and Public Health*, 18(11), 5780. https://doi.org/10.3390/IJERPH18115780

Chapter 5

An ensemble machine learning-based intelligent system for human activity recognition using sensory data

Shahan Abdulla[1,2], *Mohammed Diykh*[1,2,3], *Siuly Siuly*[4] *and Mumtaz Ali*[1]

Abstract

Human activity recognition (HAR) from sensor data has become an attractive research topic due to its application in areas such as healthcare, human–computer interaction and smart environments. Although several research works have been done on HAR, the current studies are not enough to produce optimised performance balancing efficiency and accuracy. To solve this issue, this chapter proposes a new hybrid scheme based on permutation entropy integrated with quadratic sample entropy and ensemble classifier. First, the data are segmented into intervals, then, a combination of entropy features is extracted and evaluated. The extracted features are fed to an ensemble machine learning algorithm to classify the features vector into different activities. The proposed model is tested on a publicly available sensor dataset. The results demonstrate that the proposed scheme is efficient to improve HAR while obtaining an accuracy between 95% and 96% and outperforms considerably the existing methods.

Keywords: HAR; Permutation entropy; Quadratic sample entropy; Ensemble machine

5.1 Introduction

Human activity recognition (HAR) uses motion data acquired from different sources, such as cameras and sensors, to better understand human behaviours [1–3].

[1]UniSQ College, University of Southern Queensland, Australia
[2]University of Thi-Qar, College of Education for Pure Science, Iraq
[3]Information and Communication Technology Research Group, Scientific Research Centre, Al-Ayen University, Iraq
[4]Institute for Sustainable Industries & Liveable Cities, Victoria University, Australia

Those data are normally analysed using data intelligent models to classify the human activity [4]. Recognising daily human activities accurately is also an important aspect in various real-time applications, such as personal healthcare, disease diagnosis based on smart home devices [5,6].

Most of the recent approaches of HAR focus on classifying several human activities based on machine learning algorithms [7,8]. Such activities include running downstairs or upstairs and walking. Generally, human activities can be classified into three groups based on the complexity and the duration: simple activities, short activities, and complex activities. *Short activities* are comprised of those with a very short duration, such as switching from sitting to standing. *Simple activities* consist of basic activities such as ascending and descending stairs, walking, and reading. Whereas *complex activities* include a combination of basic activities interacting with other objects or individuals [3,4]. This chapter focusses on the development of an intelligent predictive system to recognise some of these basic human activities.

Many predictive methodologies have been developed to address various HAR-related problems. Those methodologies can be classified into two types: vision-based approaches and sensor-based approaches [9]. For example, Jethanandani *et al.* [10] designed a multi-label classification ensemble mode to categorise human activities. Four classification approaches including a *k*-nearest neighbour, logistic regression, Bernoulli, and decision tree were used on ARAS dataset in that paper. Jain *et al.* [11] suggested two descriptors named histogram of gradient and centroid features-based Fourier system. A multiclass support vector machine (SVM) and a *k*-nearest classifiers were also adopted to classify the extracted features. Jalloul *et al.* [12] developed a model based on a complex network technique for HAR using random forest classifier to classify the network features into different human activities. Gupta *et al.* [13] applied a body-worn wireless accelerometer to investigate HAR with two feature selection approaches named a relief-F and sequential forward floating to identify the most powerful features. They used Naïve Bayes and *k*-nearest classifiers for classification purposes.

Neural networks-based approaches have also been utilised widely in HAR due to their ability to automatically learn the important features from the signals without the need for human domain knowledge. Garcia *et al.* [7] designed a multi-class algorithm called the ensemble of auto-encoders. Xiao *et al.* [14] proposed a federated learning system which used a perceptive extraction network and. tested on four datasets. Altuve *et al.* [15] applied a bidirectional long short-term memory neural network. In this study, the data were decomposed into intrinsic mode functions based on an improved complete ensemble empirical mode to extract different features. The extracted features were fed into the proposed long short-term memory neural network.

This chapter proposes a new model for HAR using entropy features coupled with an ensemble classifier. The main contributions of this chapter are summarised as follows: (1) propose a new model for HAR system involving entropy features with the ensemble classifier; (2) propose permutation entropy, and quadratic sample entropy for extracting efficient attributes from sensor data; (3) design an ensemble model to classify the human activities.

5.2 Methodology

The proposed method consists of two major stages: (1) segmentation and features extraction stage, which includes segmentation of HAR data and extracting the desired features; (2) the classification stage, which classifying the extracted features into different human activities. Figure 5.1 shows the workflow of the proposed method with a detailed description of each stage.

5.2.1 Data segmentation

In this study, we adopted our published work in [15–22] to partition the HAR data into a set of intervals based on the sliding window technique. The segmented data were used as an input to the proposed model. Two window sizes were used to partition the data with an overlap of 50%. For the UCI-HAR dataset, a window size of 128 was applied to segment the data. Due to the length of each activity in OPPORTUNITY dataset was short, we considered a short window size to obtain more samples. As a result, a window size of 24 was considered to segment the OPPORTUNITY dataset. Suppose X is a time series data of HAR based on proposed technique, X can be partitioned a sequence of intervals $S = \{s_1, s_2, s_3, \ldots, s_n\}$, where n representing the total number of intervals, with each interval containing m datapoints.

5.2.2 Empirical mode decomposition

To extract features from HAR data, the empirical mode decomposition (EMD) is adopted in this paper [23,24]. Each segment is passed through EMD. EMD technique decomposes a time series data into a set of intrinsic mode functions (IMFs)

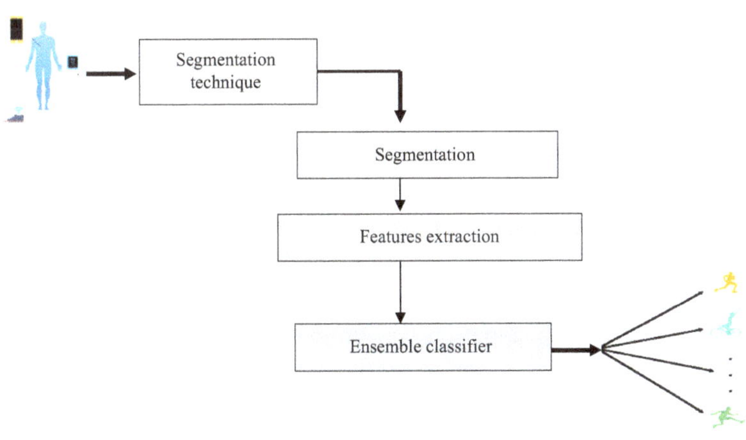

Figure 5.1 The proposed model for HAR

using a shifting process. Each set of IMFs should fulfil the following two conditions:

- The number of extrema and the number of zero crossings are the same by at most one.
- The mean of upper and lower envelopes at any instance should be zero. A time series is represented as a summation of residual components and corresponding IMFs:

$$x(f) = \sum_{k=1}^{N} imf_m(f) + r_m(f) \tag{5.1}$$

where N denotes the number of IMF and r_m refers to the final residue.

5.2.2.1 Permutation entropy

In this chapter, two entropy features are proposed and extracted from each IMF. Permutation entropy has been used widely to analyse time series. We extracted Permutation entropy from each MFI to classify human activities [25]. Suppose a time series $x = \{x_i\}_{i=1}^{M}$ with an embedding dimension n and time delay τ. The embedding vectors can be represented as

$$X(i) = \left[x_i, x_{i+1}, \ldots\ldots, x_{i+(m-1)\tau}\right] \tag{5.2}$$

Where $j = 1, 2, \ldots, N - (m-1)\tau$. Each of $-(m-1)\tau$ subsectors is arranged in the ascending order as

$$x(j + (p_1 - 1)\tau) \leq x(j + (p_2 - 1)\tau) \leq \ldots\ldots \leq x(j + (p_m - 1)\tau) \tag{5.3}$$

An ordinal pattern is assigned to each $x(j)$ and it is described as

$$\pi(j) = o_1, (o_2, \ldots. o_m) \tag{5.4}$$

The probability distribution p of each ordinal pattern is computed, then the normalised permutation entropy is defined as

$$PE(m, \tau) = \left[\frac{\sum_{j=1}^{h} p_j * \ln(p_j)}{\ln(m!)} \right. \tag{5.5}$$

where $h \leq m!$ *and* $\ln(m!)$ refer to the maximum and minimum PE values, respectively.

5.2.2.2 Quadratic entropy

Different types of entropies are developed to analyse time series. One of the most powerful entropies is quadratic entropy [26]. This entropy considers a random variable X as finite discrete with complete probability collection as mentioned in (5.6):

$$p_j \geq 0, \text{where } j = 1, 2, 3, \ \ldots \ n \tag{5.6}$$

$$\sum_{j=1}^{m} p_j = \mathbf{1} \tag{5.7}$$

The probability of event is represented as p_j. The quadratic entropy is computed using (5.8)

$$Qe\ k(x) = \sum\nolimits_{j=1}^{m} p_j(1 - p_j) \tag{5.8}$$

where k identifies the entropy information of samples. The entropy information is identified by (5.9)

$$entropy\ (k, H) = \sum_{g \in G} \left(\frac{|k_g|}{|k|} \right) * entropy\ (k_g) \tag{5.9}$$

The attribute's entropy H is denoted by entropy $entropy(k, H)$, where H refers to the attribute value.

5.2.3 Classification based on ensembles model

In this paper, an ensemble model is suggested to classify human activities. The suggested model designed a combined set of clustering and classification algorithms in one model [28]. We aim at tackling the limitations in clustering and classification ensembles. In this paper, to maximise the number of clustered data and minimise the number of un-clustered data, the resulted clustered data from ensemble clustering model is utilised as the training set to train the classification ensemble and the un-clustered data is used as the testing set. Figure 5.2 depicts the proposed model in which ensemble of clustering integrated with classification ensemble.

5.2.3.1 Ensemble clustering algorithm

In this paper, three clustering algorithms namely k-means, fuzzy clustering algorithm and k-nearest are selected to form the ensemble clustering [27,28]. We selected the maximum and the minimum value of k for each clustering algorithm.

We adopted the weighted kappa and Cohen's unweighted metrics to evaluate the ensemble clustering to assess the degree of agreement between the number of clusters generated by different clustering algorithms. The indexes return a value between 0 and 1 where 0 refers to absolute disagreement and 1 indicates to complete agreement. We adopted this procedure to reveal the unchanging cluster. The clustering results of different algorithms are combined. To maximise the agreement and obtain core groups of data belonging to the same label, the group labels are aligned. We found that some data belong to different groups. These data are labelled as un-clustered and used as a testing set to evaluate the proposed ensemble classification.

5.2.3.2 Ensemble classification algorithm

We suggested the ensemble classification to tackle the un-clustered data. The ensemble classification is trained using the previously clustered data while the

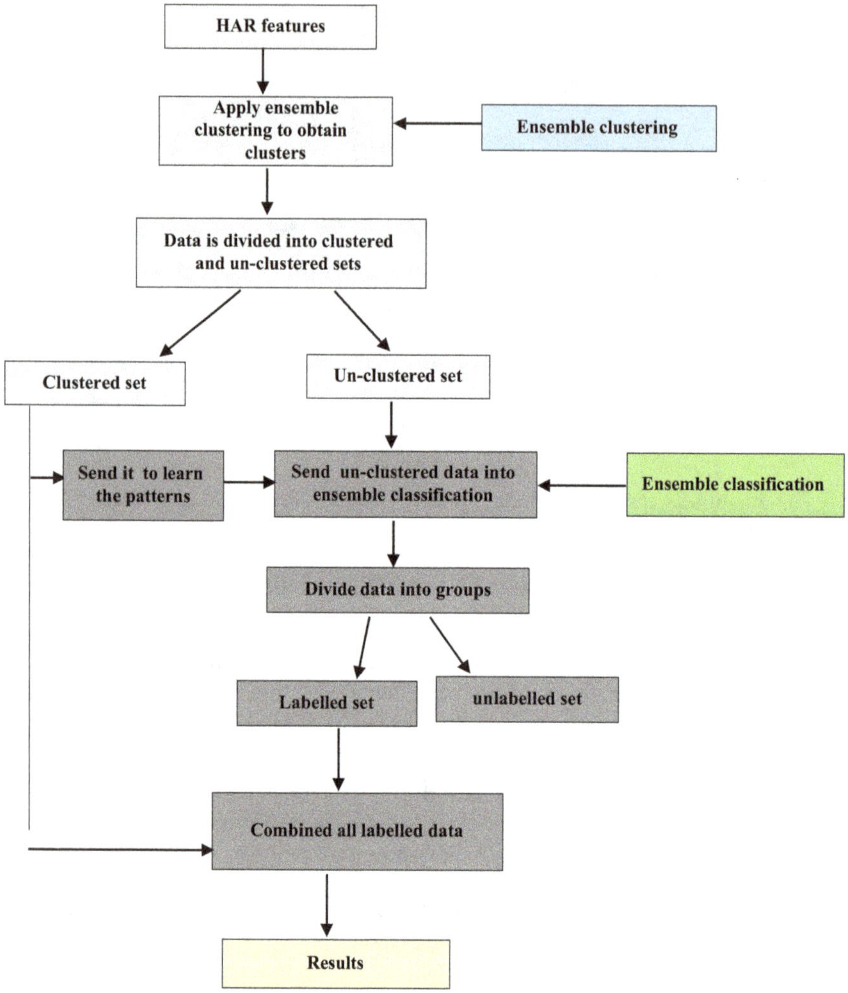

Figure 5.2 The proposed ensemble classifier for neonatal detection

un-clustered data are used as the test set. The voting strategy is adopted to integrate the classification results of classification models. Three algorithms named a support vector machine, artificial neural networks (ANN), and Naïve Bayes are used to form the classification ensemble.

We calculated a weight for each classifier model based on its performance. The error ratio is used as a metric to obtain the weight for each classifier model. As a result, the classifier with low error rate is allocated a high weight. Equation (5.10) is used to assign a weight p for each classification model:

$$p = \log \frac{1 - error}{error\ (Classifer)} \tag{5.10}$$

Table 5.1 Details of two datasets used for HAR

Dataset	Activities	Number of classes	Sample rate	Number of samples
WISDM	Upstairs Standing, sitting, walking downstairs, walking, jogging	6	20 Hz	1,098,207
UCI HAR_2012	Walking, upstairs, downstairs, standing, sitting, and laying	6	50 Hz	10,299

5.2.4 Experimental dataset

In this paper, two publicly available datasets were used to evaluate the proposed model for HAR. The datasets are named: WISDM and UCI HAR. The details of these datasets are presented in Table 5.1 [29–31].

Dataset 1 – WISDM dataset: This dataset was acquired from 36 subjects. Each subject was asked to perform certain daily activities including standing, sitting, walking upstairs and downstairs, jogging. All subjects carried an Android phone in their pants leg pocket. A graphical user interface application was installed in the phone to control the data collection process. This application records several information including user's name, activity type, label of the activity, and start and stop the data collection.

Dataset 2 – UCI-HAR dataset: The recordings were collected at University of California Irvine. 30 subjects aged from 19 to 48 were involved to collect the dataset. Subjects were asked to wear Samsung Galaxy S2 around the wrest to record activities. A total of six activities including walking, standing, sitting, laying down walking downstairs and upstairs were recorded. Each subject was given 5 s rest between each activity. Each subject repeats the activities twice. On the first trial, the smartphone was placed on the left side of the belt while on the second trial the smartphone was fixed by the subjects as preferred.

5.3 Experimental results

In this paper, an HAR model was designed based on entropy features. The data ware segmented and then sent to empirical mode decomposition technique. Two entropy features were extracted named quadratic entropy and permutation entropy were extracted from each IMF. The extracted features were sent to the proposed ensemble model. The performance of the proposed model was evaluated on two datasets. Tables 5.2 and 5.3 report the confusion matrixes results for HAR.

Table 5.2 lists the confusion matrix for the proposed model on WISDM dataset. The proposed model showed a high classification rate with activities standing, walking. However, the lower accuracy rate was obtained with upstairs. We noticed that some misclassification happened among the activities: the *Upstairs, Downstairs, and Walking* classes. Our investigation showed that data points per activity was not sufficient to learn important features of each activity. To solve this issue, we tried to increase the window size for this dataset.

The performance of the proposed model was also tested with UCI HAR_2012 dataset. The proposed model achieved a high recognition rate, however, it misclassified between some classes. The results in Tables 5.2 and 5.3 showed that the proposed model based on entropy features coupled with the ensemble classifier has a high potential to be implemented in real-time system for HAR.

5.3.1 Comparisons with other machine learning algorithms

To evaluate the proposed ensemble classifier, it was compared with other several machine learning algorithms. Table 5.4 reports the comparison results. The

Table 5.2 Average confusion matrix for the proposed model on WISDM dataset

Predict/ground truth	Downstairs	Jogging	Walking	Upstairs	Sitting	Standing
Downstairs	**0.94**	0.01	0.01	0.04	0.00	0.00
Jogging	0.02	**0.095**	0.00	0.01	0.01	0.01
Walking	0.00	0.01	**0.096**	0.00	0.02	0.01
Upstairs	0.04	0.01	0.02	**0.92**	0.01	0.00
Sitting	0.02	0.01	0.00	0.00	**0.96**	0.01
Standing	0.01	0.01	0.00	0.01	0.01	**0.96**

Note: Bold values indicate an improvement in results.

Table 5.3 Average confusion matric for the proposed model on UCI HAR_2012 dataset

Predict/ground truth	Laying	Walking	Sitting	Standing	Upstairs	Downstairs
Laying	**0.96**	0.02	0.01	0.00	0.01	0.00
Walking	0.00	**0.95**	0.000	0.01	0.02	0.01
Sitting	0.01	0.01	**0.96**	0.01	0.01	0.01
Standing	0.01	0.01	0.02	**0.96**	0.01	0.01
Upstairs	0.00	0.01	0.01	0.00	**0.95**	0.03
Downstairs	0.00	0.01	0.00	0.02	0.02	**0.95**

Note: Bold values indicate an improvement in results.

Table 5.4 Comparisons with machine learning algorithms

Technique/dataset	WISDOM	UCI HAR_2012
	ACC	ACC
The proposed model	**96**	**95**
Ensemble classification	87	85
SVM	83	82
Ensemble clustering	86	83

Note: Bold values indicate an improvement in results.

obtained results showed that the proposed ensemble model scored a higher accuracy for two datasets compared with other classifiers. In this experiment, 10-cross validation was used to evaluate the proposed model. The extracted features were sent to the four classifiers named the SVM, ensemble classification, ensemble clustering. All results were recorded and reported in Table 5.4.

5.3.2 Comparison with previous studies

In this experiment, the proposed model was compared with the state of the art. The selected previous studies were used the same dataset to obtain a fair comparison. These studies applied LSTM [32,34], CNN [35], ensemble of EN, a lightweight CNN [33]. The comparison results are reported in Table 5.5. We can observe that the proposed model outperformed all studies. It achieved 96% and 95% with two datasets.

5.3.3 Comparisons of computation time

We evaluate the computation time of the proposed mode in this section. Figure 5.3 reports the computation time of the proposed compared with the SVM, ensemble classification, and ensemble clustering based on different sample sizes. Each

Table 5.5 Comparisons with previous work

Authors	**Method**	**WISDM**	**UCI HAR_2012**
Zhao *et al.* [32]	LSTM	72.61%	X
Tang *et al.* [33]	CNN based on Lego filters	92.82%	96.90%
Wu *et al.* [34]	CNN	78.14%	X
Dong *et al.* [35]	CNN, MP	81.05	X
Xia *et al.* [14]	LSTM	95.85	95.78
The proposed model	Ensemble classifier-based entropy features	96	95

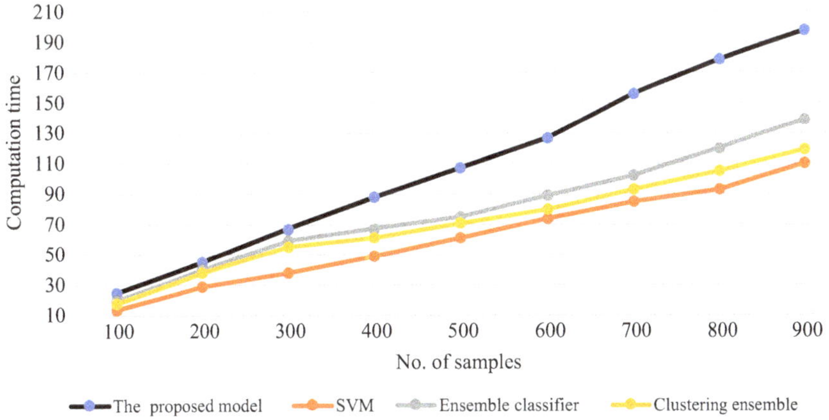

Figure 5.3 Comparisons of complexity time

segment represents one window of HAR data. A total of 900 segments were used in this experiment to evaluate the proposed model. Figure 5.3 shows that proposed model scored higher execution time compared with ensemble classification and ensemble clustering. However, the SVM recorded the lowest execution time. The noticeable increase in the proposed model's execution time is acceptable as it combined two ensembles in one model.

5.4 Conclusion

This paper proposes a model for HAR. This model adopted entropy feature coupled with an ensemble classifier. The efficiency of the proposed model was tested with two publicly available datasets. The experimental results showed the superiority of the proposed system to some of the other methods. The proposed ensemble classifier showed a high recognition rate than traditional classifiers. Our future research will focus on investigating the application of the proposed model to other datasets, such as aesthetic time series or blood pressure estimation.

References

[1] Wang, J., Y. Chen, S. Hao, X. Peng, and L. Hu. "Deep learning for sensor-based activity recognition: a survey." *Pattern Recognition Letters* 119 (2019): 3–11.

[2] Zhang, M. and A.S. Alexander. "Human daily activity recognition with sparse representation using wearable sensors." *IEEE journal of Biomedical and Health Informatics* 17(3) (2013): 553–560.

[3] Pecchia, L., L.L. Chen, C. Nugent, and J. Bravo (eds.). *Ambient Assisted Living and Daily Activities: 6th International Work-Conference, IWAAL 2014*, Belfast, UK, December 2–5, 2014, Proceedings. Vol. 8868. Springer, 2014.

[4] Dobbins, C., R. Rawassizadeh, and E. Momeni. "Detecting physical activity within lifelogs towards preventing obesity and aiding ambient assisted living." *Neurocomputing* 230 (2017): 110–132.

[5] Perera, C. and A.V. Vasilakos. "A knowledge-based resource discovery for Internet of Things." *Knowledge-Based Systems* 109 (2016): 122–136.

[6] Abdel-Basset, M., W. Ding, and L. Abdel-Fatah. "The fusion of Internet of Intelligent Things (IoIT) in remote diagnosis of obstructive sleep apnea: a survey and a new model." *Information Fusion* 61 (2020): 84–100.

[7] Garcia, K.D., T. Carvalho, J. Mendes-Moreira, J.M.P. Cardoso, and A.C.P.L. F. de Carvalho. "A study on hyperparameter configuration for human activity recognition." In *International Workshop on Soft Computing Models in Industrial and Environmental Applications*. Cham: Springer, 2019.

[8] Aggarwal, J.K., and M.S. Ryoo. "Human activity analysis: a review." *ACM Computing Surveys (CSUR)* 43(3) (2011): 1–43.

[9] Hussain, Z., Q.Z. Sheng, and W.E. Zhang. "A review and categorization of techniques on device-free human activity recognition." *Journal of Network and Computer Applications* 167 (2020): 102738.

[10] Jethanandani, M., A. Sharma, T. Perumal, and J.R. Chang. "Multi-label classification based ensemble learning for human activity recognition in smart home." *Internet of Things* 12 (2020): 100324.

[11] Jain, A. and V. Kanhangad. "Human activity classification in smartphones using accelerometer and gyroscope sensors." *IEEE Sensors Journal* 18(3) (2017): 1169–1177.

[12] Jalloul, N., F. Poree, G. Viardot, *et al.* "Activity recognition using complex network analysis." *IEEE Journal of Biomedical and Health Informatics* 22 (4) (2017): 989–1000.

[13] Gupta, P. and T. Dallas. "Feature selection and activity recognition system using a single triaxial accelerometer." *IEEE Transactions on Biomedical Engineering* 61(6) (2014): 1780–1786.

[14] Xiao, Z., X. Xu, H. Xing, *et al.* "A federated learning system with enhanced feature extraction for human activity recognition." *Knowledge-Based Systems* 229 (2021): 107338.

[15] Altuve, M., P. Lizarazo, and J. Villamizar. "Human activity recognition using improved complete ensemble EMD with adaptive noise and long short-term memory neural networks." *Biocybernetics and Biomedical Engineering* 40(3) (2020): 901–909.

[16] Diykh, M., S. Abdulla, R.C. Deo, S. Siuly, and M. Ali. "Developing a novel hybrid method based on dispersion entropy and adaptive boosting algorithm for human activity recognition." *Computer Methods and Programs in Biomedicine* 229 (2023) 107305.

[17] Diykh, M., Y. Li, P. Wen, and T. Li. "Complex networks approach for depth of anesthesia assessment." *Measurement* 119 (2018): 178–189.

[18] Diykh, M., S. Abdulla, K. Saleh, and R. Deo. "Fractal dimension undirected correlation graph-based support vector machine model for identification of focal and non-focal electroencephalography signals." *Biomedical Signal Processing and Control* 54 (2019): 101611.

[19] Diykh, M., F. S. Miften, S. Abdulla, K. Saleh, and J. H. Green. "Robust approach to depth of anaesthesia assessment based on hybrid transform and statistical features." *IET Science, Measurement & Technology* 14.1 (2019): 128–136.

[20] Diykh, M., F.S. Miften, S. Abdulla, *et al.* "Texture analysis based graph approach for automatic detection of neonatal seizure from multi-channel EEG signals." *Measurement* 190 (2022): 110731.

[21] Miften, F.S., M. Diykh, S. Abdulla, *et al.* "A new framework for classification of multi-category hand grasps using EMG signals." *Artificial Intelligence in Medicine* 112 (2021): 102005.

[22] Alsafy, I., and M. Diykh. "Developing a robust model to predict depth of anesthesia from single channel EEG signal." *Physical and Engineering Sciences in Medicine* 45 (2022): 1–16.

[23] Biju, K., H.A. Hakkim, and M. Jibukumar. "Ictal eeg classification based on amplitude and frequency contours of IMFS." *Biocybernetics and Biomedical Engineering* 37(1) (2017): 172–183.

[24] Djemili, R., H. Bourouba, and M.A. Korba. "Application of empirical mode decomposition and artificial neural network for the classification of normal and epileptic eeg signals." *Biocybernetics and Biomedical Engineering* 36 (1) (2016): 285–291.

[25] Cura, O.K., S.K. Atli, H.S. Türe, and A. Akan. "Epileptic seizure classifications using empirical mode decomposition and its derivative." *BioMedical Engineering OnLine* 19(1) (2020): 1–22.

[26] Chen, Z., Y. Li, H. Liang, and J. Yu. "Improved permutation entropy for measuring complexity of time series under noisy condition." *Complexity* 2019 (2019): 1–2.

[27] Khare, S.K., N.B. Gaikwad, and V. Bajaj. "VHERS: a novel variational mode decomposition and Hilbert transform-based EEG rhythm separation for automatic ADHD detection." *IEEE Transactions on Instrumentation and Measurement* 71 (2022): 1–10.

[28] Lafta, R., J. Zhang, X. Tao, Y. Li, M. Diykh, and J.C.W. Lin. "A structural graph-coupled advanced machine learning ensemble model for disease risk prediction in a telehealthcare environment." In *Big Data in Engineering Applications*, Springer, Singapore, 2018, pp. 363–384.

[29] Weiss, G.M., K. Yoneda, and T. Hayajneh. "Smartphone and smartwatch-based biometrics using activities of daily living." *IEEE Access* 7 (2019): 133190–133202.

[30] Reiss, A. and D. Stricker. "Introducing a new benchmarked dataset for activity monitoring." In *2012 16th International Symposium on Wearable Computers*,2012, pp. 108–109. IEEE.

[31] Anguita, D., A. Ghio, L. Oneto, X. Parra Perez, and J.L. Reyes Ortiz. "A public domain dataset for human activity recognition using smartphones." In *Proceedings of the 21th International European Symposium on Artificial Neural Networks, Computational Intelligence and Machine Learning*, 2013, pp. 437–442.

[32] Zhao, Y., H. Haddadi, S. Skillman, *et al.* "Privacy-preserving activity and health monitoring on databox." In *Proceedings of the Third ACM International Workshop on Edge Systems, Analytics and Networking*, 2020.

[33] Tang, Y., Q. Teng, L. Zhang, F. Min, and J. He. "Layer-wise training convolutional neural networks with smaller filters for human activity recognition using wearable sensors." *IEEE Sensors Journal 21.1* (2020): 581–592.

[34] Wu, Q., K. He, and X. Chen. "Personalized federated learning for intelligent IoT applications: a cloud-edge based framework." *IEEE Open Journal of the Computer Society* 1 (2020): 35–44.

[35] Dong, Y., X. Chen, L. Shen, and D. Wang. "EaSTFLy: efficient and secure ternary federated learning." *Computers & Security* 94 (2020): 101824.

Chapter 6

Challenges in the acquisition of non-invasive brain signals – electroencephalographic signals (EEG)

K. Gopika Gopan[1] *and Neelam Sinha*[1]

Abstract

The most prominent issue in the acquisition of biomedical signals, apart from the problems associated with the sensors for capturing these signals, is the presence of artifacts. In this chapter, these artifacts are discussed mainly with respect to electroencephalographic (EEG) signals. The chapter introduces EEG and the different methods employed to capture the EEG signals. In particular, the advantages of non-invasive EEG and the challenges associated with the captured signals are highlighted. The artifacts present in EEG can be broadly classified as (a) physiological and (b) non-physiological. The physiological artifacts such as ocular cardiac artifacts and non-physiological artifacts such as power line interference are elaborated along with the reason for such artifacts. In this chapter, three physiological artifacts, namely, ocular, cardiac, and electromyogenic, are explained along with two non-physiological artifacts, namely, electrode movement and power line interference. To effectively utilise EEG, the signals need to undergo artifact removal. Multiple methodologies are employed to remove the artifacts from EEG, and the most widely used methods involve Independent Component Analysis (ICA) and Wavelets. The ocular and electromyogenic artifacts can be handled using a range of methods such as adaptive filtering, blind source separation, wavelet transform, Riemannian Geometry, and deep learning approaches. Cardiac artifacts are commonly eliminated using adaptive filtering as well as blind source separation and deep learning approaches. One commonly used method to tackle the electrode placement artifact is independent component analysis. Power line interference is mainly eliminated using notch filters, however, adaptive filtering and hybrid methods of wavelet and blind source separation are also explored in multiple works. These methods are explained and analysed in this chapter, highlighting each one's advantages and disadvantages. Additionally, the challenges in employing a commercial EEG headset compared to a medical EEG headset are also explored.

[1]International Institute of Information Technology – Bangalore, India

The chapter concludes with an overview of various methods employed for artifact removal in EEG and associated active research areas.

Keywords: Electroencephalographic signals; Artifacts; Artifact removal; Commercial EEG headset

6.1 Introduction to nervous system

The human body consists of multiple organs with specific functions. All organs need to carry out their functions in coordination with other organs for the proper functioning of the body. The coordination of the functions along with cognition, emotion, regulation of the biological clock, and body temperature is carried out by the nervous system [1]. There are two parts to the nervous system, namely, Central Nervous System (CNS) and Peripheral Nervous System (PNS) [2]. CNS consists of brain and the spinal cord functions as the command centre for the nervous system. CNS receives and analyses the signals from various parts of the body. After interpreting the signals, necessary commands are sent back to the body. Cranial nerves and spinal nerves which are part of PNS obtain signals from various sensory receptors of the body and transfer them to CNS, and relay the commands from CNS back to the body. Any damage to the nervous system can result in serious health problems. For instance, a person having a spinal injury may get paralysed since the spinal cord is damaged and signals from the brain cannot be relayed to the muscles for their movement [3]. Though the disabling effect of spinal cord injury can be overcome by using brain computer interface (BCI) which transfers information to muscles directly from the brain, in most cases, brain injury cannot be reversed [2]. Hence, the brain forms the most complex and vital organ both for the analysis of information from sensory receptors and for relaying necessary commands. Additionally, the brain is also involved in emotion, sleep regulation, control of body temperature, cognition, and creativity [1]. Understanding brain function is necessary to fully comprehend the various neurological disorders, to assess mental health, and also to utilise brain signals to control machines.

6.2 Types of brain signals

The electrical signals generated by the brain provide insight into the functioning of the brain. The signals can be decomposed into frequency components or brain waves. In most analyses of brain signals, 0–60 Hz is considered as the bandwidth with useful information. This can be further divided as: Delta (0–4 Hz), Theta (4–8 Hz), Alpha (8–14 Hz), Beta (14–30 Hz), and Gamma (30–60 Hz). Often analyses of specific brain waves are required for understanding the cognitive functions or detection of disorders.

Brain signals can be obtained either invasively or non-invasively. The invasive method involves using a surgical procedure to either implant electrodes on the specific surfaces of brain regions or inside the grey matter of the brain for capturing the extracellular activities of neurons [4]. The signals obtained using invasive methods have a high signal-to-noise ratio (SNR) but require the expertise for surgical implantation of electrodes without causing damage to the brain. The non-invasive method involves placing the electrodes (wet or dry) on the scalp to obtain the electroencephalographic (EEG) signals. This method is inexpensive and involves the least risk to the subjects but the obtained signals have low SNR and are affected by noise and artifacts. Thus, EEG signals need to be pre-processed to remove noise and artifacts before they can be utilised for any analysis.

6.2.1 Neuronal recordings

The neuronal recordings refer to the measurement of brain activity invasively in the grey region of the brain using bio-compactable electrodes. These recordings capture extracellular activities associated with neurons and local field potentials [5]. Recording from a single electrode placed extracellular to neurons provides information on a limited number of cells while multi-electrode arrays provide information on the interactions between neurons on a larger scale. The main advantage is that SNR for these signals is high as the signals are picked up from as near the source as possible. The electrodes are placed surgically inside the brain which requires a highly specialised surgeon. Additionally, there is a risk of the body reacting to a foreign object (electrodes) which in some cases results in rejection. BCI using implanted electrodes is used in studies involving paralysed patients [6] and visually impaired people [7]. The interface aims to aid the patients in the movement of limbs or in restoring vision.

6.2.2 Electrocorticographic (ECoG) signals

Electrocorticographic signals are signals obtained by placing electrodes inside the skull on the cerebral cortex which enables capturing cortical potentials [8]. These electrodes do not penetrate the brain but are placed on its surface. The placement of electrodes inside the skull requires surgery but is not as complex as that required for neuronal recordings. Since the potential at the surface of the brain is measured, SNR is less than that of neuronal recordings. However, the spatial resolution obtained using ECoG is in the scale of millimeters. Additionally, the high-frequency bandwidth enables analyses of the high Gamma band of frequency 70–180 Hz. ECoG is used in partially invasive BCI [9], which offers the advantage that the brain need not be penetrated for electrode placement as in invasive BCI. Unlike non-invasive BCI, higher SNR as well as high spatial and temporal resolution is obtained using ECoG. Hence, the use of ECoG in BCI provides a good trade-off between invasiveness and the quality of the signal obtained.

6.2.3 EEG signals

EEG signals are obtained using scalp electrodes and reflect mainly the electrical activity of pyramidal neurons [10]. Though invasive and partially invasive methods of obtaining brain signals provide higher SNR and high spatial as well as temporal resolution, the surgical risk to the patient is still high including the reaction of the body to the electrodes. The use of EEG eliminates the surgical risk while providing the flexibility to change the location of electrodes as required for the analysis. This is however at the cost of low SNR and low spatial resolution. Additionally, EEG is more affected by artifacts [11,12] than ECoG and neuronal recordings. This necessitates the need for an efficient pre-processing of raw EEG. With the availability of easy-to-use commercial EEG headset, EEG has been used in a wide range of applications such as sleep stage analysis and sleep disorder detection [13,14], emotion analysis [15], control of robotic arm using thoughts [16], control of the movement of wheelchair [17], assessment of visual creativity [18], detection of neurological disorders [19–21], assessment of psychiatric disorders [22,23], stress detection [24], and migraine analysis [25].

6.3 Artifacts in EEG

Artifacts, are electrical signals captured by the electrode, whose origin is not the brain. The electrical signals captured by EEG are the result of signals that travel from neurons to the scalp through the brain–blood barrier and the skull. This provides ample means for the signal to be corrupted with artifacts from other body functions as well as from external electrical sources. The analysis of artifact-contaminated EEG can lead to misinterpretation of signals leading to wrong results.

The EEG artifacts can be categorised into two based on the source, namely, physiological artifacts and non-physiological artifacts.

6.3.1 Physiological artifacts

These artifacts are caused due to signals arising during normal biological functions in the body of the subject (e.g., heartbeat and eye movement). These signals contaminate the electrical signals corresponding to cortical activities and often can mimic abnormal cerebral activity such as seizures [26]. For instance, chewing artifact is known to mimic ictal activity in generalised seizures [27].

The three main physiological artifacts commonly found in EEG are ocular, cardiac, and electromyogenic artifacts.

6.3.1.1 Ocular artifacts

The large differences in the potential between the cornea of the eyes and the retina caused by changing electric fields of eye regions result in ocular artifacts in EEG during eye movements or eye blinks. This mechanism is known as cornea-retinal dipole movement [28], where the cornea and the retina are considered to be positively and negatively charged, respectively. In general, the ocular artifacts tend to have larger amplitudes than EEG, especially eye blinks. They are prominent in the frontal region of the brain, especially in the pre-frontal region and their effect reduces further the distance of the electrodes from the frontal region. The properties of the scalp, skull, and tissues determine the extent of ocular artifacts and hence these artifacts affect less in the posterior regions of the brain.

The two main types of ocular artifacts observed in EEG are:

1. *Eye blinks*: These artifacts have low frequency, typically less than 4 Hz, and very large amplitude. These artifacts are prominently evident in the frontopolar

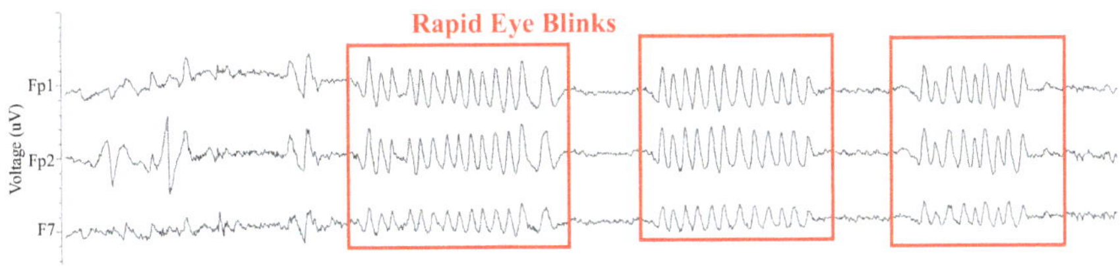

Figure 6.1 EEG of a person sitting still with intermittent rapid eye blinks resulting in low-frequency waves

electrodes, Fp1 and Fp2. Eye blinks cause a sharp increase in potential followed by a decrease over a duration of 0.25–0.3 s in these electrodes. In the bipolar montage, eye blinks appear as large negative potential prominent in electrodes Fp1–F7, Fp1–F3, Fp2–F8, and Fp2–F4. Rapid eye blinking, as seen in Figure 6.1, produces artifacts with a low frequency similar to Theta or Delta waves.

2. *Lateral eye movement*: The movement of the eyes changes the electric field of the eye region, which in turn affects the electric field of the scalp. This induces eye movement artifacts in EEG. Based on the direction of the movement of the eyes, that is, whether towards the left or the right side, the frontal electrode potential of one side increases while the potential of the other side decreases [29]. For instance, in the bipolar montage, when a person looks to the left, electrode Fp1–F7 (placed on the left frontal region of the brain) has positive potential while Fp2–F8 (placed on the right frontal region of the brain) has a negative potential.

Apart from these two types of artifacts, rotation of the eyes and eye twitching also cause artifacts in EEG. The position of the cornea with respect to the electrodes decides the location and also whether positive or negative potential is created.

Electro-oculographic signals (EOG) can be used to assess the artifacts and eliminate them. Since ocular artifacts are prominent in EEG, many studies [30,31] focus on the elimination or suppression of ocular artifacts. Adaptive filtering, blind source separation (BSS), wavelet transform, and deep learning approaches have been explored in literature for ocular artifact removal.

6.3.1.2 Cardiac artifact

The electrical potential associated with cardiac activity causes artifacts in EEG. These artifacts are clearly visible in EEG montages with ear electrodes (A1 and A2) due to the close proximity of these electrodes to the heart. The most commonly observed cardiac artifact occurs as a sharp waveform and can be easily identified due to the temporal synchronisation of cardiac artifacts with the QRS complex of ECG as seen in Figure 6.2. Another type of cardiac artifact is observed when the electrodes are placed on an artery. These artifacts, known as cardio-ballistic artifacts, appear as slow waves and have a fixed time relation to QRS complex. Adaptive filtering, BSS, template matching, spectral analyses, and deep learning approaches are some methods used in literature [32–34] for cardiac artifact removal.

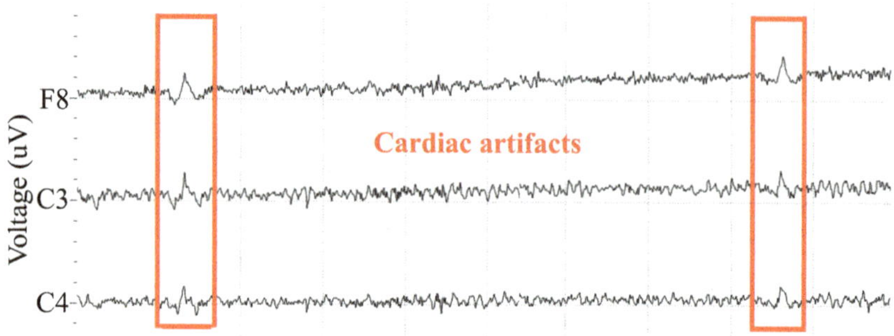

Figure 6.2 *EEG of a person sitting still with closed eyes. Cardiac artifacts are observed as synchronised sharp waveforms in all three channels. This artifact is clearly visible in channels with references being A1 and A2, the ear electrodes.*

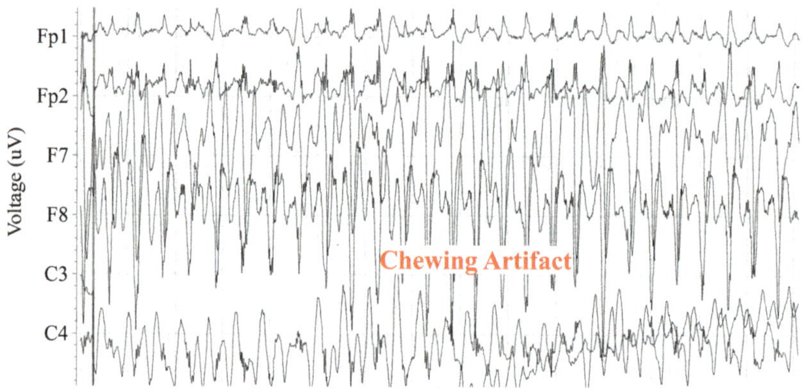

Figure 6.3 *EEG of a person sitting still while chewing a candy. Generalised high-frequency bursts of activity are observed in EEG.*

6.3.1.3 Electromyogenic (EMG) artifact

EMG artifacts or muscle artifacts are mainly caused due to the movement of facial, head, and neck muscles. These artifacts occur at various frequencies based on muscle activity and have overlapping spectrum with EEG. The duration of EMG artifact varies according to the duration of muscle activity. One of the sources of EMG artifacts is chewing, which utilises the temporalis muscles. This artifact, as seen in Figure 6.3, is identifiable due to its generalised high-frequency bursts of activity in EEG [29]. Tongue movement is another source of EMG artifact. The tip of the tongue is considered to be more electro-negative than the base [29]. Hence the movement of the tongue causes changes in the electric field that affects EEG. This artifact is also known as glossokinetic artifact. This artifact is characterised by slow activity and its polarity depends on the movement of the tongue. Jaw clenching, sniffling, swallowing, and

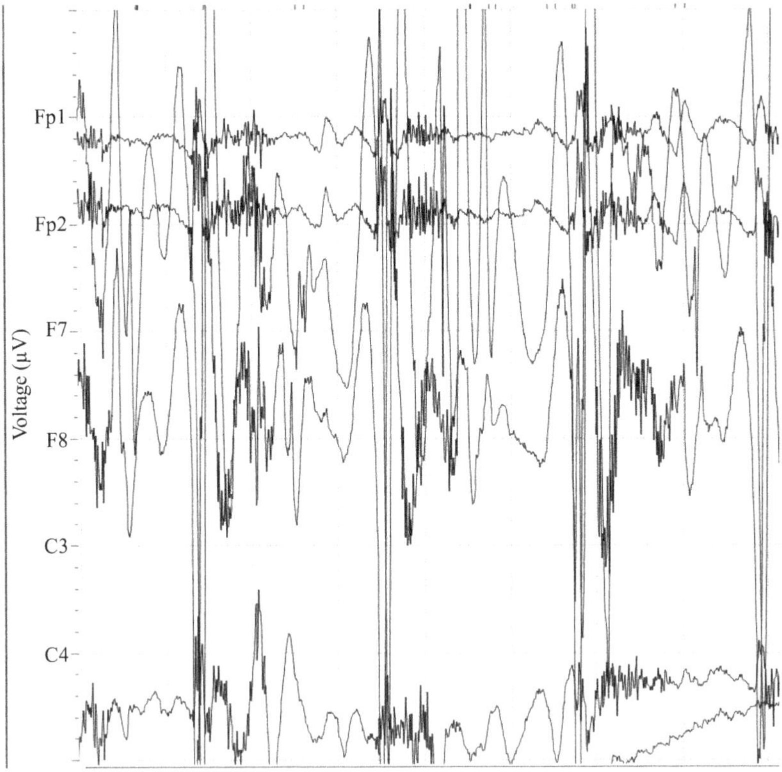

Figure 6.4 EEG of a person sitting still with intermittent clenching of the jaw. Generalised high-frequency muscle artifacts can be observed.

teeth grinding are some of the other sources of EMG artifacts in EEG. Figure 6.4 shows artifacts occurring due to jaw clenching. Similar to ocular artifacts, methods such as adaptive filtering, BSS, hybrid methods involving wavelets and BSS, and deep learning approaches are utilised in literature [35–37] to remove EMG artifacts.

6.3.2 Non-physiological artifacts

These artifacts are caused by electrodes or electrical interference such as power line interference, ventilators, and intravenous drips. Electrode-related artifacts occur due to disconnection of the electrode, spontaneous discharges, or movement of electrodes. Two frequently occurring non-physiological artifacts in EEG are electrode movement and power line interference.

6.3.2.1 Electrode movement

The movement of electrodes introduces artifacts having no fixed morphology. They may occur as disorganised unusually high amplitude signals with low frequency. The electrode movement artifacts are restricted to the electrode concerned and in

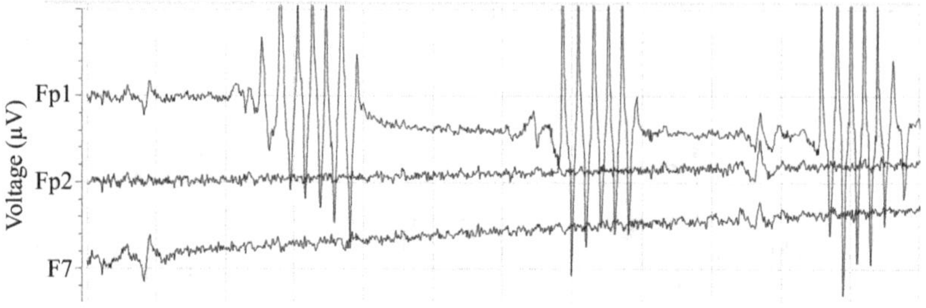

Figure 6.5 EEG of a person sitting still with the intermittent movement of FP1 electrode characterised as low frequency and high amplitude signal

bipolar montage, a phase reversal may be observed. Figure 6.5 shows a sample EEG with the intermittent movement of Fp1 electrode. Ideally, this artifact needs to be checked before the EEG signals are recorded for the analysis task by observing the readings. However, in cases where the tasks of movement of limbs like walking are required for the experiment, electrode movement can occur due to head movement when using dry electrodes. In such cases, Independent Component Analysis [38] can be used for the removal of electrode movement artifacts or the epochs detected with this artifact can be rejected [39].

6.3.2.2 Power line interference

Power line interference is caused by electromagnetic interference due to a power line, electrical equipment, or improper grounding of signal recording equipment. Since the power line has a 50 Hz frequency in most parts of the world except in the United States of America (USA) where it is 60 Hz, the interference occurs at 50 Hz (or 60 Hz in the USA). Hence, such noise can be eliminated by using a notch filter of 50 Hz (or 60 Hz). Figure 6.6 shows a sample EEG with power line interference. The use of a notch filter leads to the distortion of the frequency spectrum. Another method is to use an adaptive filtering method [40] such as Normalised Least Mean Squares algorithm. However, this method requires a reference signal with power line noise characteristics for filtering out the noise from the corrupted signal. Though methods such as Wavelet-Independent Component Analysis [41] are also explored for the removal of power line interference, the most commonly used method is a notch filter. This is because a notch filter is easy to implement, does not require any reference channel, and is computationally inexpensive. Additionally, the distortion caused by the notch filter for power line interference removal in most cases can be ignored for EEG analysis.

6.3.3 Gait-related artifacts

EEG recorded during the movement of subjects, such as walking or exercising makes the signal prone to multiple artifacts [42]. Hence, most studies use tasks that can be performed in static settings or use motor imagery/minimal limb movements. The analysis of EEG while the subject is mobile is crucial for a high-performing BCI. Currently, such BCI uses invasive electrodes or partially invasive electrodes due to excessive artifacts associated with EEG in a mobile subject. For instance, the control of a bionic arm using invasive electrodes enables the subject to carry on activities involving whole-body movements.

Artifacts observed during gait arise from different sources. Head displacement is a common artifact observed during whole-body movement. For instance, walking causes displacement of the head horizontally, vertically, and from back to front. This displacement increases with the speed of walking. This can be observed in Figures 6.7 and 6.8 showing EEG while walking and running,

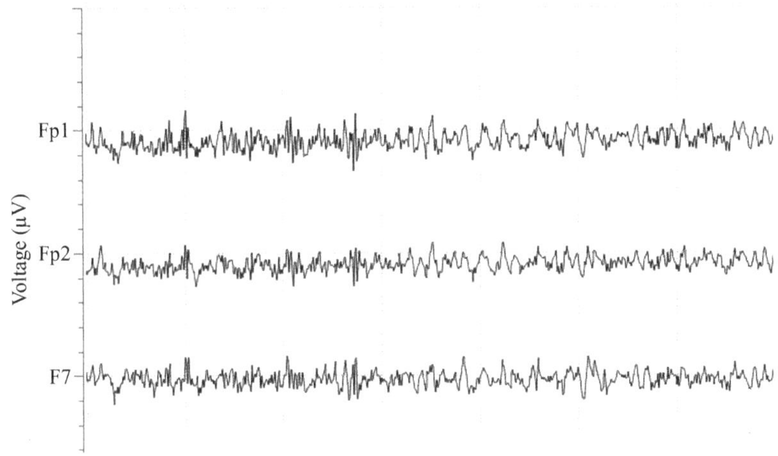

Figure 6.6 EEG of a person sitting still without using 50 Hz notch filter. Noise is observed in all electrodes characterised by high frequency.

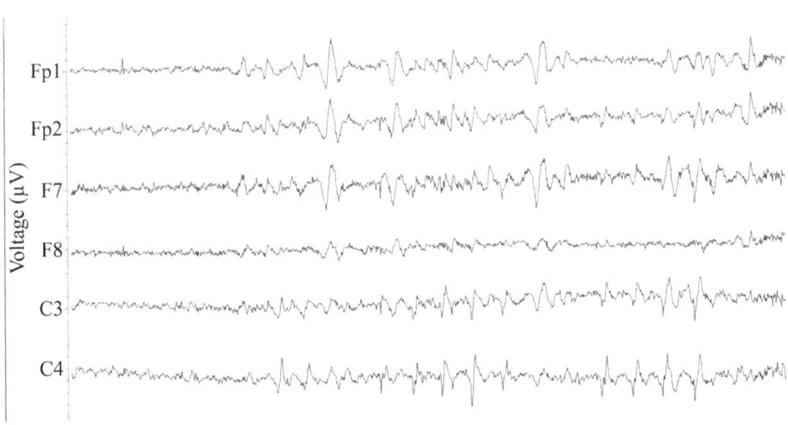

Figure 6.7 EEG of a person standing still initially and then starting walking. EEG observed during walking includes artifacts from multiple sources. The change in EEG when walking is performed can be observed as the change in EEG with higher amplitude and affected by multiple artifacts.

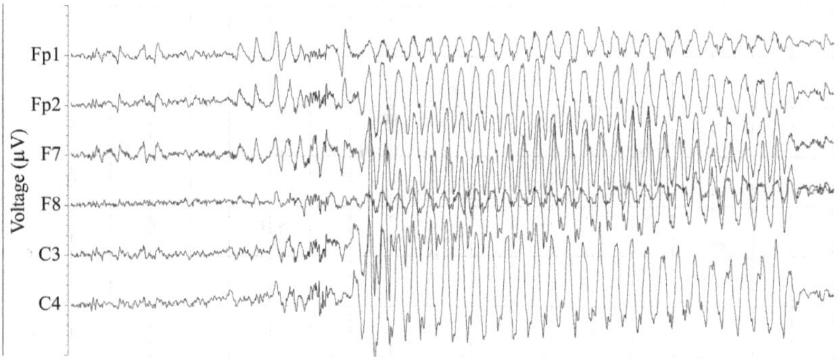

Figure 6.8 EEG of a person initially walking and suddenly starting running. EEG observed during running shows a profound change from the walking EEG. This is because the displacement of the head and other movements increase when a person runs compared to walking.

respectively. Other artifacts associated with head displacement are caused by eye movement and neck muscle movement. Additionally, electrode movement artifacts are also observed based on the type of electrodes used for recording. Though multiple methods have been developed to address the artifacts during whole body movement [43,44], artifact removal is still a challenge in tasks involving gait-related analysis.

6.4 Methods to remove artifacts

The raw EEG obtained using scalp electrodes is corrupted with artifacts. In literature, the artifacts can be dealt with in three ways:

- Ignore the artifact if it would not affect the analysis and carry on the analysis
- Remove the parts of EEG corrupted with artifacts and utilise the remaining EEG for analysis
- Suppress or remove artifacts using various methods to obtain signals with minimal artifacts without affecting the actual EEG signals needed for analysis

In this section, various methods used for removing the artifacts from raw EEG signals ranging from filtering to deep learning approaches are explored. Six main commonly used methods are discussed here, namely,

1. Adaptive filtering
2. BSS – independent component analysis (ICA)
3. Wavelet transform approach
4. Hybrid approaches: ICA wavelet
5. Riemannian geometry
6. Deep learning approaches

6.4.1 Adaptive filtering

In adaptive filtering, the obtained EEG is assumed to be a mixture of pure EEG and noise/artifacts but they are uncorrelated:

$$EEG_{raw} = EEG_{pure} + Artifacts \tag{6.1}$$

Adaptive filtering is used to eliminate artifacts that have overlapping spectra with pure EEG. The adaptive part of these filters is that they can modify their weight parameters iteratively based on the features of the signals being analysed [45]. Figure 6.9 shows a generic methodology for adaptive filtering used in EEG artifact removal. The adaptive filtering method contains two inputs, namely, the raw EEG with artifacts ($EEG_{raw}(n)$) and a reference signal ($A(n)$). The reference signal can be a sine wave of 50 Hz or 60 Hz corresponding to power line interference or artificial/real ECG or artificial/real EOG depending on the type of artifact that needs to be filtered from raw EEG. The main objective of adaptive filtering is to iteratively update the weights/coefficients of the filter, thereby changing its frequency response to generate the signal ($A_{est}(n)$) similar to the artifact to be removed from raw EEG and then subtract this signal from the raw EEG:

$$EEG_{filtered}(n) = EEG_{raw}(n) - A_{est}(n) \tag{6.2}$$

$$EEG_{filtered}(n) = EEG_{pure}(n) + Artifacts(n) - A_{est}(n) \tag{6.3}$$

Squaring (6.3) results in

$$\begin{aligned} EEG^2_{filtered}(n) &= EEG^2_{pure}(n) + (Artifacts(n) - A_{est}(n))^2 \\ &\quad - 2EEG_{pure}(n)(Artifacts(n) - A_{est}(n)) \end{aligned}$$

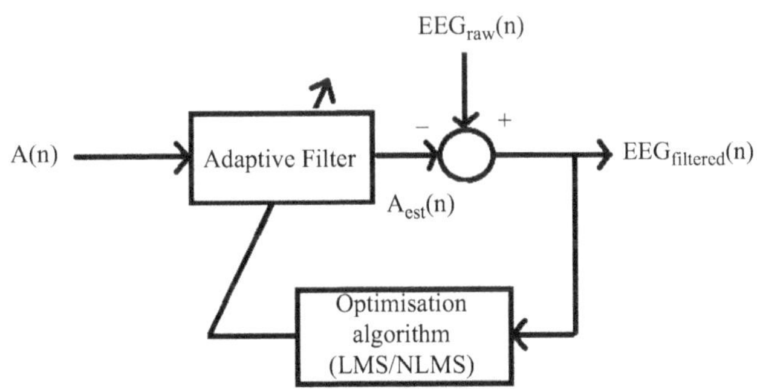

Figure 6.9 A generic block diagram of adaptive filtering. This method requires a reference signal of the artifact that is to be filtered out from raw EEG. Hence, at a time only a single artifact can be removed. However, a cascade of adaptive filters would enable multiple artifact removal.

Since EEG_{pure} and artifacts are assumed to be uncorrelated, the Expectation of the above equation is

$$E\left[EEG_{filtered}^2(n)\right] = E\left[EEG_{pure}^2(n)\right] + E\left[(Artifacts(n) - A_{est}(n))^2\right] \tag{6.4}$$

The aim here is to make $EEG_{filtered}$ as close to EEG_{pure} as possible by adjusting the filter weights. This requires minimisation of $E[(Artifacts(n) - A_{est}(n))^2]$. Since $E[EEG_{pure}^2(n)]$ is not affected by the changes in the filter weights, the minimisation of $E[(Artifacts(n) - A_{est}(n))^2]$ is equivalent to the minimisation of $E[EEG_{filtered}^2(n)]$.

The most commonly used optimisation algorithm for adaptively updating the weight is the least mean square (LMS) algorithm.

6.4.1.1 LMS algorithm

LMS algorithm is a robust optimisation algorithm with low computational complexity. Hence, this algorithm is widely used in the adaptive filtering of artifacts from raw EEG [46]. The objective here is to update the filter weights such that $E[(Artifacts(n) - A_{est}(n))^2]$ is minimised or the estimated artifacts are as close to the artifacts present in raw signal as possible. $A_{est}(n)$ can be written in terms of filter weights as

$$A_{est}(n) = W^T(n)A(n) \tag{6.5}$$

The weight update equation for LMS algorithm is

$$W(n+1) = W(n) + \Delta W(n) \tag{6.6}$$

$$= W(n) + 2\mu EEG_{filtered}A(n) \tag{6.7}$$

It can be observed that the LMS weight update equation does not require any prior information on artifact or signal statistics. It requires instantaneous weights and coefficient μ which controls the rate of updation. μ is also known as step-size or learning rate. For stability of the filter and for convergence, μ must satisfy $0 < \mu < \frac{2}{\lambda_{max}}$, where λ_{max} is the largest eigenvalue of the correlation matrix of the reference signal. Practically, this condition is modified as $0 < \mu < \frac{2}{||A(n)||^2}$, where $A(n)^2$ is the squared norm of reference signal.

6.4.1.2 Normalized LMS algorithm

The fixed learning rate for every iteration is the main disadvantage of the LMS algorithm. To overcome this, normalisation of the learning rate is introduced and the extended LMS algorithm is known as normalised least mean square (NLMS) algorithm [47]. NLMS has the advantage that the rate of convergence is faster than LMS. The normalised learning rate is calculated as

$$\eta(n) = \frac{\mu}{||\ A(n)\ ||^2 + \delta} \tag{6.8}$$

where $||\ A(n)\ ||^2$ is the squared norm of the reference signal and $\delta > 0$ is a regularisation term introduced to overcome any numerical difficulties when the squared norm is very small. The normalised learning rate $\eta(n)$ is dimensionless while the dimension of μ is that of inverse power.

The weight update equation for NLMS is

$$W(n+1) = W(n) + 2\eta(n)EEG_{filtered}A(n) \tag{6.9}$$

$$= W(n) + 2\frac{\mu}{\| A(n) \|^2 + \delta} EEG_{filtered}A(n) \tag{6.10}$$

where 0<μ<2 in equation without regularisation term and $0 < \mu < 2 + \frac{2\delta}{\|A(n)\|^2}$ in equation with regularisation term.

6.4.2 BSS – ICA

BSS refers to a set of algorithms that separate the source signals from a linear mixture of signals given only the signal mixture and no prior knowledge of the mixing matrix or source signals are known. Assumptions in this methodology are

- The sources are statistically independent
- The mixing matrix has linearly independent columns

Let there be N sources producing signals $s_1(k)$, $s_2(k)$,..., $s_N(k)$ forming source signal vector $s(k)$ and M number of signals observed as $x_1(k)$, $x_2(k)$,..., $x_M(k)$ forming the observed signal vector $\mathbf{x}(k)$. The observed signal vector can be written in terms of the source signal vector as

$$\mathbf{x}(k) = \mathbf{A}s(k) \tag{6.11}$$

where $\mathbf{A}$ is the mixing matrix and it is assumed to be a square matrix ($M = N$). The goal of BSS is to estimate $s(k)$ and $\mathbf{A}$ from $\mathbf{x}(k)$ alone. The BSS problem is illustrated in Figure 6.10.

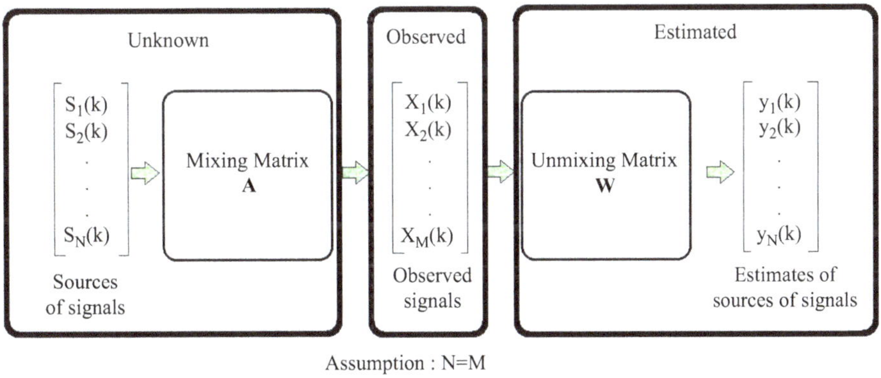

Figure 6.10 BSS problem. The number of observed signals M *is assumed to be equal to the number of sources N. The aim is to estimate the unmixing matrix and source signals given only the observations.*

The most commonly used BSS algorithm is ICA. Apart from the assumption of statistically independent sources, another assumption is that the independent components have non-Gaussian distribution [48]. This is because when independent variables are Gaussian, the mixing matrix cannot be determined as the ICA model can be estimated up to an orthogonal transformation only. But if only one independent variable is Gaussian, estimation of the model can be still carried out.

For convenience, (6.11) can be written in terms of a random variable as

$$\mathbf{x} = \mathbf{A}\mathbf{s} \tag{6.12}$$

The objective of the ICA is to estimate the mixing matrix **A**, and use its inverse to estimate the independent sources as

$$\mathbf{s} = \mathbf{W}\mathbf{x} \tag{6.13}$$

where $\mathbf{W} = \mathbf{A}^{-1}$. The multi-channel data need to be pre-processed before ICA can be carried out. The main pre-processing to be carried out are:

1. *Centering*: Centering means to make the variable zero-mean. That is, the mean vector of the observed variables is subtracted from the observed variables. This implies that estimates of source variables are zero mean. The mean vector of the source can be determined after the mixing matrix is estimated from zero-mean data as **Wm**, where **W** is the de-mixing matrix (inverse of mixing matrix) and **m** is the mean vector subtracted during pre-processing from observation variable. The estimated source mean vector can then be added to the centered estimates of sources to complete the estimation.

2. *Whitening*: This process is to make the components of the observed vector uncorrelated with unit variances. In other words, instead of estimating the larger number of parameters of the mixing matrix, a lesser number of parameters of the orthogonal mixing matrix need only be estimated. Eigen value decomposition (EVD) is often used for whitening. Let EVD of the covariance matrix of observation variable yield

$$cov(\mathbf{x}, \mathbf{x}^T) = E\{\mathbf{x}\mathbf{x}^T\} = \mathbf{Q}\Lambda\mathbf{Q}^T \tag{6.14}$$

where **Q** is the orthogonal matrix of eigenvectors of covariance matrix $E\{\mathbf{x}\mathbf{x}^T\}$ and Λ is the diagonal matrix with diagonal entries being eigenvalues. The whitening process can now be described as

$$\widetilde{x} = \mathbf{Q}\Lambda^{-1/2}\mathbf{Q}^T\mathbf{x} \tag{6.15}$$

The whitening process also transforms the mixing matrix into an orthogonal mixing matrix as follows:

$$\widetilde{\mathbf{x}} = \mathbf{Q}\Lambda^{-1/2}\mathbf{Q}^T\mathbf{A}\mathbf{s} = \widetilde{\mathbf{A}}\mathbf{s} \tag{6.16}$$

Additionally, a reduction in the dimension of data can be carried out by discarding those eigenvalues that are too small. This helps in eliminating noise as well as prevents ICA from overlearning. In the case of EEG, band-pass filtering may also be carried out to retain only those frequencies relevant to the application.

The process of centering and whitening together is known as sphering. The observation vector is now zero-mean with zero covariance having unit variances. The next step is to estimate the orthogonal transformation. Among the many algorithms for estimating orthogonal transformation, FastICA [49] is the most commonly used due to its fast and reliable convergence and its lack of adjustable parameters. It is a fixed point iterative algorithm used to maximize non-Gaussianity of $\mathbf{w}^T\mathbf{x}$, where $\mathbf{w}$ is one of the rows of the inverse of $\mathbf{A}$. $\mathbf{w}^T\mathbf{x}$ corresponds to one of the independent components:

$$y = \mathbf{w}^T\mathbf{x} = \mathbf{w}^T\mathbf{A}\mathbf{s} \tag{6.17}$$

Equation (6.17) shows the equation for retrieving a single independent component. As more components are added, the y becomes more Gaussian as per Central Limit Theorem. Hence, the aim here is to find $\mathbf{w}^T\mathbf{A}$ that leads to maximum non-Gaussian y.

There are mainly three measures of non-Gaussianity used in ICA [48]:

1. Kurtosis expressed as $E\{y^4\} - 3(E\{y^2\})^2$ is a fourth-order cumulant whose value becomes zero for the Gaussian variable, where y is a random variable that is centered with unit variance. This measure is sensitive to outliers.
2. Negentropy is defined as $J(\mathbf{y}) = H(\mathbf{y}_{gauss}) - H(\mathbf{y})$, where $\mathbf{y}$ and $\mathbf{y}_{gauss}$ are random vector and Gaussian random variable with same covariance matrix. If $\mathbf{y}$ is Gaussian, then Negentropy becomes zero, else it is positive. This computation of this measure is extremely difficult.
3. Approximation of Negentropy is defined as $J(y) \approx \sum_{i=1}^{p} k_i[E\{G_i(y)\} - E\{G_i(v)\}]^2$, where y is centered random variable with unit variance, k_i correspond to positive constants, G_i are non-quadratic functions and v is standardised Gaussian variable. This measure is computationally fast and robust and becomes zero only when y is Gaussian, else is positive always. For a single non-quadratic function, the equation becomes $J(y) \propto [E\{G(y)\} - E\{G(v)\}]^2$.

FastICA utilizes approximation of negentropy, $J(\mathbf{w}^T\mathbf{x})$, as the measure of non-Gaussianity [49]. The norm of $\mathbf{w}$ is constrained to unity. The most commonly used choices of $G(r)$ are $\frac{1}{c}\log\cosh(cr)$ and $-\exp(-r^2/2)$, where $1 \leq c \leq 2$. Let g be the derivative of G, then the FastICA algorithm to estimate one of the independent components is as follows:

1. Initialize $\mathbf{w}$ randomly
2. $\mathbf{w}^+ = E\{\mathbf{x}g(\mathbf{w}^T\mathbf{x})\} - E\{g'(\mathbf{w}^T\mathbf{x})\}\mathbf{w}$
3. $\mathbf{w} = \frac{\mathbf{w}^+}{\|\mathbf{w}^+\|}$
4. If convergence does not occur, go to step 2.

To estimate several independent components, FastICA algorithm needs to be run using multiple w vectors.

The main disadvantage of using ICA is that the order of independent components is not preserved and their variances cannot be determined.

6.4.3 Wavelet transform approach

Wavelet transform results in better time–frequency localization than Fourier transform and hence is useful in analyzing non-stationary signals such as EEG. Wavelet transform [50] decomposes a signal into a set of functions which are scaled and translated versions of a basis function, known as mother wavelet ψ. Mathematically, it is represented as

$$\Psi_{j,k}(t) = 2^{\frac{j}{2}}\psi(2^j t - k) \tag{6.18}$$

where j is the scaling factor and k is the translation factor. It can be observed that by dilating the mother wavelet with 2^j and translating the function on a grid whose interval is proportional to 2^{-j}, any function can be represented by wavelets. This is possible since high-frequency components of the signal can be represented by the compressed version of the mother wavelet while low-frequency components are represented by dilated versions of the mother wavelet. Hence, wavelet transform is a multi-resolution analysis [51] and can be implemented using a filter bank as shown in Figure 6.11. Each level consists of a high pass filter and a low pass filter that separate the input into "detail coefficients" and "approximate coefficients" corresponding to high-frequency and low-frequency components, respectively.

The assumption in artifact removal using wavelet transform is that the artifacts can be isolated when the EEG is represented in terms of wavelet basis or mother wavelet and the wavelet coefficients corresponding to the artifacts can be eliminated using a threshold. The choice of threshold is crucial and a universal threshold [52], defined as follows, is used commonly,

$$th = \sigma\sqrt{2lnN} \tag{6.19}$$

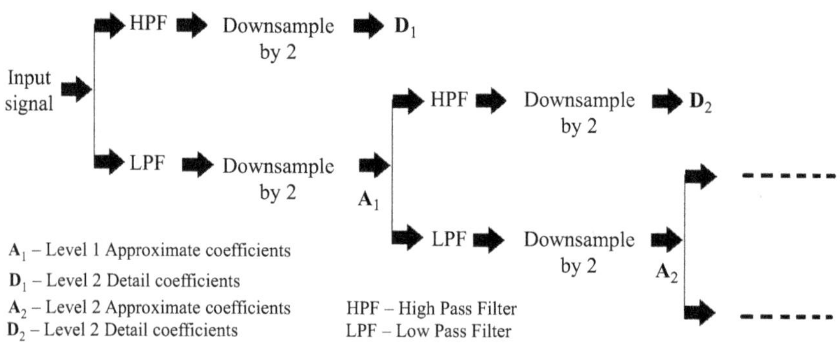

Figure 6.11 Wavelet transform implemented using feature banks. At each level, the input signal is split into high-frequency components and low-frequency components followed by down-sampling by two. The resultant high-frequency components are called detail coefficients and low-frequency components are called approximate coefficients. The approximate coefficients are given as input to the next level of decomposition.

where N and σ correspond to the number of wavelet coefficients and their standard deviation, respectively. This thresholding is carried out at each level of decomposition to eliminate artifacts.

In artifact removal of EEG of seizure patients, additional care must to taken to avoid seizure components as artifacts. An artifact removal method using wavelet transform for seizure EEG which preserves the seizure component introduced by Islam *et al.* [53] is summarized below:

- Stage 0: Seizure templates are prepared from simulated seizures or seizure databases for comparison in the later stages with the detected artifacts to ensure they are not seizure components.
- Stage 1a: The EEG signal is divided into non-overlapping segments of length N.
- Stage 1b: Pass the segments through a high pass filter (30 Hz). This step results in signals containing Gamma waves and artifacts of high frequency. This signal is considered to be free of seizures as they appear within 30 Hz. Hence, artifact removal of signals above 30 Hz can be carried out without consideration of the seizures.
- Stage 2: The signal from Stage 1a undergoes a Stationary Wavelet Transform (SWT) of 8-Level. Detail coefficients corresponding to frequencies of 0.5–30 Hz need to be analysed for seizures while other detail coefficients can be denoised using a modified threshold directly before reconstruction.
- Stage 3: The detail coefficients corresponding to 0.5–30 Hz undergo similarity measurement with templates of Stage 0. Two threshold limits are obtained using similarity measures and based on these thresholds, the epochs are confirmed to contain artifacts. The high pass filtered EEG epochs of Stage 1b along with its calculated universal threshold is also passed to this step to confirm whether the epochs are indeed artifacts. The artifact contained epochs undergo denoising utilising "non-negative garrote shrinkage function."
- Stage 4: The denoised epochs, seizure epochs, and epochs free of artifact are then used to reconstruct EEG resulting in an artifact-free EEG signal.

The major drawback of artifact removal using wavelet transform is that if the artifact spectral content is not well localised, this method would not be able to separate those artifacts from the EEG.

6.4.4 Hybrid approaches: ICA-wavelet

Artifact removal using wavelet transform cannot be used to remove those artifacts with overlapping spectra with the EEG while one of the limitations of the ICA is that the number of sources must be equal to the number of observations. Wavelet transform and ICA have been combined to form a hybrid method to overcome these shortcomings. There are numerous ways to combine wavelet and ICA. One of the hybrid methods involves using wavelet transform initially to extract various wavelet sub-bands and coefficients of each sub-band undergo artifact detection based on some threshold. The wavelet sub-bands with detected artifacts are subjected to ICA for the elimination of artifacts. The EEG signal is then reconstructed after artifact removal by inverse ICA and inverse wavelet transform. An example of a hybrid algorithm following this methodology is summarized below [54]:

1. DWT is carried out on EEG channels to extract EEG sub-bands for each channel.
2. Renyi's entropy and Kurtosis are calculated for wavelet sub-bands for each channel to detect the channel sub-bands affected by artifacts.
3. Channel sub-bands with artifacts undergo ICA to obtain independent components.
4. The independent components are divided into non-overlapping segments, and Renyi's entropy and Kurtosis are calculated for each segment of independent components.
5. The independent components that exceed a fixed threshold for Renyi's entropy and Kurtosis are selected.
6. The segments of the selected independent component that exceed the threshold for Renyi's entropy and Kurtosis are set to zero. This ensures that rather than cancelling the entire independent component, only artifact-affected segments are corrected.
7. Inverse ICA is carried out to obtain artifact-corrected independent components. This is followed by an inverse wavelet transform to reconstruct the artifact-corrected EEG signal.

Another version of the hybrid method [55,56] involves performing ICA on the raw EEG followed by wavelet transform. Here the EEG signal undergoes ICA initially to obtain the independent components and the mixing matrix. The wavelet transform is applied to this and corresponding wavelet coefficients are obtained. These wavelet coefficients undergo thresholding to remove artifacts. The reconstruction of EEG involves inverse wavelet transform followed by multiplication of the mixing matrix. One of the benefits of this method is that it reduces the rejection of other cerebral activities in EEG as artifacts, thereby preserving information pertaining to spectral and coherence characteristics of the neural activity.

Some of the hybrid methods involve a trained classifier to identify the artifacts in wavelet independent components rather than using thresholding. An example of this method using support vector machine (SVM) as the pre-trained classifier is summarised below [57]:

1. EEG signal undergoes wavelet decomposition to eliminate high-frequency noise and obtain the relevant EEG sub-bands.
2. The wavelet decomposed signal undergoes ICA to obtain independent components.
3. SVM classifier is pre-trained with independent components with artifacts (eye blinks) and independent components free of any artifactual content. The features input to SVM are Variance, Shannon's entropy, Kurtosis, and range of amplitude. The pre-trained classifier is used to evaluate the incoming independent components after ICA for artifact detection.

4. Independent components with artifacts as identified by SVM classifier undergo another DWT followed by thresholding to eliminate artifacts.
5. The artifact eliminated wavelet components are reconstructed using inverse DWT to form independent components which undergo inverse ICA to obtain artifact removed EEG.

6.4.5 Riemannian geometry

Riemannian geometry (RG) approach for EEG artifact rejection [58] is a relatively new method that came into existence along with the use of RG in analysing EEG for classification [59–61]. EEG analysis using RG utilises spatial covariance matrices of multichannel EEG. Covariance matrices are symmetric positive definite matrices and hence reside on a Riemannian manifold equipped with the inner product where Euclidean metrics cannot be applied. However, each point on the Riemannian manifold has an associated vector space known as Tangent Space where Euclidean measures are valid. Figure 6.12 shows a sample Riemannian manifold of covariance matrices.

The covariance matrix of EEG $X \in \mathbb{R}^{r \times n}$ with r channels and n samples can be calculated as

$$Covmat = \frac{1}{n-1} XX^T \tag{6.20}$$

where $Covmat \in \mathbb{R}^{r \times r}$.

Each point on the Riemannian manifold is a covariance matrix. The distance between two points, C_1 and C_2, on the manifold is calculated using Riemannian

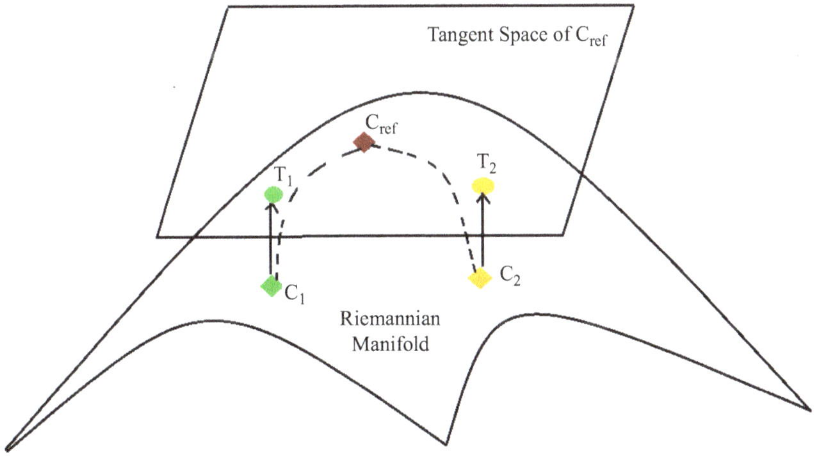

Figure 6.12 Riemannian manifold with Tangent space associated with C_{ref}. C_1 and C_2 are two points on the manifold and hence are two covariance matrices. C_{ref} is the reference point on the manifold. T_1 and T_2 are the projections of C_1 and C_2 on the Tangent space associated with C_{ref}.

distance [62] as

$$\delta(C_1, C_2) = \sqrt{\sum_{i=1}^{n} \log^2 \lambda_i} \tag{6.21}$$

where λ_i are the real eigenvalues of $C_1^{-1}C_2$.

The artifact removal of EEG using RG requires two parameters, namely, a reference point and a threshold (*th*). The reference point on the Riemannian manifold is calculated adaptively as the Riemannian mean of all EEG covariance matrices as

$$C_{ref}^{i+1} = (C_{ref}^{i})^{1/2}\left[(C_{ref}^{i})^{-1/2}C(C_{ref}^{i})^{-1/2}\right]^{1/\alpha}(C_{ref}^{i})^{1/2} \tag{6.22}$$

where C is the current covariance matrix, α is the learning parameter, and (C_{ref}^{i}) is the reference point of the previous iteration. The reference point is updated only when a signal free of artifacts is detected. The threshold *th* is calculated using the mean μ and standard deviation σ of the distances between the points on the Riemannian manifold and the reference point as

$$th = \mu + 0.25\sigma \tag{6.23}$$

The artifact rejection is carried out by calculating the distance between all points on the Riemannian manifold with a reference point and eliminating those points as artifacts whose distance exceeds the threshold value.

An extended version of this method [63] utilises spectral clustering which does not depend on the choice of threshold or reference point. The number of clusters is determined by the eigen-gap heuristic [64]. In this method, a fully connected similarity graph is constructed for covariance matrices with pairwise Riemannian distance as edge weights. Spectral clustering is carried out on the nodes (covariance matrices) of the graph and only the cluster with the largest number of elements is retained. The rest of the clusters are considered outliers and rejected.

6.4.6 Deep learning approaches

In recent years, EEG artifact removal using deep learning methods are explored. One of the works [65] utilise a stacked sparse auto-encoder as the deep learning network for ocular artifact removal from raw EEG. This method trains the network in offline mode with training samples free of ocular artifacts to enable the network to reconstruct the EEG and hence learn the representation and higher order statistical features of EEG. Once the network is trained, it can be used in online mode to eliminate ocular artifacts from EEG. To provide training samples free of ocular artifacts, the training dataset undergoes interception for the selection process. This is carried out using kurtosis. The raw EEG is divided into segments with the same time period as the occurrence of ocular artifacts (0.2–0.4 s). Kurtosis is calculated for each segment and those segments whose kurtosis value exceeds the threshold are eliminated from the offline training sample selection. The selected samples are normalised before training. During the online analysis, the input raw EEG is segmented to match the length of the segments of training samples of offline mode. The EEG segment inputs in the online mode need to be standardised. The normalised training sample from offline mode is used as a reference. After standardisation, it can be observed that the EEG are restricted to (0,1) interval while ocular artifacts are greater than one. Since the sigmoid activation is used in the network, the ocular artifacts get eliminated as the output of the neurons would be restricted to [0,1] interval.

Another method [66] of removing EOG artifacts utilises U-Net architecture. A standard 2D U-Net architecture is used with the activation function being Exponential Linear Unit. A 1×1 convolution maps the 64 feature vector to the classes. The EEG data is collected from participants with their eyes closed to avoid ocular artifacts. EOG signals are obtained from the same participants using eyes open state. EOG is used to simulate EEG with artifacts. Hence artifact-free EEG and EEG contaminated with ocular artifacts are used here. Artifact-free EEG is used to assess the performance of the network. Analysis using U-Net is carried out in three ways and in all the three methods, the U-Net is trained to result in EEG free of ocular artifacts. The three methods are:

1. Each channel is analysed using a U-Net, resulting in n number of U-Nets corresponding to n channels of EEG data. The greyscale image input is obtained by low-pass filtering channel data vector to remove high-frequency noise and then resampling followed by Min–Max normalisation and reshaping to convert the vector to greyscale image of 80×80 pixels.
2. Single U-Net model is used and the greyscale images of all channels are input to this network.
3. The greyscale image input is obtained by low-pass filtering each channel data and then forming a matrix with each column corresponding to low-pass filtered channel data. If r is the length of each channel data and there are n number of channels, then the matrix is of size $r \times n$. The resultant matrix is reshaped to 512×256 and converted to an 8-bit greyscale image. A single U-Net is used in this method and the input for a single subject is a single greyscale image containing all channel data.

Other methods explored in literature include Deep Convolutional Neural Networks (DCNN) [67], 1D Residual Convolutional Neural Networks [68], Modified U-Net CNN [69], Depthwise separable CNN [70], and Generative Adversarial Networks (GAN) [71]. There are also methods that combine traditional methods with deep learning approaches [72,73].

Though deep learning approaches are able to overcome the limitations of traditional artifact removal methods such as high computational complexity, need for reference channel, and difficulty in separating the artifacts that have an overlapping spectrum with neural information resulting in the elimination of neural information, most deep learning approaches use simulated data or data obtained in a controlled setting. Thus, the efficacy of deep learning approaches in artifact removal for long-duration real EEG, especially EEG associated with disorders, still needs to be explored.

6.5 Challenges in using commercial EEG headsets

Commercial EEG headsets with dry electrodes provide a means to capture the scalp EEG without a skilled neuro-technician. They provide the advantage of portability and easy-to-handle equipment for capturing brain signals. However, a number of challenges arise when using a commercial headset. One of the issues associated with using commercial EEG headsets is that the standard single-size of headsets does not fit comfortably for all people. This causes discomfort or electrode misalignment and the EEG obtained may be inaccurate. Another commonly found issue is the frequent electrode disconnection when the electrodes of commercial headsets are placed on the scalp of people with thick hair. While some commercial headsets offer the choice of electrodes to be used for EEG acquisition, most commercial headsets have fixed electrodes. This limits the freedom to choose the appropriate electrodes using domain knowledge based on the experiment being conducted. Additionally, the headsets are expensive as the number of electrodes provided in the headset increases. Often on the basis of experimental tasks, the experiment requires a sufficiently appropriate number of electrodes in the specified electrode locations.

Apart from these, EEG picked up by these commercial headsets would often be contaminated with artifacts such as ocular or muscle artifacts, much more than those used clinically. This is because when EEG is taken by a skilled neuro-technician in a clinical setting, in the majority of the cases, the patients would be lying down and staying still. External interferences are kept at a minimum. They might be exposed to photic stimulation to induce any abnormal EEG but overall the amount of movement is less and hence artifacts are minimised. In the case of commercial headsets, though some proprietary algorithms are used by headsets to eliminate artifacts, the possibility of artifacts being present in the captured EEG is high and hence appropriate artifact removal methods need to be employed.

Most commercial headsets currently available are used (1) to assess the attention level [74], (2) to utilise P300 waves in simple applications such as cursor control [75], and (3) to play video games like Invaders Reloaded [76]. However, they are far from being used with high performance for BCI such as driving a car using thoughts. Hence, currently, the general public has limited use of commercial EEG headsets.

6.5.1 Single channel headsets

The most commonly used single-channel commercial EEG headset is the *Neurosky Mindwave Mobile*. This headset consists of a single channel at FP1 with reference at the left ear or A1. The headset contains in-built proprietary artifact removal algorithms, however, external artifact removal algorithms would ensure that the artifacts are indeed minimised. The approach using wavelet transform and the hybrid method of wavelet and ICA, would be a reasonable choice for artifact removal for a single channel EEG headset. Adaptive filtering requires a reference channel, which is not possible in this headset. ICA and Riemannian Geometry approaches require multiple channels and hence cannot be employed for this headset. However, deep learning approaches can be employed with sufficient data.

The main drawback of such an EEG headset is that the signal captured using a single channel has limited information and hence has limited applications. The most commonly used application is to assess the attention level of users in educational settings [77]. Attempts have been made to assess various cognitive functions [18,78–80] using a single channel commercial EEG headset and while the results are promising, the information from a single channel seems to be insufficient for producing a deployable system. Hence, single-channel commercial headsets are currently used in educational and research settings only.

6.5.2 Multi-channel headsets

Numerous multi-channel commercial EEG headsets are available with a varying number of electrodes. Some of these headsets have fixed electrode placement while others provide flexibility in the choice of electrodes. Apart from this, headsets with dry and wet electrodes are available.

Muse is a low cost commonly used headset consisting of four dry electrodes and is often used to assess relaxation in the user. *Insight* and *Epoc* are two commercial EEG headsets offered by Emotiv having five and 14 fixed electrodes, respectively. Insight has dry electrodes while Epoc has wet electrodes. They are used with various BCI applications [81–83]. Neuroelectric's *Enobio* headset provides flexibility in electrodes up to 32 electrodes. However, the available option for the 32 electrode positions is fixed. This headset also has the option for the user to buy just 8, 20, or full 32 electrodes, thereby providing flexibility in price at the cost of the number of electrodes. This headset has been used for fatigue detection [84], BCI [85], stress analysis [86], and assessment of effectiveness of meditation [87]. ANT Neuro provides high-density EEG devices with *eego mylab* providing up to 256 electrodes. These high-density devices have been used in the analyses of cognitive functions and disorders like insomnia [88].

The most commonly used algorithm for artifact removal in multi-channel EEG use ICA and hybrid ICA-wavelet approaches. Adaptive filtering and Riemannian Geometry approaches are also feasible but used less commonly due to the need for the reference channel in adaptive filtering, and threshold along with a reference point in Riemannian Geometry approaches. Deep learning approaches are promising but are yet to be assessed on long-duration real EEG and disorder EEG.

6.6 Conclusion

The electrical signals generated in the brain obtained through non-invasive scalp electrodes have the advantage that complex surgical procedures are not required and the changes in electrode position as required by the task can be carried out easily. Additionally, with the availability of multiple commercial headsets, these signals can be obtained with ease for varied applications. However, these electrical signals, known as EEG signals suffer from poor spatial resolution and low SNR. EEG is also highly affected by artifacts, both physiological and non-physiological. Artifact removal or elimination is one of the major challenges in EEG acquisition.

Artifacts such as ocular, electromyogenic, cardiac, electrode movement, and power line interference are the commonly occurring artifacts in EEG. Artifacts in EEG, when a person is in locomotion, occur due to multiple sources and it is a challenging task to remove all the artifacts in gait-related analysis. Six commonly used methods of artifact removal are discussed in this chapter. These methods are summarised in Table 6.1 based on the need for a reference channel, use in single channel scenarios, and drawbacks or assumptions of each method. Table 6.2 summarises the most commonly used artifact removal method for each artifact considered in this chapter. Challenges in using commercial EEG headsets for analysis are also discussed in this chapter with examples from single-channel and multi-channel headsets.

Artifact removal in EEG is an active research area since it is still difficult to completely isolate or remove all artifacts, especially, in tasks involving the analysis of a person's gait. Deep learning approaches seem to be promising but real-time analyses of long-duration EEG or EEG of patients with disorders are yet to be assessed. The use of commercially available EEG headsets has the advantage of

Table 6.1 Summary of artifact removal methods for EEG

Method	Reference channel	Single/multi channel EEG	Comments
Adaptive filtering	Required	Both	Requires reference channel of the artifact being removed such as EOG channel
ICA	Not required	Multi-channel	Main assumptions: (1) Statistical independence of sources and non-gaussianity. (2) No. of sources = No. of observation
Wavelet	Not required	Both	Requires the artifact spectral content to be well localised, hence not effective if artifacts and EEG have overlapping spectrum. Requires appropriate threshold
Hybrid wavelet-ICA	Not required	If ICA carried out first then multi-channel only else both	Involves combinations of multiple methods. High computational complexity
Riemannian Geometry	Not required	Multi-channel	Requires finding reference point on manifold and appropriate threshold
Deep learning	Usually not required	Both	Assessed in simulated or semi-simulated EEG with artifacts. Need further exploration on long-term real EEG and disorder EEG

Table 6.2 Artifacts and commonly used artifact removal methods

Artifacts	Artifact removal methods
Ocular	Adaptive filtering, ICA, Wavelet, Hybrid method, RG, Deep learning approaches
Cardiac	Adaptive filtering, ICA, Deep learning approaches
Electromyogenic	Adaptive filtering, ICA, Wavelet, Hybrid method, RG, Deep learning approaches
Electrode movement	ICA
Power line Interference	Notch filters, Adaptive filtering, Hybrid method

portability and ease of use, however, multiple challenges as discussed in this chapter hinder the use of such headsets for high-performance tasks such as bionic limb movements where currently invasive electrodes are used. The use of deep learning approaches for real-time artifact removal with high-density commercial EEG headsets such as those from ANT Neuro needs to be explored to understand the extent of artifact removal that can be achieved currently with available resources.

References

[1] Kandel ER, Schwartz JH, Jessell TM, *et al. Principles of Neural Science*, vol. 4. McGraw-Hill, New York, NY; 2000.

[2] Noback CR, Ruggiero DA, Strominger NL, *et al. The Human Nervous System: Structure and Function*, vol. 744. Springer Science & Business Media; 2005.

[3] Somers MF. *Spinal Cord Injury: Functional Rehabilitation.* Prentice Hall; 2001.

[4] Kipke DR, Vetter RJ, Williams JC, *et al.* Silicon-substrate intracortical microelectrode arrays for long-term recording of neuronal spike activity in cerebral cortex. *IEEE Transactions on Neural Systems and Rehabilitation Engineering.* 2003;11(2):151–155.

[5] Buzsáki G, Stark E, Berényi A, *et al.* Tools for probing local circuits: High-density silicon probes combined with optogenetics. *Neuron.* 2015;86(1):92–105.

[6] Milekovic T, Sarma AA, Bacher D, *et al.* Stable long-term BCI-enabled communication in ALS and locked-in syndrome using LFP signals. *Journal of Neurophysiology.* 2018;120(7):343–360.

[7] Lewis PM, Ackland HM, Lowery AJ, *et al.* Restoration of vision in blind individuals using bionic devices: A review with a focus on cortical visual prostheses. *Brain Research.* 2015;1595:51–73.

[8] Newton HB. *Handbook of Neuro-oncology Neuroimaging.* Academic Press; 2016.

[9] Zhang Y. Invasive BCI and noninvasive BCI with VR/AR technology. In: *International Conference on Artificial Intelligence, Virtual Reality, and Visualization (AIVRV 2021)*, vol. 12153. SPIE; 2021. p. 186–192.

[10] Olejniczak P. Neurophysiologic basis of EEG. *Journal of Clinical Neurophysiology*. 2006;23(3):186–189.

[11] Mahmood D, Nisar H, and Voon YV. Removal of physiological artifacts from electroencephalogram signals: A review and case study. In: *2021 IEEE 9th Conference on Systems, Process and Control (ICSPC 2021)*. IEEE; 2021. p. 141–146.

[12] Mahmood D, Leong HJ, and Nisar H. Effectiveness of online ocular artifact removal from electroencephalogram signal during neurofeedback training. In: *2022 International Conference on Emerging Trends in Smart Technologies (ICETST)*. IEEE; 2022. p. 1–6.

[13] Taran S and Bajaj V. Sleep apnea detection using artificial bee colony optimize hermite basis functions for EEG signals. *IEEE Transactions on Instrumentation and Measurement*. 2019;69(2):608–616.

[14] Gopan KG, Prabhu SS, and Sinha N. Sleep EEG analysis utilizing inter-channel covariance matrices. *Biocybernetics and Biomedical Engineering*. 2020;40(1):527–545.

[15] Lin YP, Wang CH, Jung TP, *et al.* EEG-based emotion recognition in music listening. *IEEE Transactions on Biomedical Engineering*. 2010;57(7):1798–1806.

[16] Bousseta R, El Ouakouak I, Gharbi M, *et al.* EEG based brain computer interface for controlling a robot arm movement through thought. *Irbm*. 2018;39(2):129–135.

[17] Tanaka K, Matsunaga K, and Wang HO. Electroencephalogram-based control of an electric wheelchair. *IEEE Transactions on Robotics*. 2005;21 (4):762–766.

[18] Gopan KG, Reddy SA, Rao M, *et al.* Analysis of single channel electroencephalographic signals for visual creativity: A pilot study. *Biomedical Signal Processing and Control*. 2022;75:103542.

[19] Durongbhan P, Zhao Y, Chen L, *et al.* A dementia classification framework using frequency and time-frequency features based on EEG signals. *IEEE Transactions on Neural Systems and Rehabilitation Engineering*. 2019;27 (5):826–835.

[20] Gopan KG, Harsha A, Joseph LA, *et al.* Adaptive neuro-fuzzy classifier for 'Petit Mal' epilepsy detection using Mean Teager Energy. In: *2013 International Conference on Advances in Computing, Communications and Informatics (ICACCI)*. IEEE; 2013. p. 752–757.

[21] Shah D, Gopan KG, and Sinha N. Analysis of EEG for Parkinson's disease detection. In: *2022 IEEE International Conference on Signal Processing and Communications (SPCOM)*. IEEE; 2022. p. 1–5.

[22] Newson JJ and Thiagarajan TC. EEG frequency bands in psychiatric disorders: A review of resting state studies. *Frontiers in Human Neuroscience*. 2019;12:521.

[23] Siuly S, Khare SK, Bajaj V, *et al.* A computerized method for automatic detection of schizophrenia using EEG signals. *IEEE Transactions on Neural Systems and Rehabilitation Engineering.* 2020;28(11):2390–2400.

[24] Sharma LD, Bohat VK, Habib M, *et al.* Evolutionary inspired approach for mental stress detection using EEG signal. *Expert Systems with Applications.* 2022;197:116634.

[25] Frid A, Shor M, Shifrin A, *et al.* A biomarker for discriminating between migraine with and without aura: Machine learning on functional connectivity on resting-state EEGs. *Annals of Biomedical Engineering.* 2020;48(1):403–412.

[26] McKay JH and Tatum WO. Artifact mimicking ictal epileptiform activity in EEG. *Journal of Clinical Neurophysiology.* 2019;36(4):275–288.

[27] McKay JH and Tatum WO. Artifact mimicking ictal epileptiform activity in EEG. *Journal of Clinical Neurophysiology.* 2019;36(4):275–288.

[28] Kaya I. A brief summary of EEG artifact handling. In *Brain-Computer Interface*. London: IntechOpen, 2019.

[29] Tatum IV WO. *Handbook of EEG Interpretation*. New York, NY: Springer Publishing Company, 2021.

[30] Ranjan R, Sahana BC, and Bhandari AK. Ocular artifact elimination from electroencephalography signals: A systematic review. *Biocybernetics and Biomedical Engineering.* 2021;41(3):960–996.

[31] Croft RJ and Barry RJ. Removal of ocular artifact from the EEG: A review. *Neurophysiologie Clinique/Clinical Neurophysiology.* 2000;30(1):5–19.

[32] Tamburro G, Stone DB, and Comani S. Automatic removal of cardiac interference (ARCI): A new approach for EEG data. *Frontiers in Neuroscience.* 2019;13:441.

[33] Jiang JA, Chao CF, Chiu MJ, *et al.* An automatic analysis method for detecting and eliminating ECG artifacts in EEG. *Computers in Biology and Medicine.* 2007;37(11):1660–1671.

[34] Hamaneh MB, Chitravas N, Kaiboriboon K, *et al.* Automated removal of EKG artifact from EEG data using independent component analysis and continuous wavelet transformation. *IEEE Transactions on Biomedical Engineering.* 2013;61(6):1634–1641.

[35] Hu J, Wang Cs, Wu M, *et al.* Removal of EOG and EMG artifacts from EEG using combination of functional link neural network and adaptive neural fuzzy inference system. *Neurocomputing.* 2015;151:278–287.

[36] Li Y, Wang PT, Vaidya MP, *et al.* Electromyogram (EMG) removal by adding sources of EMG (ERASE): A novel ICA-based algorithm for removing myoelectric artifacts from EEG. *Frontiers in Neuroscience.* 2021;14:597941.

[37] Chen X, Xu X, Liu A, *et al.* Removal of muscle artifacts from the EEG: A review and recommendations. *IEEE Sensors Journal.* 2019;19(14):5353–5368.

[38] Iriarte J, Urrestarazu E, Valencia M, *et al.* Independent component analysis as a tool to eliminate artifacts in EEG: a quantitative study. *Journal of Clinical Neurophysiology.* 2003;20(4):249–257.

[39] Saab M and Gotman J. A system to detect the onset of epileptic seizures in scalp EEG. *Clinical Neurophysiology*. 2005;116(2):427–442.

[40] La Rosa AB, Pereira PT, Ücker P, *et al.* Exploring NLMS-based adaptive filter hardware architectures for eliminating power line interference in EEG signals. *Circuits, Systems, and Signal Processing*. 2021;40(7):3305–3337.

[41] Kaushal G, Jain V, and Singh A. Removal of power line interference from EEG using wavelet-ICA. *International Journal of Computer Applications*. 2015;975:8887.

[42] Jacobsen NSJ, Blum S, Witt K, *et al.* A walk in the park? Characterizing gait-related artifacts in mobile EEG recordings. *European Journal of Neuroscience*. 2021;54(12):8421–8440.

[43] Snyder KL, Kline JE, Huang HJ, *et al.* Independent component analysis of gait-related movement artifact recorded using EEG electrodes during treadmill walking. *Frontiers in Human Neuroscience*. 2015;9:639.

[44] Gwin JT, Gramann K, Makeig S, *et al.* Removal of movement artifact from high-density EEG recorded during walking and running. *Journal of Neurophysiology*. 2010;103(6):3526–3534.

[45] Urigüen JA and Garcia-Zapirain B. EEG artifact removal—state-of-the-art and guidelines. *Journal of Neural Engineering*. 2015;12(3):031001.

[46] Correa AG, Laciar E, Patiño H, *et al.* Artifact removal from EEG signals using adaptive filters in cascade. *Journal of Physics: Conference Series*. 2015; 90:012081.

[47] La Rosa AB, Pereira PT, Ücker P, *et al.* Exploring NLMS-based adaptive filter hardware architectures for eliminating power line interference in EEG signals. *Circuits, Systems, and Signal Processing*. 2021;40(7):3305–3337.

[48] Hyvärinen A and Oja E. Independent component analysis: Algorithms and applications. *Neural Networks*. 2000;13(4–5):411–430.

[49] Hyvärinen A. Fast ICA by a fixed-point algorithm that maximizes non-Gaussianity. *Independent Component Analysis: Principles and Practice*. 2001.

[50] Daubechies I. *Ten Lectures on Wavelets*. SIAM; 1992.

[51] Jawerth B and Sweldens W. An overview of wavelet based multiresolution analyses. *SIAM Review*. 1994;36(3):377–412.

[52] Donoho DL and Johnstone IM. Minimax estimation via wavelet shrinkage. *The Annals of Statistics*. 1998;26(3):879–921.

[53] Islam MK, Rastegarnia A, and Yang Z. A wavelet-based artifact reduction from scalp EEG for epileptic seizure detection. *IEEE Journal of Biomedical and Health Informatics*. 2015;20(5):1321–1332.

[54] Mammone N and Morabito FC. Enhanced automatic wavelet independent component analysis for electroencephalographic artifact removal. *Entropy*. 2014;16(12):6553–6572.

[55] Castellanos NP and Makarov VA. Recovering EEG brain signals: Artifact suppression with wavelet enhanced independent component analysis. *Journal of Neuroscience Methods*. 2006;158(2):300–312.

[56] Mahajan R and Morshed BI. Unsupervised eye blink artifact denoising of EEG data with modified multiscale sample entropy, kurtosis, and wavelet-ICA. *IEEE Journal of Biomedical and Health Informatics.* 2014;19(1):158–165.

[57] Sai CY, Mokhtar N, Arof H, *et al.* Automated classification and removal of EEG artifacts with SVM and wavelet-ICA. *IEEE Journal of Biomedical and Health Informatics.* 2017;22(3):664–670.

[58] Barachant A, Andreev A, and Congedo M. The Riemannian Potato: An automatic and adaptive artifact detection method for online experiments using Riemannian geometry. In: *TOBI Workshop IV*; 2013. p. 19–20.

[59] Navarro-Sune X, Hudson A, Fallani FDV, *et al.* Riemannian geometry applied to detection of respiratory states from EEG signals: The basis for a brain–ventilator interface. *IEEE Transactions on Biomedical Engineering.* 2016;64(5):1138–1148.

[60] Gaur P, Pachori RB, Wang H, *et al.* A multi-class EEG-based BCI classification using multivariate empirical mode decomposition based filtering and Riemannian geometry. *Expert Systems with Applications.* 2018;95:201–211.

[61] Gopan KG, Sinha N, Jayagopi DB, *et al.* Alcoholic EEG analysis using Riemann geometry based framework. In: *2019 27th European Signal Processing Conference (EUSIPCO).* IEEE; 2019. p. 1–5.

[62] Lee JM. *Introduction to Riemannian Manifolds,* vol. 176. New York, NY: Springer; 2018.

[63] Yamamoto MS, Sadatnejad K, Tanaka T, *et al.* Detecting EEG outliers for BCI on the Riemannian manifold using spectral clustering. In: *2020 42nd Annual International Conference of the IEEE Engineering in Medicine & Biology Society (EMBC).* IEEE; 2020. p. 438–441.

[64] Von Luxburg U. A tutorial on spectral clustering. *Statistics and Computing.* 2007;17(4):395–416.

[65] Yang B, Duan K, Fan C, *et al.* Automatic ocular artifacts removal in EEG using deep learning. *Biomedical Signal Processing and Control.* 2018;43:148–158.

[66] Mashhadi N, Khuzani AZ, Heidari M, *et al.* Deep learning denoising for EOG artifacts removal from EEG signals. In: *2020 IEEE Global Humanitarian Technology Conference (GHTC).* IEEE; 2020. p. 1–6.

[67] Zhang H, Wei C, Zhao M, *et al.* A novel convolutional neural network model to remove muscle artifacts from EEG. In: *ICASSP 2021-2021 IEEE International Conference on Acoustics, Speech and Signal Processing (ICASSP).* IEEE; 2021. p. 1265–1269.

[68] Sun W, Su Y, Wu X, *et al.* A novel end-to-end 1D-ResCNN model to remove artifact from EEG signals. *Neurocomputing.* 2020;404:108–121.

[69] Onners J, Alam M, Cichy B, *et al.* U-EEG: A deep learning autoencoder for the detection of ocular artifact in EEG signal. In: *2021 IEEE Signal Processing in Medicine and Biology Symposium (SPMB).* IEEE; 2021. p. 1–5.

[70] Khatwani M, Rashid HA, Paneliya H, *et al.* A flexible multichannel eeg artifact identification processor using depthwise-separable convolutional

neural networks. *ACM Journal on Emerging Technologies in Computing Systems (JETC)*. 2021;17(2):1–21.
[71] Sawangjai P, Trakulruangroj M, Boonnag C, *et al.* EEGANet: Removal of ocular artifact from the EEG signal using generative adversarial networks. *IEEE Journal of Biomedical and Health Informatics*. 2021;2022:4875399.
[72] Lee SS, Lee K, and Kang G. EEG artifact removal by Bayesian deep learning & ICA. In: *2020 42nd Annual International Conference of the IEEE Engineering in Medicine & Biology Society (EMBC)*. IEEE; 2020. p. 932–935.
[73] Lopes F, Leal A, Medeiros J, *et al.* Ensemble deep neural network for automatic classification of EEG independent components. *IEEE Transactions on Neural Systems and Rehabilitation Engineering*. 2022;30:559–568.
[74] Peng S, Chen L, Gao C, *et al.* Predicting students' attention level with interpretable facial and head dynamic features in an online tutoring system (Student Abstract). In: *Proceedings of the AAAI Conference on Artificial Intelligence*, vol. 34; 2020. p. 13895–13896.
[75] Sun Y, Sheng W, Khemri NA, *et al.* Evaluation of P300-based BCI using a non-invasive commercial EEG sensor. In: *2017 IEEE 7th Annual International Conference on CYBER Technology in Automation, Control, and Intelligent Systems (CYBER)*. IEEE; 2017. p. 115–120.
[76] Tunstel E, Cobo MJ, Herrera-Viedma E, *et al.* Systems science and engineering research in the context of systems, man, and cybernetics: Recollection, trends, and future directions. *IEEE Transactions on Systems, Man, and Cybernetics: Systems*. 2020;51(1):5–21.
[77] Sezer A, Inel Y, Seçkin A, *et al.* An investigation of university students' attention levels in real classroom settings with NeuroSky's MindWave mobile (EEG) device. In: *International Educational Technology Conference*. Istanbul, Turkey. vol. 27; 2015. p. 88–101.
[78] Permana K, Wijaya S, and Prajitno P. Controlled wheelchair based on brain computer interface using Neurosky Mindwave Mobile 2. In: *AIP Conference Proceedings*, vol. 2168. AIP Publishing LLC; 2019. p. 020022.
[79] Rusanu OA, Cristea L, Luculescu MC, *et al.* Experimental model of a robotic hand controlled by using NeuroSky Mindwave Mobile headset. In: *2019 E-Health and Bioengineering Conference (EHB)*. IEEE; 2019. p. 1–4.
[80] Gopan KG, Reddy S, Krishnan K, *et al.* EEG analysis of mathematical cognitive function and startle response using single channel electrode. *CSI Transactions on ICT*. 2020;8(4):367–376.
[81] Prince D, Edmonds M, Sutter A, *et al.* Brain machine interface using Emotiv EPOC to control robai cyton robotic arm. In: *2015 National Aerospace and Electronics Conference (NAECON)*. IEEE; 2015. p. 263–266.
[82] Liu Y, Jiang X, Cao T, *et al.* Implementation of SSVEP based BCI with Emotiv EPOC. In: *2012 IEEE International Conference on Virtual Environments Human-Computer Interfaces and Measurement Systems (VECIMS) Proceedings*. IEEE; 2012. p. 34–37.

[83] Ahad R, Rahman K, Fuad N, *et al.* Body motion control via brain signal response. In: *2018 IEEE-EMBS Conference on Biomedical Engineering and Sciences (IECBES)*. IEEE; 2018. p. 696–700.

[84] Jain A, Abbas B, Farooq O, *et al.* Fatigue detection and estimation using auto-regression analysis in EEG. In: *2016 International Conference on Advances in Computing, Communications and Informatics (ICACCI)*. IEEE; 2016. p. 1092–1095.

[85] Vourvopoulos A, Niforatos E, and Giannakos M. EEGlass: An EEG-eyeware prototype for ubiquitous brain-computer interaction. In: *Adjunct Proceedings of the 2019 ACM International Joint Conference on Pervasive and Ubiquitous Computing and Proceedings of the 2019 ACM International Symposium on Wearable Computers*; 2019. p. 647–652.

[86] Rajendran V, Jayalalitha S, Adalarasu K, *et al.* Analysis and classification of stress among students using EEG as biomarker. *ECS Transactions.* 2022;107 (1):1857.

[87] Manjunatha Siddappa DK. A cognitive approach towards measuring effectiveness of meditation using Enobio-8 EEG device. *European Journal of Molecular & Clinical Medicine*;7(08):2020.

[88] Ling J, Lin X, Li X, *et al.* Altered brain activity related to inhibitory processing in youth with insomnia. *Journal of Sleep Research.* 2021;30(6): e13398.

Chapter 7

Cognitive task and workload classification using EEG signal

Debanjali Bhattacharya[1] *and Neelam Sinha*[1]

Abstract

Cognitive workload is defined as the load generated within our brain as a result of cognitive and psychological effort which is given to accomplish the desired task. Analysis of mental workload and cognitive stress monitoring is extremely crucial as it directly impacts humans' performance in executing the tasks as well as in mental health. Recent studies in neuroscience have differentiated various cognitive states by analyzing local changes in neuronal activity in human brain. Given cost-effectiveness and high sensitivity to detect brain activity fluctuations for varieties of cognitive tasks, electroencephalography (EEG) analysis has been widely studied by researchers in order to understand the information about brain functionality that EEG signal contains. Here we present a comprehensive overview on how EEG can be utilized to identify different cognitive states for assessing mental workload. A systematic review of the literature was performed on EEG signal for different mental stress tasks, where the subjects undergo with varying degrees of task complexity. In the present study, the results are analyzed based on the type of tasks, different EEG preprocessing methods and classification models. The study summarizes the recent practices and performance outcomes for mental workload classification using EEG signal.

Keywords: EEG; Mental workload; Cognitive tasks; Machine learning; Deep learning

7.1 Introduction

Recent advances in AI and modern neuroimaging techniques have substantially increased the number of research in studying the functioning of human brain for a variety of cognitive tasks such as emotion recognition, motor imagery tasks, seizure detection, quantifying drowsiness and mental fatigue, attention, working memory,

[1]International Institute of Information Technology – Bangalore, India

effect of mental stress, sleep analysis, and meditation. In this chapter, we will focus on mental workload (MW) classification task using electroencephalogram (EEG) signals. Quantification of workload can be performed by assessing cognitive and psychological effort, required in order to complete the desired tasks. Too high workload may increase mental stress and reduces the work efficiency; on the other hand too small workload may disconnect the person from the given task [1]. Thus, continuous evaluation and assessment of MW is pivotal for keeping workload in an optimal level so that human performance with which tasks need to be executed can improve. There are three methods by which MW can be assessed [2,3]:

1. *Subjective workload* – Here, users report their perceived levels of MW that is assessed through questionnaire responses at various levels. So subjective assessment is primarily based on the rankings or scales for quantifying the amount of MW a person is feeling. There exist varieties of subjective workload measurement techniques that can allow individuals to evaluate or rate their present workload. These measurements are usually administered in prior, during, and after successful completion of the given task. Some examples of subjective workload include:
   (i) Self-assessment: In this type of assessment, an 'n'-point scale is used to label MW. It was first designed for air traffic controllers where workload classes are scaled in the range of 1 (underutilized workload) to 5 (excessive workload).
   (ii) Subjective workload assessment technique (SWAT): It is primarily designed in order to measure and rank the MW of pilot in relation to three factors: time, psychological effort, and mental stress. Participants are required to rate these three factors on a scale that ranges from 1 to 3, based on which an overall workload score is computed.
   (iii) National Aeronautics and Space Administration-Task Load Index (NASA-TLX): This is an extensively used assessment method for quantifying MW [4], measured during the execution of a given task by the participants. It is a multidimensional rating tool by which overall workload is computed. Here, the computation is based on a weighted mean of six different types of workload sub-scale ratings which are effort, performance, frustration, temporal demand, physical, and mental aspects of a task.
2. *Behavioral workload* – It relies on task performance measures like accuracy, response times, error rate by which the capacity of an individual is examined to characterize the mental states. The MW estimated in this type is based on how well a person accomplishes the given task with increasing workload.
3. *Objective workload* – Although subjective and behavioral workload measures are used in many literature, however using these methods it is difficult to provide the real-time measurements of MW. This gap can be filled by instrumental or objective method which is capable to provide real-time correlates of MW by exploring physiological and neuronal activity measures which can obtained through functional MRI (fMRI), functional near-infrared spectroscopy (fNIRS), electroencephalography (EEG), magnetoencephalogram (MEG), electrocardiography (ECG), and galvanic skin response. Since the

signals obtained from EEG, ECG, fMRI can directly measure the physical reactions of the body to the amount of mental work, a person is experiencing and does not involve any direct response from the person (unlike subjective measures), making this objective method the best way to quantify MW.

There are five main areas in human body from where signals are decoded to measure MW. These are speech signals, brain activity, eye activity, cardiac activity, and respiratory activity. Among all these, nowadays, the majority of the active research has focused on analyzing the brain activity; how brain functions during the execution of a specific cognitive task. Being simple, compatible, having high temporal resolution and improved dry electrode technologies, EEGs have gained popularity over other available non-invasive devices. Recent studies have considered EEG as an appropriate tool that can be used to capture the dynamically changing complex brain wave patterns in any cognitive tasks. In the current study, we have considered the EEG signals only that measure changes in brain activity for MW classification task.

In this chapter, we present the systematic review of the literature on cognitive task and workload classification using EEG. In this regard, the study attempts to address some critical questions. The three main research questions that are addressed in this chapter are:

RQ1. Which kind of cognitive tasks are executed for MW classification using EEG? Are there any specific cognitive tasks, mostly used for MW classification using EEG?
RQ2. What types of EEG features and classifiers are used for MW classification? With advances of deep learning (DL) in past few years, there exist many studies that have preferred DL models for classification instead of using traditional ML classifiers. The third research question, addressed in this study is:
RQ3. Is there any particular deep neural network (DNN) architecture, appropriate for MW classification using EEG?

To address these RQs, all peer-reviewed published articles on cognitive tasks and MW classification strategies using EEG are compiled. Different cognitive tasks, EEG preprocessing, EEG feature extraction and selection methods, ML classification model, DNN architecture, and performance measures are reviewed thoroughly. The overall trend analysis reveals that MW classification tasks using EEG can be performed more effectively with specific ML and DL network architecture selection. This information is collected and presented in Table 7.1, which can help in studying the future applications of MW classification using EEG signals.

7.1.1 Neural basis of EEG

Firing of neurons within the human brain produced electrical signals which are measured by EEG. Cortical neurons are electrically excitable cells where the transmission and processing of information are realized by electrochemical signaling through synapses [21]. A typical neuron consists of an axon, cell body, and

Table 7.1 Summary of studies that used ML classifiers in classifying MW using EEG

Study	Ch.	Subj.	Freq. range (Hz)	Feature	Classification	LC[a]	Acc. (%)
[2]	8	48	1–45 M:F 25:23 27 ± 6.6 yrs	PSD, PMSC, AM, PMSC of AM	RF	2	97
[5]	11	17	4–48	PSD	Gaussian process regression	2	91
[6]	6	8	1–40	PSD	Canonical correlation	2	–
[7]	14	10 M:F 10:0 20–30 yrs	–	RMS	NN, sub-band energy PSD, EI	3	90
[8]	32	13 M:F 11:2 22–25 yrs	1–30	Energy	LDA, BPNN, KNN, SVM	3	79.8 (avg)
[9]	14	8 M:F 8:0 21–24 yrs	–	First-order stats Hjorth para. FFT, wavelet	NB, DT SVM, KNN XGBoost	3	88.01 (avg)
[10]	40	8 M:F 8:0 24.8 ± 0.8 yrs	0.5–30	Total power of FFT	BPNN SVM, KNN RF	3	85.7
[11]	14	8 M:F 6:2 19–30 yrs	1.5–40	PSD RMS	SVM	2	98.5[b], 95.5[c]
[12]	12	12 M:F 5:7 23.5 ± 3 yrs	2–25	SV	LR, GMM, LDA	3	74 (avg)
[13]	29	31 M:F 9:12 25.9 ± 4.9 yrs	1–30	PSD	LDA	3	95
[14]	19	20 M:F 20:0 22.8 yrs (avg)	3–35	PSD	SVM, DT, KNN	2	85.4
[15]	19	11 M:F 11:0 22 yrs (avg)	–	PSD	SVM, DT, RF, KNN	3	91
[16]	14	11 M:F 6:5	1–4	HOC	KNN	7	89.6
[17]	64	47 M:F 23:24 24.5 ± 1.7 yrs	1–50	PSD, neural complexity, connectivity measures	LR, RF XGBoost ANN	6	r^2 stats
[18]	8	20	–	α, θ power eyes closed, pupil diameter, correct-ness	SVM, RF, KNN	3	0.33 (F1-score)
[19]	14	12 M:F 12:0	2–42	PSD, FD first-order, stats	SVM, KNN	4	93.2
[20]	32	13 M:F 11:2 22–25 yrs	1–30	IC of EEG, PSD	Cross-session subspace alignment with SVM	2	84.7

[a]LC – level of complexity.
[b]98.5% accuracy – n-back task.
[c]95.5% accuracy – arithmetic task.

several dendrites. Axon sends the signals from one neuron to the dendrite of another neuron through cell body [22]. Two kinds of electrical activities are related to cortical neurons from which EEG signals are generated. These are (i) action potentials and (ii) postsynaptic potentials. Action potentials are voltage spikes, having limited potential field and induce very small current in the axon (typically, less than 10 ms) [23]. Postsynaptic potentials are produced by the binding of neurotransmitters of postsynaptic neuron to their membrane receptors. This voltage is generated with a significantly longer duration (typically, 45–200 ms) and having larger potential field as compared to action potential. Due to this reason, postsynaptic potentials are generally considered as the primary source of EEG signal [23]. The electrical activity of cortical neuron is captured in EEG and is recorded at the brain scalp. However, there are several conditions that need to be satisfied while recording EEG. One such condition is, there should be enough number of cortical neurons which must be located closer to the brain scalp. Moreover, with a certain geometric configuration, these neurons need to be activated in synchronized manner [24]. The primary composition of cerebral cortex is pyramidal neurons. It is seen that these pyramidal neurons are located very near to scalp and are extremely polarized with the dominant orientation perpendicular to the brain surface [24]. Hence, it is the cortical pyramidal neurons that provide EEG signal [24].

7.1.2 EEG measurement

The key element of EEG measurement system is electrodes which are categorized into three types: (i) several active electrodes, (ii) one reference electrode, and (ii) one ground electrode, based on their diverse roles in the recording. Reference electrode ideally needs to be set at zero or near-zero potential at remote location. Thus, the differences in potential between active and reference electrode mirror the electrical activity near active electrode. The general purpose of ground electrode is to reduce noise, generated in electrical circuit.

In order to obtain low contact impedance with good electrical conductivity, the gap between the scalp and electrodes is filled by the conductive medium, generally, gel or saline. It is seen that gel is better conductive medium as compared to the saline (less stable conductive connection), as it can lessen artifacts generated from the skin surface and from the movements of subjects [24]. However, these conventional wet electrodes possess some limitations in EEG data acquisition: (i) time taken for applying gel on scalp is too long and also it becomes harder to clean the gel residuals, (ii) skin exfoliation is essential to limit scalp-electrode junction impedance, (iii) application of wet electrodes may cause allergic infection, and (iv) over-consumption of gel has to be monitored cautiously in order to prevent short-circuit between electrodes which are close to each other, so that it cannot affect signal transduction. These problems can be solved by utilizing dry electrodes that consist of a single metal (usually stainless steel) and acts as a conducting medium between the electrode and the skin. Satisfactory performance is seen to be achieved by using these dry electrodes in EEG recording.

For EEG studies, signals are recorded from multiple active electrodes, located at different positions on scalp. Thus the total number of electrodes needs to be decided in routine EEG recordings. The use of high-density multichannel configuration of arrays of active electrodes can improve the spatial resolution. However, data acquisition and analysis using multiple electrodes is time-consuming and expensive. Moreover, it is seen that the increased number of electrodes may degrade the quality of EEG signal [24]. In the year 1958, International Federation in EEG and clinical neurophysiology [25] formulated a standard protocol for placing EEG electrodes, which is named as 'International 10–20 system' and was revised repeatedly by the American Electroencephalographic Society later [26,27]. According to this system, a total of 75 electrodes are placed at specific standard locations on the scalp. The 10–20 system electrode placement is shown in Figure 7.1. Since electrodes are located at 10% and 20% points along latitude and longitude, it is called 10–20 system [24].

The spontaneous electrical activity obtained from EEG shows certain characteristic waveforms over a wide frequency ranges. This is classified into five distinct frequency bands: (i) delta ($\delta < 4$ Hz), (ii) theta (4 Hz $< \theta <$ 8 Hz), (iii) alpha (8 Hz $< \alpha <$ 13 Hz), (iv) beta (13 Hz $< \beta <$ 30 Hz), and (v) gamma ($\gamma >$ 30 Hz) [24]. The electrical signal pattern at each of these frequency bands shows specific distribution, having specific biological significance. For example, it is seen that α wave which is typically measured at the occipital lobe in an awake person with the eyes

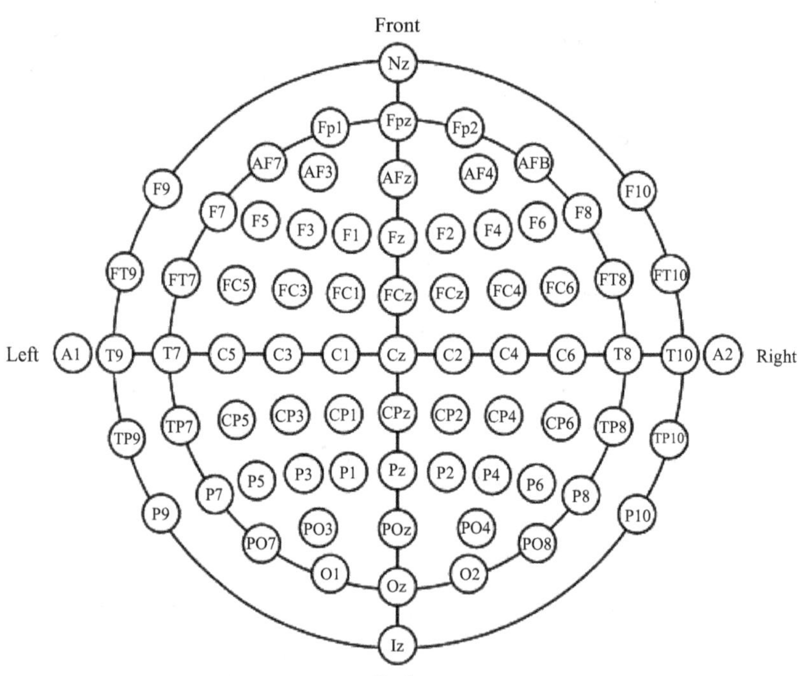

Figure 7.1 EEG electrode placements according to International 10–20 system [24,27]

closed, attenuates with eyes opening or other mental activity [24]. When a task is given to a subject, the task-related specific cognitive event disturbs the spontaneous EEG signals. With the help of advanced signal processing techniques, different features or neural responses are extracted from EEG signal for the analysis. The mean responses are termed as event-related potentials (ERPs), related to the particular events.

7.2 Study selection

Literatures for cognitive task and workload classification using EEG signal are identified by following a methodical review procedure. Literature search was conducted on June 10, 2022 mainly within PubMed databases using the following keywords: 'EEG' AND 'mental workload', 'cognitive workload', 'cognitive functions', 'cognitive task classification', 'mental workload classification', 'mental workload assessment', 'mental states' AND 'Deep Neural Networks', OR 'Deep Learning', AND 'Machine Learning'. The articles that did not satisfy the inclusion criteria (as described in Section 7.2.1) were excluded. Thereafter, available full texts of the remaining articles were studied thoroughly. The temporal distribution of studies which are reviewed exhaustively is plotted in Figure 7.2.

7.2.1 Inclusion and exclusion criteria

The current study includes only published full-text articles with the following criteria and excludes other literature that did not meet the inclusion criteria.

- *Healthy subjects:* Articles related to neurological disorders, clinical research, subjects undergoing some medical treatments were kept outside of the scope of this chapter.

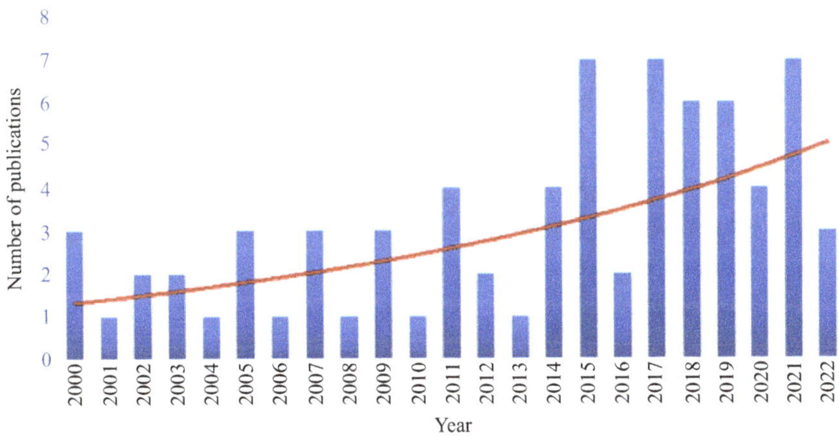

Figure 7.2 Temporal distribution of research publications in classifying MW using EEG with an increased number of publications from 2017 to 2021. The red line shows the trend of research publication on MW using EEG from the year 2000 to 2022.

- *Electroencephalography:* The literature which are reviewed for this chapter included only EEG-based studies on healthy participants. Research that combine EEG with other medical image modalities, for example, fMRI, fNIRS, MEG, etc. are excluded from analysis. Similarly articles that include MW classification using physiological features such as respiration measures, ECG measures, ocular measures, speech measures, neuroendocrine measure, are also excluded.
- *MW classification task only:* The present study focused exclusively on MW classification using EEG. Research on non-human studies, power analysis, assessment of cognitive tasks other than MW and the studies that did not consider any cognitive tasks are excluded in the current review. Moreover any studies that include cross-task cognitive workload recognition based on EEG are also not considered since cross-task experiments consider cognitive tasks which are quite different from the tasks that generate MW.
- *Time:* In order to have a comprehensive study on this topic, literature within the past two decades (2000–2022) are included in this chapter.
- *English language articles:* For this chapter, we have studied articles from peer-reviewed journals, conference proceedings, dissertations and textbooks, which are published only in English language.
- *Peer-reviewed full-text articles:* The chapter considers articles only if it was published in reputed peer-reviewed journals and conference proceedings. This results a comprehensive review of a total 72 articles.

From each of the studies, relevant information is extracted that include experiments of cognitive tasks for generating MW, artifact removal methodologies, EEG feature extraction, feature selection, and classification techniques using machine learning (ML) and deep neural network (DNN) architecture. The general framework for MW classification is shown in Figure 7.3.

7.2.2 Data summary measures

The data categories, collected for this chapter, are as follows:

A. Number of subjects for MW classification:
    - Ratio of male and female participants
    - Mean age/age range, as reported in literature
B. Total number of EEG channels used
C. Artifact removal technique
    - Manual removal
    - Automatic removal
    - No removal
D. Frequency range used for analysis
E. EEG feature extraction types:
    - Frequency domain features
    - Time domain features

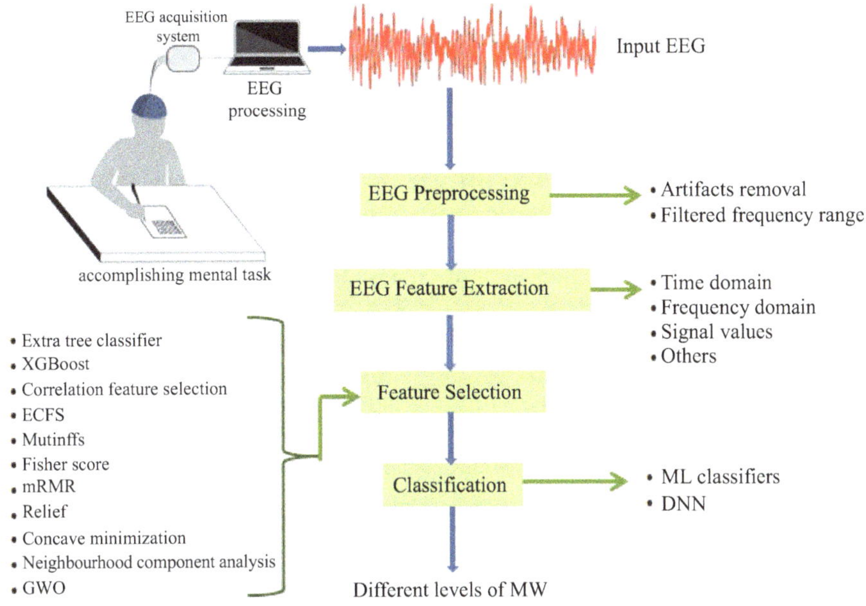

Figure 7.3 Mental workload classification framework using EEG

- functional connectivity features
- Raw EEG signal values

F. Classification model
- Traditional ML models
- Deep learning neural network models

G. Level of task complexity
H. Performance metrics
- Percentage of accuracy
- R^2 statistic and correlation measure

7.3 Cognitive tasks and workload classification

This section first details the brief description of different cognitive tasks which are performed using EEG in order to generate MW. Then, for analyzing and classifying different MW labels, the EEG pre-processing, feature extraction, and selection methods are detailed. For MW classification, different ML classifiers and the recent trends in DNN architecture are studied. The present section is concluded with a case-study, performed on a publicly available dataset; thus allowing us for a direct comparative analysis on the particular ML/DL architecture selections to classify MW.

7.3.1 Cognitive tasks for generating workload

There are several cognitive tasks that are carried out to quantify MW. This section briefly illustrates the type of MW tasks as reported in literature.

7.3.1.1 Forward digit span (FDS)

In this type, a sequence of numbers is sequentially presented on screen and the subjects need to memorize the whole sequence for entering the numbers according to the given order within stipulated time. Different difficulty levels are included by incrementally increasing the length of the series [5,6].

7.3.1.2 Backward digit span (BDS)

This is similar to the FDS, except that here the participants are asked to enter the numbers in the reverse manner from the one presented, within time. Similar to FDS, difficulty labels can be varied by increasing the length of the sequence [5,6].

7.3.1.3 Logical tasks

The inferential skills of the participants on information series are tested in this type of task. For example, in [5], the subjects were instructed to find the missing number in the given sequence according to the logic within 30 sec.

7.3.1.4 Learning activity

In order to assess MW during learning of a particular course, different learning activities with varying labels can be given. In [5], this learning activity comprises three phases (i) a pre-test session (typically yes/no-response), (ii) a learning session on a trigonometry course, and (iii) a quantitative aptitude session where numerical problems with an increasing difficulty level of the same order are given to all the participants. Prior information on self-assessment in a particular subject can be collected from each participant (low, moderate or high) for accurate estimation of classification performance.

7.3.1.5 NASA MATB-II

The Multi-Attribute Task Battery (MATB-II) is used to estimate the performance of the operator and the workload [2,7,8,28–30]. MATB provides a benchmark set of computer-based tasks that are similar to the activities performed by aircraft crew members usually perform in flight. In this type, the participants are required to complete the four simultaneous sub-tasks which are system monitoring, tracking, scheduling, and resource management, under different mental loads in MATB-II. The difficulty levels of this task are defined by the frequency of occurrence of the four sub-tasks.

7.3.1.6 Solidworks 3D modeling

In this type of task for generating MW, the participants were instructed to use different tools and features in order to construct 3D models in such a way so that a lot of information is embedded into the model [9]. The subjects having no past history and experience of Solidworks usage were selected to create the 3D models having different levels of model complexity.

7.3.1.7 Real multimachine assignment simulation of oceanauts

For this task, the test is conducted on the platform, which is a simulation scene of the manned submersible 3D operating system in order to simulate the real multimachine assignment [10]. Here, the task usually divided into three levels: spatial positioning, grasping, and comprehensive operational tasks. This three-task experiment can be used to construct the MW level (low, medium, and high) of the oceanauts, where the low MW is labeled for the spatial positioning task, medium MW is labeled for grasp task, and the high MW is labeled for comprehensive operational tasks.

7.3.1.8 N-back task

Several literature have used N-back tasks for classifying MW [1,11–13,31,32]. Here, 0-back (N=0) and 2-back (N=2) tasks represent low and high workloads, respectively. For example, in the 0-back condition if the targ*et al*phabet is 'T', in the 2-back condition, subject needs to decide if the alphabet displayed at a specific time is identical as the alphabet displayed 2 sequences earlier. Thus, at each time, the subjects need to memorize 'N' numbers (for N-back task) of previous alphabets as the sequence progresses.

7.3.1.9 Arithmetic task

Here, subjects are instructed to perform mental arithmetic calculations [6,11, 14–17,33]. The result of each arithmetic operation is displayed on screen after certain time. If the result is true, then such number is considered as the valid number, otherwise it is invalid. The level of MW can be increased by increasing the number of digits and number of operations to be performed and decreasing the required time. In [15], 17 different arithmetical tasks were carried out by participants with a total six intervals. Raven's Advanced Progressive Matrices Set II problems were performed in [17] with 36 items in increasing difficulty level.

7.3.1.10 Pilot pattern classifier

This is used for distinguishing the MW level of the pilot, based on EEG signal and eye-tracking information. In order to incorporate cognitive interference, a secondary task can be introduced which is needed to be performed along with the primary one. One example of this secondary task is reported in [18] where the participants are instructed to convert passive phrase to active phrase and vice versa. In [18], each participant went through three different sets of twenty primary visual search tasks followed by the secondary transformation tasks.

7.3.1.11 SIMKAP simultaneous capacity test

This type of cognitive task consists of three phases. In the first phase, the participants are required to complete baseline task of item matching (items can be numbers, letters, and shapes) in three trials [19,34–39] within stipulated time. Followed by this, in the second phase, the subjects are required to perform an auditory based problem-solving task. Finally, in the third phase, the subjects are asked to execute the multitasking activity within specific duration. This multitasking incorporates execution of previous phase-1 and phase-2 activities in addition of little variations to the auditory task. After each task, subjects are also instructed to rate their MW level on 1 to 9 scale along with a questionnaire. Four possible MW levels can be identified from SIMKAP tasks: level 1 – No task state, level 2 – Visual-matching task, level 3 – Auditory task and level 4 – Multitasking.

7.3.1.12 Automatic enhanced cabin air management system (ACAMS)-based simulated human–machine task

The ACAMS platform provided a micro-world in a spaceflight [40–45]. In this task, whenever a fault occurs in any parameter of the automatic control system, the participants are required to fix the error manually. The variations in the difficulty levels of each session are designed via a cyclic-loading schedule of eight consecutive phases.

7.3.1.13 Stroop color test

It is a well-known task to study attention. In [46], authors have used this neuropsychological test in order to assess MW. The Stroop color and word test (SCWT) is widely used in assessing capability to constrain cognitive distraction that arises when processing of a specific stimulus attribute delays the simultaneous processing of another stimulus feature. This is well known as Stroop effect. Apart from assessing the ability to inhibit cognitive interference SCWT is also used to measure other cognitive functions such as attention, processing speed, cognitive flexibility, and working memory.

7.3.1.14 Driving simulation platform

The platform consists of (i) operational instruments for driving stimulation such as seat, steering wheel, loudspeaker, LCD, and projector; (ii) the devices for physiological signal collection, such as 64 electrodes EEG neuro-scanner, camera for blinking-eye detection, and heart-rate sensor; (iii) computer to record all the parameters during driving; and (iv) additional computer in order to collect alert tasks and to process physiological signals. Post-training, subjects are instructed to drive at the particular track for two or more times without deviating from the track for safety purpose. In addition to EEG brain wave data, in [47], authors have also collected facial expression (with the help of camera) and the heart rate simultaneously using the electrodes which are attached on the wrist of the subjects.

7.3.2 EEG preprocessing and signal analysis

7.3.2.1 Artifacts removal

Artifacts are unwanted signals which are generated due to the presence of environmental noise and experimental error. Since these types of artifacts originate from external factors, they are also referred as extrinsic artifacts. There is another type of artifacts, termed as, physiological artifacts that is generated from the body itself. Physiological artifacts are also termed as intrinsic artifact. Some common examples of physiological artifacts include (i) ocular artifacts: caused as a result of movement and blinking of eyes that propagate over the scalp and recorded by EEG;

(ii) muscle artifacts: originated due to contraction and stretch of muscle, talking, etc.; and (iii) cardiac artifacts: produced by placing EEG electrodes near the blood vessel that causes recording of heartbeat by EEG. While environment artifacts can removed by simple filtering technique as the frequency of environment artifacts are very distinct from desired signal frequency, the physiological artifacts are little difficult to remove. The methods for detection and deletion of such EEG artifacts are explored extensively in several literature [48–51]. The methodologies of artifact removal generally fall into one of the following two categories: (A) manual removal [52–56], (B) automatic removal [1,2,4,9,13,15–17,19,40,54,57–59]. However, it is seen that some studies on MW classification did not report any particular type of artifact removal techniques [58–61]. Studies that utilized shared public dataset [34,36,37,39] did not report any specific artifact removal techniques since they have used already cleaned dataset. The most common artifact removal algorithm which is used in literature is ICA, followed by DWT. Based on the variation in task complexity, many studies used filters in order to reduce the EEG bandwidth. It is seen that low pass and band pass filters are widely used for this purpose. The frequency ranges utilized in the studies are shown in Figure 7.4. The mean frequency was found to be in the range from 1.91 ± 1.3 Hz to 41.74 ± 13.04 Hz.

7.3.2.2 Feature extraction

Due to a high inter-subject variability of EEG signals, EEG feature extraction is crucial in order to estimate the level of MW. There exists a wide range of methods for extracting suitable features for classification. Some examples of feature extraction methodologies include tunable Q wavelet transform, empirical mode decomposition, empirical wavelet transform, variational mode decomposition, optimal allocation of samples, fast Fourier transform, filtering [62–66]. In context of MW classification, the extracted features which are reported in literature are broadly categorized into four groups. This is shown in Figure 7.5. These are

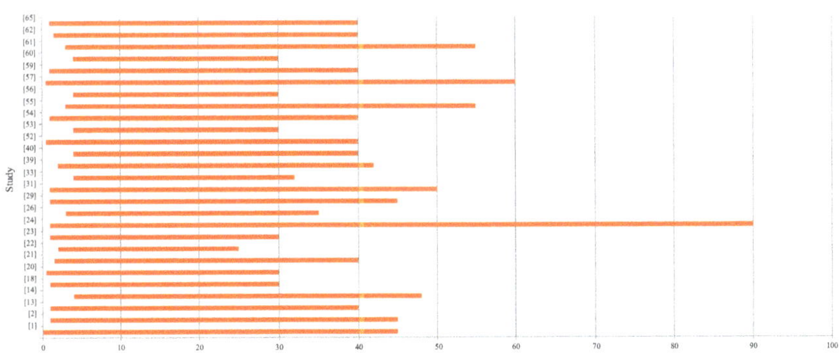

Figure 7.4 The range of frequency of EEG signal, considered for MW classification task for each identified study

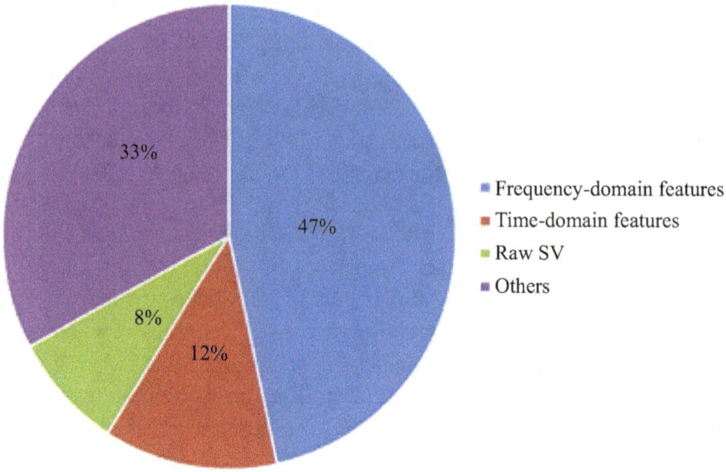

Figure 7.5 Types of EEG features used across studies in classifying MW using EEG. Examples of other features include functional connectivity measures, engagement index (EI), task load index (TLI), amplitude of ERP components.

frequency-domain features (used by 47% of the studies), time domain features (used by 12% of the studies), the raw EEG signal values (SV) (used by 8% of the studies), and other features (12% of the studies). Some examples of other features include function connectivity measures, engagement index (EI), task load index (TLI), amplitude of ERP components. As seen from Figure 7.5, majority of the studies preferred to analyze EEG data in frequency domain because of their strong association with behavioral patterns. Among all extracted frequency-domain features, it was observed that majority of the studies have reported the utility of power spectral density (PSD) for workload assessment and classification that measured signal power across different sub-band frequencies [2,4,6,7,9–11,13–15,17, 19,35,36,40,55,57,59,61,67]. Besides PSD, several other types of time and frequency domain features like sub-band energy [7], first-order statistics [9,36], Hjorth parameter [9], wavelet decomposition [9], RMS [7,11], fractal dimension [19], higher order crossing (HOC) [16], non-linear entropy features such as slope entropy, dispersion entropy, permutation entropy, sample entropy, approximation entropy [36,68], spatial and spectral filter based features [31], autoregressive coefficient [36], Hurst exponent [36], mean degree, clustering coefficient, oblique visibility graphs (OVG) degree distribution [39] are extracted in order to classify MW for different cognitive tasks. It is noted that some neural networks (NN), especially convolutional neural networks (CNN), utilized spectrograms as its inputs [53,56,58]. Traditionally spectrograms are the useful post-processing tool for data visualization. Since CNN's has the ability to learn from the input images, spectrograms can be used as an input feature to the classifier.

There are many literatures that used raw EEG SVs as features or inputs to the NN [12,52,54,69]. As NN are capable to learn complex features automatically from large amounts of data, it inspires researchers to feed directly raw EEG SV into the NN without deriving hand-crafted features from it.

Different from time/frequency-domain feature and raw EEG SV, there are several other EEG features reported in literature. Specific cortical and sub-cortical regions are activated for different cognitive tasks and the brain dynamically coordinate to transfer the information. Quantification of neuronal interactions can be performed by exploring functional connectivity of neurons [17]. Different techniques are reported in literature that quantifies these interactions using EEG [1,2,17]. In [17], authors used three different neural complexity metrics like mean Lempel-Ziv complexity across channels, multi-scale entropy, and detrended fluctuation analysis. In [17], authors used mutual information for quantifying the inter-dependency between the obtained time-series EEG data. Phase locking value is used to quantify the phase synchronization of EEG signals, acquired from different brain regions. The information transfer between brain regions is quantified by phase transfer entropy to estimate the causal influence of one brain region on another brain region. In [2], authors used phase and magnitude spectral coherence (PMSC), amplitude modulation rate-of-change (AMRC), and PMSC-AM measuring the connectivity between cortical regions that tell about the interactions between different brain processes. For assessing pilot's psycho-physiological state, EEG and eye-tracking information were used to differentiate different states of MW [18]. These authors have used α-power, θ-power, left and right eye closure percentage, left and right pupil diameter and correctness. Apart from this, task engagement index (EI) [7,70–76], task load index (TLI) [33,77,78], and amplitude of ERP components [33,79–86] are also used as features for MW classification.

7.3.2.3 Feature selection

The signals or features obtained from EEG are usually high-dimensional features. Thus, it is important to reduce the complexity of such high-dimensional signals. Feature selection for dimensionality reduction is a common step in classification tasks so that redundant, correlated and irrelevant features can be removed in order to improve classification performance. Feature selection becomes essential not only because it decreases the number of features for further processing but it also increases the computation speed. Different ranking methods exist that allow arranging the features based on their accurate prediction ability. The various feature selection methods reported in the literature are extra trees classifier, extreme gradient boosting, correlation feature selection [9], feature selection by Eigen vector centrality (ECFS), relief, mutinffs-mutual information feature selection method, Fisher score [14,15], minimum redundancy maximum relevance filter method (mRMR) [2,14], supervised infinite feature selection, concave minimization [14], neighborhood component analysis [35], and Grey wolf optimization (GWO) [36].

7.3.3 Workload classification

The classification of MW using EEG is mainly performed by two approaches, one with the traditional ML methods and second, with the help of deep learning (DL) architectures. The distribution of studies that have used different ML and DL classification models for assessing MW is shown in Figure 7.6. Apart from classification, there are some studies that have reported statistical measures like *r*2 statistic, canonical correlation [6,17], subjective, and objective rating score [5,6,87] in order to assess the pattern of MW, detected during the execution of varying levels of cognitive activities as performed by the participants. For subjective ratings, NASA-TLX [4,87,88] technique has been used that depends on conscious perceived experience by the participants with respect to the effort produced and the complexity of the task.

7.3.3.1 MW classification using ML model

For the classification, it is seen that 63% studies have used different ML models to predict workload label. It was found that 15% of the literature have reported the utility of multiclass support vector machine (SVM) [8–11,14,15,18,19,35,39], followed by K-nearest neighbor (KNN) (13%) [8–10,14–16,18,19], and linear discriminant analysis (LDA) (7%) [8,12,13,31]. Along with SVM, KNN, and LDA

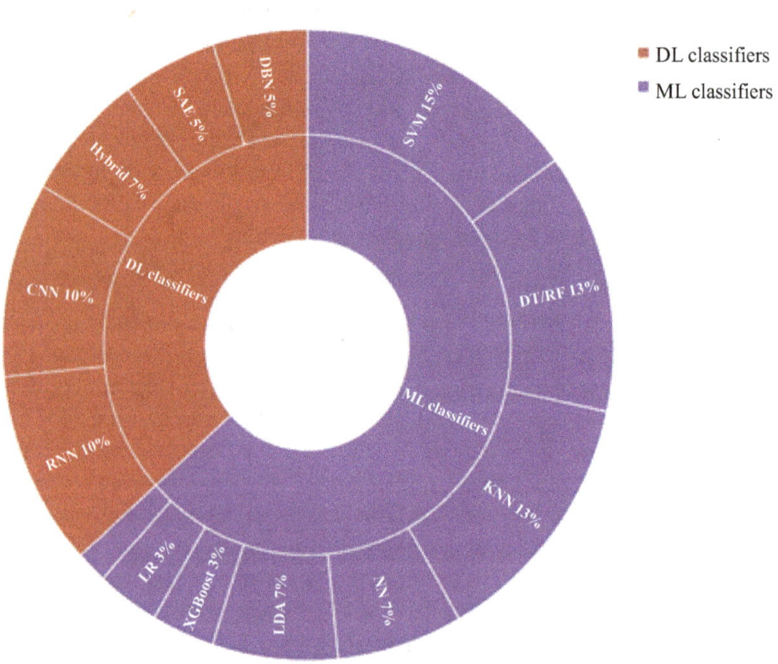

Figure 7.6 MW classification model across studies. The inner circle demonstrates the in-general categories of classifiers (red region: studies that used DL classifiers; blue region: studies that used DL classifiers) while the outer circle demonstrates more specific ML/DNN architecture choices from each category. The outer circle also depicts the percentages of studies that used a particular ML/DL classifier.

authors also used Naive Bayes (NB) [9], Gaussian process regression model [5], decision tree (DT) [9,14,15,39], random forest (RF) [2,10,15,17,18], logistic regression [12,17], Gaussian mixture model [12], and XGBoost [9,17] to classify the workload using various EEG features. The different EEG features, level of task complexity, and ML classification model used across studies are summarized in Table 7.1 along with average or highest achieved accuracy by the study, as reported in the literature.

7.3.3.2 Trends in deep learning for MW classification

The review of the literature on understanding the trends in forming deep learning architecture for workload classification using EEG is illustrated in this section. Table 7.2 shows the chosen DNN architecture along with the highest accuracy achieved by study. It is found that 37% of the total studies have used various DL architectures in classifying MW as shown in Figure 7.6. The most widely used DL frameworks for MW classification are CNN's (used by 10% of DL studies), the network which consists of alternating convolution with pooling layers (mostly, max-pooling layers) [1,34,52–54] and RNN's (used by 10% of DL studies) [1,34,36,61,68]. RNN includes a sequence of recurrent layers followed by fully-connected layers. After CNN and RNN, the second most prevalent choice is hybrid architecture, consisting of 7% of the total DL studies. It is seen that all the studies that utilized hybrid architecture used hybrid CNN for workload classification [55,56,58,69]. Hybrid CNN's incorporates a series of recurrent layers in addition to traditional convolutional and pooling layers, seen in CNN (e.g., CNN+RNN, CNN

Table 7.2 Summary of data that used DL strategy in classifying MW using EEG

Study	Subject	Frequency range (Hz)	DL strategy	Features	Accuracy (%)
[1]	19	0.1–45	CNN, LSTM	Functional connectivity	97.92
[40]	8	4–40	SDAE, En-SDAE[a]	PSD	92
[52]	120	0.5–40	CNN	SV	80
[53]	15	4–30	CNN	Spectrogram	90
[54]	10	1–40	CNN	SV	93
[55]	8	3–55	Hybrid (CNN+RNN)	PSD	87
[56]	13	4–30	Hybrid (CNN+RNN)	Spectrogram	91
[89]	8	–	DBN	Entropy	77.4
[60]	6	0.5–40	DBN	MAD	93
[58]	13	4–30	Hybrid (CNN+RNN)	Spectrogram	93
[61]	13	4–30	RNN-LSTM	PSD	93
[59]	8	1.5–40	SAE	PSD	74.3
[90]	15	0.5–60	DBN	PSD	97
[67]	7	1–40	SAE	PSD	95.5
[68]	36	–	LSTM, BLSTM, GRU	Entropy	99
[69]	37	–	Hybrid (CNN+DBN)	SV	82.8

[a]En-SDAE – ensemble SDAE.

+DBN). The next group of DNN architecture is DBN (used by 5% of DL studies) [57,60,89], which consists of stacked restricted Boltzmann machines followed by a series of fully-connected layers. The final group of studies for MW classification utilized stack autoencoder (SAE) [40,59,67] and transformer network model [37], having fully connected layers.

Activation functions are also noted for all studies that utilized DL architectures. For convolutional layers, majority of DL strategies employed ReLU activation function [1,34,53–56,58]. The second prevalent activation function includes Sigmoid [40,59,67,68]. There is only one study that used tanh activation function in their bidirectional LSTM (BLSTM) architecture [34]. It is also seen that most of DNN classification frameworks employed softmax activation function in the fully connected layer.

7.3.4 Case study on publicly available MW dataset

Most of the articles which are reviewed in the current chapter used EEG data which were generated from a wide variety of cognitive tasks performed by different participants of varying gender and age groups. These datasets are not publicly available. Since these datasets were generated from wide varieties of tasks having unique data procurement procedures, comparative analysis between these studies is not valid. Contrary to this, studies that have performed MW classification using same dataset could provide a more meaningful comparison. Five such studies are identified that used one common publicly available dataset for classifying MW [34–37,39]. For MW classification, the widely used dataset which is publicly available is simultaneous task EEG workload (STEW) dataset [35].

STEW dataset consists of raw EEG data of participants (n=48) who have performed a multitasking workload experiment, called, the ‘SIMKAP multitasking test’ in order to assess the capability of individual subjects in multitasking and in stress tolerance. The task details of SIMKAP is illustrated in Section 3.1.11. The order of tasks and questions in SIMKAP are fixed for all 48 participants. The brain activity of the subjects at resting state was also recorded before starting of the test for analysis. The Emotiv EPOC device, with 14 channels and sampling frequency of 128 Hz, was used to obtain the EEG data. As discussed in Section 3.1.11, the SIMKAP represents rating-based multi-tasking activity. In STEW dataset two different types of classes are introduced; Type-I: ‘No task Vs. SIMKAP task’ experiment having 2 MW levels: low and high (binary class) and Type-II: ‘SIMKAP-based multitask’ – which is further divided into 3 MW levels (low, moderate, and high) for ternary classification. The rating scale was set 1–9, with 1–3 represents ‘low scale’, 4–6 represents ‘moderate scale’, and 7–9 represents ‘high scale’. The device comprises of 14 electrodes located at 14 distinct brain locations according to the standard 10–20 international system (Figure 7.1). The collected raw EEG data is then pre-processed in order to remove artifacts resulting from muscle movement. On STEW data, artifact and noise removal was performed by high-pass filtering at 1 Hz followed by artifact subspace reconstruction (ASR) to remove the large amplitude artifact from specific channel.

Table 7.3 A comparison of chosen architecture and EEG features across MW classification studies using the publicly available STEW dataset [35]

Study	Architecture	Feature	Accuracy (%)
[34]	1D CNN-BLSTM	SV	96.77 (2-class), 95.36 (3-class)
[35]	SVR	PSD	69.2 (3-class)
[36]	LSTM, BLSTM	PSD, AR coeff, approx. entropy, Hurst exponent first-order statistics	86.33 (2-class), 82.57 (3-class)
[37]	Transformer network	SV	95.28 (2-class), 88.72 (3-class)
[39]	OVG, DT SVM	Mean degree, clustering coeff. Degree distribution	89.6 (2-class), 79.5 (3-class)

Table 7.3 shows the result of comparative analysis of chosen architecture and EEG features across the studies that have performed MW classification task using publicly available STEW dataset. In [35], the authors performed ASR, PSD feature extraction, and neighborhood component analysis for MW classification. Support vector regression (SVR) model yields the accuracy of 69.20% for the 'SIMKAP multitask' classification. In [34], a cascaded one-dimensional CNN and BLSTM (1DCNN-BLSTM) model has been proposed and tested over STEW data for binary and ternary classification of MW. Here, instead of handcrafted feature extraction, authors have used an end-to-end DL over all 14 channel EEG signals for classification purpose. In binary and ternary classification, the accuracy of 96.77% and 95.36% have been achieved respectively, with 7-fold cross-validation. In [36], a deep hybrid model based on BLSTM and LSTM has been proposed that achieves 86.33% and 82.57% classification accuracy for binary and ternary classification, respectively. In [37], the efficacy of transformer network in classifying MW using was explored. The network has achieved an accuracy of 95.28% (binary) and 88.72% (ternary) using raw EEG data without feature extraction. In [39], oblique visibility graphs (OVG) was constructed and features like mean degree, clustering coefficient, and degree distribution were extracted. Here, DT and SVM were used for classification, where the better result was achieved by SVM as compared to DT with 89.60% and 79.50% accuracy for binary and tertiary classification, respectively. Comparing all the literature that used STEW dataset, the best result for binary and ternary classification was obtained using 1D CNN-BLSTM (96.77% and 95.66%) model in terms of accuracy, as reported in [34]. It was also observed that with STEW dataset, the DL architectures [34,36,37] were found to yield better result in MW classification as compared to traditional ML classifiers [35,39].

7.4 Discussion

There are several literature that reported the potential of using EEG in studying varieties of cognitive tasks such as emotion recognition, mental states and sleep stage classification, detection of diseases like schizophrenia and seizures [91–94]. The current chapter is aimed to understand how brain EEG signals can be utilized for the classification of MW task. As a multidimensional paradigm, workload has been defined in terms of the resources available in order to meet demands for a specific task. Several studies have defined the term MW [95] and reported the possible solutions for quantifying it. While too high workload reduces the performance in executing the task, a too low workload also lessens the incentive for executing the specific task [96]. Thus, a reasonable workload is always desirable for maintaining a prolific working environment. Continuous assessment of mental state is possible by exploring the functioning of human brain during the execution of a given task [95,97]. Many studies have focused on differentiating workloads levels using of EEG where workload is generated during the execution of a specific cognitive task. This section describes recommendations for choosing workload classification methodologies based on ML and DNN architecture, as collected from literature, utilizing different EEG features.

7.4.1 Are there any specific cognitive tasks, mostly used for MW classification using EEG?

There exists a large variety of cognitive tasks, performed in order to assess MW using EEG. The most prevalent cognitive tasks included in 20% of the total studies is arithmetic tasks, followed by N-back tasks (15%) and NASA-MATB-II (15%). There also exist literature that have reported subjective measures or self-report measures to assess MW [95] where, based on personal experience during the execution of a task, the participants provide the quantitative and/or qualitative assessment. It is seen that using EEG, studies have utilized NASA-TLX multidimensional rating for MW assessment since this index provides both the diagnostic information about the particular sources as well as a global summary of MW [4,7].

7.4.2 Choice of EEG features and ML classifiers

A number of features were extracted from raw EEG in order to perform workload classification using traditional ML and DNN classifiers. Figures 7.5 and 7.6 show the recommendations for EEG features and ML/DL classifiers, respectively. Among all frequency domain features, PSD has been extensively utilized for assessing MW. It is seen that PSD of the parietal-α bands is reduced and frontal-θ bands is increased with an increase in task difficulty [87,96,98–106]. Similar results have been observed with wavelet EEG features [99]. Using PSD as features, 98.5% and 95.5% classification accuracies were achieved with N-back task and arithmetic tasks, respectively using SVM [11]. Using KNN as a classifier maximum accuracy of 85.4% [14] and 91% was obtained with extracted PSD feature [15]. In [15], ANOVA for repeated measures reveals significantly stronger power of EEG waves for low and high cognitive workload, as compared to relax condition. However, neither statistical significant differences nor strong correlations between low and high cognitive workload level have been found from ANOVA. In [7], EI was best identified feature for NN classifiers. The EEG-

EI, $\beta/(\theta + \alpha)$ (proposed by Prinzel *et al.* in [71]) is usually used to identify the decline in task engagement. Typically EI is defined as 'the ratio between β power and the sum of α and θ power associated with certain EEG measurement channels' [107]. The direct correlation which was found between the EI and task load [108], making EI a promising measure to quantify workload status and thus highly recommended in order to design adaptive systems [73]. The adaptive system is able to provide situational alertness by keeping the reasonable workload. When EI increases, the system will be shifted to automatic mode, the result of which reduces the value of EEG EI. Conversely, when EI decreases, the system will shifted back to manual mode, resulting in increase in MW that in turn increases the EI index [70–76]. Similar to EI, some literatures reported the utility of measuring TLI which is defined as 'the ratio of the mean frontal midline θ energy to the mean parietal α energy'. Like EI, an upsurge of TLI over time has been observed with increase in cognitive task difficulty levels [33,77,78]. While comparing different EEG channels, it was seen that EEG channels in anterior frontal region are most appropriate for assessing MW, provided the maximum classification accuracy of 90% using NN [7]. Also RMS value in γ sub-band was able to differentiate MW with statistically significant difference at p-value. In [9], the highest accuracy of 83.33% and 88.01% was reported using XGBoost classifier with and without feature selection, respectively. In [10], multi-class SVM with particle-swarm optimization algorithm yields an average classification accuracy of 85.7% for assessing workload of the oceanaut. In [19], 90.39% and 80.09% accuracy using SVM was obtained for two-class and four-class MW classification, respectively, by the combination of statistical and FD feature. In [16], KNN classifier was used for the classification of workloads with seven labels and achieved 89.61% accuracy using HOC feature. Here, the effect of the workload found at the frontal and temporal lobe of brain. Also, the reaction time (time required to answer by the participant after displaying MCQs) was observed to be increased with increase in load levels. In [8], the highest accuracy of 79.8% was obtained using SVM with ICA-energy feature. In [6], the decline in performance of participants was observed with an increase in task difficulty. Comparing all literature that used classical ML approaches, the minimum accuracy of 74% (across all channels) was found in [12] where raw EEG data was utilized as features. A few studies on MW have investigated the amplitude of ERP. It was observed that the value of certain ERP components such as mismatch negativity, late positive potential, early and late slow-wave component (SW1 and SW2, respectively), P2, P3, N1, and N2 decreases as MW increases. Conversely, the latency of the ERP component was found to increase with an increase in workload [79–85]. Some experiments have combined features like PSD and P300 ERP components [33,80,86]. They have found a decline both in amplitude of P300 and the PSD of α band. However, a rise in the PSD of θ bands was observed when MW increases. While comparing spectral indicators and spatially filtered ERPs, in [109], authors concluded that as compared to spectral domain, the spatially filtered ERP technique is more precise and effective in distinguishing different workload levels.

7.4.3 Is there any particular DNN architecture, appropriate for MW classification using EEG?

The detailed review of DL studies reveals the recommendation of DBN or RNN's or CNN's for MW classification. In [69], authors reported that as compared to classical CNN, DBN significantly outperformed MW classification task, whereas authors in [53] found several CNN variations outperformed MW classification as compared to typical DBN. Three studies were published by Yin *et al.* based on one common MW dataset that showed the potential efficacy of SAE as compared to traditional SVM in classifying MW [59,67,89]. Using ensemble stacked denoising auto-encoder (SDAE), 92% accuracy was achieved that outperforms several classical shallow and deep classifiers [40]. With entropy as features, classification accuracy of 99% was reported in [68] using various RNNs. In [2], classification accuracy of 97% was achieved by Harvard automated processing pipeline for electroencephalography (HAPPE) enhancement and adaptive filtering techniques. The authors suggested that a relationship among attention, sensory motor integration, and workload may exist due to presence of an upsurge in γ activity, seen in parietal cortex region. In [1], the best performance with an accuracy of 97.92% was achieved by the combinations of PLV with Conv-LSTM, and PLV with CNN, that highlighted the efficacy in using the combination of EEG-based functional connectivity measures and DL architecture to classify MW. Using PSD as feature, the highest accuracy of 97% was obtained from DBN [57], followed by an accuracy of 95% which was obtained from SAE [67].

Apart from using single DNN architecture for MW classification, five studies were found that applied hybrid DL architectures. Hybrid DNN structures use a combination of two or more standard DL algorithms. Four of these five studies combine different layers (e.g. one or more RNN layers) along with the typical convolutional layers [55,56,58,69]. There is one study that used RNN-LSTM modules [61]. A few of these studies have compared classification performance of these hybrid architectures and the architectures based on traditional DNN in terms of accuracy. Comparing all hybrid architecture, it is seen that the hybrid structures that include recurrent layers along with convolutional layers outperformed when compared against standard DNN algorithms. For example, with spectrogram as extracted feature, hybrid CNN-RNN [58] gives 93% accuracy as compared to the result reported in [53] that used single CNN and yields 90% accuracy in classifying MW. However, it is seen that the hybrid CNN-DBN architecture could able to bring only 82.8% accuracy when EEG raw SVs were fed to its input. Thus, it is recommended to explore more in-depth research of these hybrid network structures. Apart from these DL network design, further research needs to be conducted in understanding how DNN interprets raw EEG vs. denoised EEG, since this has not been addressed clearly in the recent literature.

7.5 Conclusion

In this chapter, a systematic review of literature on the applications of EEG for quantifying human MW is explored based on research studies conducted between

2000 and 2022 on this topic. The presented study allowed a comprehensive analysis of the research trends relevant to cognitive tasks and MW classification. An increasing trend of publication in this field was observed over the last 5 years. While most of studies applied PSD and other statistical EEG features to evaluate cognitive performance using traditional ML algorithms, with advances of DL, there are many recent studies that used different DNN architectures to classify MW. In this chapter, both ML and DL techniques for classifying MW are explored. Three main critical questions are addressed in our study that include (i) types of specific cognitive tasks, mostly used for MW classification using EEG; (ii) different types of EEG features that have been used for classification of MW using both ML and DL algorithms; and (iii) types of specific DL network structures which are mostly suitable for MW classification. All peer-reviewed published articles on MW classification using EEG are screened, followed by a detailed review of different EEG pre-processing methods, feature extraction and selection, DL network structure and performance analysis of the ML/DL classifiers. We have also explored publicly available STEW dataset which was seen to be utilized in multiple studies for MW classification using EEG, permit us to compare results directly based on different feature extraction and classification strategies. After comparing the different NN architecture and classification strategy, reported in individual studies, it is found that MW can be classified more efficiently with specific EEG features and ML/DL design choices. The data-related key information is summarized into recommendation tables that can serve as a starting step in choosing specific EEG features and classification models for future applications of MW classification using EEG.

References

[1] Gupta A, Siddhad G, Pandey V, *et al.* Subject-specific cognitive workload classification using EEG-based functional connectivity and deep learning. *Sensors.* 2021;21:6710.

[2] Rosanne O, Albuquerque I, Cassani R, *et al.* Adaptive filtering for improved EEG-based mental workload assessment of ambulant users. *Front Neurosci.* 2021;15:611962.

[3] Sea R. EEG based cognitive workload assessment for maximum efficiency. *IOSR J Electron Commun Eng*. 2021:34–38.

[4] Hart S and Staveland L. Development of NASA-TLX (Task Load Index): results of empirical and theoretical research. In *Human Mental Workload.* Amsterdam: Elsevier, 1988.

[5] Chaouachi M, Jraidi I, and Frasson C. *Modeling Mental Workload Using EEG Features for Intelligent Systems,* vol. 6787. Berlin, Heidelberg: Springer, 2011.

[6] Berka C, Levendowski D, Lumicao M, *et al.* EEG correlates of task engagement and mental workload in vigilance, learning, and memory tasks. *Aviat Space Environ Med.* 2007;78:B231–B244.

[7] Sushil Cea. EEG based cognitive workload classification during NASA MATB-II multitasking. *Int J Cogn Res Sci Eng Educ*. 2015;3:35–41.

[8] Qu H, Shan Y, Liu Y, *et al.* Mental workload classification method based on EEG independent component features. *Appl Sci.* 2020;10(9):3036.

[9] Vishal A, Sreeja S, and Debasis S. *An automated approach for task evaluation using EEG signals*, CoRR abs/1911.02966, 2019.

[10] Xiaoguang L, Lu S, Cong Y, *et al.* Mental workload classification of oceanauts from EEG data using multiclass support vector machines, 2022. bioRxiv 2022.03.08.483450.

[11] Quadri A and George P. Mental workload estimation using wireless EEG signals, 2019. bioRxiv 755033.

[12] Coffey E, Brouwer A, and Van J. Measuring workload using a combination of electroencephalography and near infrared spectroscopy. In *Proceedings of the Human Factors and Ergonomics Society Annual Meeting*, 2012, vol. 56, no. 1.

[13] Yichuan L, Hasan A, and Patricia A. Multisubject "Learning" for mental workload classification using concurrent EEG, fNIRS, and physiological measures. *Front Hum Neurosci.* 2017;11:389.

[14] Plechawska-Wojcik M, Borys M, Tokovarov M, *et al.* Classifying cognitive workload based on brain waves signal in the arithmetic tasks' study. In *Proceeding of the International Conference on Human System Interaction*, 2018.

[15] Ałgorzata PW, Mikhail T, Monika K, *et al.* A three-class classification of cognitive workload based on EEG spectral data. *Appl Sci.* 2019;9(24):5340.

[16] Jadhav N, Manthalkar R, and Joshi Y. Analysis of effect of meditation on cognitive load using higher order crossing features. In *Proceedings of the International Conference on Communication and Signal Processing (ICCASP 2016)*, 2016.

[17] Nir F, Tomer F, Kobi G, *et al.* EEG-based prediction of cognitive load in intelligence tests. *Front Hum Neurosci.* 2019;13:00191.

[18] Lobo J, Javier D, Simone F, *et al.* Cognitive workload classification using eye-tracking and EEG data. In *Proceedings of the International Conference on Human-Computer Interaction in Aerospace*, 2016, vol. 16.

[19] Lim W, Sourina O, Liu Y, *et al.* EEG-based mental workload recognition related to multitasking. In *Proceedings of the International Conference on Information, Communications and Signal Processing (ICICS)*, 2015.

[20] Qu H, Zhang M, and Pang L. Mental workload classification method based on EEG cross-session subspace alignment. *Mathematics.* 2022;10(11):1875.

[21] Hildebrandt J and Smith D. *The Nervous System: Neurons, Networks, and the Human Brain*. Colorado Springs: Great Pacific Media, 2008.

[22] Squire L. *Fundamental Neuroscience*, third ed. Amsterdam/Boston: Elsivier, 2008.

[23] Rowan A and Tolunsky E. *Primer of EEG: With a Mini-Atlas.* Philadelphia, PA: Butterworth-Heinemann, 2003.

[24] Li Hea. *EEG Signal Processing and Feature Extraction*. Amsterdam/Boston: Springer, 2019.

[25] RoJasper H. The ten-twenty electrode system of the international federation. *Electroencephal. Clin Neurophysiol.* 1958;10:3–6.

[26] Acharya J, Hani A, Cheek J, *et al.* American Clinical Neurophysiology Society Guideline 2: guidelines for standard electrode position nomenclature. *J Clin Neurophysiol.* 2016;33(4):308–311.

[27] Society ACN. Guideline 5: guidelines for standard electrode position nomenclature. *J Clin Neurophysiol.* 2006;23:2.

[28] Ryan M and Wea Z. Workload classification across subjects using EEG. In *Proceedings of the Human Factors and Ergonomics Society 55th Annual Meeting*, 2011.

[29] Wang Z, Hope R, Wang Z, *et al.* An EEG workload classifier for multiple subjects. In *Proceedings of the IEEE Engineering in Medicine and Biology Society (EMBC '11)*, 2011.

[30] Hefron R, Borghetti B, Christensen J, *et al.* Deep long short-term memory structures model temporal dependencies improving cognitive workload estimation. *Pattern Recogn Lett.* 2017;94:96–04.

[31] Christian M, Camille J, and Fabien L. EEG based workload estimation across affective contexts. *Front Neurosci.* 2014;8:00114.

[32] Ryan H, Brett B, Christine S, *et al.* Cross-participant EEG-based assessment of cognitive workload using multi-path convolutional recurrent neural networks. *Sensors.* 2018;18(5):1339.

[33] Anu H, Kristian L, Jussi K, *et al.* Estimating brain load from the EEG. *Sci World J.* 2009;9:639–651.

[34] Sharma V and Ahirwal M. Quantification of Mental Workload Using a Cascaded Deep One-dimensional Convolution Neural Network and Bi-directional Long Short-Term Memory Model; 2021. TechRxiv. Preprint.

[35] Lim W, Olga S, and Lipo W. STEW: simultaneous task EEG workload dataset. *IEEE Dataport.* 2018.

[36] Chakladar D, Shubhashis D, Roy P, *et al.* EEG-based mental workload estimation using deep BLSTM-LSTM network and evolutionary algorithm. *Biomed Signal Process Control.* 2020;60:101989.

[37] Siddhad G, Anmol G, Dogra D, *et al.* Efficacy of transformer networks for classification of raw EEG *data.* 2022, 1–9. doi:10.48550/arXiv.2202.05170.

[38] Dan W, Yier L, Liang H, *et al.* Detection of the driver's mental workload level in smart and autonomous systems using physiological signals. *Math Prob Eng.* 2022;2022:1–13.

[39] Zhu G, Zong F, Zhang H, *et al.* Cognitive load during multitasking can be accurately assessed based on single channel electroencephalography using graph methods. *IEEE Access.* 2021;9:33102–33109.

[40] Shuo Y, Zhong Y, Yagang W, *et al.* Assessing cognitive mental workload via EEG signals and an ensemble deep learning classifier based on denoising autoencoders. *Comput Biol Med.* 2019;109:159–170.

[41] Zhang J, Yin Z, and Wang R. Recognition of mental workload levels under complex human–machine collaboration by using physiological features and adaptive support vector machines. *IEEE Trans Human–Mach Syst.* 2015;45 (2):200–214.

[42] Zixuan C, Zhong Y, and Jianhua Z. Recognition of cognitive load with a stacking network ensemble of denoising autoencoders and abstracted neurophysiological features. *Cogn Neurodynam*. 2021;15:425–437.
[43] Zhong Y and Jianhua Z. Cross-subject recognition of operator functional states via EEG and switching deep belief networks with adaptive weights. *Neurocomputing*. 2017;260:266–281.
[44] Zhong Y and Jianhua Z. Cross-session classification of mental workload levels using EEG and an adaptive deep learning model. *Biomed Signal Process Control*. 2017;33:30–47.
[45] Jianhua Z and Sunan L. A deep learning scheme for mental workload classification based on restricted Boltzmann machines. *Cogn Tech Work*. 2017;19:1–25.
[46] Fatemeh F, Zhuo Z, Goh W, *et al.* Inter-subject transfer learning with an end-to-end deep convolutional neural network for EEG-based BCI. *J Neural Eng*. 2019;16:0226007.
[47] Hong Z, Chen Y, Guojun D, *et al.* EEG classification of driver mental states by deep learning. *Cogn Neurodynam*. 2018;12:597–606.
[48] Nathan K and Contreras-Vidal J. Negligible motion artifacts in scal electroencephalography (EEG) during treadmill walking. *Front Hum Neurosci*. 2016;9:708.
[49] Delorme A, Sejnowski T, and Makeig S. Enhanced detection of artifacts in EEG data using higher-order statistics and independent component analysis. *NeuroImage*. 2007;34:1443–1449.
[50] Kline J, Huang H, Snyder K, *et al.* Isolating gait-related movement artifacts in electroencephalography during human walking. *J Neural Eng*. 2015;12:046022.
[51] Kilicarslan A, Grossman R, and Contreras-Vidal J. A robust adaptive denoising framework for real-time artifact removal in scalp EEG measurements. *J Neural Eng*. 2016;13:026013.
[52] Ang K and Guan C. Inter-subject transfer learning with end-to-end deep convolutional neural network for EEG-based BCI. *J Neural Eng*. 2019;16:026007.
[53] Jiao Z, Gao X, Wang Y, *et al.* Deep convolutional neural networks for mental load classification based on EEG data. *Pattern Recognit*. 2018;76:582–595.
[54] Zeng H, Yang C, Dai G, *et al.* Classification of driver mental states by deep learning. *Cogn Neurodyn*. 2018;12(6):597–606.
[55] Hefron R, Borghetti B, Schubert K, *et al.* Cross-participant EEG-based assessment of cognitive workload using multi-path convolutional recurrent neural networks. *Sensors*. 2018;1:1339.
[56] Bashivan P, Rish I, Yeasin M, *et al.* Learning representations from eeg with deep recurrent convolutional neural networks, 2015. ArXiv:1511.06448.
[57] Fea L. Deep models for engagement assessment with scarce label information. *IEEE Trans Hum-Mach Syst*. 2017;47:598–605.

[58] Kuanar S, Athitsos V, Pradhan N, *et al.* Cognitive analysis of working memory load from EEG, by a deep recurrent neural network. In *Proceedings of the IEEE International Conference on Acoustics, Speech and Signal Processing (ICASSP)*, 2018.

[59] Yin Z and Zhang J. Recognition of cognitive task load levels using single channel EEG and stacked denoising autoencoder. In *Chinese Control Conference*, 2016, vol. 2016.

[60] Zhang J and Li S. A deep learning scheme for mental workload classification based on restricted Boltzmann machines. *Cogn Technol Work.* 2017;19:607–631.

[61] Hefron R, Borghetti B, Christensen J, *et al.* Deep long short-term memory structures model temporal dependencies improving cognitive workload estimation. *Pattern Recognit Lett.* 2017;94:96–104.

[62] Khare S, Varun B, and Upadhyay N. Classification of emotions from EEG signals using time-order representation based on the S-transform and convolutional neural network. *Electron Lett.* 2020;56:1359–1361.

[63] Khare S and Varun B. A hybrid decision support system for automatic detection of Schizophrenia using EEG signals. *Comput Biol Med.* 2022;141:105028.

[64] Khare S and Varun B. A self-learned decomposition and classification model for schizophrenia diagnosis. *Comput Biol Med.* 2021;211:106450.

[65] Khare S and Varun B. A CACDSS for automatic detection of Parkinson's disease using EEG signals. In *Power and Signal Processing (CAPS)*, 2021.

[66] Khare S, Varun B, and Gaikwad NB. VHERS: a novel variational mode decomposition and Hilbert transform-based EEG rhythm separation for automatic ADHD detection. *IEEE Trans Instrum Meas.* 2022;71:4008310.

[67] Yin Z and Zhang J. Cross-session classification of mental workload levels using EEG and an adaptive deep learning model. *Biomed Signal Process Control.* 2017;33:30–47.

[68] Varshney A, Ghosh S, Padhy S, *et al.* Automated classification of mental arithmetic tasks using recurrent neural network and entropy features obtained from multi-channel EEG signals. *Electronics.* 2021;10:1079.

[69] Hajinoroozi M, Mao Z, and Huang Y. Prediction of driver's drowsy and alert states from EEG signals with deep learning. In *Proceedings of the IEEE International Workshop on Computational Advances in Multi-Sensor Adaptive Processing*, 2015.

[70] Prinzel L, Freeman F, Scerbo M, *et al.* Effects of a psychophysiological system for adaptive automation on performance, workload, and the event-related mation on performance, workload, and the event-related potential P300 component. *Hum Factors.* 2003;45(4):601–614.

[71] Prinzel L, Freeman F, Scerbo M, *et al.* A closed-loop system for examining psychophysiological measures for adaptive task allocation. *Int J Aviat Psychol.* 2000;10(4):393–410.

[72] Freeman F, Scerbo M, Mikulka P, *et al.* An evaluation of an adaptive automation system using a cognitive vigilance task. *Biol Psychol.* 2004;67 (3):283–297.

[73] Bailey N, Freeman F, Scerbo M, *et al.* Comparison of a brain-based adaptive system and a manual adaptable system for invoking automation. *Hum Factors.* 2006;48(4):693–709.

[74] Freeman F, Scerbo M, and Mikulka P. Effects of a biocybernetic system on vigilance performance. *Hum Factors.* 2002;44(4):654–664.

[75] Freeman F, Scerbo M, and Pea M. Evaluation of a psychophysiologically controlled adaptive automation system, using performance on a tracking task. *Appl Psychophysiol Biofeedback.* 2000;25(2):103–115.

[76] Freeman F, Scerbo M, and Pea M. A closed loop system for examining psychophysiological measures for adaptive task allocation. *Int J Aviat Psychol.* 2000;10(4): 393–410.

[77] Hockey G, Nickel P, Roberts A, *et al.* Sensitivity of candidate markers of psychophysiological strain to cyclical changes in manual control load during simulated process control. *Appl Ergon.* 2009;40(6):1011–1018.

[78] Gevins A and Smith M. Neurophysiological measures of cognitive workload during human–computer interaction. *Theor Issues Ergon Sci.* 2003;4 (1–2):113.

[79] Giraudet L, Imbert J, Berenger M, *et al.* The neuroergonomic evaluation of human machine interface design in air traffic control using behavioral and EEG/ERP measures. *Behav Brain Res.* 2015;294:246–253.

[80] Kathner I, Wriessnegger S, Muller-putz G, *et al.* Effects of mental workload and fatigue on the P300, alpha and theta band power during operation of an ERP (P300) brain–computer interface. *Biol Psychol.* 2014;102:118–129.

[81] Lei S, Welke S, and Roetting M. Driver's mental workload assessment using EEG data in a dual task paradigm. In *Proceeding of the International Technical Conference on the Enhanced Safety of Vehicle*, 2009.

[82] Raabe M, Rutschmann R, Schrauf M, *et al.* Neural correlates of simulated driving: auditory oddball responses dependent on workload. In *Foundations of Augmented Cognition*. Mahwah, NJ: Lawrence Erlbaum Associates, Inc. Publishers, 2005.

[83] Allison B and Polich J. Workload assessment of computer gaming using a single-stimulus event related potential paradigm. *Biol Psychol.* 2008;77 (3):277–283.

[84] Lean Y, Shan F, Xuemei Q, *et al.* Effects of mental workload on long-latency auditory evoked-potential, salivary cortisol, and immunoglobulin. *Neurosci Lett.* 2011;80(1):31–34.

[85] Miller M, Rietschel J, Mcdonald C, *et al.* A novel approach to the physiological measurement of mental workload. *Int J Psychophysiol.* 2011;80:75–78.

[86] Brouwer A, Hogervorst M, Van J, *et al.* Estimating workload using EEG spectral power and ERPs in the n-back task. *J Neural Eng.* 2012;9 (4):045008.

[87] Fallahi M, Motamedzade M, Heidarimoghadam R, *et al.* Assessment of operators' mental workload using physiological and subjective measures in cement, city traffic and power plant control centers. *Heal Promot Perspect.* 2016;6(2):96–103.

[88] Noyes J and Bruneau D. A self-analysis of the NASA-TLX workload measure. *Ergonomics*. 2007;50(4):514–519.

[89] Yin Z and Zhang J. Cross-subject recognition of operator functional states via EEG and switching deep belief networks with adaptive weights. *Neurocomputing*. 2017;260:349–366.

[90] Fea L. Deep models for engagement assessment with scarce label information. *IEEE Trans Hum-Mach Syst*. 2017;47:598–605.

[91] Siuly S, Khare S, Bajaj V, *et al.* A computerized method for automatic detection of schizophrenia using EEG signals. *IEEE Trans Neural Syst Rehab Eng*. 2020;28(11):2390–2400.

[92] Khare S and Bajaj V. Time–frequency representation and convolutional neural network-based emotion recognition. *IEEE Trans Neural Networks Learn Syst*. 2021;32(7):2901–2909.

[93] Khare S, Varun B, Sachin T, *et al.* Multiclass sleep stage classification using artificial intelligence based time-frequency distribution and CNN. In *Artificial Intelligence-Based Brain-Computer Interface*.

[94] Khare S, Varun B, Sachin T, *et al.* Classification of mental states from rational dilation wavelet transform and bagged tree classifier using EEG signals. In *Artificial Intelligence-Based Brain-Computer Interface*, 2022.

[95] Longo L, Wickens C, Hancock P, *et al.* Human mental workload: a survey and a novel inclusive definition. *Front Psychol*. 2022;13:883321.

[96] Ryu K and Rohae M. Evaluation of mental workload with a combined measure based on physiological indices during a dual task of tracking and mental arithmetic. *Int J Ind Ergon*. 2005;35:991–1009.

[97] Alexander C, Yongtian H, and Jose C. Deep learning for electro-encephalogram (EEG) classification tasks: a review. *J Neural Eng*. 2019;16 (031001):031001.

[98] Borghini G, Astolfi L, Vecchiato G, *et al.* Measuring neurophysiological signals in aircraft pilots and car drivers for the assessment of mental workload, fatigue and drowsiness. *Neurosci Biobehav Rev*. 2014;44:58–75.

[99] Murata A. An attempt to evaluate mental workload using wavelet transform of EEG. *Hum Factors*. 2005;47(3):498–508.

[100] Mathan S, Feyereisen T, and Whitlow S. WorkSense: exploring the feasibility of human factors assessment using electrophysiological sensors. In *Proceedings of the ICACS*, 2007.

[101] Wilson G. An analysis of mental workload in pilots during flight using multiple psychophysiological measures. *Int J Aviat Psychol*. 2002;12 (1):63–77.

[102] Gea B. A neurophysiological training evaluation metric for Air Traffic Management. In *Proceedings of the International Conference of the IEEE Engineering in Medicine and Biology Society*, 2014.

[103] Lei S and Roetting M. Influence of task combination on EEG spectrum modulation for driver workload estimation. *Hum Factors*. 2011;53(2):168–179.

[104] Pea A. Air traffic-controllers (ATCO): neurophysiological analysis of training and workload. *Ital J Aerosp Med*. 2015;12:35.

[105] Pea A. Reliability over time of EEG-based mental workload evaluation during Air Traffic Management (ATM) tasks. In *Proceedings of the International Conference of the IEEE Engineering in Medicine and Biology Society (EMBC)*, 2015.

[106] Smith M, Gevins A, Brown H, *et al.* Monitoring task loading with multivariate EEG measures during complex forms of human-computer interaction. *Hum Factors*. 2001;43(3):366–380.

[107] Zhang J, Yin Z, and Wang R. Design of an adaptive human-machine system based on dynamical pattern recognition of cognitive task-load. *Front Neurosci*. 2017;11:1–18. https://doi.org/10.3389/fnins.2017.00129.

[108] Cea B. EEG correlates of task engagement and mental workload in vigilance, learning, and memory tasks. *Aviat Space Environ Med*. 2007;78(5): B231–B244.

[109] Roy R, Charbonnier S, Campagne A, *et al.* Efficient mental workload estimation using task using task-independent EEG features. *J Neural Eng*. 2016;13:026019.

Chapter 8

Automatic detection of Parkinson's disease using non-linear signal decomposition and machine learning techniques

Smith K. Khare[1], Nikhil B. Gaikwad[1], Hesam Akbari[2], Varun Bajaj[3] and G.R. Sinha[4]

Abstract

The manual techniques for detecting Parkinson's disease (PD) include computer examination tests, interviews, and activity-based recognition. However, such methods are burdensome, error-prone, and time-consuming. Therefore, there is a need for an automatic decision-making support system for the timely detection of PD. This chapter explores the utility of electroencephalogram (EEG) signals for accurate PD detection. The EEG provides detailed information about distinct neural characteristics during PD. But, the nonstationary and nonlinear nature of EEG makes its analysis difficult, resulting in reduced system performance. Therefore, this chapter presents an empirical wavelet transform-based nonlinear analysis of EEG signals. Various hidden characteristics (statistical features) are extracted from the decomposed multi-component EEG and classified using nonlinear multiple machine learning techniques. The effectiveness of the proposed model is tested by evaluating various performance measures and comparing them with existing state-of-the-art methods. The proposed model has obtained the highest accuracy of 98.22% and 98.65% in classifying EEG events of healthy control (HC) subjects vs. PD patients with the absence of medication (PDAM) and HC vs. PD with medication (PDOM). In addition to this, five performance matrices are also evaluated to test system performance. The developed model automatically detects PD from different brain regions. The brain's frontal area provides insightful information in PD detection with the highest accuracy. In contrast, the temporal part of the brain has obtained the least accuracy of PD detection. The developed model is ready to be deployed for PD detection by neuro experts in real-time scenarios.

[1]Department of Electrical and Computer Engineering, Aarhus University, Denmark
[2]Department of Biomedical Engineering, Islamic Azad University, South Tehran Branch, Iran
[3]Electronics and Communication Discipline, Indian Institute of Information Technology Design and Manufacturing, India
[4]International Institute of Information Technology Bangalore (IIITB), India

Keywords: Electroencephalogram; Parkinson's disease detection; Empirical wavelet transform; feature extraction; Machine learning techniques

8.1 Introduction

Parkinson's disease (PD) is a neurodegenerative condition of the brain characterized by number of motory and non-motory symptoms. It is the second most common neurodegenerative disorder after Alzheimer's disease [1]. It is a result of weakening or dying of the neurons that control the motory movement. The non-motory symptoms include disorders related to mental health, cognitive impairment, abnormality in sleep, pain, and disturbances of sensors while the motory symptoms are rigidity in movement, tremors, imbalanced walking, and slow movement. In addition, involuntary movements and painful involuntary muscle contractions resulted in speech mobility and difficulty in carrying routine activities [2]. The reports published by Global Burden of Disease has reported 1.02 million cases of PD in the year 2017 [1]. In addition, 6.1 million cases have been reported in 2016, which shows an increase of 21.7% from 1990 to 2016 [3]. The risk of developing PD increases with the age. It is most common that people after the age of 50 years develop PD. The probability of developing PD in men is 1.5 times higher than that of women [3]. The increasing number of PD cases demands timely detection and proper medication for PD to perform routine activities.

Many methods have been proposed over the years to detect and diagnose PD which include interviewing, hand-writing tests, imaging, and signaling. The interviewing and hand-writing techniques are burdensome, biased, and prone to human errors. The imaging techniques like positron emission tomography scan, computerized tomography scan, and magnetic resonance imaging are costly, require additional recordings, and are bulky when compared with signaling techniques like electroencephalogram (EEG) signals [4,5]. In addition, EEG has been extensively used in different applications like mental state detection, drowsiness, attention deficit hyperactivity disorder, and emotions [6–9]. Non-linear decomposition techniques play a vital role in extracting meaningful information from non-stationary EEG signals. Various methodologies have been developed to detect PD using EEG signals. Analysis of different EEG rhythms has been explored for detecting PD using fast Fourier transform (FFT) [10]. A machine learning and analysis of variance method has been developed to present the preliminary results to detect PD [11]. A technique combining smoothed pseudo-Vigner-Wille distribution-based time–frequency representation and a convolutional neural network approach have been developed to detect PD [12]. In [13], a wavelet and short-time Fourier transform (STFT) technique has been developed in combination with various classifiers to detect PD. A wavelet-based decomposition to extract different rhythms and subbands from the EEG signals has been used to detect PD. Various features from these rhythms and subbands are extracted and classified using support vector machine (SVM), Naive Bayes, artificial neural networks (ANN), random under-sampling boosted tree, and linear discriminant analysis classifiers [14–16].

A non-linear techniques combing recurrent quantification analysis and statistical features of lower and higher order have been analyzed to detect PD in [17,18]. A tunable Q wavelet transform technique has been explored to detect PD [2,19]. Detection of PD has also been explored using autoregressive model and wavelet packed entropy features [20]. The combination of wavelet decomposition and FFT techniques has been explored for detecting PD [21]. In [22,23], Mel Frequency Cepstral Coefficients along with wavelet techniques have been used using SVM classifier to separate health control (HC) and PD EEG segments. A convolutional neural network has been explored to detect the HC and PD EEG segments [24]. A model composed of minimum average maximum tree and singular value decomposition has been explored in [25]. A variational mode decomposition combined with extreme learning machine classifier has been used to detect PD [26].

This chapter presents a combination of non-linear decomposition techniques and minimal features, which are combined to provide input to different non-linear machine learning techniques to detect PD. At first, the EEG signals are combined to form brain regions. The combined EEG signals are then fed to the empirical wavelet transform (EWT) technique to obtain different subbands (SBs) levels. Five features are identified as significant features extracted from the SB, classified using various classifiers, and evaluated the performance of the best performing classifier. Contributions to the chapter are as follows:

1. Exploring the analysis of different brain regions to find the strength of separation of EEG signals.
2. Existing techniques like FFT, STFT, and wavelet transform require a predefined basis function for representing the EEG signals. In this chapter, an EWT technique has been explored to extract meaningful information from the EEG.
3. Identifying the minimal number of features required to be fed to classifiers to obtain the highest performance of the model.
4. Lastly, find the best machine learning model that provides the optimal performance for classifying the HC and PD EEG segments.

The remaining of chapter is organized as follows: Section 8.2 provides material and methods, in Section 8.3 results of the methodology are discussed, and finally, Section 8.4 provides the gist of the chapter.

8.2 Material and methods

This section provides the details about the proposed model. The model composed of four sections namely dataset, empirical wavelet transform, features, and machine learning techniques. The pictorial flow of the proposed model is shown in Figure 8.1.

8.2.1 Dataset

The method is implemented on a publicly available EEG dataset of PD and HC. The dataset was recorded at the University of San Diego and curated at the University of Oregon. The EEG records were collected using a 32-channel ActiveTwo biosemi

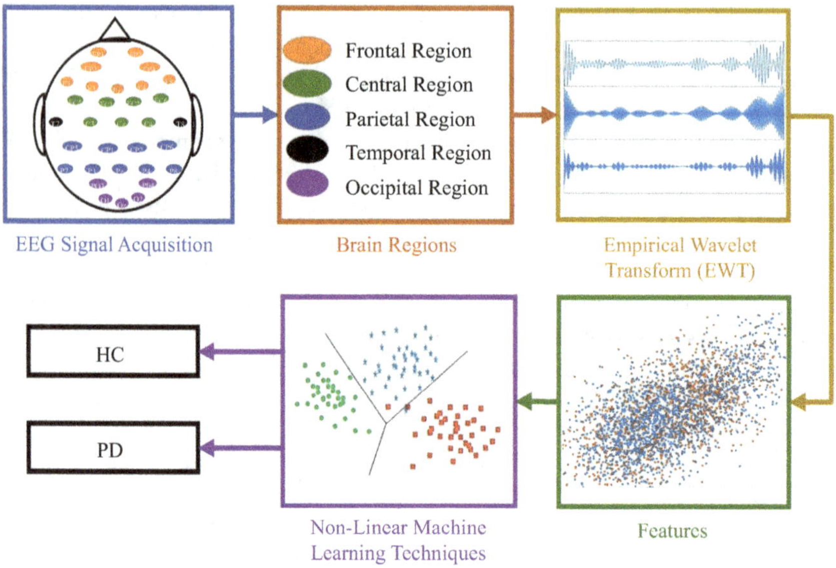

Figure 8.1 Pictorial view of the proposed PD detection model

instrumentation system. The sampling frequency for recording the EEG data was set at 512 Hz, and data were recorded for 3 min. The recordings were composed of EEG signals from 15 PD (seven males and eight females) subjects and 16 HC (nine females and seven males) subjects. The mean age of HC subjects was 63.5 $\pm$ 9.667 years, and the mean age of PD subjects was 63.26 $\pm$ 8.198 years, respectively. The EEG recordings of PD subjects were recorded in two sessions, one with medication (PDOM) and one without medication (PDAM). During the PDAM recordings, the patients refrained from taking medication for at least 12 h before the signal acquisition. While collecting the data, the subjects were asked to sit comfortably. The subjects from both the classes were aged, matched and righthanded. This chapter uses the 2s epochs of EEG recordings. Therefore, a final signal comprises 1,024 samples in each EEG segment of HC, PDAM, and PDOM subjects. Before proceeding toward signal analysis, we partitioned the 32 channels into five brain regions. The brain regions were categorized as frontal (FR), central (CR), parietal (PR), occipital (OR), and temporal (TR). The FR composed of channels 1, 2, 3, 4, 27, 28, 29, 30, 31, CR composed of 5, 6, 8, 23, 25, 26, 32 channels, PR with channel numbers 9, 10, 11, 12, 13, 19, 20, 21, 22, OR having the channel configuration of 14, 15, 16, 17, 18, and finally, TR with channel numbers 7, 24. The details of the dataset are available in [27–30]. The brain regions composed of FR, CR, PR, TR, and OR with their corresponding channels are shown in Figure 8.2.

8.2.2 Empirical wavelet transform (EWT)

EWT is an effective tool for analyzing a non-stationary EEG signal. It is an approach that builds adaptive wavelets capable of generating or extracting

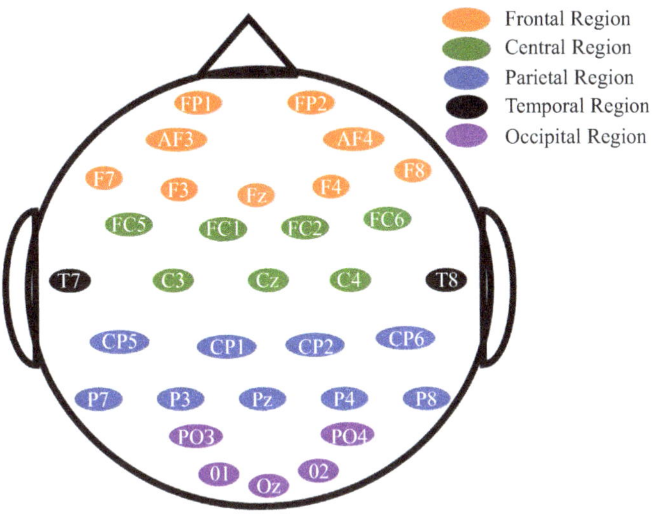

Figure 8.2 Constitution of various brain regions and their corresponding channels

amplitude-modulated (AM)–frequency-modulated (FM) components. The developed AM–FM components have a compact support Fourier spectrum. These components are Fourier spectrum equivalent segments and applying the filtering to detect each corresponding support [31,32]. The detailed coefficients are represented as inner products with empirical wavelets given as

$$\gamma_f^\delta(n,t) = \langle f, \theta_n \rangle = \int f(\tau)\overline{\theta_n(\tau - t)}d\tau = (\widehat{f}(\omega)\overline{\widehat{\theta}_n(\omega)})^\nu \tag{8.1}$$

The approximation coefficient by the inner product with the scaling function is given as

$$\gamma_f^\delta(0,t) = \langle f, \psi_1 \rangle = \int f(\tau)\overline{\psi_1(\tau - t)}d\tau = (\widehat{f}(\omega)\overline{\widehat{\psi}_1(\omega)})^\nu \tag{8.2}$$

where the empirical scaling function and empirical wavelet functions are represented by $\widehat{\theta}_n(\omega)$ and $\widehat{\psi}_n(\omega)$. The function $f(t)$ is the non-stationary EEG signal, τ is the shift. The reconstruction of the signal is obtained by the following relation represented as

$$\begin{aligned} f(t) &= \gamma_f^\delta(0,t) * \psi_1(t) + \sum_{n=1}^{N} \gamma_f^\delta(n,t) * \theta_n(t) \\ &= (\widehat{\gamma}_f^\delta(0,t) * \widehat{\psi}_1(t) + \sum_{n=1}^{N} \widehat{\gamma}_f{}^\delta(n,t) * \widehat{\theta}_n(t))^\nu \end{aligned} \tag{8.3}$$

The illustrative subbands obtained after EWT decomposition are shown in Figure 8.3.

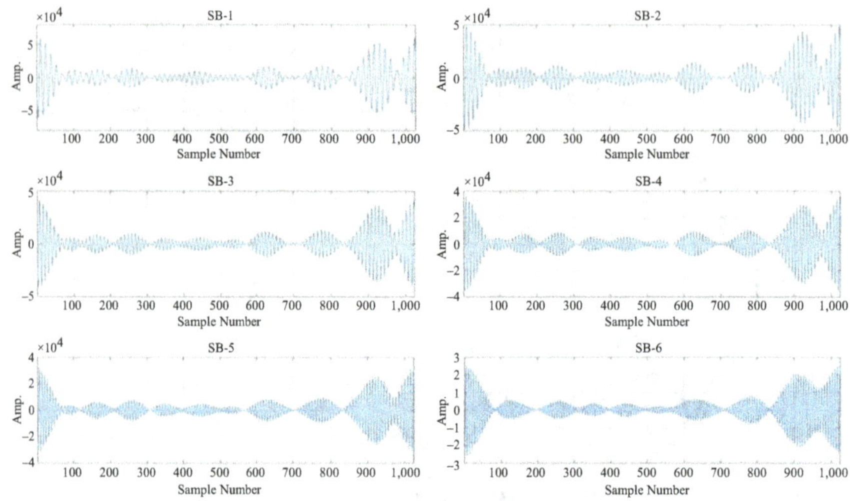

Figure 8.3 Exemplary subbands obtained from EWT decomposition

8.2.3 Features

This chapter uses five features as statistical measures obtained from the subbands of different brain regions. Features play a key role in reducing the dimensionality of the EEG records and smaller input dimension to the classifier. In this work, five features were used as significant measures for classification. These features are namely zero crossing rate (ZCR), wavelet entropy (WAVENT), spectral flatness (SPEC FLAT), Kraskov Entropy (KEN), and permutation entropy (PERM ENT). ZCR is a measure of sign change (negative to positive and vice versa) in an array between two adjacent samples. It returns 1 when one sample is positive and another is negative, while it returns zero for other cases. Kraskov entropy is an estimator of the Shannon entropy, WAVENT indicates the energy distribution of the signal, PERM ENT is a measure of complexity of a dynamic signal which captures order relations between values of a time series and extracts a probability distribution of the ordinal patterns. At the same time, SPEC FLAT is the ratio of geometric and arithmetic mean of subbands in a power spectrum (also known as Weiner entropy). The feature details are available in [2, 26, 33, 34].

8.2.4 Non-linear machine learning techniques

Machine learning has brought a significant revolution in healthcare technologies by automating the decision-making process. This chapter uses supervised machine-learning algorithms to draw a boundary between the features of HC and PD subjects. Since no standard available can tell which algorithm needs to be used for the desired application, we have deployed multiple classifiers to test the model performance. These models are support vector machine (SVM), ensemble bagged tree, k-nearest neighbor (KNN), narrow neural network (narrow-NN), and bilayered NN classifiers. However, these classifiers work on different cost functions to draw the decision boundaries. If the features are not scaled to a standard range, the decision may be biased, resulting in degraded system performance. Therefore, a *Z*-score normalization has been used before applying an input to the feature matrix. The details about the classifiers are available in [35–37].

8.3 Results

As mentioned earlier, the dataset comprises 15 PD and 16 HC subjects, respectively. The data is acquired at a sampling rate of 512 Hz, and a 2s epoch has been used to analyze the EEG signals. Therefore, each EEG segment is composed of 1,024 samples. Five brain regions have been used for analysis: FR, PR, CR, OR, and TR, each with a combination of 9, 9, 7, 5, and 2 channels. Simulations have been carried out on an Intel(R) Core(TM) i7-8550U CPU @ 1.80 GHz with 16 GB RAM 64-bit operating system. A MATLAB® 2021a version has been used to perform the simulations. A common experimental platform is maintained to evaluate the performance of the model. The boundary for EWT has been set uniformly and empirically at [4,8,13,30]. Distinct values of subbands are generated while decomposing the signals using EWT. However, we have taken only 6 SBs for our analysis (minimum values of SBs obtained). We have performed a 10-fold cross-validation evaluation to evaluate the machine learning models. In this, an input data matrix is divided into ten equal parts. Nine parts are used for training, and the remaining is used for testing, repeated 10 times. The final evaluation of the performance matrices is obtained by taking the mean of testing results obtained for each iteration. The default parameters are used for each classifier to evaluate the performance. Data is standardized for SVM, KNN, and NN. For SVM, the Gaussian kernel is used with automatic scaling of the kernel, while for KNN, the number of neighbors set at 10, Euclidean matrix with squared inverse evaluation is used. For narrow NN and Bilayered NN, three layered models are used with a neuron combination of [10 10 2] with maximum iterations set at 1,000. Finally, for the ensemble bagged tree classifier, the number of learners was 10, a number of splits at 3030, and a learning rate of 0.01 was used.

In the first part, we performed the statistical test to check the discrimination ability of the features. For this, we have evaluated the mean and standard deviation (STD) for each feature of HC, PDAM, and PDOM classes in different brain regions. Tables 8.1–8.5 show the statistical properties of the parts of CR, FR, OR, PR, and TR regions. It is evident from these tables that features of each class and every SB has non-overlapping mean and STD, reflecting their better discrimination ability. The results of Tables 8.1–8.5 motivate us to use these feature sets for further evaluation.

In the second step, the most discriminant feature set is identified, and a feature matrix is formed by combining it for each brain region. This feature matrix is normalized by using *Z*-score normalization to scale all the features to a standard range to avoid bias in decision-making. The normalized feature matrix is applied to various classifiers to test the classification accuracy of each SB. Figures 8.4–8.8

Table 8.1 Statistical properties (mean ± STD) of the CR features for HC, PDAM, and PDOM classes

Central	Mean ± STD				
Features Class	**KEN**	**WAVENT**	**SPEC FLAT**	**ZCR**	**PERM ENT**
	HC				
SB1	15.33 ± 0.01	913.25 ± 0.81	0.363 ± 0.001	0.235 ± 0.001	0.689 ± 0.0012
SB2	8.01 ± 0.048	5837.74 ± 45.38	0.01 ± 0.002	0.576 ± 0.018	0.6928 ± 0.0005
SB3	8.87 ± 0.046	3284.95 ± 28.48	0.01 ± 0.003	0.542 ± 0.025	0.6925 ± 0.0008
SB4	10.09 ± 0.071	1934.84 ± 16.75	0.023 ± 0.003	0.593 ± 0.025	0.6915 ± 0.0013
SB5	11.76 ± 0.046	1192.46 ± 6.9	0.038 ± 0.003	0.608 ± 0.025	0.691 ± 0.0016
SB6	13.85 ± 0.027	774.85 ± 2.15	0.142 ± 0.006	0.625 ± 0.006	0.6917 ± 0.0006
Class	PDOM				
SB1	14.85 ± 0.031	861.06 ± 3.02	0.268 ± 0.007	0.234 ± 0.004	0.6886 ± 0.0029
SB2	8.44 ± 0.048	6110.46 ± 47.28	0.007 ± 0.002	0.563 ± 0.02	0.6928 ± 0.0004
SB3	9.13 ± 0.06	3380.5 ± 29.61	0.014 ± 0.006	0.518 ± 0.027	0.6926 ± 0.0007
SB4	10.02 ± 0.076	1895.89 ± 17.84	0.021 ± 0.007	0.532 ± 0.029	0.6918 ± 0.0015
SB5	11.3 ± 0.099	1123.53 ± 11.84	0.033 ± 0.008	0.574 ± 0.039	0.6908 ± 0.0024
SB6	13.28 ± 0.09	725.57 ± 5.4	0.095 ± 0.011	0.65 ± 0.026	0.6883 ± 0.0019
Class	PDAM				
SB1	14.91 ± 0.019	902.16 ± 0.77	0.43 ± 0.001	0.202 ± 0.01	0.6836 ± 0.0023
SB2	8.07 ± 0.049	5875.49 ± 51.85	0.01 ± 0.002	0.571 ± 0.019	0.6929 ± 0.0014
SB3	8.91 ± 0.05	3303.37 ± 28.52	0.012 ± 0.003	0.547 ± 0.03	0.6925 ± 0.0009
SB4	10.01 ± 0.062	1914.47 ± 15.38	0.02 ± 0.002	0.53 ± 0.026	0.6921 ± 0.0011
SB5	11.44 ± 0.076	1144.03 ± 8.84	0.053 ± 0.004	0.57 ± 0.029	0.6905 ± 0.002
SB6	13.22 ± 0.023	721.4 ± 2.96	0.104 ± 0.005	0.545 ± 0.021	0.6898 ± 0.0014

Table 8.2 Statistical properties (mean ± STD) of the FR features for HC, PDAM, and PDOM classes

Frontal	**Mean ± STD**				
Features	**KEN**	**WAVENT**	**SPEC FLAT**	**ZCR**	**PERM ENT**
Class	**HC**				
SB1	15.55 ± 0.013	930.06 ± 0.55	0.425 ± 0.001	0.143 ± 0.003	0.6711 ± 0.0038
SB2	8.47 ± 0.052	6228.51 ± 48.98	0.01 ± 0.002	0.563 ± 0.02	0.6928 ± 0.0005
SB3	9.19 ± 0.067	3430.54 ± 33.32	0.011 ± 0.004	0.529 ± 0.027	0.6926 ± 0.0007
SB4	10.22 ± 0.074	1964.55 ± 17.56	0.02 ± 0.004	0.565 ± 0.03	0.692 ± 0.0014
SB5	11.84 ± 0.056	1200.13 ± 8.36	0.035 ± 0.003	0.589 ± 0.032	0.6901 ± 0.0019
SB6	13.82 ± 0.032	770.33 ± 3.33	0.132 ± 0.005	0.671 ± 0.012	0.6858 ± 0.0015
Class	PDOM				
SB1	15.28 ± 0.024	896.44 ± 0.73	0.363 ± 0.006	0.208 ± 0.002	0.6849 ± 0.0034
SB2	8.84 ± 0.064	6530.86 ± 56.95	0.008 ± 0.003	0.546 ± 0.02	0.6928 ± 0.0005
SB3	9.46 ± 0.06	3563.69 ± 32.68	0.017 ± 0.007	0.513 ± 0.028	0.6925 ± 0.0008
SB4	10.38 ± 0.067	2002.64 ± 18.8	0.028 ± 0.009	0.512 ± 0.03	0.6923 ± 0.0012
SB5	11.61 ± 0.088	1169.8 ± 11.45	0.042 ± 0.01	0.602 ± 0.046	0.6891 ± 0.0035
SB6	13.6 ± 0.055	747.52 ± 3.75	0.096 ± 0.01	0.65 ± 0.021	0.6925 ± 0.0005
Class	PDAM				
SB1	15.16 ± 0.024	930.71 ± 0.59	0.616 ± 0.001	0.184 ± 0.003	0.6861 ± 0.0017
SB2	8.64 ± 0.048	6407.12 ± 49.05	0.012 ± 0.002	0.539 ± 0.02	0.6927 ± 0.0005
SB3	9.34 ± 0.058	3522.29 ± 31.23	0.015 ± 0.003	0.524 ± 0.032	0.6926 ± 0.0007
SB4	10.35 ± 0.067	1996.42 ± 18.38	0.023 ± 0.003	0.495 ± 0.031	0.6919 ± 0.0015
SB5	11.84 ± 0.071	1198.29 ± 10.15	0.057 ± 0.004	0.594 ± 0.037	0.6917 ± 0.0015
SB6	13.55 ± 0.041	745.91 ± 3.78	0.132 ± 0.007	0.555 ± 0.029	0.6894 ± 0.0018

Table 8.3 Statistical properties (mean ± STD) of the OR features for HC, PDAM, and PDOM classes

Occipital	**Mean ± STD**				
Features	**KEN**	**WAVENT**	**SPEC FLAT**	**ZCR**	**PERM ENT**
Class	**HC**				
SB1	15.38 ± 0.019	907.5 ± 0.71	0.391 ± 0.001	0.241 ± 0.001	0.6749 ± 0.0011
SB2	8.21 ± 0.058	6060.34 ± 53.81	0.011 ± 0.002	0.563 ± 0.02	0.6929 ± 0.0004
SB3	9.02 ± 0.048	3363.97 ± 27.09	0.016 ± 0.003	0.526 ± 0.026	0.6925 ± 0.0008
SB4	10.27 ± 0.057	1972.49 ± 15.43	0.024 ± 0.003	0.562 ± 0.029	0.6919 ± 0.0014
SB5	11.87 ± 0.055	1210.6 ± 8.01	0.056 ± 0.004	0.582 ± 0.025	0.6911 ± 0.0014
SB6	13.97 ± 0.035	787.31 ± 3.21	0.21 ± 0.007	0.628 ± 0.011	0.6908 ± 0.0018
Class	PDOM				
SB1	15.13 ± 0.021	884.75 ± 2.73	0.428 ± 0.007	0.269 ± 0.009	0.6914 ± 0.0015
SB2	8.77 ± 0.054	6426.49 ± 50.18	0.011 ± 0.003	0.546 ± 0.02	0.6929 ± 0.0004
SB3	9.39 ± 0.061	3502.69 ± 31.76	0.02 ± 0.008	0.516 ± 0.028	0.6925 ± 0.0008
SB4	10.33 ± 0.079	1970.82 ± 18.74	0.031 ± 0.011	0.49 ± 0.03	0.6922 ± 0.0013
SB5	11.6 ± 0.11	1166.88 ± 13.55	0.057 ± 0.012	0.552 ± 0.038	0.69 ± 0.0025
SB6	13.65 ± 0.068	750.49 ± 4.89	0.093 ± 0.007	0.614 ± 0.019	0.6896 ± 0.0024
Class	PDAM				
SB1	15.24 ± 0.016	914.72 ± 0.86	0.435 ± 0.001	0.174 ± 0.005	0.6849 ± 0.0051
SB2	8.29 ± 0.044	6138.83 ± 45.91	0.01 ± 0.002	0.539 ± 0.021	0.6929 ± 0.0004
SB3	9.02 ± 0.054	3374.51 ± 30.92	0.013 ± 0.005	0.541 ± 0.029	0.6925 ± 0.0009
SB4	10.16 ± 0.055	1950.28 ± 16.72	0.017 ± 0.003	0.516 ± 0.03	0.6924 ± 0.001
SB5	11.57 ± 0.059	1160.14 ± 9.03	0.051 ± 0.003	0.553 ± 0.034	0.6914 ± 0.0017
SB6	13.19 ± 0.025	724.22 ± 3.56	0.143 ± 0.008	0.591 ± 0.026	0.6867 ± 0.0027

Table 8.4 Statistical properties (mean ± STD) of the PR features for HC, PDAM, and PDOM classes

Parietal	**Mean ± STD**				
Features	**KEN**	**WAVENT**	**SPEC FLAT**	**ZCR**	**PERM ENT**
Class	**HC**				
SB1	15.27 ± 0.017	904.75 ± 1.41	0.345 ± 0.002	0.268 ± 0.001	0.6908 ± 0.0011
SB2	8.08 ± 0.051	5935.79 ± 51.3	0.011 ± 0.002	0.594 ± 0.02	0.6928 ± 0.0004
SB3	8.95 ± 0.045	3333.07 ± 28.18	0.014 ± 0.003	0.522 ± 0.025	0.6927 ± 0.0006
SB4	10.18 ± 0.066	1948.89 ± 14.34	0.022 ± 0.003	0.56 ± 0.027	0.6922 ± 0.0011
SB5	11.72 ± 0.052	1194.46 ± 7.51	0.057 ± 0.004	0.575 ± 0.027	0.6897 ± 0.002
SB6	13.91 ± 0.028	782.99 ± 2.55	0.202 ± 0.007	0.595 ± 0.01	0.6897 ± 0.0016
Class	PDOM				
SB1	15.05 ± 0.024	874.09 ± 1.99	0.331 ± 0.006	0.278 ± 0.011	0.6895 ± 0.0022
SB2	8.57 ± 0.05	6241.13 ± 45.85	0.009 ± 0.003	0.561 ± 0.02	0.6928 ± 0.0004
SB3	9.26 ± 0.062	3440.04 ± 31.4	0.014 ± 0.006	0.516 ± 0.028	0.6926 ± 0.0008
SB4	10.15 ± 0.078	1930.93 ± 19.66	0.024 ± 0.008	0.511 ± 0.033	0.6921 ± 0.0012
SB5	11.44 ± 0.108	1148.9 ± 12.42	0.044 ± 0.009	0.572 ± 0.036	0.6919 ± 0.0015
SB6	13.43 ± 0.075	736.57 ± 5.2	0.094 ± 0.01	0.645 ± 0.025	0.6885 ± 0.0025
Class	PDAM				
SB1	15.07 ± 0.02	896.83 ± 1.14	0.36 ± 0.002	0.19 ± 0.005	0.6835 ± 0.0026
SB2	8.1 ± 0.046	5944.38 ± 47.11	0.01 ± 0.002	0.56 ± 0.019	0.6928 ± 0.0004
SB3	8.93 ± 0.053	3328.89 ± 30.62	0.012 ± 0.003	0.549 ± 0.027	0.6926 ± 0.0007
SB4	10.05 ± 0.054	1920.12 ± 15.79	0.02 ± 0.003	0.546 ± 0.029	0.6923 ± 0.0011
SB5	11.52 ± 0.07	1154.36 ± 9.16	0.045 ± 0.003	0.59 ± 0.028	0.6905 ± 0.0016
SB6	13.27 ± 0.034	728.69 ± 3.62	0.128 ± 0.009	0.579 ± 0.025	0.6901 ± 0.0007

Table 8.5 Statistical properties (mean ± STD) of the TR features for HC, PDAM, and PDOM classes

Temporal	**Mean ± STD**				
Features Class	**KEN HC**	**WAVENT**	**SPEC FLAT**	**ZCR**	**PERM ENT**
SB1	14.95 ± 0.02	905.12 ± 0.35	0.488 ± 0.002	0.176 ± 0.056	0.6927 ± 0
SB2	8.27 ± 0.051	6133.28 ± 48.42	0.011 ± 0.002	0.569 ± 0.02	0.6929 ± 0.0003
SB3	9.04 ± 0.045	3377.33 ± 26.72	0.013 ± 0.002	0.521 ± 0.027	0.6926 ± 0.0008
SB4	10.37 ± 0.058	1982.5 ± 15.51	0.023 ± 0.004	0.588 ± 0.027	0.6925 ± 0.0009
SB5	11.75 ± 0.054	1198.36 ± 8.78	0.048 ± 0.004	0.576 ± 0.032	0.6908 ± 0.0017
SB6	13.73 ± 0.033	774.55 ± 2.97	0.153 ± 0.008	0.524 ± 0.014	0.6893 ± 0.0021
Class	PDOM				
SB1	15.04 ± 0.013	902.18 ± 1.56	0.41 ± 0.002	0.191 ± 0.008	0.6885 ± 0.0018
SB2	8.61 ± 0.054	6331.11 ± 52.82	0.009 ± 0.002	0.546 ± 0.02	0.6928 ± 0.0004
SB3	9.23 ± 0.059	3464.8 ± 30.5	0.012 ± 0.004	0.519 ± 0.028	0.6922 ± 0.0011
SB4	10.13 ± 0.072	1945.9 ± 20.08	0.022 ± 0.007	0.494 ± 0.028	0.6924 ± 0.0011
SB5	11.52 ± 0.092	1169.56 ± 10.36	0.052 ± 0.013	0.626 ± 0.045	0.691 ± 0.0021
SB6	13.48 ± 0.05	745.97 ± 2.85	0.124 ± 0.011	0.67 ± 0.012	0.6828 ± 0.0036
Class	PDAM				
SB1	15.1 ± 0.018	913.42 ± 0.24	0.376 ± 0.001	0.103 ± 0.004	0.6797 ± 0.0005
SB2	8.36 ± 0.047	6171.53 ± 48.42	0.008 ± 0.002	0.542 ± 0.021	0.6928 ± 0.0004
SB3	9.05 ± 0.051	3372.19 ± 30.6	0.007 ± 0.003	0.542 ± 0.029	0.6926 ± 0.0008
SB4	10.1 ± 0.055	1945.72 ± 16.09	0.013 ± 0.002	0.536 ± 0.025	0.6925 ± 0.0009
SB5	11.5 ± 0.064	1160.35 ± 8.12	0.037 ± 0.01	0.596 ± 0.027	0.6925 ± 0.0008
SB6	13.17 ± 0.072	719.93 ± 3.11	0.076 ± 0.005	0.534 ± 0.007	0.6927 ± 0.0003

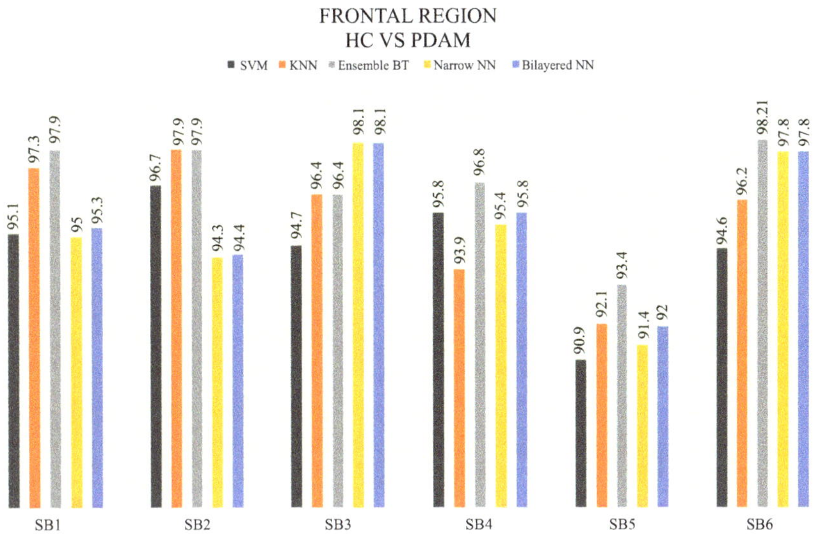

Figure 8.4 Accuracy obtained using different classifiers for SB1–SB6 in FR for HC vs. PDAM classification

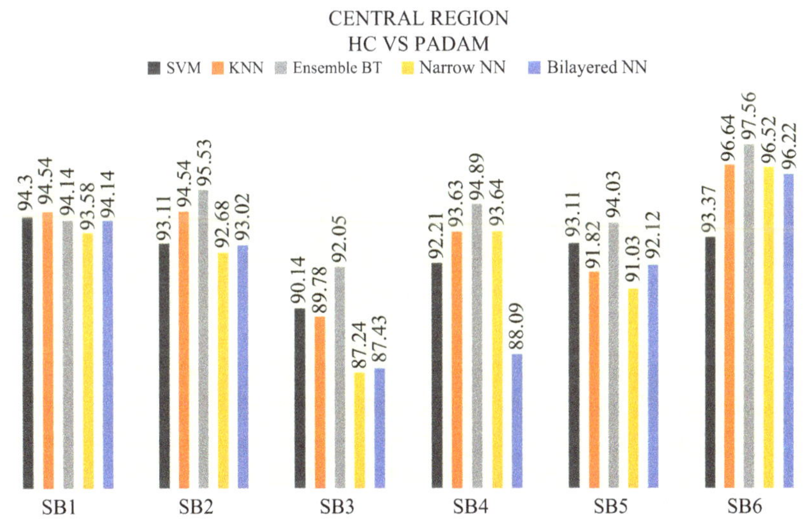

Figure 8.5 Accuracy obtained using different classifiers for SB1–SB6 in CR for HC vs. PDAM classification

show the accuracy obtained using different classification techniques for HC vs. PDAM classification. It can be evident from Figures 8.4 to 8.8 that the FR is the best among all for detecting the changes happening in the brain during HC and PDAM states. The EEG signals of FR show more discriminant properties, due to

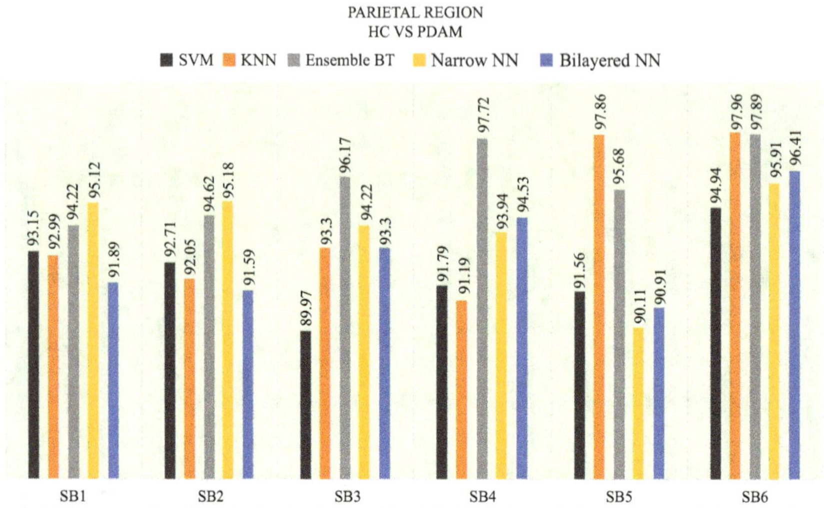

Figure 8.6 Accuracy obtained using different classifiers for SB1–SB6 in PR for HC vs. PDAM classification

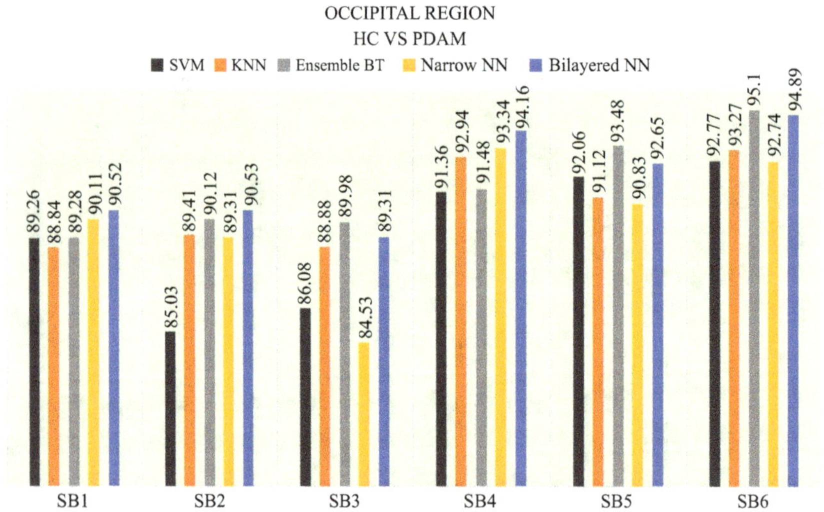

Figure 8.7 Accuracy obtained using different classifiers for SB1–SB6 in OR for HC vs. PDAM classification

which the highest accuracy has been obtained in the classification of HC vs. PDAM. The voting scheme to detect the best classification algorithm shows that the ensemble bagged tree classifier has the highest votes for each brain region. The second and third-best classification algorithms are Bilayered NN and

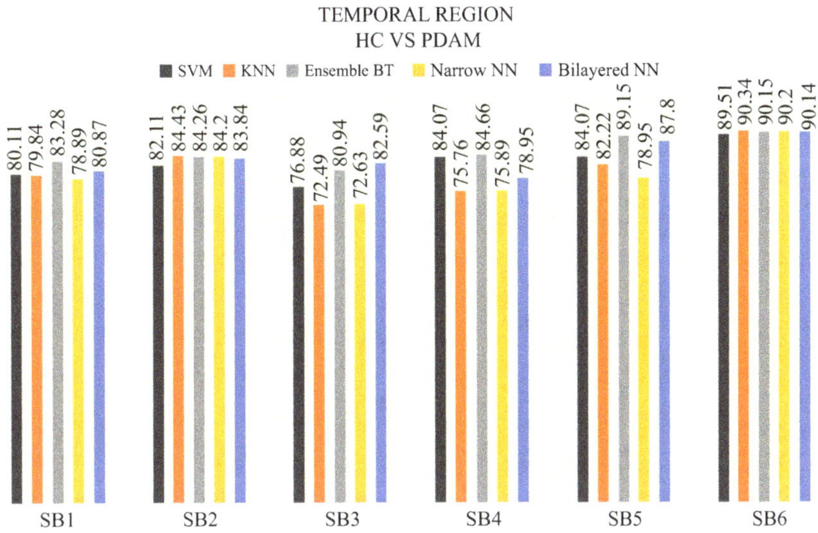

Figure 8.8 Accuracy obtained using different classifiers for SB1–SB6 in TR for HC vs. PDAM classification

narrow NN, followed by the KNN classifier, while SVM is the worst-performing classification technique. SB-6 has provided the best classification performance for each FR, CR, PR, TR, and OR, among all the SB, respectively. In HC vs. PDAM classification for FR, SB-3 and SB-2 are second and third best, while the least classification accuracy has been obtained for SB-5. For CR, OR, and TR, SB-3 has turned out to be the worst-performing SB. In the case of PR, SB-2, and SB-3 collectively show the worst performance in classifying HC vs. PDAM cases.

Similarly, Figures 8.9–8.13 show the accuracy obtained using different classification techniques for HC vs. PDOM classification. It can be evident from Figures 8.9 to 8.13 that the FR is the best among all for detecting the changes happening in the brain during HC and PDAM states. The EEG signals of the frontal part have more distinguishable patterns for HC and PDOM classes reflecting its better classification performance. The ensemble bagged tree provides the highest accuracy among all other classifiers, whereas the SB-6 provides the highest classification among all different subbands. In HC vs. PDOM classification for FR, CR, and PR, SB-1 delivers the worst performance, while for OR and TR, SB-3 and SB-4 show the worst performance in the classification of HC vs. PDOM cases.

Figures 8.4–8.13 show that the ensemble bagged tree has provided the best classification rate for classifying HC vs PDAM and HC vs. PDOM EEG segments in SB-6. This motivates us to use this classifier to evaluate system performance in SB-6. For this, we have considered six performance parameters for

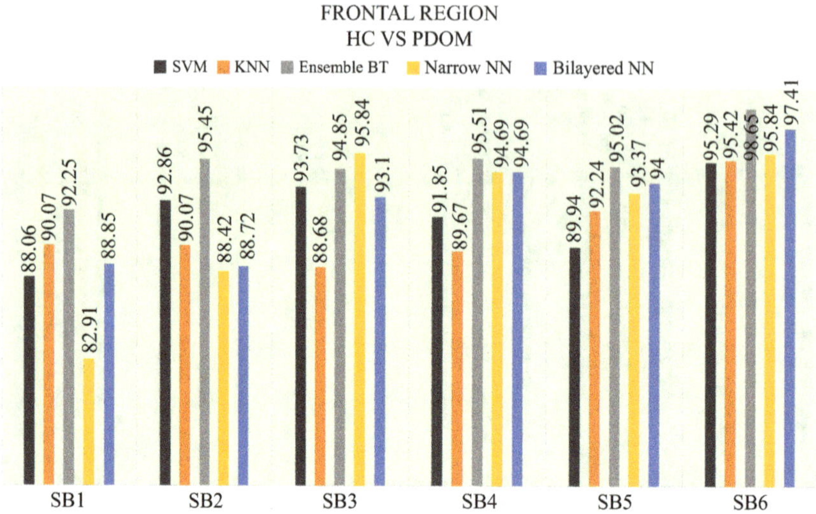

Figure 8.9 Accuracy obtained using different classifiers for SB1–SB6 in FR for HC vs. PDOM classification

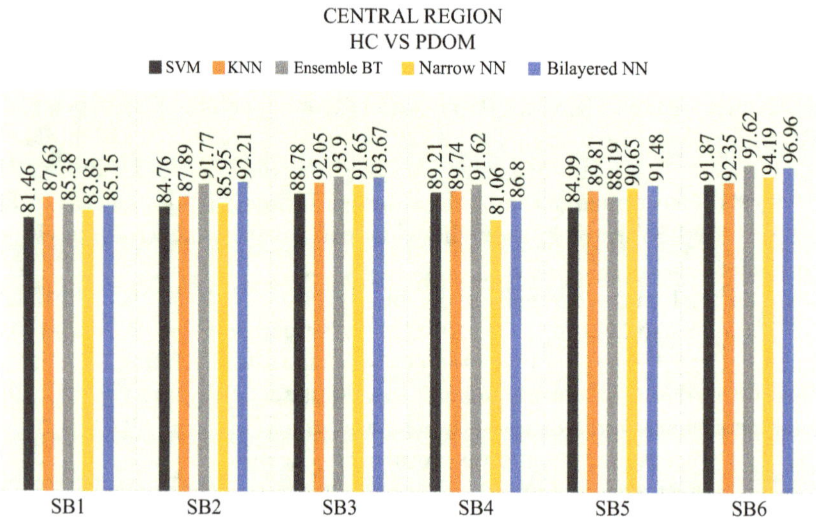

Figure 8.10 Accuracy obtained using different classifiers for SB1–SB6 in CR for HC vs. PDOM classification

the binary classification of HC vs. PDAM and HC vs. PDOM, which are accuracy (ACC), F1-score (F1), negative predictive value (NPV), sensitivity (SEN), precision (PRC), and specificity (SPE). The mathematical representation of these performance measures is represented as

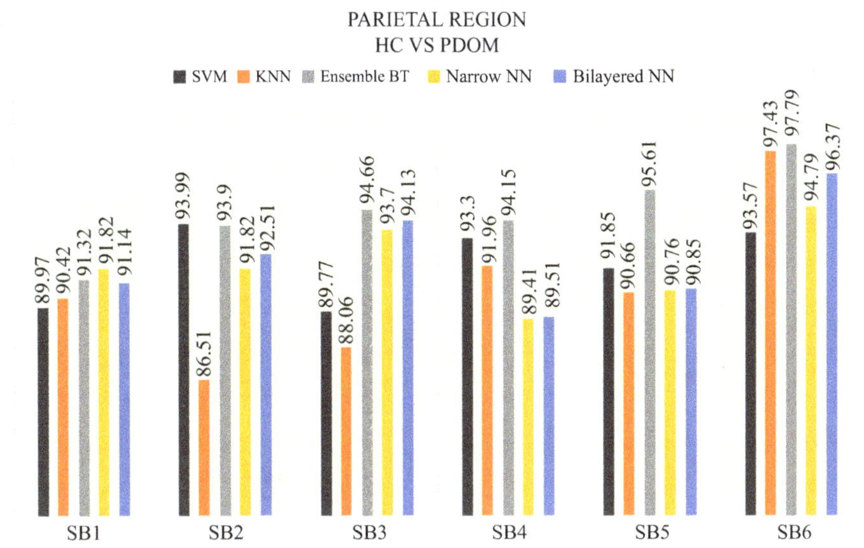

Figure 8.11 Accuracy obtained using different classifiers for SB1–SB6 in PR for HC vs. PDOM classification

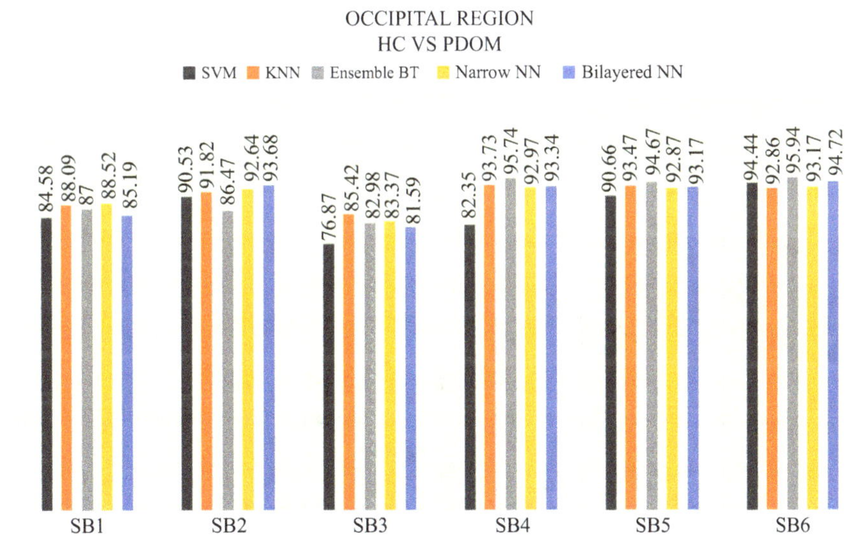

Figure 8.12 Accuracy obtained using different classifiers for SB1–SB6 in OR for HC vs. PDOM classification

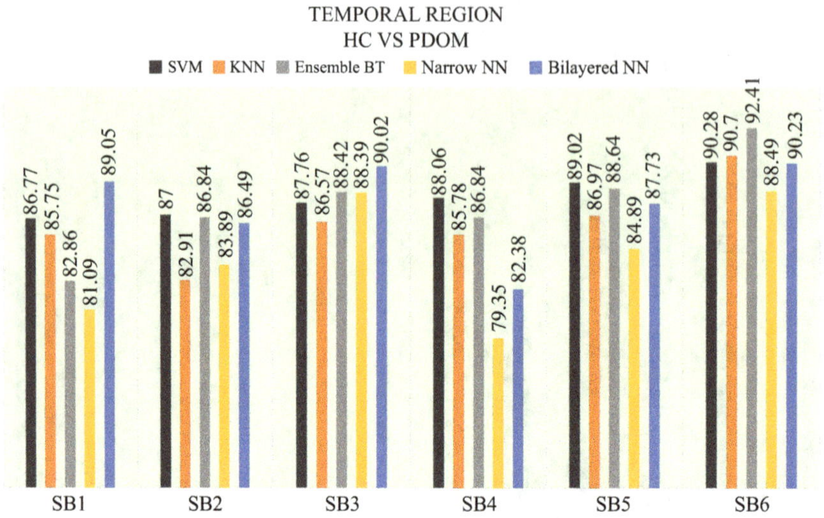

Figure 8.13 Accuracy obtained using different classifiers for SB1–SB6 in TR for HC vs. PDOM classification

$$\begin{aligned} ACC &= \frac{P_T + N_T}{P_T + N_T + P_F + N_F} \\ PRC &= \frac{P_T}{P_T + P_F} \\ SEN &= \frac{P_T}{P_T + N_F} \\ SPE &= \frac{N_T}{N_T + P_F} \\ F1 &= \frac{2(PRC * SEN)}{(PRC + SEN)} \\ NPV &= \frac{N_T}{N_T + N_F} \end{aligned} \tag{8.4}$$

where P_T, N_T, P_F, and N_F are true positive, true negative, false positive, and false negative events.

Table 8.6 shows the performance obtained for both cases using the ensemble bagged tree classifier. It can be seen from the table that the highest ACC, SEN, SPE, PRC, F1, and NPV obtained for FR in HC vs. PDAM classification are 98.22%, 99.86%, 96.71%, 96.53%, 98.17%, and 99.87%, while for HC vs. PDOM, it is again provided highest in FR with a score of 98.65%, 99.32%, 98%, 97.93%, 98.62%, and 99.35%, respectively.

Finally, the receiver operating characteristics (ROC) and the area under the curve (AUC) have obtained to check the binary classification ability of the

Table 8.6 Performance measure of HC vs. PDAM and HC vs. PDOM obtained for ensemble bagged tree classifier

Case	**PDAM**					**PDOM**				
Region	**FR**	**CR**	**PR**	**OR**	**TR**	**FR**	**CR**	**PR**	**OR**	**TR**
ACC	98.22	97.56	97.89	95.12	90.86	98.65	97.62	97.79	95.94	92.41
SEN	99.86	97.09	98.45	95.61	90.74	99.32	97.16	97.54	95.68	91.17
SPE	96.71	98.02	97.35	94.65	90.99	98.00	98.09	98.03	96.20	93.69
PRC	96.53	98.00	97.27	94.47	90.80	97.93	98.07	98.00	96.13	93.73
F1	98.17	97.54	97.85	95.04	90.77	98.62	97.61	97.77	95.91	92.43
NPV	99.87	97.13	98.50	95.76	90.93	99.35	97.19	97.58	95.76	91.12

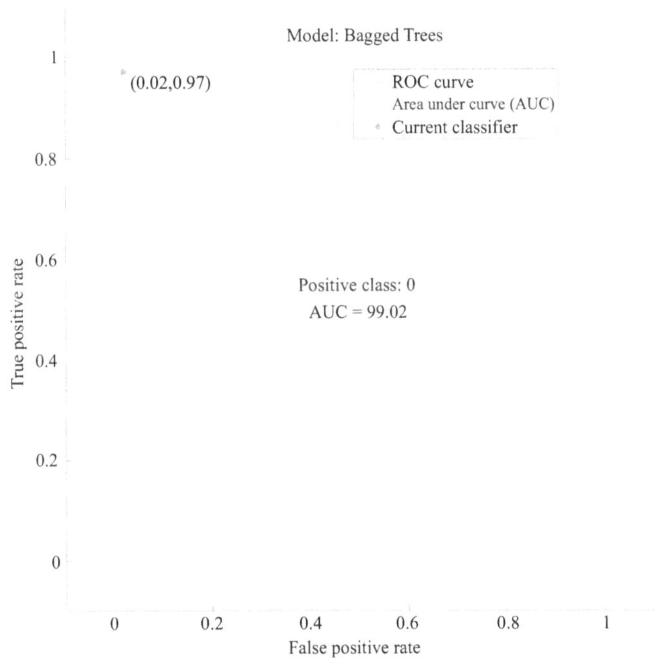

Figure 8.14 AUC and ROC obtained for HC vs. PDAM classification in SB-6

ensemble bagged tree classifier for HC vs. PDAM and HC vs. PDOM. Figures 8.14 and 8.15 provide the ROC and AUC curves for HC vs. PDAM and HC vs. PDOM classification obtained in SB-6. An AUC of 99.02% and 99.86% has been obtained for the classification of HC vs. PDAM and HC vs. PDOM showing the strength of binary classification.

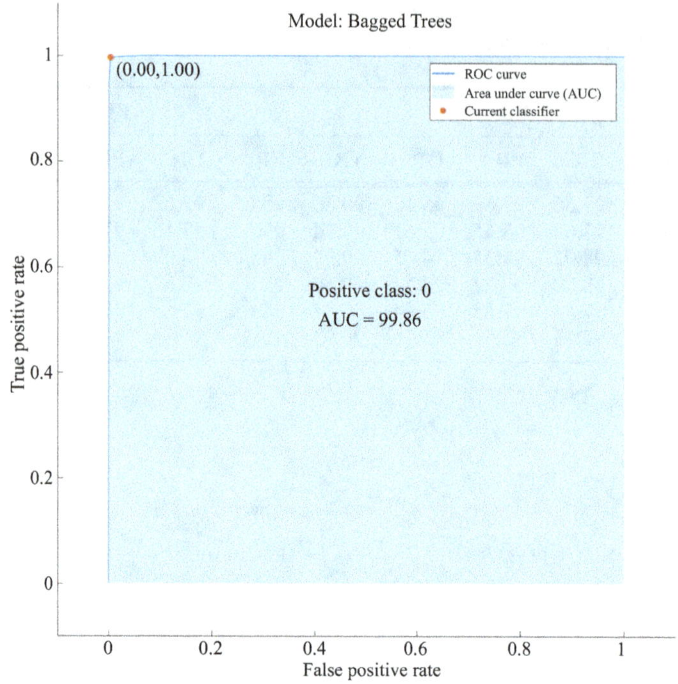

Figure 8.15 AUC and ROC obtained for HC vs. PDOM classification in SB-6

8.4 Conclusion

This chapter explores the utility of EWT-based non-linear decomposition and non-linear machine learning classifier techniques for automatically detecting PD using EEG signals. This chapter concludes that the brain's frontal region provides EEG's most distinguishable pattern, which has obtained the highest performance. In contrast, the temporal region has provided the least performance depicting its overlapping nature in the EEG pattern of HC and PD subjects. Out of many subbands obtained using EWT decomposition, SB-6 has provided the most representative characteristics and discriminable features, with the highest classification performance for all the classification techniques. Finally, the ensemble classifier has turned out to be the best-performing classifier among SVM, KNN, narrow NN, and bilayered NN classifiers. Due to this, it has provided the highest classification rate of 98.22% and 98.65% in the classification of HC vs. PDAM and HC vs. PDOM classes. The EEG patterns of HC and PD patients with medication are more distinguishable than those generated for HC and PD patients without medication. This model is ready to be deployed for the real-time detection of PD in complex scenarios with a simplistic approach.

References

[1] Ou Z, Pan J, Tang S, *et al.* Global trends in the incidence, prevalence, and years lived with disability of Parkinson's disease in 204 countries/territories from 1990 to 2019. *Frontiers in Public Health*. 2021;9:776847.

[2] Khare SK, Bajaj V, and Acharya UR. Detection of Parkinson's disease using automated tunable Q wavelet transform technique with EEG signals. *Biocybernetics and Biomedical Engineering*. 2021;41(2):679–689.

[3] The facts about Parkinson's Disease [homepage on the Internet]. Massachusetts, USA: Harvard Health Publishing; Harvard Medical School. https://www.health.harvard.edu/diseases-and-conditions/the-facts-about-parkinsons-disease.

[4] Khare SK and Bajaj V. A hybrid decision support system for automatic detection of Schizophrenia using EEG signals. *Computers in Biology and Medicine.* 2022;141:105028.

[5] Khare SK and Bajaj V. A self-learned decomposition and classification model for schizophrenia diagnosis. *Computer Methods and Programs in Biomedicine.* 2021;211:106450.

[6] Khare SK, Bajaj V, Sengur A, *et al.* Classification of mental states from rational dilation wavelet transform and bagged tree classifier using EEG signals. In: *Artificial Intelligence-Based Brain-Computer Interface*. New York, NY: Elsevier; 2022. p. 217–235.

[7] Khare SK, Bajaj V, and Sinha G. *Automatic Drowsiness Detection Based on Variational Non-Linear Chirp Mode Decomposition Using Electroencephalogram Signals*. Vol. 1. IOP Science; 2020.

[8] Khare SK, Gaikwad NB, and Bajaj V. VHERS: a novel variational mode decomposition and Hilbert transform-based EEG rhythm separation for automatic ADHD detection. *IEEE Transactions on Instrumentation and Measurement.* 2022;71:1–10.

[9] Khare S, Nishad A, Upadhyay A, *et al.* Classification of emotions from EEG signals using time-order representation based on the S-transform and convolutional neural network. *Electronics Letters*. 2020;56(25):1359–1361.

[10] Pezard L, Jech R, and Ržička E. Investigation of non-linear properties of multichannel EEG in the early stages of Parkinson's disease. *Clinical Neurophysiology*. 2001;112(1):38–45.

[11] Betrouni N, Delval A, Chaton L, *et al.* Electroencephalography-based machine learning for cognitive profiling in Parkinson's disease: preliminary results. *Movement Disorders.* 2019;34(2):210–217.

[12] Khare SK, Bajaj V, and Acharya UR. PDCNNet: an automatic framework for the detection of Parkinson's disease using EEG signals. *IEEE Sensors Journal.* 2021;21(15):17017–17024.

[13] Obukhov YV, Gabova A, Zaljalova Z, *et al.* Electroencephalograms features of the early stage Parkinson's disease. *Pattern Recognition and Image Analysis*. 2014;24(4):593–604.

[14] Baby MS, Saji A, and Kumar CS. Parkinsons disease classification using wavelet transform based feature extraction of Gait data. In: *2017 International Conference on Circuit, Power and Computing Technologies (ICCPCT)*. IEEE; 2017. p. 1–6.
[15] Jeong DH, Kim YD, Song IU, *et al.* Wavelet energy and wavelet coherence as EEG biomarkers for the diagnosis of Parkinson's disease-related dementia and Alzheimer's disease. *Entropy*. 2015;18(1):8.
[16] Liu G, Zhang Y, Hu Z, *et al.* Complexity analysis of electroencephalogram dynamics in patients with Parkinson's disease. *Parkinson's Disease*. 2017;2017:8701061.
[17] Khoshnevis SA and Sankar R. Classification of the stages of Parkinson's disease using novel higher-order statistical features of EEG signals. *Neural Computing and Applications*. 2021;33(13):7615–7627.
[18] Murugappan M, Khare SK, Alshuaib W, *et al.* Electroencephalogram signals based emotion classification in Parkinson's disease using recurrence quantification analysis and non-linear classifiers. In: *Computer-Aided Design and Diagnosis Methods for Biomedical Applications*. London: CRC Press; 2021. p. 1–34.
[19] Murugappan M, Alshuaib W, Bourisly AK, *et al.* Tunable Q wavelet transform based emotion classification in Parkinson's disease using electroencephalography. *PLoS ONE*. 2020;15(11):e0242014.
[20] Han CX, Wang J, Yi GS, *et al.* Investigation of EEG abnormalities in the early stage of Parkinson's disease. *Cognitive Neurodynamics*. 2013;7(4):351–359.
[21] Rizvi SQA, Wang G, and Xing X. Early detection of Parkinson disease using wavelet transform along with Fourier transform. In: *International Conference on Smart City and Informatization*. New York, NY: Springer; 2019. p. 323–333.
[22] Soumaya Z, Taoufiq BD, Nsiri B, *et al.* Diagnosis of Parkinson disease using the wavelet transform and MFCC and SVM classifier. In: *2019 4th World Conference on Complex Systems (WCCS)*. New York, NY: IEEE; 2019. p. 1–6.
[23] Drissi TB, Zayrit S, Nsiri B, *et al.* Diagnosis of Parkinson's disease based on wavelet transform and mel frequency cepstral coefficients. *International Journal of Advanced Computer Science and Applications*. 2019;10(3):125–132.
[24] Oh SL, Hagiwara Y, Raghavendra U, *et al.* A deep learning approach for Parkinson's disease diagnosis from EEG signals. *Neural Computing and Applications*. 2020;32(15):10927–10933.
[25] Tuncer T, Dogan S, and Acharya UR. Automated detection of Parkinson's disease using minimum average maximum tree and singular value decomposition method with vowels. *Biocybernetics and Biomedical Engineering*. 2020;40(1):211–220.
[26] Khare SK and Bajaj V. A CACDSS for automatic detection of Parkinson's disease using EEG signals. In: *2021 International Conference on Control,*

Automation, Power and Signal Processing (CAPS). New York, NY: IEEE; 2021. p. 1–5.

[27] George JS, Strunk J, Mak-McCully R, *et al.* Dopaminergic therapy in Parkinson's disease decreases cortical beta band coherence in the resting state and increases cortical beta band power during executive control. *NeuroImage: Clinical.* 2013;3:261–270.

[28] Swann NC, de Hemptinne C, Aron AR, *et al.* Elevated synchrony in Parkinson disease detected with electroencephalography. *Annals of Neurology.* 2015;78(5):742–750.

[29] Jackson N, Cole SR, Voytek B, *et al.* Characteristics of waveform shape in Parkinson's disease detected with scalp electroencephalography. *eNeuro.* 2019;6(3):ENEURO.0151-19.2019.

[30] Rockhill AP, Jackson N, George J, *et al.* UC San Diego Resting State EEG data from patients with Parkinson's disease. *OpenNeuro*; 2020.

[31] Khare SK, Bajaj V, Siuly S, *et al.* Classification of schizophrenia patients through empirical wavelet transformation using electroencephalogram signals. *Modelling and Analysis of Active Biopotential Signals in Healthcare.* 2020;1:1–26.

[32] Gilles J. Empirical wavelet transform. *IEEE Transactions on Signal Processing.* 2013;61(16):3999–4010.

[33] Khare SK and Bajaj V. Entropy-based drowsiness detection using adaptive variational mode decomposition. *IEEE Sensors Journal.* 2020;21(5):6421–6428.

[34] Khare SK, Gaikwad N, and Bokde ND. An intelligent motor imagery detection system using electroencephalography with adaptive wavelets. *Sensors*. 2022;22(21):8128.

[35] Ullo SL, Khare SK, Bajaj V, *et al.* Hybrid computerized method for environmental sound classification. *IEEE Access.* 2020;8:124055–124065.

[36] Siuly S, Khare SK, Bajaj V, *et al.* A computerized method for automatic detection of schizophrenia using EEG signals. *IEEE Transactions on Neural Systems and Rehabilitation Engineering.* 2020;28(11):2390–2400.

[37] Gaikwad NB, Khare SK, Satpute N, *et al.* Hardware implementation of high-performance classifiers for edge gateway of smart automobile. In: 2022 *1st International Conference on the Paradigm Shifts in Communication, Embedded Systems, Machine Learning and Signal Processing (PCEMS)*; 2022. p. 74–77.

Chapter 9

A review on gait kinematics acquisition sensors and its advancements in IoT and machine learning

Saaveethya Sivakumar[1,2], Alpha Agape Gopalai[2], King Hann Lim[1] and Darwin Gouwanda[2]

Abstract

The increasing need for analyzing lower limb kinematics urges the application of Internet of Things (IoT) and machine learning for advancement in sensors, towards accurate, real-time, fast, and easy access measurements. Lower limb kinematics such as joint angles and motion trajectories are necessary to distinctly explain the functionality of the locomotion. In clinical gait analysis, the details of locomotor function are important to identify abnormalities such as cerebral palsy, hip osteoarthritis and stroke. In sports, lower limb kinematics are used to evaluate the performances of sports' athletes. On the other hand, robotic prosthesis devices use joint kinematics for movement control. Among the kinematics sensors, laboratory systems are the gold standard which attaches reflective markers on the wearer and capture movement using infrared camera systems. However, laboratory systems are expensive, time consuming, require experienced professionals for data recording, and limited in space which has a tendency to alter natural walking patterns. Furthermore, laboratory systems are not real time analyzed with IoT or advanced protocols. Applications of wearable kinematics sensors (e.g., inertial measurement units and force-sensitive resistor instrumented foot insoles), enabled with IoT are recognized as an emerging field for outdoor gait monitoring. Current research trends towards applications of machine learning to enhance and improve the quality of kinematics data acquisition and analysis of wearable sensors. As such, this chapter presents a detailed systematic review of kinematics systems (from traditional to the latest advancements), its potentials, limitations, and future trends especially in the IoT and machine learning field. A detailed understanding of available methodologies is important to investigate possible directions and areas to be improved.

Keywords: Gait; Kinematics; IoT; Machine learning

[1]Faculty of Engineering and Science, Curtin University Malaysia, Malaysia
[2]School of Engineering, Monash University, Malaysia

9.1 Introduction

Gait explains the manner of walking. Gait is the primary feature of survival and is considered as a health index in the medical field. Clinically, gait quantifies lower limb pathology conditions, monitor progression of treatments, applied for rehabilitation and prosthesis designs [1,2]. Walking is generally sequential and synchronized coordination of lower limb postures that are dynamic and cyclic in nature. The common term used when studying the dynamic postures during walking is referred as Gait analysis. Kinematics are one of the primary parameters used for quantifying dynamic gait motion.

Kinematics are the motion of joints and body segments which are not related to forces acted internally and externally. Gait kinematics study lower limb acceleration, position, velocity, higher derivatives of time-based properties of motion and angles. Kinematics are investigated clinically for monitoring and diagnosing gait disorders, and in sports to analyze improvements in sports practices. Clinically, kinematics in the lower limb such as joint angles and segment displacements are crucial measurements that distinctly explain the functionality of the lower limb for identifying gait abnormalities [1].

Kinematics are classified based on (1) laboratory methods and (2) wearable sensors. Figure 9.1 elaborates the classifications of gait kinematics methods. Laboratory methods are considered the gold standard, which follows transitional protocols for kinematics measurements. Laboratory methods are further classified into inverse kinematics (IK) and direct kinematics (DK). In DK, joint angles are calculated directly from the measured trajectories. IK calculates joint angles by removing the residual artefacts of the markers, through the introduction of joint constraints. Wearable sensors are widely applied in outdoor experiments and activities.

Past literature [2–5] documented the potential of gait kinematics in identifying and quantifying disorders such as fall risk, cerebral palsy, hip osteoarthritis, and stroke. Wearable robotic devices such as exoskeletons and robotic prosthesis utilize the kinematics information for its control mechanisms. Literature [6–8] reports the importance of kinematics model in lower exoskeletons in order to ensure the movements that are in the desired trajectories. Furthermore, joint kinematics information is also used in robotic prosthesis devices to control movements [9,10].

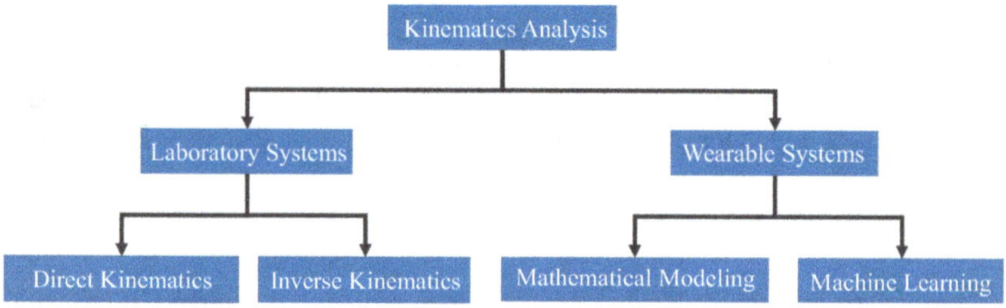

Figure 9.1 Sensor-based method for gait kinematics analysis

Sports utilize the joint kinematics information to monitor athlete performances. Joint angles are analyzed in running gait [11,12]. Phinyomark *et al.* [13] classified healthy runners using knee angle data. The growing need for gait analysis drives advancements in sensors and Internet of Things (IoT), for accurate and efficient kinematics measurements.

Each of the above discussed methods has its drawbacks that are been treated with various algorithms and techniques over the years. Research trends towards machine learning (ML) applications are powered by IoT that is applied to improve and enhance kinematics measurements. As such, this chapter presents a systematic review of different kinematics measurement sensors and systems, its potential, drawbacks, and future direction.

9.2 Search methodology and review process

The papers are selected for the review through an electronic search in cooperation with Curtin University and Monash University Library databases. The databases searched are Web of Science, SpringerLink, Google Scholar, ScienceDirect, SpringerLink, IEEE Xplore, and manual search of cited references. The procedure is shown in Figure 9.2.

The selected papers are grouped under two categories namely laboratory systems and wearable systems. Papers under laboratory systems are further grouped based on DK and IK. Similarly, papers related to wearable systems are categorized further into mathematical modeling and machine learning. The papers that are grouped as per the above processed are then reviewed and documented as per its contributions, accuracy, limitations, and future potential.

9.3 Laboratory systems

Laboratory-based systems are the gold standard for gait analysis, which are equipped with optical cameras for kinematics measurements and force platform(s) for kinetic measurements. Figure 9.3 depicts an example of a gait laboratory system. In general, a 3-dimensional (3D) optical motion capture setup is utilized to measure the kinematics of gait. Reflective markers are attached to the lower limb anatomical landmarks [14]. These markers are attached to the skin subject's skin surface with the assumption that it is directly attached to the bone/segment of interest.

The reflections of the markers are tracked by the optical camera system to map the trajectories of the movements. The number of cameras is dependent on the scale of the movement and the camera volume. However, a minimum of two cameras are required and the best performance can be acquired using a five-six camera setup [1]. The markers emit a reflection for the infrared lights released by the optical camera system. These reflections are then identified as motion trajectories that are mapped in 3D space. The position and geometry of the anatomical landmarks are identified using the captured positions and motion of the reflective markers [15].

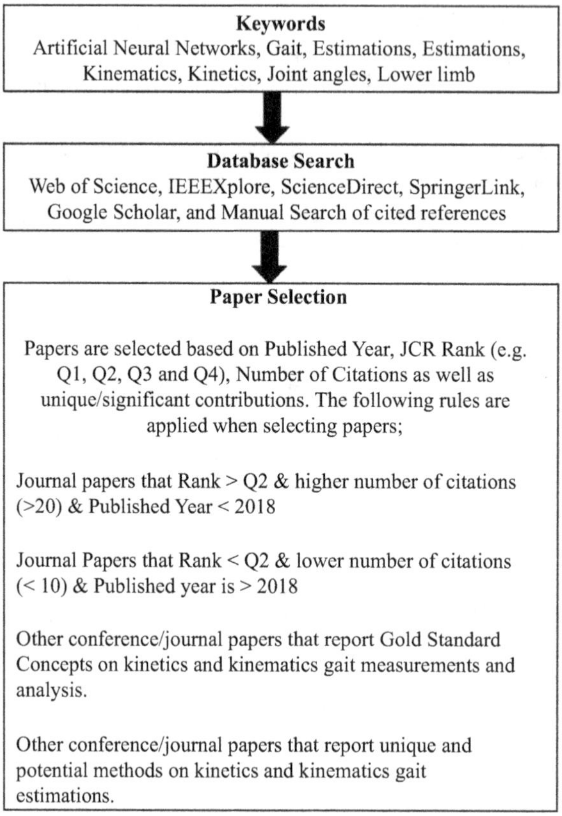

Figure 9.2 Systematic search methodology

Marker placement is performed in two methods: (1) landmark specific and (2) cluster specific. Only the main bony landmarks are considered when the markers are placed based on landmark-specific method. Helen Hayes marker set and the Newington Children's marker set are two such examples of the landmark-specific markers as shown in Figure 9.4(a) and (b) respectively. On the other hand, markers placed according to the cluster system are for segment tracking. Unlike landmark-specific markers, cluster marker systems are focused on segments and not to a specific anatomical landmark. Three-four markers are attached in a non-collinear manner on the segment of interest. Cleveland Clinical system is one of the widely used cluster marker setups in gait analysis (see Figure 9.4(a)). As seen in Figure 9.3, cluster markers are mainly attached to thigh and shank.

Vicon (Oxford, UK), Qualisys (Gothenburg, Sweden), and OptiTrack are examples of motion capture systems. A summary of laboratory-based systems is shown in Table 9.1. These systems are integrated with compatible hardware such as

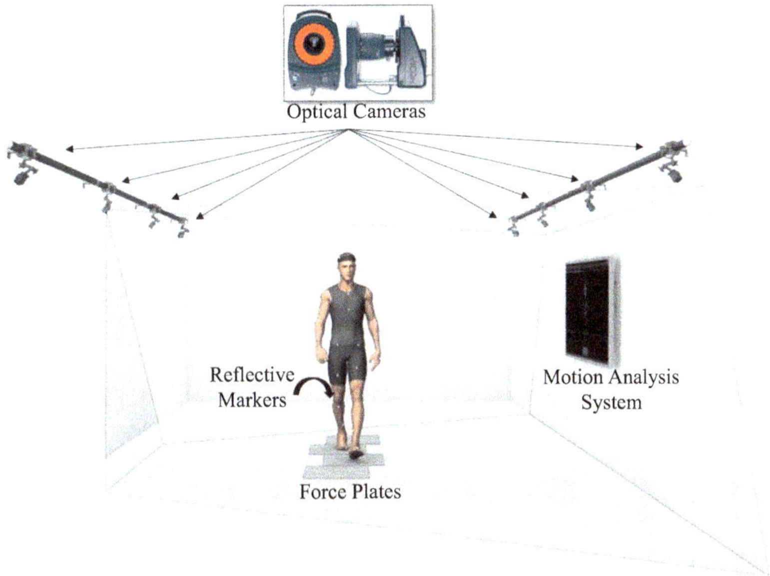

Figure 9.3 Gait laboratory setup [16]

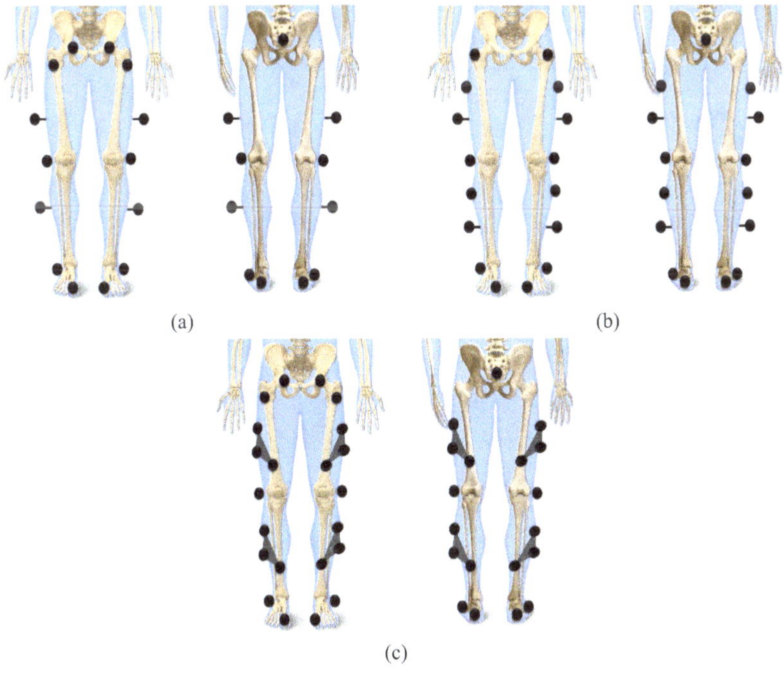

Figure 9.4 Gait marker sets. (a) Helen Hayes, (b) Newington Children's, and (c) Cleveland Clinic.

Table 9.1 Summary of motion capture systems namely: Optitrack PRIMEx, Qualisys, and Vicon [19]

Feature	**Optitrack PRIMEx**	**Qualisys**	**Vicon**
Type of system	Marker-based (passive and active)	Marker-based (passive and active)	Marker-based (passive and active)
Number of cameras	Minimum – 2 Maximum – 200+	Minimum – 2 Maximum – 200+	Minimum – 2 Maximum – 200+
Camera dimensions Height × weight × depth (cm)	Optitrack PRIMEx 13: 6.9 × 6.9 × 5.3 Optitrack PRIMEx 22: 12.6 × 12.6 × 10.7 Optitrack PRIMEx 41: 12.6 × 12.6 × 13.2	Qualisys Arqus: 13 × 14 × 13 Qualisys Miqus: 14 × 9 × 8 Qualisys Miqus Hybrid: 14 × 9 × 8	Vicon Vantage: 16.2 × 12.5 × 13.4 Vicon Vero: 8.3 × 8 × 13.5
Lense type	Prime	Narrow, standard and wide	Vicon Vantage: : Standard/Wide Vicon Vero: : Varifocal
Minimum–maximum filed view	Optitrack PRIMEx 13: 56° Optitrack PRIMEx 22: 79° Optitrack PRIMEx 41: 51°	Qualisys Arqus: 29 × 22°–82 × 48° Qualisys Miqus: 25 × 25°–80 × 53° Qualisys Miqus Hybrid: 46 × 27°–83 × 53°	Vicon Vantage: 54.7 × 54.7°–76.4 × 76.4° Vicon Vero: 44 × 23.6°–98.1× 50.1°
Maximum Sampling Frequency	Optitrack PRIMEx 13: 240 Hz Optitrack PRIMEx 22: 360 Hz Optitrack PRIMEx 41: 180 Hz	Qualisys Arqus: 10,000 fps Qualisys Miqus: 650 fps Qualisys Miqus Hybrid: 340 fps	Vicon Vantage: 2,000 fps Vicon Vero: 250 fps
Resolution	Optitrack PRIMEx 13: 1.3 MP Optitrack PRIMEx 22: 2.2 MP Optitrack PRIMEx 41: 4.1 MP	Qualisys Arqus: 26 MP Qualisys Miqus: 4 MP Qualisys Miqus Hybrid: 2 MP	Vicon Vantage: 16 MP Vicon Vero: 2.2 MP
Data collection, visualization, and platform connection	Software, connected by Ethernet	Software connected by Ethernet	Software: Evoke, Shogun, Nexus, and Tracker Connected by Ethernet
Accuracy: 3D position	Optitrack PRIMEx 13: 0.2 mm Optitrack PRIMEx 22: 0.15 mm Optitrack PRIMEx 41: 0.10 mm	Qualisys Arqus: 0.03 mm Qualisys Miqus: 0.07 mm Qualisys Miqus Hybrid: 0.11 mm	0.0017 mm

electromyography (EMG) sensors and force platforms to measure external force and muscle parameters. The motion trajectories captured from the reflective markers are processed to calculate gait kinematics such as joint angles and angular velocities. As a result, marker-based kinematic analysis can be classified into two main categories namely: (1) DK [17] and (2) IK [18].

9.3.1 Direct kinematics

DK is used by the conventional methods to compute joint angles and its derivatives. DK models can be further classified into two sections [20]: (1) landmark specific and (2) cluster specific. In landmark-specific DK, the geometry and positions of the anatomical sections are calculated by considering only the major bony landmarks of the lower limb. Examples of DK models are Newington Children's Hospital Model and Helen Hayes Model [1]. The joint kinematics of the Helen Hayes model is calculated using the Eular and Cardan angle concepts that are embedded with the motion capture systems such as Vicon Plug-in-Gait (PIG) [1]. However, landmark-specific DK methods are only focused on the main anatomical bony landmarks and it only captures movements of the specific positions the markers are attached. Therefore, limited landmarks could be tracked. These drawbacks of the DK methods are addressed using the cluster-based DK analysis.

DK are calculated by tracking motions of cluster markers that are attached on the segments and joints [21]. DK can be calculated up to 6 degrees of freedom (DOF) [21]. A manual calibration is conducted at each anatomical landmark to identify the associated segments of interest [22]. Cleveland Clinic (Motion Analysis Corp, Santa Rosa, USA) [20] is one of the widely applied cluster-based DK methods. In the Cleveland Clinic method, manual calibration is conducted to identify the anatomical landmarks by normalizing the cluster to a known/defined marker location (e.g. normalizing the thigh marker cluster to knee location). Horsak *et al.* [17] reported the reliability of the Cleveland Clinic system using Vicon PIG, where ankle and knee locations are utilized for calibrations.

DK is the fundamental method for kinematics analysis and has been widely studied over the years. The potential of DK for gait analysis is clinically proven clinically especially in quantifying stroke, obesity, and cerebral palsy [23,24]. DK calculates gait kinematics with the assumption that the markers are rigidly attached to the skin. As such, DK eliminates the requirement of marker scaling, which results in a fast and much straightforward process for joint kinematics calculations. However, identifying the non-palpable landmarks [e.g. hip joint center (HJC)] is a challenge. The joint center locations of the non-palpable landmarks are estimated using predictive and functional methods. However, an offset is generated when estimating non-palpable joint centers that could result in distortions in the measurements and joint angles with an expanded range of motions (ROMs). Literature also reports that introducing joint constraints can upsurge the sensitivity of errors [18,25].

The joint center locations are calculated based on the marker positions that locate the anatomical landmarks of the lower limb. However, the markers are placed with a general rule for all subjects, and not subject specific. DK further assumes the markers are rigidly attached to the skin. But in reality, markers tend to generate inconsistent movements on the skin generating soft tissue artefacts [26]. As a result, marker positions can be dislocated generating faulty measurements [27]. An indirect method known as "IK" is developed to eliminate the marker position error caused by the DK methods.

9.3.2 Inverse kinematics

In IK, the marker position errors are eliminated by introducing joint constraints to the predefined skeletal-joint model (see Figure 9.5). Generally, this method is known as the Global Optimization Algorithm [20], and in biomechanics, it is named as IK. Joint kinematics are calculated by first adding physical joint constraints to the model, to minimize the mean square error (MSE) calculated between the model joints and surface marker positions [20]. As such, accurate construction of the skeletal-joint model is important in IK.

The skeletal-joint model is a multilink system that comprises of multiple kinematics chains connected by joints. The human body segments and joints are denoted by the model's kinematic chains and the joints, respectively [27]. The joint states are defined by its DOFs. For example an unconstrained joint has 6 DOFs. When the joint is unconstrained, the joint motion is larger compared to the realistic range, which result in measurement defects such as soft tissue artefacts and marker position error [27]. Hence, in IK, joint constraints are introduced to restrict the relative motion between segments to minimize the error [20].

Charlton *et al.* [20] added joint constraints to the ankle, knee, and hip, which resulted in reduced standard deviation of the inter-marker distance and estimated segment lengths, with a clinically significant accuracy. Cerveri *et al.* [28] used IK to reduce the soft tissue artefacts of a model that was initially affected by surface marker obstructions and its false identifications. Furthermore, the IK algorithm uses a weighting factor to quantify the average soft tissue artefact of each segment. This weighting factor together with the joint constraints helps to generate a practically realistic skeletal-joint model that contributes to accurate kinematics calculations with a significant reduction in the surface marker error compared with DK [29].

Many user-friendly software tools for IK analysis are commercially available [25]. OpenSim [30] is an open sources software tool that uses multibody dynamics engine Simbody [31] for IK computations. OpenSim constructs the models as kinematics chains using anthropometric information which are only specific to each subject such as segment length, bone geometry, and joint center axes. AnyBody (AnyBody Technology, Aalborg, Denmark) is another software that computes IK, personalization capabilities. Visual3D (C-Motion, Germantown, MD, USA) is also another widely utilized clinically approved IK analysis software and has the ability to be interfaced with motion capture systems such as Vicon (Oxford, UK) and Qualisys (Gothenburg, Sweden). These software tools compute IK in real time. Furthermore, these software tools eliminate the need for additional joint kinetic sensors as they can also estimate joint kinetic parameters such as muscle-tendon

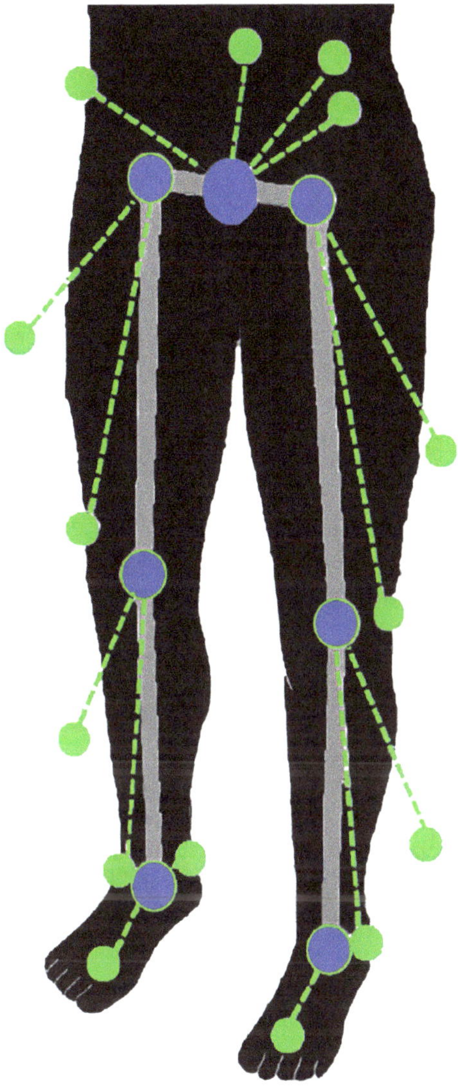

Figure 9.5 Example of a kinematic chain based on Helen Hayes marker set. Blue circles depict model joints and green circles depict model markers. Green arrows point to the corresponding joints of each marker. This image is modified from Charlton et al. [20].

forces [32], induced accelerations [33], and joint contact forces [34] using measured external forces (e.g. ground reaction forces (GRFs)) and calculated IK.

Kainz *et al.* [25] highlighted the importance of the joint kinetic parameters that are calculated using IK for treatments and planning of cerebral palsy. In his study,

Kaize emphasized that the joint kinetics calculated from IK provides reliable information to quantify musculoskeletal disorders such as cerebral palsy. However, the accuracy of IK depends largely on the locations of the joint centers (e.g. HJC) [18]. Kainz *et al.* [18] suggested to use medical images as the first priority in identifying HJCs. Other alternative approaches to minimize errors in the HJC would be functional methods and scaling methods [18]. IK is proven as a reliable algorithm for removing skin artefacts caused by markers. Nevertheless, marker scaling is crucial for accurate IK analysis.

Laboratory-based IK analysis is considered the widely applied gold standard method for gait data acquisition and analysis, that is clinically practiced with the introduction of user-friendly software tools. However, laboratory gait analysis is expensive and time consuming, as such daily analysis is not practical and feasible. Moreover, the camera and sensors are mounted to a fixed volume within a limited laboratory space. Subjects are required to walk within the stipulated camera visibility range ensuring each walking step is placed correctly on the force plates that are mounted on the laboratory floor. As a result, the subject's natural walking style is altered, generating faulty readings.

Marker-based motion analysis systems are not practical to be used with IoT. Furthermore, the force platforms and motion cameras are triggered at once, to ensure parallel data logging. As such, laboratory systems are not powered by IoT. Therefore, it is always required to have a trained individual to perform and monitor the gait measurement process. The data processing and analysis are performed locally. As such, the analysis and reporting need to be manually shared across other devices and systems. Furthermore, as the sensors are fixed to the laboratory systems, it is not capable of measuring gait in outdoor environments [35]. Outdoor gait measurements on a daily basis, are useful to quantify abnormalities and monitor improvements in prescribed treatments. The introduction of wearable sensors provides the advantage of user-friendly, inexpensive, and long-term gait measurements in outdoor environments that has gained a significant attention among clinicians over the years.

9.4 Wearable systems

Wearable sensors are widely used in clinical and sports applications for ambulatory gait measurements [36]. Wearable sensors are user-friendly and portable and provide qualitative and quantitative results that are repeatable and can be extended for longer duration with minimal cost [37]. Accelerometers, gyroscopes, and magnetometers are the three main wearable sensors used for kinematic analysis. The movement of the attached sensors is measured to estimate the motion such as joint and segment angles [38]. Measurements can be performed using a single wearable sensor or combining multiple wearable sensors. Wearable sensors can be attached to the joints and segments of the lower limb such as foot, shank, thing, knee, hip, and ankle. Figure 9.6 illustrates an example of a wearable sensor setup for measuring segment angles.

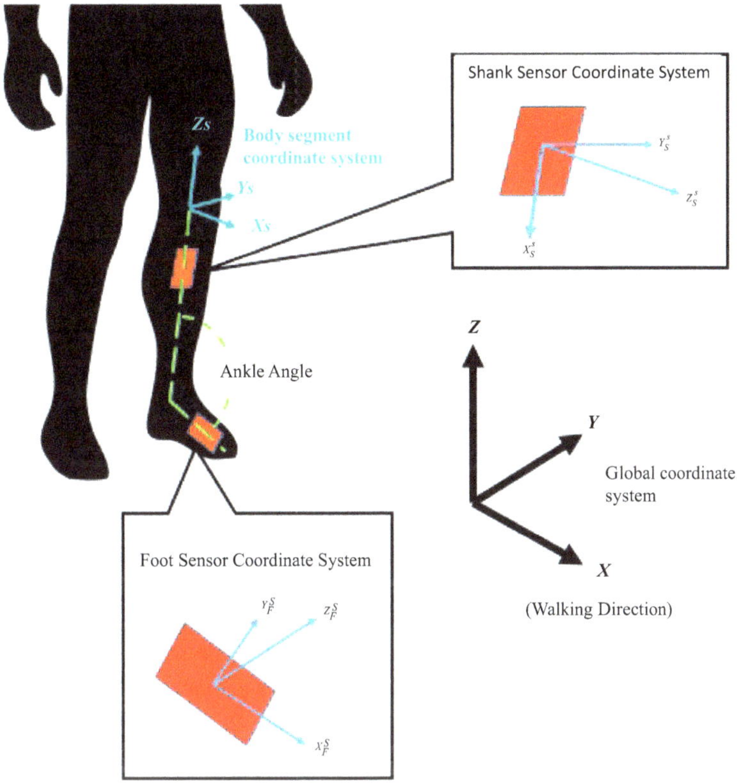

Figure 9.6 Gait measurements using kinematics wearable sensors.

9.4.1 Literature review – wearable systems with mathematical modeling

As wearable sensors are capable of measuring data on long-term basis, it is more suitable to study gait characteristics and related abnormalities such as identification of walking trajectories [39], auto-detection of daily gait activities of elderly [40], Parkinson's gait characteristics [41], fall detection [42,43], and post-operative gait characteristics [44].

Wearable sensor commercialization for gait analysis has taken place rapidly over the past years [36]. Such examples of wearable kinematics sensors are IMUs such as Xsens Motion Technologies (Enschede, The Netherlands) and APDM Wearable Technologies (Portland). Furthermore, wearable kinematics sensors are further expanded to more advanced applications in 3D motion tracking such as Polhemus Liberty ETS (Polhemus Incorporated, Colchester, VT, USA) and M3D TEC GIHAN Co. (Kyoto, Japan). These wearable 3D motion tracking systems require combining multiple wearable sensors to improve their performance. The data directly measured from these sensors need to be further processed to compute the parameters of interest. Kinematics analyses based on wearable sensors are classified into (1) mathematical modeling and (2) ML [37].

Accelerometer is the first wearable sensors utilized for gait assessments [36]. The micro-electro-mechanical systems (MEMS) introduced miniature, low-powered, and cost-effective accelerometers which are recognized as ideal for gait measurements. MEMS accelerometers are used to measure acceleration by integrated miniature mass systems [37,45]. These miniature masses spring systems move with respect to the change of their inertia, that is been transformed in to electric signals. MEMS acceleration data comprised of rotation, translation, and gravitation acceleration. However, these devices are affected by sensor noise that requires additional signal processing to remove gravitational effect-related errors [37,46]. As such, accelerometers are combined with magnetometers and gyroscopes to reduce gravitational errors [47].

Gyroscope sensors measure the angular rate by capturing the response of the rotating moving mass also known as the Coriolis force [37]. The raw gyroscope angular rate data are then integrated to compute the joint angles [48]. Gyroscopes are not affected by gravitational errors, sensor vibrations, and are less sensitive to motion changes [37]. On the other hand, magnetometers are suitable to acquire additional information such as the orientation of the segment in relation to the axis of its magnetic field [36]. Roetenberg *et al.* [49] presented a method to compute orientation by combining magnetometers and accelerometers. Magnetometers identified the heading orientation which a driving source for the system to be less affected by external infrastructure and be self-contained [50]. IMUs are combined sensors that comprise of accelerometer, gyroscopes, and magnetometers to calculate joint angles and orientations and are used in applications such as instrumented shoes and exoskeletons for gait analysis. The trunk inclination angles are measured using IMU sensors in a study conducted by Farris *et al.* [51]. Bao *et al.* [52] built a smart insole with IMU sensors for gait recognition. The acceleration data of the IMU sensor are integrated to compute joint angles of the lower limb during gait [48]. However, integrating the acceleration data could lead to drift error that needs to be treated with advanced algorithms [48].

Literature reports many methods to eliminate the drift offset and sensor noise error associated with wearable inertial sensors. Weijun *et al.* [36] present a detailed review of wearable inertial sensors in gait analysis and reviewed various methods applied for drift and sensor error reduction. Tong *et al.* [53] used a high pass filter to remove the shank and thigh angle drift. Cikajlo *et al.* [54] applied a Kalman filter to remove the gyroscope drift if the shank angle. Rebula *et al.* [55] used an improved Kalman filtering method to identify foot position, by utilizing an accelerometer as an inclinometer sensor. Zhang *et al.* [56] filtered the wearable sensor data using a Gaussian Particle Filter to calculate hip angles. Although the above discussed methods treated the sensor drift, the data were affected by skin artefacts caused by the movements between the sensor and the skin surface. Skin artefacts are generated by the movements of the muscles against the sensor, resulting in high frequency errors in the signals [57]. LeMoyne *et al.* [58] constructed an orientation filter based on a gradient descent algorithm. In this study, LeMoyne *et al.* [58]

enhanced the sensitivity of the algorithm towards the computational intensity and sampling rate. However, data filtering not only eliminates the sensor drift but also removes low-frequency components of the signal, which might eliminate important information of the signal characteristics.

Wearable inertial sensor-based gait data acquisition is a fast-growing field where many attempts are made to benchmark the wearable sensor method with gold standard methods for gait kinematic analysis. Han *et al.* [59] estimated joint angles from gyroscope data incorporating a 2-point error estimation algorithm. The study by Han [59] was benchmarked against Vicon Optical Motion Capture System. Abhayasinghe *et al.* [60] reported a cost effect and less complex orientation filter that is based on a gyroscope integration technique known as GIOF algorithm to estimate forward and backward thigh swigging angle. In Abhayasinghe's study [60], the GIOF algorithm was validated against the gold standard motion capture systems. Bidabadi *et al.* [61] validated the ankle pitch angle computed based on IMU sensor with that of the Vicon motion capture system. However, IMU sensors need to be calibrated to align its local axis coordinates with the segment axis. IMU calibrations are conducted by performing certain defined movements and postures [46]. In a review, Seel *et al.* [46] highlighted that the IMU calibration based on movements and postures could affect the accuracy of sensor readings due to subject's motion and posture precision. Furthermore, identifying the locations to which the IMU sensor to be attached is a tedious task, as the IMU sensors local coordinates and body segment axis are required to be measured manually. The manual measurements of IMU sensor coordinates and segment axis are performed in 3D space that could extend the sensor setup duration [62].

Ambulatory gait measurements in outdoor environments became a reality with the introduction of wearable sensors. However, for accurate gait measurements, the wearable sensors need to be attached to multiple positions of the hip, shank, thigh, and foot. Qiu *et al.* [48] used four IMUs attached to the lower limb to monitor fluctuations in the joint angles and foot elevations. Roetenberg *et al.* [49] designed a motion capture suite using 16 IMU sensors. The number of sensors required depends on the number of segments to be investigated. When the number of segments that need to be analyzed increases, the required number of IMU sensors as well increases, which consequently increases the computational parameters and associated cost [37]. These IMU sensors are attached to the segments using a stretchable belt or adhesive tapes to control to assess movements during gait motion. It is important to sensors the sensors are attached firmly for accurate measurements, as IMU data are highly prone to skin artefacts that are caused by relative movements between the sensor and the skin surface [63]. Moreover, sensor attachment on patients with abnormal gait is a cumbersome task that needs to be performed carefully. For example, patients with knee fracture might experience discomfort when a sensor is attached for gait assessments. As a result, the subject might alter his/her normal gait, leading to faulty sensor readings that would mislead the clinical diagnosis [64,65]. As such, the current research in gait analysis aims to address the above discussed challenges associated with wearable inertial sensors using ML algorithms [66].

9.4.2 Wearable systems with ML

ML is a subbranch of Artificial Intelligence (AI), which has the capability of self-learning tasks through observing a set of examples [67]. Figure 9.7 shows different classes and concepts of ML. ML can be subdivided mainly into statistical methods, model-based methods, and deep learning methods. Statistical methods follow simple mathematical operations to predict, classify, or analyze data. For example, support vector machines (SVM) construct discriminatory hyperplanes between different classes by projecting input data into a high-dimensional feature space [68]. Decision tree represents a graph-like structure, where events are denoted by nodes and edges of the graph denote rules/conditions [69]. K-Nearest neighbor (KNN) compares the similarities between the testing and training data through the property values given to each data point by the k nearest neighbors [70].

Model-based methods are mainly artificial neural networks (ANNs) which are advanced ML concepts, inspired by biological neuron system of the human brain [71–74]. Many ANN architectures are presented over the years. The simplest ANN architecture is the multi-layer perceptron (MLP) [75] which is also defined as the three-layer feed-forward neural network. The MLP architecture is depicted in Figure 9.8. MLP follows a feed-forward topology, where the inputs and outputs are connected through a hidden layer. The desired outputs are calculated by self-adjusting the network weights and bias between layers repeatedly, until the network error or cost function reaches a minimum or desired target [71,76,77]. MLPs with one hidden layer hold universal approximation properties, which are powered by its

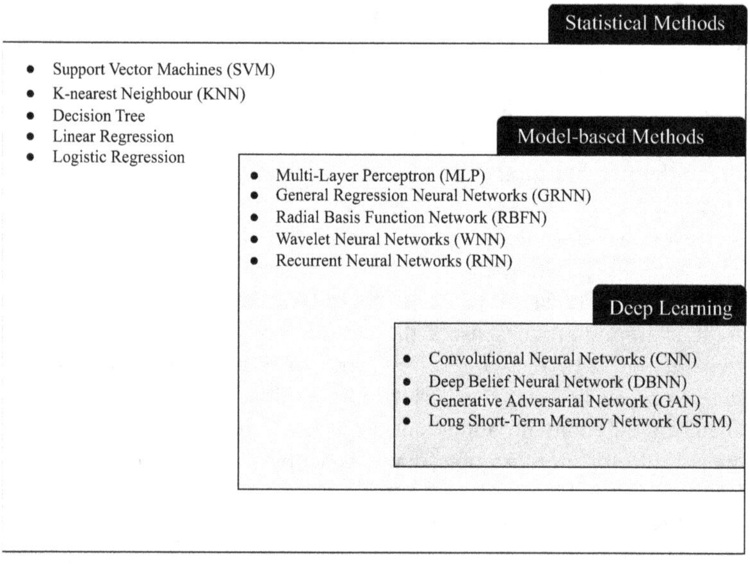

Figure 9.7 ML classes and concepts

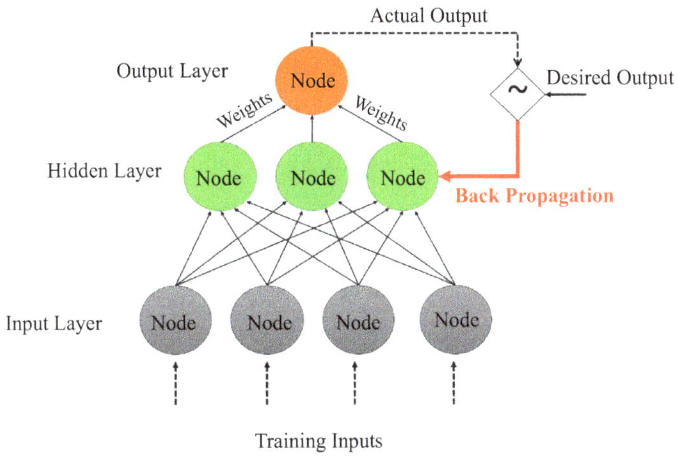

Figure 9.8 MLP architecture

hidden layer activation functions. The hidden layer of the MLPs is activated by global activation functions for example sigmoid and hyperbolic tangent [78]. The concept of the MLP network is used as the base when constructing other advanced neural networks.

Other ANN architectures are general regression neural networks (GRNN), radial basis function network (RBFN), wavelet neural networks (WNN), and recurrent neural networks (RNN). RBFN network that is an extended version of the MLP is activated by radial basis functions. Radial basis function has the ability to capture localized nonlinear data structures [70] which is useful for applications related to biomechanics. WNNs are with a similar mechanism as the RBFN network but its hidden layer activation is performed using wavelets which is used for applications that require time-frequency decomposition [79]. GRNN follows the kernel density estimation theorem [70]. RNNs are applied for sequential data that identify time-series data patterns to predict future scenarios. RNN has an internal memory that remembers input data patterns and uses the memory of previous input to predict the next output [80].

When the number of hidden layers increases, the network is defined as a deep neural network. Examples of such deep neural networks which are powered by deep learning concepts are convolutional neural network (CNN), deep belief neural network (DBNN), generative adversarial network (GAN), and long short term memory network (LSTM). CNN follows a hierarchical feature learning concept through a series of layers. CNNs are mainly applied for computer vision applications, where data is represented as spatial relationships [68]. GANs are generative models that learn the probability distribution of training data to generate new data samples similar in nature with random and minor variations [68]. Multiple ML models can also be combined to create hybrid models to achieve advanced performance. For example, DBNN is a hybrid model that stacks multiple RBFNs such that the output of the previous RBFN is the input of the next RBFN [81]. LSTM is a hybrid network which is an expanded version of RNN with extended memory that eliminates the long-term dependencies observed in RNN [80]. Another such example of a hybrid model is the CNN LSTM that is specifically designed to predict time series sequence data with spatial features such as images or videos [82].

ML is applied by drafting a functional relationship between the given set of inputs and outputs (e.g. mapping GRFs with acceleration). The end goal is to generate an accurate ML model that can estimate accurate outputs for fresh or unseen set of input features. Nevertheless, explainable ML has also gained interest especially among clinicians and medical experts. Explainable ML is useful when interpreting how a model operates and generates the output. In other words, a model's output is explained using statistics and domain knowledge. Among the available ML models, statistical methods in general are interpretable [83]. For example, linear regression and logistic regression are explainable using equations represented by model coefficients [84]. Decision trees are explainable using if-then rules [84]. However, majority of model-based and deep learning concepts are in general "black box" systems that are not interpretable. Models that are not interpretable cause trust issues, especially in medical research and healthcare. For example, a system that is powered by ML only warns the medical expert of the risk of a particular sickness without the indication of proper reasoning for such alerts, may results in confusion and require further assessment that may delay the treatment process. As such, new advancements are required in developing explainable ML, despite model type, for healthcare to foster trust and acceptance of ML concepts in the medical domain.

Sandhu *et al.* [85] discussed in detail the potential of ANNs in gait analysis. Table 9.2 depicts the summary of ANN methods currently available in gait estimations mainly classified as (1) kinematics estimations and (2) kinetics estimations. Tables 9.2 and 9.3 summarize ML methods applied in gait estimations. Table 9.2 summarizes ML methods that estimated kinematics from kinematics sensor inputs, whereas Table 9.3 summarizes ML-based kinematics–kinetics estimation methods and vice versa. It is noted that ML is rarely used to estimate joint angles [66]. Sivakumar *et al.* [78] presented a systematic review of existing methods associated with ML-based gait estimations. ANN is the widely utilized ML model for gait estimations.

9.4.2.1 Importance of feature selection

The dimensionality of each sample affects networks' training efficiency and estimation accuracy. Features selection reduces input data dimensionality by selecting salient features that best resemble overall data structures. Feature selection is performed to either select the primary input variable or to select salient data points within each selected variable. Dimensionality reduction techniques were seldom applied in present methods related to gait estimations. Oh *et al.* [89] and Choi *et al.* [90] followed SOM GA-GRNN which is a supervised ML technique used in

Table 9.2 Summary of ML-based kinematics estimations with inputs from wearable kinematics sensor data. Accuracies are stated in root mean square error (RMSE), correlation coefficient (ρ), mean absolute deviation (MAD), percentage of variance unexplained (PVU), and normalized root mean square error (NRMSE)

Author	Type	Feature selection	Input	Output	Results
Goulermas *et al.* [70]	MLP, kNN RBFN, GRNN	–	Foot acceleration angular velocity	Joint angles	$\rho > 0.9$ and RMSE < 4.36° MAD < 3.13°
Finlow *et al.* [86]	GRNN	–	Foot & shank acceleration & angular velocity	Joint angles	Intra-subject: $\rho > 0.9$, MAD < 2.3°, PVU < 14.50°; inter-subject: $0.7 < \rho < 0.89$, 4.918° < MAD < 9.068° 22.39° < PVU < 68.31°
Goulermas *et al.* [87]	GASI	–	Foot & shank acceleration & angular velocity	Joint angles	Intra-subject $\rho > 0.9$ and 1.42° < PVU < 12.78°; inter-subject $0.89 < \rho < 0.95$ and 9.52° < PVU < 21.53°
Wouda *et al.* [88]	MLP	–	ANN1 – lower-body orientations	ANN1 knee joint angles	Knee joint angle – RMSE < 5° & $\rho > 0.9$

selecting input variables with high correlations to its outputs. However, supervised ML methods pre-require training samples to learn data characteristics for accurate variable selection. On the other hand, PCA is a widely applied, unsupervised feature selection approach, which locates orthogonal projections of input variables that holds the highest variance to its outputs [100]. PCA algorithm is dependent on the complete data set (with all signals included), where a slight change in the data set leads to a different group of principal components. The complete data set should be evaluated for PCA algorithm to identify highest variance components. Such techniques are not applicable to be used in real-time as most real-time applications take in one signal at a time.

SOM GA-GRNN and PCA methods focus on reducing the number of input variables, but not on reducing the number of data points within each variable. The majority of available methods used full-length signals representing multiple variables as inputs to the network. Even though all points within each signal collectively the transfer overall pattern, not all data points are important for ANN modeling. Sivakumar *et al.* [97–99] presented knowledge-based feature selection methods to reduce the number of data points in gait signals. In this study, the authors used gait intervals of gait events to select prominent features of the gait cycle. Acquired data, in general, comprises of both useful and redundant data. Training the network with redundant data results in excessive training duration.

Table 9.3 Summary of ANN-based gait kinematics-kinetics estimation methods and vice versa. Accuracies are stated in root mean square error (RMSE), correlation coefficient (ρ), and normalized root mean square error (NRMSE)

Author	Type	Feature selection	Input	Output	Results
Oh *et al.* [89]	MLP	Prior knowledge, SOM, GA-GRNN	Trajectory/acceleration	GRF & moments	GRF $\rho >$: 0.841 & rRMSE $<$ 10.9% Moments $> \rho$: 0.841 rRMSE!25.5%
Choi *et al.* [90]	MLP	SOM & GA-GRNN	Trajectory & velocity	3D GRFs	$\rho >$0.47 & RMSE $<$ 0.61 N/kg
Ardestani *et al.* [91]	WNN	–	EMG & GRFs	Joint moments	NRMSE $<$ 10% & $> \rho = 0.94$
Ardestani *et al.* [92]	TDNN	–	Kinetics/kinematics	Contact pressure	Pressure reduction (25%)
Ardestani *et al.* [93]	WTDNN	–	Knee kinematics & forces	Contact pressure	RMSE $<$ 2.6 MPa, NRMSE $<$ 9.3% and $\rho > 0.96$
Jacobs and Ferris [94]	MLP	–	Pressure and tendon	GRF, COP & ankle joint torques	NRMSE $<$ 10% for walking at 1.0 m/s and 1.5 m/s
Ardestani *et al.* [75]	MLP	–	Displacement & acceleration	GRF & moments	$\rho > 0.98$ (training data) $\rho = 0.97$ (validation data)
Ngoh *et al.* [95]	MLP	–	Foot acceleration	vGRF	RMSE $<$ 0.017 & $> \rho = 0.99$
Wouda *et al.* [88]	MLP	–	ANN2 – knee joint angles	ANN2-vGRF	GRF-RMSE $<$ 0.27 of BW & $\rho > 0.9$
Komaris *et al.* [96]	MLP	–	Shank acceleration	3D GRF	RMSE $<$ 0.042
Sivakumar *et al.* [97]	MLP	Gait intervals	GRFs	Ankle joint angle	NRMSE $<$ 9.112% $\rho > 0.94$
Sivakumar *et al.* [98]	MLP	Gait intervals and force characteristics	GRFs	Ankle joint angle	Left ankle NRMSE 5.745 $\pm$ 1.642% Right ankle NRMSE 6.536 $\pm$ 0.9798% $\rho > 0.95$
Sivakumar *et al.* [99]	WNN	Gait intervals	GRFs	Ankle, knee & hip angle	Ankle RMSE 2.860 $\pm$ 1.097 Knee RMSE 3.978 $\pm$ 1.369 Hip RMSE 2.071 $\pm$ 0.924, $\rho > 0.945$

Hence, suitable dimensionality reduction methods should be utilized to remove redundant data and select salient data points that best represent input data patterns.

9.4.2.2 Machine learning-based kinematics estimation

Literature reports four studies that focused on ML-based estimating lower limb gait kinematics [70,86–88]. Goulermas *et al.* [70] used ML to estimate hip, knee, and ankle joint angles from transnational acceleration and foot rotational velocity as inputs. In his study, Goulermas tested seven ML models namely MLP, GRNN, KNN, and RBFN. GRNN reported the best performance with RMSE <4.36°, correlation coefficient ($\rho > 0.9$, and MAD < 3.13°). Finlow *et al.* [86] tested the GRNN network for estimating joint angles during gait from linear acceleration and angular velocity inputs of the foot and shank. The accuracies are analyzed with respect to intra- (training data) and inter-subject data (testing data). Intra-subject accuracies reported PVU < 14.50°, MAD < 2.3° and $\rho > 0.9$. Inter-subject accuracies were comparatively poorer reporting 22.39° < PVU < 68.31°, 4.918° < MAD < 9.068° and $0.7 < \rho < 0.89$. In another study by Goulermas *et al.* [87], a unique method known as GASI was presented, which combined GRNN with auxiliary similarity information (ASI) theory to estimate lower body joint angles. GASI uses adaptive weighting theory by considering the trajectory slopes and proximity. The GASI method has indicated improved accuracies only for intra-subject data (1.42° < PVU < 12.78° and $\rho > 0.9$). The inter-subject accuracies are poorer with (9.52° < PVU < 21.53° and $0.89 < \rho < 0.95$). Overall, GRNN estimations are with higher accuracies compared to that of GASI. Furthermore, the above discussed methods collectively have reported estimations with lower accuracies for inter-subject data. This explains the requirement for a larger training dataset for network performance validation. As such, the above reported methods were not fully validated in estimating lower limb joint angles from foot acceleration data, Furthermore, the authors highlighted the sensor noise associated with wearable kinematics sensors as one of the main drawbacks for the poor accuracies [86].

Wouda *et al.* [88] used ML to estimate joint angles from lower limb segment orientations as inputs. Results were promising as the author reports $\rho > 0.9$ and RMSE < 5°. The inputs were measured from three IMUs, that were attached to multiple lower limb joint segments. However, the sensitivity of wearable sensors on different joint locations needs to be tested. This is because in reality, the sensors are attached and detached repeatedly, and, during this process, there is a higher tendency for sensor dislocation [46]. Furthermore, attaching the sensor in different locations but carefully positioning is a time-consuming process that also has a tendency to generate skin artefacts that can affect the overall accuracy of the measured data and its associated estimations [62].

Feature selection is a vital step in ML modeling. However, the above methods in kinematics estimations did not apply feature selection methods to their algorithms [70,86–88]. Feature selection is useful for removing data affected by sensor noise, and also to extract salient data point by eliminating redundant data. As such, with the application of feature selection, the data dimensionality is reduced, which consequently reduces the training duration and enhances model accuracy. Overall, the reported literature reports the potential of ML, in particularly ANNs for estimating gait kinematics from wearable sensor inputs from the foot. However, the further research to support and improve the proposed direction is recommended to validate its applicability in clinical research [70,86,87].

In a recent review conducted by Caldas *et al.* [66], the potential of ML for gait kinematics estimations is highlighted. In his review, Caldas *et al.* [66] pointed out that a limited number of studies are currently reported in ML-based gait kinematics estimations. These ML-based kinematics estimation methods, used only kinematics data as inputs (e.g. joint angles). In other words, the ML models were constructed for kinematics inputs (e.g. foot accelerations) to kinematics output (e.g. joint angles) estimations.

In gait, the relationship between the changes in joint angles with respect to changes in foot kinetics (e.g. GRFs) can be explained by Newton's second law of motion [88]. Many mathematical methods in gait estimations, followed the Newton second law to model the relationship between gait kinetics and kinematics [101,102]. Sivakumar *et al.* [99] recently reported a novel approach in estimating lower limb joint angles from GRF data using an ML approach. In this study, the authors utilized a WNN for the estimations and validated its performance against the widely used MLP networks. The study reports improved accuracies with 50% than the MLPs and is being validated against the gold standard qualisys motion capture system. The method presented a future direction in eliminating the reliance on kinematics sensors and camera-based laboratory systems and its associated drawbacks such as drift and sensor noise.

9.4.2.3 ML-based kinetics estimation

Many studies were reported in estimating kinetics from kinematics inputs for gait applications, using ML models. MLPs were used by Choi *et al.* [90] to estimate GRFs from velocity and inertia forces. However, accuracies report poor correlations $\rho < 0.47$. Oh *et al.* [89] used comparatively simple MLP architecture (14 inputs, 3 hidden nodes, and 6 output nodes) to estimate GRFs and moments from 14 acceleration inputs. Results reported by Oh *et al.* [89] state improved correlations $\rho > 0.9$ when compared that with Choi *et al.* [90]. Furthermore, the estimation accuracies of the method [90] are not tested with abnormal gait data. The normal walking patterns get altered when an abnormality is present in gait. For example, the peak of the knee flexion angle will be decreased in patients with cerebral palsy [103]. During minimum toe clearance, the forefoot gets trapped that expands and alters the gait phase duration [104]. As such, it is vital to validate the ML approaches in gait estimations on its clinical value.

Ardestani *et al.* [75] tested the method proposed by Oh *et al.* [89], against pathological gait data. The study reported promising estimations with $\rho > 0.9$. In this study, Ardestani *et al.* [75] only involved four subjects to collect the required pathological gait data. The gait database is then constructed by artificially expanding the collected data into a probabilistic kinematic database using Principal Component Analysis (PCA). However, as this database [75] is artificially constructed, it does not truly represent the natural pathological gait characteristics.

Therefore, the methods [75,89] require to be further validated against the pathological data originally collected from patients with gait abnormalities.

ML is also utilized to reduce the number of sensors utilized in gait analysis. Traditionally, force platforms are used to measure GRFs, which are typically expensive and require a laboratory space to mount it on to the floor. Force platforms are not portable and constrict the laboratory space to a limited area. Wearable foot insoles with instrumented force sensors are getting attention that is useful in eliminating the expensive laboratory settings and associated drawbacks. Jacob and Ferris [94] designed a foot insole with pressure sensors, to estimate GRFs, center of pressure (COP) and ankle joint torques. The study [94] resulted in normalized root mean square error (NRMSE) <10%. In this study [94], the authors highlighted the significance of calculating joint torques by inverse dynamics approach through combining ankle angles with pressure data for accuracy improvements. In practical settings, joint angles are measured either from camera settings in the laboratory which are expensive and not portable or through the wearable kinematics sensors, that are affected by sensor noise, skin artefacts and drift. As such, the study presented by Jacob and Ferris [94] showed the future direction for ambulatory gait measurements using instrumented foot insoles and ML.

In another study, Ardestani *et al.* [92] used time-delay neural networks (TDNNs) and genetic algorithm (GA) for estimating foot pressure from joint angles and contact forces as inputs. In comparison with finite element analysis (FEA) and multi-body dynamics, TDNN produced estimations with better accuracies, by reducing the contact pressure in the knee by 25%. The authors in [92] suggested TDNN as an efficient ML model compared to MLPs. The tapped delay connection at the input layer enables the TDNN network to store short-term memory of the input data patterns. Nevertheless, the TDNNs clinical significance in estimating gait parameters is yet to be investigated.

Activation functions are important for modeling and performance of ANNs. The above discussed ANN architectures in gait applications [75,89,90,92,94] are activated by hyperbolic tangent sigmoid function which is also defined as one of the global activation functions. During the activation process, the global activation functions tend to discard the local data patterns [105]. Furthermore, the global activation function also tends to converge the network's cost function to intermediate local minimums that could mislead the training process by not converging to the expected target [91].

In another study, Ardestani *et al.* [91] used ANNs to avoid converging in local minima traps. The hidden layer of the WNNs is activated by wavelet functions, in which a time-frequency decomposition is conducted. In this study by Ardestani *et al.* [91], WNNs are used to estimate joint moments from GRF and EMG inputs. The study [91] reported $\rho > 0.94$ and NRMSE <10%. Authors also compared the WNNs with MLPs and proved that WNNs can be used for gait estimations with better accuracies than the traditionally used MLP networks. Furthermore, the study also highlighted the rapid convergence capability of WNNs and its significance in real-time applications. In another study, Ardestani *et al.* [93] used wavelet time delay neural network (WTDNN) to estimate maximum contact pressure from knee joint reaction forces and angles for real-time rehabilitation. WTDNN is an advanced version of WNN which combines TDNN with wavelet functions in its hidden layer. WTDNN reported improved estimation accuracies with $\rho > 0.9$ and 9.3% NRMSE for trunk sway and 6.2% NRMSE for medial trust. However, further testing is required as this method [91] was only validated against four subjects, that is insufficient for clinical acceptance. Wavelet activation functions thoroughly localize data in time–frequency domain. As such wavelet activation functions reduce training duration, converge faster, and avoid local minima convergence [79].

The above discussed methods used multi-variable inputs to estimate gait parameters. However, input from multiple variables increases the dimensionality of data, which eventually increases training duration and complexity of the network architecture. Feature selection is generally incorporated with ML algorithms to reduce the data dimensionality, by selecting salient features from the data map that best represents the data characteristics. Feature selection could be conducted to select data types (variable types) or to select salient data points within a single data type (variable type). Oh *et al.* [89] and Choi *et al.* [90] used SOM-GA-GRNN [106–108] to reduce data dimensionality. Oh *et al.* [89] selected 14 data types (variables) from 825, whereas Choi *et al.* [90] selected 13 data types (variables) from 1098 sensor data. Majority of the gait estimation methods use single-variable types as inputs for its ML models. Ngoh *et al.* [95] used only foot acceleration to estimate vGRFs. Wounda *et al.* [88] used only knee joint angles to estimate vGRFs. Komaris *et al.* [96] published a recent method that used shank acceleration to estimate 3D GRFs. Sivakumar *et al.* [99] used vGRFs to estimate joint angles of the knee, hip, and ankle. Current literature proves the potential of using single variable inputs for accurate gait estimations (Table 9.2).

The number of input and output nodes of any network depends on the data points of the input and output variables. It is vital to utilize a simpler network architecture with a minimal number of input and output nodes, to achieve fast network convergence and accurate estimations. However, majority of current approaches in gait estimations did not utilize any feature selection algorithms to reduce the number of input data points. Wounda *et al.* [88], Ngoh *et al.* [95], and Komaris *et al.* [96] used the full-length signal (100 points) to model an MLP, that resulted in a network architecture with 100 inputs and 100 outputs nodes. In such scenarios, the network architecture gets further complicated as multiple number of variables in full-length is used for modeling the network. Komaris *et al.* [96] used full-length GRFs in 3D (*X–Y–Z*) for their ANN model, which resulted in high dimensional input data structures and therefore converging the network with a higher number of hidden nodes (250 hidden nodes) and computational cost. Further, literature also used sequential inputs to model networks for gait estimations [75,89–94]. Sequential inputs feed data points to the network one at a time. As such, sequential input concept does not require separate input/output nodes for each data point. When the data is inserted in a sequential manner, the number of input/output nodes follows the number of input/output variables. As such, a sequential input concept is a good alliterative to reduce the number of nodes in the input and output layers. Moreover, the reported literature also reports a higher number of nodes in the hidden layer (15–30 nodes per hidden layer). It is vital to consider reducing the dimensionality of the input data through variable and data point selection, which consequently achieve simpler network architecture and less computational cost.

Most of the reported literature on ML-based gait estimations, focused on estimating foot kinetics from lower limb kinematics. Foot kinetics are estimated to eliminate the costly laboratory-based force platforms and measurements. Only limited literature focused on estimating joint kinematics based on ML techniques. Available ML-based kinematics estimation methods mainly looked into reducing the number of wearable kinematics sensors and the simplifying its sensors layouts, associated with gait data acquisition. Practically, wearable kinematics sensors cause skin artefacts, drift, and noise errors. Additional signal processing is required to remove these sensor errors (skin artefacts, drift, and noise errors). Furthermore, the placement of kinematic sensors needs to be carefully performed by identifying right locations of joints of the lower limb segments. Pre/post calibration is also a mandatory requirement associated with wearable kinematics sensors. As such, wearable kinematics sensors are not practical to be self-used by patients/subjects on a daily basis for gait monitoring. As such, alternative approaches are necessary to be investigated to identify possible solutions for the drawbacks associated with wearable kinematics sensors for outdoor gait measurements. Recent trends now focusing on eliminating the full use of kinematics sensors by using ML-based estimation methods, through estimating joint kinematics from wearable foot kinetic sensor input.

9.5 Wearable sensors-based IoT systems

IoT connects multiple devices and sensors over the Internet to communicate and interact through a network system [109]. Recently, advancement in sensor technologies promoted cost-effective (reducing human cost) wearable sensors powered by IoT. IoT-based wearable kinematics sensors are becoming increasingly popular especially for personalized gait analysis. IoT-based systems connect multiple sensors to create connections between physicians and patients/users through a network for gait assessments. IoT connects devices and users through four stages: (1) sensing layer, (2) network layer, (3) analysis layer, and (4) application layer [109]. At the sensing layer, the data is collected from wearable kinematics sensors and transferred to the network layer. Wireless or wired networks such as Wi-Fi, Bluetooth, or ZigBee are adopted in the network layer. The sensors are connected to the IoT system via the network layer. The sensor measurements collectively enter the network layer and prepare to be processed in the analysis layer. During the analysis layer, the sensor data is pre-processed and filtered to remove the drift, noise, and skin artefact errors followed by the applications of gait analysis algorithms. User context information such as location and state could also be used for more advanced and personalized gait analysis. Lastly, the outcome of the analyzed gait is presented to users through the user interface. IoT cloud networks-based gait assessments are useful for clinical applications, so that clinicians and patients could connect without physical presence for gait monitoring and diagnosis.

Recently, walking gait identification via WiFi, radio frequency (RFID), and millimeter wave is also gaining interest [110]. Available studies assumed that the unique walking patterns and body shapes could trigger the sensor signals distinctively, which could be used for applications such as user verification [111]. Several methods that are focused on user verification based on WiFi signals are discussed in literature such as WiFiU [112], WiWho [113], WiFiID [114], and FreeSense [115]. Among which FreeSense [115] has produced promising results as it is capable of detecting signals and performing analysis even at disturbed WiFi channel. FreeSense is capable of capturing disturbed waveforms, and pre-processes the signals using discrete wavelet transform and principal component analysis to capture distinguished signal characteristics. Luo *et al.* used RFID to recognize gait by monitoring the interruptions of the RFID signals [110].

Nweke *et al.* [116] and Llamas *et al.* [117] explained about the different applications of wearable sensors powered by IoT for human activity monitoring. Multiple sensors are attached to different joints and segments on the human body, and are connected through the network, providing data from different joints and segments. The consolidated data from all sensors are analyzed and an alert is generated to the user if any unusual activity is detected. Llamas *et al.* [118] introduced an IoT sensor-based setup for gait recognition using open-source hardware. The system collects the gait behavior of the user, through the wearable sensors that are connected via the IoT network and performs analysis for user identification (authentication). Terada *et al.* [119] used a sensor on the ankle to collect acceleration and angular velocity data to identify users through an IoT-based scoring algorithm that uses predefined threshold values for classifications. Muaaz *et al.* [120] use gait acceleration data collected from a pedometer sensor. The gait cycles are detected and processed on cloud using dynamic time wrapping and local maxima theory. Further the authors pre-processed the acceleration data by applying interpolation and Savitzky Golay filter. Zhang *et al.* [121] proposed a healthcare application powered by IoT, where signals from wearable sensors are acquired, processed, and predicted to generate gait report for the user. GarcÃ-a *et al.* explained the healthcare applications for identifying individuals with stroke, through analyzing gait data that are acquired from the cloud network. IoT-based wearable gait sensor networks are also utilized for the identification of Parkinson's diseases [122].

The combination of ML and IoT technology could lead to supporting reliable communications between the kinematics sensors and the cloud that will improve the analytical capacity [62–65,123–125]. These systems are formally named as AIoT (Artificial Intelligence of Things). Literature reports the significance and applications of AIoT systems to prevent fault analysis [126,127]. Gadaleta *et al.* [128] used smartphone sensors for gait monitoring and recognition, in which neural networks are used to analyze gait data and IoT network is used to communicate and alert the users. Kwapisz *et al.* [129] proposed a user authentication method from acceleration data and ML (J48 and neural networks). The acceleration data are processed in the cloud to identify a set of activities namely walking, jogging, and stair climbing. Batool *et al.* [130] introduced an IoT-based data analytic method for human activity recognition for gait data. This method [130] used ML to classify activities. However, the authors highlighted the requirement of more profound analysis for user-specific activity identification. Thang *et al.* [131] proposed an IoT system which used acceleration data for gait identification. In this study [131], data were analyzed in both time and frequency domain. The authors [131] proved that frequency domain analysis in IoT systems could generate higher identification accuracies than that of the time domain. Vhaduri *et al.* [132] proposed a continuous user identification method from physical activity data. This method connects smart phones and smart bands via IoT network. SVMs are applied on cloud for user identifications. Malik *et al.* [133] also used processed data in time and frequency domain and utilized SVM and decision tree via an IoT network for user classification and identifications.

However, prior studies on IoT gait systems are mainly tested in controlled settings [134]. The robustness is still questioned and requires further evaluations, especially in outdoor environments. In addition, prior studies were mainly focused on single person-based analysis and not generalized for larger population [134]. Past literature reports a lag when data is transferred via IoT networks [135,136]. As a result, the accuracy levels of the analysis reduce. Hence, it is important to optimize the routing model to reduce the lag to improve the accuracy. Nevertheless, it is observed that current research only focused on IoT systems to detect or classify gait conditions (e.g. fall detection and disease identification) [136]. Available IoT methods mainly focused on direct measurements and analysis of kinematics data during gait. As such, in future work, the advancements in ML-based kinematics estimation methods through wearable sensors data could be combined with IoT systems to achieve more efficient and less complex gait analysis models. Several studies also report the use of wearable foot kinetic sensors and IoT protocols for gait/user identifications. Jenkins *et al.* [137] introduced a user identification method from force sensors powered by IoT. In this study [137], GRFs are measured from force sensors during walking, and are processed in cloud to identify users, which reported acceptable accuracies. As such, in future work, it is also worth to investigate the applicability of IoT with ML-based gait kinematics estimations from foot kinetics.

9.6 Summary

The current gold standard method for gait signal analysis is conducted in laboratory setup, that are equipped with force platforms and optical cameras for motion tracking. The list of laboratory systems, its advantages, and challenges is summarized in Table 9.4. In general, laboratory systems equipped with force platforms and optical cameras are expensive and limited in space [59]. The subjects are required to walk within the limited camera volume, ensuring each step is placed on the force plates. The multiple reflective markers attached to different joints and segments cause inconvenience to the subject/wearer. Hence, laboratory systems for gait signal acquisition are not practical to be used for long-term outdoor gait measurements on daily basis [35].

Wearable sensors enable long-term ambulatory gait analysis in outdoor environments on daily basis [88]. The list of wearable systems, its advantages, and challenges are summarized in Table 9.5. Current gait measurement

Table 9.4 Summary of laboratory systems

Method	Algorithms	Devices/software tools	Advantages	Challenges
DK	Algorithms are based on marker placement; landmark and cluster specific Kinematics are calculated with respect to: Eular and Cardan angle concepts	Devices: Vicon (Oxford, UK), Qualisys (Gothenburg, Sweden), and OptiTrack DK is calculated using inbuilt software tools of respective devices	Gold standard for kinematics measurements	Marker position & skin artefact errors Offsets of joint centers & expanded ROMs High cost & time-consuming Limited space & camera visibility range Not applicable for long-term applications Not compatible for IoT-based applications Pre/post calibration requires expert assistance
IK	Global Optimization Algorithm	Devices: Vicon (Oxford, UK), Qualisys (Gothenburg, Sweden), and OptiTrack IK software tools: OpenSim, AnyBody (AnyBody Technology, Aalborg, Denmark), Visual3D (C-Motion, Germantown, MD, USA).	Gold standard for kinematics measurements Marker position errors and skin artefact errors are eliminated by introducing joint constraints Clinically applied The IK software tools can be interfaced with other motion capture systems	High cost & time-consuming Limited space & camera visibility range Not applicable for long-term applications Not compatible for IoT-based applications Pre/post calibration requires expert assistance

Table 9.5 Summary of wearable systems

Method	Algorithms	Devices/software tools	Advantages	Challenges
Mathematical modeling	Integration of accelerometer and gyroscope data to get joint angles	Devices: IMUs (accelerometer, gyroscope and magnetometer) such as Xsens Motion Technologies (Enschede, The Netherlands) and APDM wearable technologies (Portland).	User-friendly and cheaper than laboratory systems	Not gold standard
	Additional filtering to remove sensor noise such as Kalman filter and Gaussian particle filter	Software tools: 3D motion tracking such as; Polhemus Liberty ETS (Polhemus Incorporated, Colchester, VT, USA) and M3D TEC GIHAN Co. (Kyoto, Japan)	Long-term measurements in outdoor environments	Sensor noise and drift error
				Skin artefact error
			Clinically applicable	Sensor attachments in multiple positions cause discomfort on the wearer
			Applicable for IoT	Sensor placement and pre/post calibration – require expert assistance
ML	Artificial neural networks such as MLP, KNN, RBFN, GRNN, GASI, WNN, TDNN	IMUs (accelerometer, gyroscope and magnetometer) such as Xsens Motion Technologies (Enschede, The Netherlands) and APDM wearable technologies (Portland)	Reduced number of sensors & simplified sensor layout	
			Additional filtering not required	Not gold standard
			Generalize	Skin artefact error
			User-friendly and cheaper than laboratory systems	Dependent on training data accuracy and size of the training dataset
			Long-term measurements in outdoor environments	Sensor placement and pre/post calibration – require expert assistance
			Applicable for IoT	
			Estimates kinematics & kinetics	

methods mainly used wearable sensors to acquire kinematics of the lower limb. Mathematical modeling techniques are used to calculate the joint angles from the measured wearable sensor data. These mathematical methods include the integration of angular velocity measurements and signal pre-processing techniques such as filtering and removing the drift offsets. Methods also focused on using ML to estimate joint angles, aiming to eliminate the additional signal pre-processing required with mathematical modeling. ML is also being used with eliminating the need for force platforms, by estimating GRFs from IMU/wearable kinematics sensor data. However, to obtain a complete gait measurement, multiple sensors need to be attached to joint and segments. As the number of segments/joints of interest increases, the number of sensors required for the gait measurements also increases. As a result, the computational complexity of the model expands [37]. Literature reports limited studies on ML-based kinematics estimations. Out of which, majority focus on was using raw wearable kinematics sensor data to estimate joint angles (kinematics–kinematics estimations) [70,86–88]. However, these methods only focused on reducing the number of sensors and not on fully eliminating the reliance on kinematics sensors.

Wearable kinematics sensors are affected by skin-artefacts, sensor noise, and drift, which need to be removed by advanced pre-processing algorithms [36]. Placement of the sensors is time consuming and tedious as it require pre/post calibrations [62]. Furthermore, sensor placement on subjects with gait abnormalities and leg injuries could cause discomfort and alter their natural walking style, resulting in faulty gait measurements [65]. Therefore, alternative methods such as estimating joint kinematics form foot kinetics sensors that are presented to overcome the drawbacks of wearable kinematics sensors for outdoor gait analysis.

The emergence of ML and IoT systems has managed to connect multiple kinematics sensors through the cloud to perform intelligent algorithms in gait identification and analysis. ML has enabled estimating gait parameters from reduced number of sensors, whereas IoT has enabled the possibility of connecting multiple sensors and processing the analysis through cloud with shared user interfaces. These advancements have provided the opportunity for users to perform gait measurements and evaluations, without physically being present at a gait lab. However, current IoT and ML systems in gait research mainly focused on user identification (classifications) and not on regression systems. On the other hand, recent methods in IoT, ML, and gait depict promising results in foot sensor-based applications. As such, IoT and ML could be jointly used for estimating joint kinematics from foot insole data, while analyzing and presenting the processed gait information to users who are connected over the cloud network. Figure 9.9 summarizes the current trends and future directions of kinematics-based gait estimations.

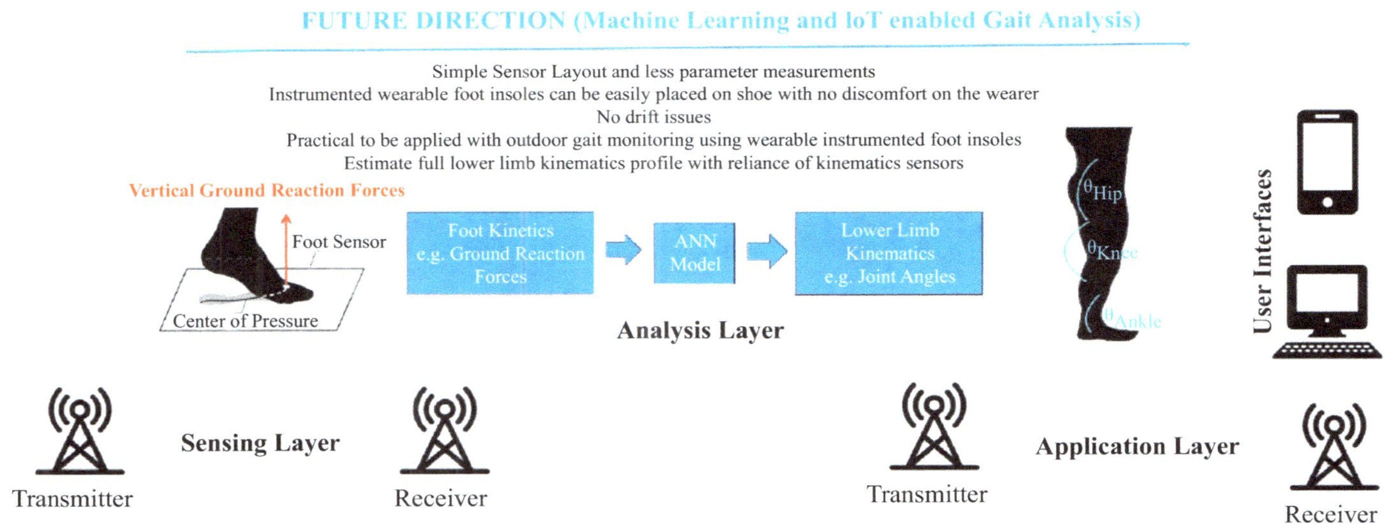

Figure 9.9 Future direction for ML and IoT-enabled gait analysis

References

[1] Sutherland DH. The evolution of clinical gait analysis: Part II. Kinematics. *Gait & Posture*. 2002;16(2):159–179. Available from: http://www.sciencedirect.com/science/article/pii/S0966636202000048.

[2] Laroche D, Tolambiya A, Morisset C, *et al.* A classification study of kinematic gait trajectories in hip osteoarthritis. *Computers in Biology and Medicine*. 2014;55:42–48. Available from: http://www.sciencedirect.com/science/article/pii/S0010482514002637.

[3] Hicks R, Durinick N, and Gage J. Differentiation of idiopathic toe-walking and cerebral palsy. *Journal of Pediatric Orthopedics*. 1988;8(2):160–163. Available from: http://europepmc.org/abstract/MED/3350949.

[4] Kaczmarczyk K, Wit A, Krawczyk M, *et al.* Gait classification in post-stroke patients using artificial neural networks. *Gait & Posture*. 2009;30(2):207–210. Available from: http://www.sciencedirect.com/science/article/pii/S0966636209001180.

[5] Verghese J, Holtzer R, Lipton RB, *et al.* Quantitative gait markers and incident fall risk in older adults. *The Journals of Gerontology: Series A*. 2009 04;64A(8):896–901.

[6] Walsh CJ, Pasch K, and Herr H. An autonomous, underactuated exoskeleton for load-carrying augmentation. In: *2006 IEEE/RSJ International Conference on Intelligent Robots and Systems*, 2006. p. 1410–1415.

[7] Yali H and Xingsong W. Kinematics analysis of lower extremity exoskeleton. In: *2008 Chinese Control and Decision Conference*, 2008. p. 2837–2842.

[8] Eskandari F, Nodoushan MN, Arshi AR, *et al.* Designing a 12 DoF lower extremity exoskeleton based on the gait kinematics data. In: *2015 22nd Iranian Conference on Biomedical Engineering (ICBME)*, 2015. p. 358–362.

[9] Hitt J, Sugar T, Holgate M, *et al.* Robotic transtibial prosthesis with biomechanical energy regeneration. *Industrial Robot: The International Journal of Robotics Research and Application*. 2009;36(5):441–447.

[10] Richter H, Simon D, Smith WA, *et al.* Dynamic modeling, parameter estimation and control of a leg prosthesis test robot. *Applied Mathematical Modelling*. 2015;39(2):559–573. Available from: http://www.sciencedirect.com/science/article/pii/S0307904X14003096.

[11] Noehren B, Scholz J, and Davis I. The effect of real-time gait retraining on hip kinematics, pain and function in subjects with patellofemoral pain syndrome. *British Journal of Sports Medicine*. 2011;45(9):691–696.

[12] Fisher D, Louw Q, Cockcroft J, *et al.* Three-dimensional kinematics of the thorax during over-ground running. *Journal of Bodywork and Movement Therapies*. 2018;22(2):300–307. Available from: http://www.sciencedirect.com/science/article/pii/S1360859217302152.

[13] Phinyomark A, Osis S, Hettinga BA, *et al.* Kinematic gait patterns in healthy runners: a hierarchical cluster analysis. *Journal of Biomechanics*. 2015;48(14):3897–3904. Available from: http://www.sciencedirect.com/science/article/pii/S0021929015005199.

[14] Perry J. *Gait Analysis: Normal and Pathological Function*. Thorofare, NJ: *SLACK*, 1992.

[15] Stagni R, Leardini A, Cappozzo A, *et al.* Effects of hip joint centre mislocation on gait analysis results. *Journal of Biomechanics.* 2000;33 (11):1479–1487. Available from: http://www.sciencedirect.com/science/article/pii/S0021929000000932.

[16] OptiTrack;. Available from: https://optitrack.com/motion-capture-movement-sciences/.

[17] Horsak B, Pobatschnig B, Schwab C, *et al.* Reliability of joint kinematic calculations based on direct kinematic and inverse kinematic models in obese children. *Gait & Posture*. 2018;66:201–207.

[18] Kainz H, Carty CP, Maine S, *et al.* Effects of hip joint centre mislocation on gait kinematics of children with cerebral palsy calculated using patient-specific direct and inverse kinematic models. *Gait & Posture*. 2017;57:154–160.

[19] Comparison of Motion Capture Systems. 2021. Available from: https://www.comparesportstech.com/compare-motion-capture-systems.

[20] Charlton IW, Tate P, Smyth P, *et al.* Repeatability of an optimised lower body model. *Gait & Posture*. 2004;20(2):213–221. Available from: http://www.sciencedirect.com/science/article/pii/S0966636203001632.

[21] Collins TD, Ghoussayni SN, Ewins DJ, *et al.* A six degrees-of-freedom marker set for gait analysis: repeatability and comparison with a modified Helen Hayes set. *Gait & Posture*. 2009;30(2):173–180. Available from: http://www.sciencedirect.com/science/article/pii/S0966636209001106.

[22] Zuk M and Pezowicz C. Kinematic analysis of a six-degrees-of-freedom model based on ISB recommendation: a repeatability analysis and comparison with conventional gait model. *Applied Bionics and Biomechanics*. 2015;2015:9. Available from: http://dx.doi.org/10.1155/2015/503713.

[23] McGinley JL, Baker R, Wolfe R, *et al.* The reliability of three-dimensional kinematic gait measurements: a systematic review. *Gait & Posture*. 2009;29 (3):360–369.

[24] Horsak B, Pobatschnig B, Baca A, *et al.* Within-assessor reliability and minimal detectable change of gait kinematics in a young obese demographic. *Gait & Posture*. 2017;54:112–118.

[25] Kainz H, Modenese L, Lloyd DG, *et al.* Joint kinematic calculation based on clinical direct kinematic versus inverse kinematic gait models. *Journal of Biomechanics.* 2016;49(9):1658–1669. Available from: http://www.sciencedirect.com/science/article/pii/S0021929016304134.

[26] Akbarshahi M, Schache AG, Fernandez JW, *et al.* Non-invasive assessment of soft-tissue artifact and its effect on knee joint kinematics during functional activity. *Journal of Biomechanics.* 2010;43(7):1292–1301.

[27] Leardini A, Belvedere C, Nardini F, *et al.* Kinematic models of lower limb joints for musculo-skeletal modelling and optimization in gait analysis. *Journal of Biomechanics.* 2017;62:77–86.

[28] Cerveri P, Rabuffetti M, Pedotti A, *et al.* Real-time human motion estimation using biomechanical models and non-linear state-space filters. *Medical and Biological Engineering and Computing.* 2003;41(2):109–123.

[29] Leardini A, Chiari L, Croce UD, *et al.* Human movement analysis using stereophotogrammetry: Part 3. Soft tissue artifact assessment and compensation. *Gait & Posture.* 2005;21(2):212–225. Available from: http://www.sciencedirect.com/science/article/pii/S0966636204000773.

[30] Delp SL, Anderson FC, Arnold AS, *et al.* OpenSim: open-source software to create and analyze dynamic simulations of movement. *IEEE Transactions on Biomedical Engineering.* 2007;54(11):1940–1950.

[31] Sherman MA, Seth A, and Delp SL. Simbody: multibody dynamics for biomedical research. *Procedia IUTAM.* 2011;2:241–261. *IUTAM Symposium on Human Body Dynamics.* Available from: http://www.sciencedirect.com/science/article/pii/S2210983811000241.

[32] Barber L, Carty C, Modenese L, *et al.* Medial gastrocnemius and soleus muscle-tendon unit, fascicle, and tendon interaction during walking in children with cerebral palsy. *Developmental Medicine & Child Neurology.* 2017;59(8):843–851.

[33] Graham DF, Carty CP, Lloyd DG, *et al.* Muscle contributions to recovery from forward loss of balance by stepping. *Journal of Biomechanics.* 2014;47 (3):667–674.

[34] Wesseling M, Derikx LC, de Groote F, *et al.* Muscle optimization techniques impact the magnitude of calculated hip joint contact forces. *Journal of Orthopaedic Research.* 2015;33(3):430–438.

[35] Villeger D, Costes A, Watier B, *et al.* An algorithm to decompose ground reaction forces and moments from a single force platform in walking gait. *Medical Engineering & Physics.* 2014;36(11):1530–1535.

[36] Tao W, Liu T, Zheng R, *et al.* Gait analysis using wearable sensors. *Sensors.* 2012;12(2):2255–2283.

[37] Rueterbories J, Spaich EG, Larsen B, *et al.* Methods for gait event detection and analysis in ambulatory systems. *Medical Engineering & Physics.* 2010;32(6):545–552. Available from: http://www.sciencedirect.com/science/article/pii/S1350453310000718.

[38] Muro-De-La-Herran A, Garcia-Zapirain B, and Mendez-Zorrilla A. Gait analysis methods: an overview of wearable and non-wearable systems, highlighting clinical applications. *Sensors.* 2014;14(2):3362–3394.

[39] Kitagawa N and Ogihara N. Estimation of foot trajectory during human walking by a wearable inertial measurement unit mounted to the foot. *Gait & Posture.* 2016;45:110–114.

[40] Ayachi FS, Nguyen HP, Lavigne-Pelletier C, *et al.* Wavelet-based algorithm for auto-detection of daily living activities of older adults captured by multiple inertial measurement units (IMUs). *Physiological Measurement.* 2016;37(3):442–461.

[41] Okuda S, Takano S, Ueno M, *et al.* Gait analysis of patients with Parkinson's disease using a portable triaxial accelerometer. *Neurology and Clinical Neuroscience*. 2016;4(3):93–97.
[42] Shibuya N, Nukala BT, Rodriguez AI, *et al.* A real-time fall detection system using a wearable gait analysis sensor and a Support Vector Machine (SVM) classifier. In: *2015 Eighth International Conference on Mobile Computing and Ubiquitous Networking (ICMU)*, 2015. p. 66–67.
[43] Rispens SM, Van Dieën JH, Van Schooten KS, *et al.* Fall-related gait characteristics on the treadmill and in daily life. *Journal of NeuroEngineering and Rehabilitation*. 2016;13(1):12.
[44] Hanly R, Doyle F, Whitehouse S, *et al.* Outpatient 3-D gait analysis one year after THA using a portable imu system. *Orthopaedic Proceedings*. 2016;98-B(SUPP_11):1–1.
[45] Brigante CM, Abbate N, Basile A, *et al.* Towards miniaturization of a MEMS-based wearable motion capture system. *IEEE Transactions on Industrial Electronics*. 2011;58(8):3234–3241.
[46] Seel T, Raisch J, and Schauer T. IMU-based joint angle measurement for gait analysis. *Sensors*. 2014;14(4):6891–6909.
[47] Shull PB, Xu J, Yu B, *et al.* Magneto-gyro wearable sensor algorithm for trunk sway estimation during walking and running gait. *IEEE Sensors Journal*. 2017;17(2):480–486.
[48] Qiu S, Liu L, Zhao H, *et al.* MEMS inertial sensors based gait analysis for rehabilitation assessment via multi-sensor fusion. *Micromachines*. 2018;9 (9):442.
[49] Roetenberg D, Luinge H, and Slycke P. Xsens MVN: full 6DOF human motion tracking using miniature inertial sensors. *Xsens Motion Technologies BV, Tech Rep*. 2009, p. 1.
[50] Wang Z, Li J, Wang J, *et al.* Inertial sensor-based analysis of equestrian sports between beginner and professional riders under different horse gaits. *IEEE Transactions on Instrumentation and Measurement*. 2018;67 (11):2692–2704.
[51] Farris RJ, Quintero HA, Murray SA, *et al.* A preliminary assessment of legged mobility provided by a lower limb exoskeleton for persons with paraplegia. *IEEE Transactions on Neural Systems and Rehabilitation Engineering*. 2013;22(3):482–490.
[52] Bao SD, Meng XL, Xiao W, *et al.* Fusion of inertial/magnetic sensor measurements and map information for pedestrian tracking. *Sensors*. 2017;17 (2):340.
[53] Tong K, and Granat MH. A practical gait analysis system using gyroscopes. *Medical Engineering & Physics*. 1999;21(2):87–94.
[54] Cikajlo I, Matjačić Z, and Bajd T. Efficient FES triggering applying Kalman filter during sensory supported treadmill walking. *Journal of Medical Engineering & Technology*. 2008;32(2):133–144.
[55] Rebula JR, Ojeda LV, Adamczyk PG, *et al.* Measurement of foot placement and its variability with inertial sensors. *Gait & Posture*. 2013;38(4):974–980.

[56] Zhang Z, Huang Z, Wu J. Ambulatory hip angle estimation using Gaussian particle filter. *Journal of Signal Processing Systems*. 2010;58(3):341–357.
[57] Ancillao A, Tedesco S, Barton J, *et al.* Indirect measurement of ground reaction forces and moments by means of wearable inertial sensors: a systematic review. *Sensors*. 2018;18(8):2564.
[58] LeMoyne R and Mastroianni T. Quantifying the spatial position representation of gait through sensor fusion. In: *Wearable and Wireless Systems for Healthcare I*. Springer, 2018. p. 105–110.
[59] Han YC, Wong KI, and Murray I. 2-Point error estimation algorithm for 3-D thigh and shank angles estimation using IMU. *IEEE Sensors Journal*. 2018;18(20):8525–8531.
[60] Abhayasinghe N, Murray I, and Sharif Bidabadi S. Validation of thigh angle estimation using inertial measurement unit data against optical motion capture systems. *Sensors*. 2019;19(3):596.
[61] Bidabadi SS, Murray I, and Lee GYF. Validation of foot pitch angle estimation using inertial measurement unit against marker-based optical 3D motion capture system. *Biomedical Engineering Letters*. 2018;8(3):283–290.
[62] Seel T, Schauer T, and Raisch J. Joint axis and position estimation from inertial measurement data by exploiting kinematic constraints. In: *2012 IEEE International Conference on Control Applications*. IEEE, 2012. p. 45–49.
[63] Cloete T and Scheffer C. Benchmarking of a full-body inertial motion capture system for clinical gait analysis. In: *2008 30th Annual International Conference of the IEEE Engineering in Medicine and Biology Society*, 2008. p. 4579–4582.
[64] Ren L, Jones RK, and Howard D. Whole body inverse dynamics over a complete gait cycle based only on measured kinematics. *Journal of Biomechanics*. 2008;41(12):2750–2759.
[65] Xiang Y, Arora JS, and Abdel-Malek K. Optimization-based prediction of asymmetric human gait. *Journal of Biomechanics*. 2011;44(4):683–693.
[66] Caldas R, Mundt M, Potthast W, *et al.* A systematic review of gait analysis methods based on inertial sensors and adaptive algorithms. *Gait & Posture*. 2017;57:204–210.
[67] Halilaj E, Rajagopal A, Fiterau M, *et al.* Machine learning in human movement biomechanics: best practices, common pitfalls, and new opportunities. *Journal of Biomechanics*. 2018;81:1–11.
[68] Janiesch C, Zschech P, and Heinrich K. Machine learning and deep learning. *Electronic Markets*. 2021;31(3):685–695.
[69] Mahesh B. Machine learning algorithms – a review. *International Journal of Science and Research (IJSR)*. 2020;9:381–386.
[70] Goulermas J, Howard D, Nester C, *et al.* Regression techniques for the prediction of lower limb kinematics. *Journal of Biomechanical Engineering*. 2005;127(6):1020–1024.

[71] Nayak R, Jain L, and Ting B. Artificial neural networks in biomedical engineering: a review. In: *Computational Mechanics – New Frontiers for the New Millennium.* 2001, p. 887–892.
[72] Bishop CM. *Neural Networks for Pattern Recognition.* Oxford University Press, 1995.
[73] Specht DF. Probabilistic neural networks. *Neural Networks.* 1990;3(1):109–118.
[74] Patra JC and Kot AC. Nonlinear dynamic system identification using Chebyshev functional link artificial neural networks. *IEEE Transactions on Systems, Man, and Cybernetics, Part B (Cybernetics).* 2002;32(4):505–511.
[75] Ardestani MM, Moazen M, and Jin Z. Sensitivity analysis of human lower extremity joint moments due to changes in joint kinematics. *Medical Engineering & Physics.* 2015;37(2):165–174.
[76] Schöllhorn W. Applications of artificial neural nets in clinical biomechanics. *Clinical Biomechanics.* 2004;19(9):876–898.
[77] Chau T. A review of analytical techniques for gait data. Part 2: neural network and wavelet methods. *Gait & Posture.* 2001;13(2):102–120.
[78] Sivakumar S, Gopalai AA, Gouwanda D, *et al.* ANN for gait estimations: a review on current trends and future applications. In: *2016 IEEE EMBS Conference on Biomedical Engineering and Sciences (IECBES)*, 2016. p. 311–316.
[79] Alexandridis AK and Zapranis AD. Wavelet neural networks: a practical guide. *Neural Networks.* 2013;42:1–27. Available from: http://www.sciencedirect.com/science/article/pii/S0893608013000129.
[80] Manaswi NK. Rnn and lstm. In: *Deep Learning with Applications Using Python.* Springer, 2018. p. 115–126.
[81] Zhao G, Zhang C, and Zheng L. Intrusion detection using deep belief network and probabilistic neural network. In: *2017 IEEE International Conference on Computational Science and Engineering (CSE) and IEEE International Conference on Embedded and Ubiquitous Computing (EUC)*, vol. 1. IEEE, 2017. p. 639–642.
[82] Elmaz F, Eyckerman R, Casteels W, *et al.* CNN-LSTM architecture for predictive indoor temperature modeling. *Building and Environment.* 2021;206:108327.
[83] Gilpin LH, Bau D, Yuan BZ, *et al.* Explaining explanations: an overview of interpretability of machine learning. In: *2018 IEEE 5th International Conference on Data Science and Advanced Analytics (DSAA).* IEEE, 2018. p. 80–89.
[84] Petch J, Di S, and Nelson W. Opening the black box: the promise and limitations of explainable machine learning in cardiology. *Canadian Journal of Cardiology.* 2021;38:204–213.
[85] Sandhu K and Kamboj VK. Role of artificial neural network for prediction of gait parameters and patterns. In: *AI Techniques for Reliability Prediction for Electronic Components.* IGI Global, 2020. p. 124–135.

[86] Findlow A, Goulermas J, Nester C, *et al.* Predicting lower limb joint kinematics using wearable motion sensors. *Gait & Posture*. 2008;28(1):120–126.

[87] Goulermas JY, Findlow AH, Nester CJ, *et al.* An instance-based algorithm with auxiliary similarity information for the estimation of gait kinematics from wearable sensors. *IEEE Transactions on Neural Networks.* 2008;19 (9):1574–1582.

[88] Wouda FJ, Giuberti M, Bellusci G, *et al.* Estimation of vertical ground reaction forces and sagittal knee kinematics during running using three inertial sensors. *Frontiers in Physiology.* 2018;9:218.

[89] Oh SE, Choi A, and Mun JH. Prediction of ground reaction forces during gait based on kinematics and a neural network model. *Journal of Biomechanics.* 2013;46(14):2372–2380.

[90] Choi A, Lee JM, and Mun JH. Ground reaction forces predicted by using artificial neural network during asymmetric movements. *International Journal of Precision Engineering and Manufacturing.* 2013;14(3):475–483.

[91] Ardestani MM, Zhang X, Wang L, *et al.* Human lower extremity joint moment prediction: a wavelet neural network approach. *Expert Systems with Applications.* 2014;41(9):4422–4433.

[92] Ardestani MM, Moazen M, and Jin Z. Gait modification and optimization using neural network–genetic algorithm approach: application to knee rehabilitation. *Expert Systems with Applications.* 2014;41(16):7466–7477. Available from: http://www.sciencedirect.com/science/article/pii/S0957417414003753.

[93] Ardestani MM, Moazen M, Chen Z, *et al.* A real-time topography of maximum contact pressure distribution at medial tibiofemoral knee implant during gait: application to knee rehabilitation. *Neurocomputing.* 2015;154:174–188. Available from: http://www.sciencedirect.com/science/article/pii/S0925231214016695.

[94] Jacobs DA and Ferris DP. Estimation of ground reaction forces and ankle moment with multiple, low-cost sensors. *Journal of Neuroengineering and Rehabilitation.* 2015;12(1):1.

[95] Ngoh KJH, Gouwanda D, Gopalai AA, *et al.* Estimation of vertical ground reaction force during running using neural network model and uniaxial accelerometer. *Journal of Biomechanics.* 2018;76:269–273.

[96] Komaris DS, Pérez-Valero E, Jordan L, *et al.* Predicting three-dimensional ground reaction forces in running by using artificial neural networks and lower body kinematics. *IEEE Access.* 2019;7:156779–156786.

[97] Sivakumar S, Gopalai A, Gouwanda D, *et al.* Estimation of joint angle from ground reaction force in human gait. In: *2018 IEEE-EMBS Conference on Biomedical Engineering and Sciences (IECBES)*. IEEE, 2018. p. 623–628.

[98] Sivakumar S, Gopalai AA, Lim KH, *et al.* Artificial neural network based ankle joint angle estimation using instrumented foot insoles. *Biomedical Signal Processing and Control.* 2019;54:101614.

[99] Sivakumar S, Gopalai AA, Lim KH, *et al.* Joint angle estimation with wavelet neural networks. *Scientific Reports.* 2021;11(1):1–15.

[100] Noori R, Karbassi AR, Moghaddamnia A, *et al.* Assessment of input variables determination on the SVM model performance using PCA, Gamma test, and forward selection techniques for monthly stream flow prediction. *Journal of Hydrology.* 2011;401(3):177–189. Available from: http://www.sciencedirect.com/science/article/pii/S0022169411001363.

[101] Faber GS, Chang C, Kingma I, *et al.* Estimating 3D L5/S1 moments and ground reaction forces during trunk bending using a full-body ambulatory inertial motion capture system. *Journal of Biomechanics.* 2016;49(6):904–912.

[102] Karatsidis A, Bellusci G, Schepers H, *et al.* Estimation of ground reaction forces and moments during gait using only inertial motion capture. *Sensors.* 2017;17(1):75.

[103] Simon SR. Quantification of human motion: gait analysis–benefits and limitations to its application to clinical problems. *Journal of Biomechanics.* 2004;37(12):1869–1880.

[104] Sato K. Factors affecting minimum foot clearance in the elderly walking: a multiple regression analysis. *Open Journal of Therapy and Rehabilitation.* 2015;3(04):109.

[105] Cordova JJ, Yu W, and Li X. Haar wavelet neural networks for nonlinear system identification. In: *2012 IEEE International Symposium on Intelligent Control.* IEEE, 2012. p. 276–281.

[106] Islam S and Kothari R. Artificial neural networks in remote sensing of hydrologic processes. *Journal of Hydrologic Engineering.* 2000;5(2):138–144.

[107] Bowden GJ, Dandy GC, and Maier HR. Input determination for neural network models in water resources applications. Part 1 – background and methodology. *Journal of Hydrology.* 2005;301(1–4):75–92.

[108] Specht DF. A general regression neural network. *IEEE Transactions on Neural Networks.* 1991;2(6):568–576.

[109] Qi J, Yang P, Waraich A, *et al.* Examining sensor-based physical activity recognition and monitoring for healthcare using Internet of Things: a systematic review. *Journal of Biomedical Informatics.* 2018;87:138–153.

[110] Luo C, Wu J, Li J, *et al.* Gait recognition as a service for unobtrusive user identification in smart spaces. *ACM Transactions on Internet of Things.* 2020;1(1):1–21.

[111] Xiao F, Guo Z, Ni Y, *et al.* Artificial intelligence empowered mobile sensing for human flow detection. *IEEE Network.* 2018;33(1):78–83.

[112] Wang W, Liu AX, and Shahzad M. Gait recognition using wifi signals. In: *Proceedings of the 2016 ACM International Joint Conference on Pervasive and Ubiquitous Computing*, 2016. p. 363–373.

[113] Zeng Y, Pathak PH, and Mohapatra P. WiWho: WiFi-based person identification in smart spaces. In: *2016 15th ACM/IEEE International Conference on Information Processing in Sensor Networks (IPSN).* IEEE, 2016. p. 1–12.

[114] Zhang J, Wei B, Hu W, *et al.* Wifi-id: human identification using wifi signal. In: *2016 International Conference on Distributed Computing in Sensor Systems (DCOSS).* IEEE, 2016. p. 75–82.

[115] Xin T, Guo B, Wang Z, *et al.* FreeSense: a robust approach for indoor human detection using Wi-Fi signals. *Proceedings of the ACM on Interactive, Mobile, Wearable and Ubiquitous Technologies*. 2018;2(3):1–23.

[116] Nweke HF, Teh YW, Al-Garadi MA, *et al.* Deep learning algorithms for human activity recognition using mobile and wearable sensor networks: state of the art and research challenges. *Expert Systems with Applications*. 2018;105:233–261.

[117] Llamas C, González MA, Hernández C, *et al.* Open source platform for collaborative construction of wearable sensor datasets for human motion analysis and an application for gait analysis. *Journal of Biomedical Informatics*. 2016;63:249–258.

[118] Llamas C, González MA, Hernández C, *et al.* Open source hardware based sensor platform suitable for human gait identification. *Pervasive and Mobile Computing*. 2017;38:154–165.

[119] Terada S, Enomoto Y, Hanawa D, *et al.* Performance of gait authentication using an acceleration sensor. In: *2011 34th International Conference on Telecommunications and Signal Processing (TSP)*. IEEE, 2011. p. 34–36.

[120] Muaaz M and Mayrhofer R. Smartphone-based gait recognition: from authentication to imitation. *IEEE Transactions on Mobile Computing*. 2017;16(11):3209–3221.

[121] Zhang Y, Yao L, Xhafa F, *et al.* Special issue editorial: wearable sensor signal processing for smart health. *Smart Health*. 2018 03;5-6.

[122] Stack E, Agarwal V, King R, *et al.* Identifying balance impairments in people with Parkinson's disease using video and wearable sensors. *Gait & Posture*. 2018;62:321–326.

[123] Qiu C, Wu F, Shi Q, *et al.* Sensors and control interface methods based on triboelectric nanogenerator in IoT applications. *IEEE Access*. 2019;7: 92745–92757.

[124] Lee GH, Park JK, Byun J, *et al.* Parallel signal processing of a wireless pressure-sensing platform combined with machine-learning-based cognition, inspired by the human somatosensory system. *Advanced Materials*. 2020;32(8):1906269.

[125] Shi Q, Dong B, He T, *et al.* Progress in wearable electronics/photonics – moving toward the era of artificial intelligence and internet of things. *InfoMat*. 2020;2(6):1131–1162.

[126] Madni AM, Madni CC, and Lucero SD. Leveraging digital twin technology in model-based systems engineering. *Systems*. 2019;7(1):7.

[127] Tao F, Cheng J, Qi Q, *et al.* Digital twin-driven product design, manufacturing and service with big data. *The International Journal of Advanced Manufacturing Technology*. 2018;94(9):3563–3576.

[128] Gadaleta M and Rossi M. Idnet: smartphone-based gait recognition with convolutional neural networks. *Pattern Recognition*. 2018;74:25–37.

[129] Kwapisz JR, Weiss GM, and Moore SA. Cell phone-based biometric identification. In: *2010 Fourth IEEE International Conference on Biometrics: Theory, Applications and Systems (BTAS)*. IEEE, 2010. p. 1–7.

[130] Batool S, Saqib NA, and Khan MA. Internet of Things data analytics for user authentication and activity recognition. In: *2017 Second International Conference on Fog and Mobile Edge Computing (FMEC)*. IEEE, 2017. p. 183–187.

[131] Thang HM, Viet VQ, Thuc ND, *et al.* Gait identification using accelerometer on mobile phone. In: *2012 International Conference on Control, Automation and Information Sciences (ICCAIS)*. IEEE, 2012. p. 344–348.

[132] Vhaduri S and Poellabauer C. Wearable device user authentication using physiological and behavioral metrics. In: *2017 IEEE 28th Annual International Symposium on Personal, Indoor, and Mobile Radio Communications (PIMRC)*. IEEE, 2017. p. 1–6.

[133] Malik MN, Azam MA, Ehatisham-Ul-Haq M, *et al.* ADLAuth: passive authentication based on activity of daily living using heterogeneous sensing in smart cities. *Sensors*. 2019;19(11):2466.

[134] Liang Y, Samtani S, Guo B, *et al.* Behavioral biometrics for continuous authentication in the internet-of-things era: an artificial intelligence perspective. *IEEE Internet of Things Journal*. 2020;7(9):9128–9143.

[135] Paraskevopoulos DC, Laporte G, Repoussis PP, *et al.* Resource constrained routing and scheduling: review and research prospects. *European Journal of Operational Research*. 2017;263(3):737–754.

[136] Kaur N and Sood SK. An energy-efficient architecture for the Internet of Things (IoT). *IEEE Systems Journal*. 2015;11(2):796–805.

[137] Jenkins J and Ellis C. Using ground reaction forces from gait analysis: body mass as a weak biometric. In: *International Conference on Pervasive Computing*. Springer, 2007. p. 251–267.

Chapter 10

Cognitive IoT sensors for smart industrial and biomedical applications

Shruti Sharma[1], Prafulla Vyas[2], Kavita Thakur[3] and Ashutosh Sharma[4]

Abstract

Cognitive Internet of Things (IoT) sensors are advanced sensing devices that incorporate artificial intelligence (AI) and machine learning (ML) algorithms to process and analyze real-time data from various sources. These sensors are widely used in smart industrial and biomedical applications for their ability to provide real-time information and improve operational efficiency. In industrial applications, cognitive IoT sensors can be used to monitor production processes and provide predictive maintenance. In biomedical applications, they can be used to monitor patient health and improve diagnosis and treatment. With the continued growth of the IoT and advancements in AI and ML, the use of cognitive IoT sensors is expected to increase in both industrial and biomedical applications in the coming years. This book chapter presents the innovative approach to decision making in healthcare and industrial settings, focusing on the handling of data transmission, trade, and record keeping through a cost-effective preservation method. The chapter serves as an overview of cognitive sensors and the current state of IoT in smart industries. It also delves into the impact of cognitive sensors on the operations of hospitals and industrial organizations, and explores the use of these sensors to understand the reasons behind resistance to smart products among both producers and consumers.

Keywords: Artificial intelligence; Cognitive sensors; Health monitoring; IoT devices; Machine learning; Predictive maintenance; Smart industry; Wireless sensor networks

[1]Department of Electrical and Computer Engineering, Ajou University, South Korea
[2]Disha College, India
[3]School of Studies in Electronics and Photonics, Pt. Ravishankar Shukla University, India
[4]Department of Materials Science and Engineering and Department of Energy Systems Research, Ajou University, South Korea

10.1 Introduction

Cognitive Internet of Thing (IoT) sensors are smart sensors designed to improve the efficiency and performance of industrial and biomedical applications. These sensors use artificial intelligence (AI) and machine learning (ML) algorithms to process and analyze data, and can be integrated into IoT systems to provide real-time monitoring and control. Applications of cognitive IoT sensors in industry include predictive maintenance, quality control, and process optimization, while in biomedicine, they can be used for patient monitoring, disease diagnosis and treatment, and telemedicine [1–4]. Cognitive IoT is an advanced form of IoT that uses AI and ML to process and analyze the data generated by IoT devices. The goal of cognitive IoT is to create smart, self-learning, and autonomous systems that can interact with their environment, learn from it, and make decisions based on that knowledge [5–7]. Cognitive IoT enhances the traditional IoT by incorporating cognitive computing capabilities, such as speech recognition, image and pattern recognition, and natural language processing (NLP), to improve the intelligence, efficiency and performance of IoT systems. Applications of cognitive IoT can be found in various domains such as smart homes, industrial automation, transportation, and healthcare [8–11].

The basic function of cognitive technology is to enhance human intelligence by simulating human thought processes and decision-making abilities. This is achieved by using AI techniques such as ML, NLP, computer vision, and expert systems [12–15]. Cognitive technology helps automate complex and repetitive tasks, provides insights from large amounts of data, and makes more informed decisions. It can also learn from experience and continuously improve over time, enabling it to adapt to new situations and perform increasingly complex tasks. A machine starts producing data for analysis once it is connected to a cognitive computing system. A computing scheme is a set of instructions and procedures that dictate how a computer or network of computers should process, store, and manage data. The computing scheme can determine how data is organized, how algorithms are executed, how data is transmitted, and how different components of the system interact [16–20]. There are several types of computing schemes, including: *Centralized computing*: this scheme involves a single, central computer that manages all data and processing in the system. *Distributed computing*: this scheme involves multiple computers working together in a network, each processing a portion of the data and communicating with each other to reach a common goal. *Client–server computing*: this scheme involves a central server that manages data and processing and a number of client computers that request data and services from the server [15–17,21]. *Cloud computing*: this scheme involves remote servers that provide data and processing resources over the Internet, allowing users to access the resources they need without having to manage their own infrastructure [19]. Each computing scheme has its own advantages and disadvantages, and the best scheme for a particular application depends on the specific requirements and constraints of the system. Basic function of cognitive machinery is represented in Figure 10.1.

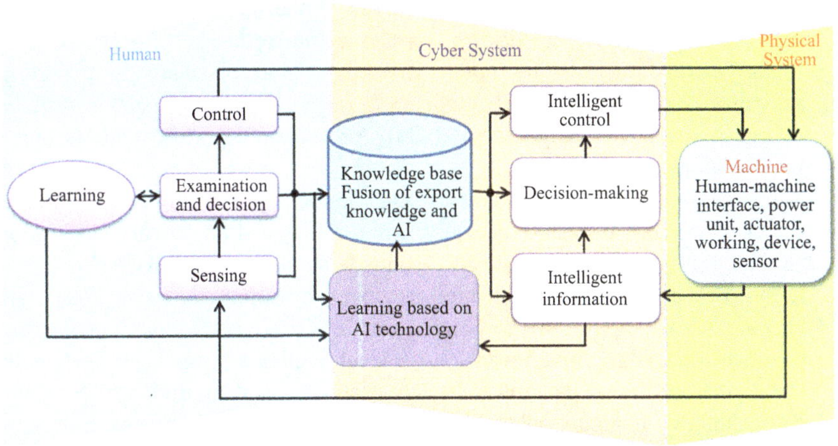

Figure 10.1 Basic function of cognitive technology [14]

A. Architecture of cognitive sensor

The architecture of a cognitive sensor is a layered structure consisting of three main classes: *Sensory class system*: This class is responsible for capturing raw data from the physical environment using various sensory modalities such as cameras and microphones [20–22]. *Preprocessing class system*: This class performs initial processing on the raw data obtained from the sensory class system. This includes tasks such as noise reduction, data normalization, and feature extraction. *Logical class system*: This class processes the preprocessed data and performs higher-level tasks such as object recognition, scene understanding, and decision making. This class is responsible for creating a semantic representation of the data and making inferences based on that representation. Each class in the architecture is designed to perform specific tasks and the overall system is designed to perform complex cognitive tasks. The three classes work together to provide a complete and integrated solution for capturing, processing and making decisions based on sensory data represents the various components and processes involved in the technology [23–27]. Some common elements that may be included in a visual illustration of cognitive technology are: *Sensors*: represented as input devices that capture data from the physical environment. *Data processing*: represented as a series of blocks that perform various data processing tasks such as data normalization, feature extraction, and semantic representation. *Machine learning models*: represented as decision making blocks that use statistical models to make predictions based on the processed data. *Knowledge base*: represented as a database of knowledge that is used by the cognitive technology to make inferences and decisions. *Output devices*: represented as devices that present the results of the cognitive technologies processing to the end-user, such as a display or a speaker [28–31]. Overall, the visual illustration of cognitive technology aims to provide a clear and concise representation of the

various components and processes involved in the technology, and how they interact with each other to perform complex cognitive tasks.
Cognitive work refers to the process of using cognitive technology to perform tasks that require human-level intelligence, such as decision making, problem solving, and pattern recognition [32–36]. *Flexibility*: cognitive work is highly flexible, allowing the technology to perform a wide range of tasks, and adapt to changing conditions and requirements. *Fully automatic*: in many cases, cognitive work is performed fully automatically, without the need for human intervention, allowing for increased efficiency and accuracy. *Accuracy*: cognitive technology is designed to perform tasks with high accuracy, by using statistical models and machine learning algorithms to process data and make predictions. *Intelligence*: cognitive work is performed with human-level intelligence, by using models that capture the underlying patterns and relationships in the data. *Adaptability*: cognitive technology is highly adaptable, allowing it to learn and improve over time, based on new data and experience [37,38]. This enables the technology to continuously improve its performance, and stay up-to-date with the latest advancements in the field. Overall, cognitive work provides a highly efficient and accurate solution for performing tasks that require human-level intelligence, by leveraging the power of advanced machine learning algorithms and statistical models. Figure 10.2 shows the visual illustration of cognitive technology.

B. Briefly on decision theory

Decision theory is a branch of mathematics and economics that deal with the study of decision making under uncertainty. It provides a framework for modeling and analyzing decisions, by taking into account the uncertainty and risk associated with different outcomes. The theory defines a set of formal methods for representing and analyzing decision problems, including methods for determining the optimal decision, given the available information and

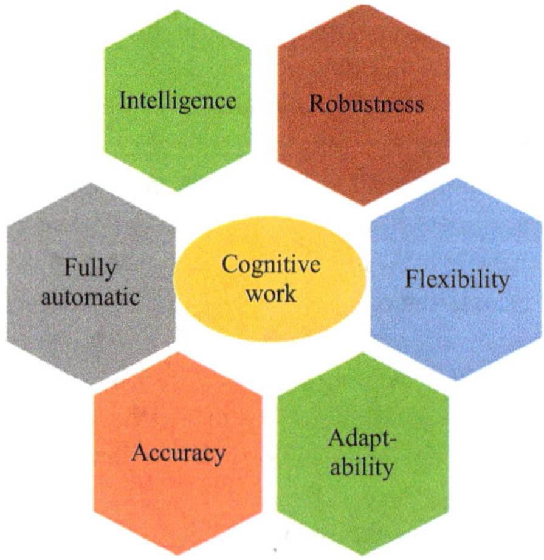

Figure 10.2 Visual illustration of cognitive technology

constraints. The goal of decision theory is to help individuals and organizations make better decisions, by providing a systematic and scientific approach to decision making. Decision theory covers a wide range of topics, including decision under uncertainty, risk assessment, utility theory, game theory, and multi-criteria decision analysis. It is widely used in a variety of fields, including finance, engineering, medicine, and operations research. Overall, decision theory provides a valuable tool for improving decision making in complex and uncertain environments, by helping individuals and organizations make informed and rational decisions [39]. A decision often involves selecting one of two or more possibilities. Since it is simpler for an individual to move from a choice to action than for an organization, decision-making theories have tended to be formed from studies of individual behavior rather than organizational behavior [40–42]. In [43], Brunsson and Hansson provide an example of a decision in their book, "Organizing Reason: Decisions, Bounded Rationality, and Organizing." The example they provide is a simple decision problem, where a manager is faced with the choice between two alternatives: investing in a new product line or not. The manager must consider the potential benefits and costs of each alternative, and make a decision based on the information available to them. In this example, Brunsson and Hansson illustrate how decision theory can be used to analyze and evaluate the different options, taking into account the uncertainties and risks involved. They show how the manager can use decision theory to determine the optimal decision, based on the available information and constraints, and how they can use it to evaluate the potential outcomes of each alternative. The example provided by Brunsson and Hansson demonstrates the power of decision theory in improving decision making, by providing a structured and systematic approach to decision making under uncertainty. It highlights the importance of considering the different factors that influence a decision, and the role of decision theory in helping individuals and organizations make informed and rational decisions [44–47]. Brunsson, a Swedish organization theorist, provides several explanations in his work on organizations and decision making. Some of the key explanations provided by Brunsson include: the role of rules and routines in organizations: Brunsson argues that rules and routines play a crucial role in shaping organizational behavior and decision making. He shows how rules and routines can both facilitate and constrain decision making, and how they can influence the way that organizations approach decision problems. The concept of organizational rationality: Brunsson introduces the concept of "organizational rationality," which refers to the way that organizations make decisions based on the available information and constraints. He argues that organizations are not always rational in the traditional sense, but instead have their own unique form of rationality that is shaped by the rules, routines, and practices that they have developed over time. The importance of symbolic actions in organizations: Brunsson highlights the role of symbolic actions in shaping organizational behavior and decision making. He shows how symbolic actions can be used to convey meaning and influence decision making, and how they can play a crucial role in shaping the way that organizations perceive and respond to decision problems [48–50]. Overall, Brunsson's work provides valuable insights into the dynamics of organizational decision making, and the role of rules, routines, and symbolic actions in shaping organizational behavior and decision making. His explanations help to deepen our understanding of the complexities of decision making in organizations, and the ways in which organizations approach decision problems.

10.2 IoT outline

A. The significance of the Internet

The progress of the Internet has been significant since its inception in the late 1960s. The following are some key milestones in the evolution of the Internet [51]. The creation of ARPANET in the late 1960s: ARPANET was the first computer network and the precursor to the modern Internet. It was created by the US Department of Defense and was designed to allow researchers to share information and collaborate on projects. The development of TCP/IP in the 1970s: Transmission Control Protocol/Internet Protocol (TCP/IP) was developed in the 1970s and became the standard communication protocol for the Internet [27,29,52–55]. It allowed computers to communicate with each other, regardless of their location or operating system. The creation of the World Wide Web (WWW) in the 1990s: WWW was created in the 1990s and transformed the Internet from a network of computers into a global network of information. The WWW allowed users to access and share information using a web browser, making it much easier to navigate the Internet. The rapid growth of the Internet in the 2000s: the early 2000s saw a rapid increase in the number of Internet users, as well as the development of new technologies and services, such as broadband, social media, and e-commerce. The Internet became an increasingly important part of everyday life, as people used it for communication, entertainment, and business. The rise of mobile Internet in the 2010s [56–59]. The rise of smartphones and tablet computers in the 2010s has dramatically changed the way that people access and use the Internet. Mobile Internet has become increasingly widespread, making it possible for people to access the Internet from anywhere, at any time [59,60]. Overall, the progress of the Internet has been rapid and transformative, and has had a profound impact on society, commerce, and communication. The Internet continues to evolve and grow, and it is likely that it will continue to play a central role in our lives for many years to come.

B. IoT sensing

IoT sensing refers to the use of sensors in IoT devices to collect data about the physical world and transmit it over the Internet. IoT sensing allows devices to interact with their environment, gather data, and share that data with other devices and systems. Examples of IoT sensing include: *Environmental sensors*: these sensors measure temperature, humidity, air quality, light levels, and other environmental factors, allowing devices to respond to changes in their environment. *Motion sensors*: these sensors detect movement, allowing devices to respond to physical activity and respond accordingly. *Location sensors*: these sensors use GPS, Wi-Fi, or other technologies to determine the device's location, allowing it to respond to its location. *Biometric sensors*: These sensors measure biometric data, such as heart rate, blood pressure, and other physiological parameters, to provide health monitoring and wellness tracking. *Image and audio sensors*: These sensors capture images and audio, allowing devices to respond to visual and auditory information. IoT sensing enables a wide range of applications, such as smart homes, wearable devices, industrial automation, and connected vehicles. It also plays a crucial role in the development of smart cities, allowing cities to monitor and respond to changes in their environment in real-time [61]. There are three types of the primary technologies that integrate "Things" into IoT ecosystems: Radio Frequency Identification (RFID): RFID is a wireless technology that uses radio waves to communicate between a tag and a reader. It enables devices to identify and track items in real-time, making it a key technology for supply chain management, inventory management, and other applications [47–49]. *Wireless Sensor Networks (WSN)*: WSNs are networks of interconnected sensors that communicate wirelessly. They enable devices to collect and transmit data about their environment, providing valuable insights for applications such as industrial automation, smart homes, and smart cities [50,51]. *Mobile computing*: Mobile computing technology enables devices to connect to the internet and communicate with other devices and systems using wireless networks. Mobile computing is a key enabler for the IoT, as it provides a platform for IoT devices to access cloud services, process data, and interact with other devices and systems [52–55]. IoT sensing plays a crucial role in various industries, including: *Manufacturing*: IoT sensors can be used to monitor production lines, collect data on machine performance, and track inventory levels in real-time. This helps manufacturers optimize their production processes and improve overall efficiency [27,29,58]. *Healthcare*: IoT sensors can be used to monitor patients' health conditions and track vital signs such as temperature, heart rate, and blood pressure. This helps healthcare providers make informed decisions about treatment and improve patient outcomes [57–59]. *Agriculture*: IoT sensors can be used to monitor soil moisture levels, crop growth, and weather conditions, providing farmers with valuable insights into the health of their crops. This information can be used to optimize irrigation and other agricultural processes [59–62]. *Energy and utilities*: IoT sensors can be used to monitor energy usage in buildings, track energy consumption. These technologies are critical for building an IoT ecosystem, as they provide the connectivity and communication capabilities required for devices to interact with each other and share data [62–65]. Development of the IoT for industry is represented in Figure 10.3. There are several ways in which the IoT can be improved for the industry: *Security*: as more and more devices are connected to the IoT, security remains a major concern. Improving the security of IoT systems is crucial to protect against hacking and other cyber threats. This can be achieved through the use of encryption, secure communication protocols, and regular software updates. *Interoperability*: ensuring that IoT devices and systems can

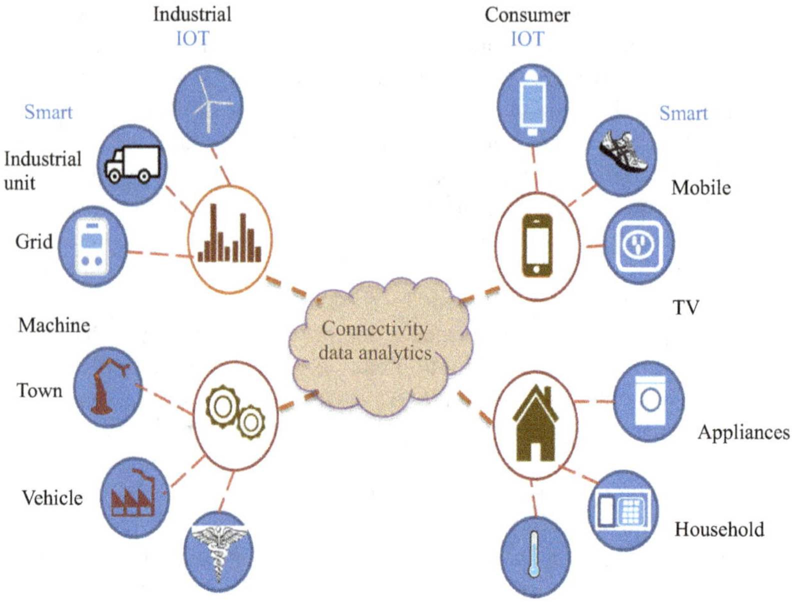

Figure 10.3 The development of the IoT for industry

communicate and work together seamlessly is crucial to the success of the IoT in industry. Improving interoperability will make it easier for organizations to integrate IoT solutions into their operations and realize the full potential of the IoT. *Scalability*: as the number of IoT devices continues to grow, it is important that IoT systems are scalable and can handle the increased load. Improving scalability will ensure that IoT systems can continue to function effectively as more devices are added. *Data management*: collecting, storing, and analyzing large amounts of IoT data requires a robust data management infrastructure. Improving data management will ensure that organizations can effectively harness the power of IoT data to drive business value [66–68]. *Cost-effectiveness*: IoT solutions can be expensive, and reducing costs is crucial to ensure that they are accessible to organizations of all sizes. Improving cost-effectiveness will make it easier for organizations to adopt IoT solutions and realize the benefits of the IoT. By addressing these areas, the IoT can be improved and become even more valuable for the industry.

10.3 Smart industrial

Smart industrial refers to the application of advanced technologies, such as AI, ML, IoT, and robotics, in the industrial sector to increase efficiency, productivity, and competitiveness. The term "SMART" refers to the ability of these technologies to make operations more "intelligent," that is, more automated, connected, and data-driven. The goal of smart industrial is to create a more flexible, adaptive, and sustainable industrial system [69,70]. "Towards Smart Factory for Industry 4.0" refers to the trend towards the integration of advanced technologies in manufacturing processes to achieve greater efficiency, flexibility, and responsiveness. The concept of the smart factory is a key aspect of Industry 4.0, the fourth industrial revolution, which aims to create a more automated, connected, and data-driven industrial system [71–74]. A smart factory leverages technologies such as the IoT, AI, ML, and robotics to create a fully integrated, real-time system for monitoring and controlling production processes. This results in a more agile and responsive manufacturing system, capable of quickly adapting to changing market conditions, customer demands, and supply chain disruptions [75–78]. The goal of a smart factory is to create a more flexible, adaptive, and sustainable industrial system, one that can respond quickly to changing market conditions and customer demands, while reducing costs, increasing efficiency, and improving product quality.

The industrial IoT (IIoT) with ML refers to the integration of ML algorithms into the IIoT ecosystem. The IIoT is a network of connected devices and machines that communicate with each other to automate industrial processes. By combining the power of machine learning with the IIoT, organizations can create more intelligent, adaptive systems that can improve efficiency, productivity, and competitiveness [79–80]. ML algorithms can analyze vast amounts of data generated by connected devices and machines to identify patterns and make predictions about

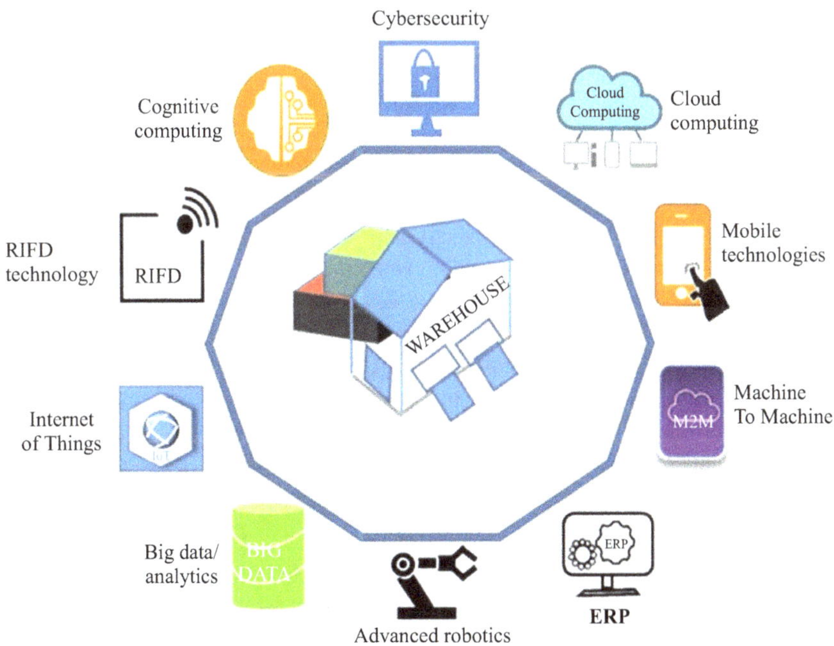

Figure 10.4 Architecture of smart industry

future events. This can help organizations to optimize production processes, improve quality control, and reduce downtime. In a smart factory, for example, ML algorithms can be used to predict maintenance needs based on sensor data, optimize production schedules based on real-time demand and supply chain data, and improve energy efficiency by adjusting processes based on real-time energy consumption data [81–84]. By integrating ML into the IIoT ecosystem, organizations can gain a competitive advantage by leveraging the power of data and analytics to create a more intelligent, adaptive, and efficient industrial system.

10.4 Biomedical

Cognitive computing in biomedical refers to the use of AI and ML algorithms to process and analyze large amounts of biomedical data to support decision making and improve healthcare outcomes [85–88]. Applications of cognitive computing in biomedical include: (1) drug discovery and development. (2) Predictive analytics for patient outcomes. (3) Medical imaging analysis. (4) Electronic medical records management. (5) Clinical trial design and optimization. (5) Personalized medicine and precision health.

These applications help to identify new targets for drug development, predict patient outcomes, improve medical imaging analysis, and optimize the delivery of care to patients.

10.5 Issues and challenges

Issues and challenges in cognitive technologies include (1) bias and fairness: ensuring that AI systems do not perpetuate existing biases and discrimination. (2) *Explainability*: understanding how AI systems make decisions and ensuring accountability. *Data quality and privacy*: ensuring that data used to train AI systems is accurate, diverse, and respects privacy rights. (3) *Regulation*: navigating the complex regulatory landscape around AI and ensuring compliance with relevant laws and ethical principles. (4) *Security*: protecting AI systems from cyber-attacks and ensuring the privacy and security of sensitive data [89]. *Talent and skills*: attracting and developing the necessary talent and skills to build, deploy, and manage AI systems. *Integration with existing systems*: integrating AI systems with existing processes, technologies, and organizational structures. *Ethical considerations*: ensuring that AI systems are aligned with ethical and moral values, such as transparency, accountability, and respect for privacy. (5) *Scalability and sustainability*: ensuring that AI systems can handle increasing amounts of data and computational complexity over time.

(A) Requirements for intelligent equipment
    The ability to learn from data and improve over time. *Real-time processing*: the ability to process large amounts of data in real-time and make decisions quickly. *Data integration*: the ability to integrate data from multiple sources and formats, such as sensors, databases, and the cloud [89–93]. The ability to

communicate and share data with other systems and devices. The ability to handle increasing amounts of data and computational complexity over time. *Human–machine interface*: a user-friendly interface that enables human operators to interact with the equipment and monitor its performance. *Security and privacy*: the ability to protect sensitive data and ensure the privacy and security of users. The ability to function reliably and consistently, even in challenging environments or under adverse conditions. The ability to provide value for money, taking into account both the initial cost of the equipment and its ongoing operational expenses. The ability to comply with relevant standards and regulations, such as safety standards, data privacy laws, and ethical guidelines.

(B) Nanotechnology-based method
Nanotechnology-based cognitive sensors are advanced sensors that utilize nanotechnology to enhance their capabilities. These sensors have the ability to self-learn, adapt, and process information in real-time, making them useful in a range of applications, such as smart homes, wearable devices, and robotics. They can detect, analyze, and respond to changes in their environment and make decisions based on that information [94–97]. The miniaturization of these sensors enables them to be integrated into smaller and more complex systems, resulting in more efficient and effective sensing capabilities. Nanotechnology-based smart industries are industries that utilize nanotechnology to improve their processes, products, and overall operations [98–100]. Some examples of nanotechnology applications in industries include: *Healthcare*: nanotechnology is used in the development of advanced medical devices, such as nanoscale sensors for monitoring vital signs and targeted drug delivery systems. *Energy*: nanotechnology is used in the development of solar panels, fuel cells, and batteries, to increase their efficiency and storage capacity. *Manufacturing*: nanotechnology is used to enhance product performance, improve material strength, and reduce production costs. *Information technology* [101–105]: nanotechnology is used to improve the performance and functionality of computer components, such as hard drives, processors, and memory chips. *Environmental protection*: nanotechnology is used to develop more efficient and effective water filtration systems, air purification systems, and to clean up toxic waste sites [106,107]. Overall, nanotechnology is helping smart industries to increase efficiency, reduce costs, and improve sustainability.

(C) Data-driven fabrication
Data-driven fabrication refers to a manufacturing process where data, such as CAD models, simulation results, and sensor readings, is used to drive the production of a product. This technology allows for the integration of data into the production process, enabling real-time adjustments and customization of products based on data analysis [108]. In industry, data-driven fabrication has the potential to revolutionize manufacturing by improving efficiency, reducing waste, and enabling mass customization. Some examples of its applications include: *Additive manufacturing (3D printing)*: using data from CAD models, 3D printers can produce complex parts and structures with high accuracy and

minimal waste [109]. *Precision manufacturing*: data from sensors and simulation results can be used to guide the production process, ensuring high precision and consistent quality. *Smart factories*: data from sensors and production equipment can be analyzed to optimize production processes and identify areas for improvement. *Personalized products*: data from customer preferences and usage patterns can be used to produce customized products that meet individual needs. Overall, data-driven fabrication is enabling industry to produce high-quality products faster and more efficiently, while reducing waste and improving sustainability.

10.6 Conclusions

In conclusion, cognitive sensing is playing a critical role in the development of smart industries. By utilizing advanced sensors that can self-learn, adapt, and process information in real-time, smart industries are able to improve efficiency, reduce costs, and enhance their overall operations. With the integration of nanotechnology and data-driven fabrication, smart industries are able to produce high-quality products faster and more efficiently, while reducing waste and improving sustainability. Cognitive sensing is a rapidly growing field with immense potential, and its continued development will likely bring about numerous new applications and advancements in the future. As such, smart industries that embrace and incorporate cognitive sensing technology will likely be well-positioned for success in the coming years. By utilizing advanced sensors that can self-learn, adapt, and process information in real-time, the biomedical field is able to improve patient outcomes and advance the field of medicine. Cognitive sensors are being used to develop innovative medical devices, such as nanoscale sensors for monitoring vital signs, targeted drug delivery systems, and implantable devices. These devices are helping to provide continuous monitoring and treatment, reducing the need for invasive procedures and improving patient comfort. In addition, cognitive sensing technology is enabling the development of personalized medicine, where treatments are tailored to the unique needs and characteristics of each patient. This is accomplished by using data from sensors and other sources to better understand patient conditions and make informed treatment decisions. Overall, cognitive sensing technology is helping to advance the biomedical field and improve patient outcomes, and its continued development will likely bring about numerous new applications and advancements in the future.

References

[1] K. Hwang and M. Chen, *Big-data Analytics for Cloud, IoT and Cognitive Computing*, New York, NY: John Wiley & Sons, 2017.

[2] M. Chen, J. Yang, Y. Hao, S. Mao, and K. Hwang, A 5G cognitive system for healthcare, *Big Data Cogn. Comput.*, 1(1), 2, 2017.

[3] Y. Chen, J. E. Argentinis, and G. Weber, IBM Watson: how cognitive computing can be applied to big data challenges in life sciences research, *Clin. Ther.*, 38(4), 688–701, 2016.
[4] A. Dhir, The emerging role of cognitive computing in healthcare: a systematic literature review, *Int. J. Med. Inform.*, 129, 154–166, 2019, doi:10.1016/j.ijmedinf.2019.04.024.
[5] A. Belle, R. Thiagarajan, S. M. Soroushmehr, F. Navidi, D. A. Beard, and K. Najarian, Big data analytics in healthcare, *Biomed Res. Int.*, 2015, p. 370194, 2015.
[6] A. Asante-Korang and J. P. Jacobs, Big data and paediatric cardiovascular disease in the era of transparency in healthcare, *Cardiol. Young*, 26(8), 1597–1602, 2016.
[7] Z. Goli-Malekabadi, M. Sargolzaei-Javan, and M. K. Akbari, An effective model for store and retrieve big health data in cloud computing, *Comput. Methods Programs Biomed.*, 132, 75–82, 2016.
[8] G. Adamson, L. Wang, and P Moore., Feature-based control and information framework for adaptive and distributed manufacturing in cyber-physical systems, *J. Manuf. Syst.*, 43, 305–315, 2017.
[9] N. Z. Aitzhan and D. Svetinovic, Security and privacy in decentralized energy trading through multi-signatures, blockchain and anonymous messaging streams, *IEEE Trans. Depend. Secur. Comput.*, 99, 1–1, 2016.
[10] S. R. Islam, D. Kwak, M. H. Kabir, M. Hossain, and K. S. Kwak, The Internet of Things for healthcare: a comprehensive survey, *IEEE Access*, 3, 678–708, 2015.
[11] J. Kim and J. W. Lee, OpenIoT: an open service framework for the Internet of Things, in *2014 IEEE World Forum on Internet of Things (WF-IoT)*, IEEE, 89–93, 2014.
[12] I. F. Akyildiz, W. Su, Y. Sankarasubramaniam, and E. Cayirci, Wireless sensor networks: a survey, *Comput. Networks*, 38(4), 393–422, 2002.
[13] R. Albert and A. Barabási, Statistical mechanics of complex networks, *Rev. Mod. Phys.*, 74(1), 47–97, 2002.
[14] I. F. Akyildiz, W. Su, Y. Sankarasubramaniam, and E. Cayirci, Wireless sensor networks: a survey, *Comput. Netw.*, 38(4), 393–422, 2002.
[15] A. Albert and A. Barabási, Statistical mechanics of complex networks, *Rev. Modern Phys.*, 74(1), 47–97, 2002.
[16] A. Amiri, Application placement and backup service in computer clustering in Software as a Service (SaaS) networks, *Comput. Oper. Res.*, 69, 48–55, 2016.
[17] B. F. Babiceanu and R. Seker, Big data and virtualization for manufacturing cyber-physical systems: a survey of the current status and future outlook, *Comput. Ind.*, 81, 128–137, 2016.
[18] A. Bahga and V. K. Madisetti, Blockchain platform for industrial internet of things, *J. Softw. Eng. Appl.*, 9(10), 533–546, 2016.
[19] Y. Bao, L. Ren, L. Zhang, *et al.*, Massive sensor data management framework in cloud manufacturing based on hadoop, in *Proceedings of the 10th*

International Conference on Industrial Informatics, IEEE Press, Piscataway, NJ, 2012, 397–401.

[20] O. F. Beyca, P. K. Rao, Z. Kong, S. T. S. Bukkapatnam, and R. Komanduri, Heterogeneous sensor data fusion approach for real-time monitoring in ultraprecision machining (UPM) process using non-parametric Bayesian clustering and evidence theory, *IEEE Trans. Autom. Sci. Eng.*, 13(2), 1033–1044, 2016.

[21] L. Bian, N. Gebraeel, and J. P. Kharoufeh, Degradation modeling for real-time estimation of residual lifetimes in dynamic environments, *IIE Trans.*, 4 (5), 471–486, 2015.

[22] T. Cheng, G. Miglaccio, J. Teizer, *et al.*, Data fusion of real-time location sensing and physiological status monitoring for ergonomics analysis of construction workers, *J. Comput. Civil Eng.*, 27(3), 320–335.

[23] S. Sharma and W. Yoon, Multiobjective optimization for resource allocation in full-duplex large distributed MIMO systems, *Adv. Electr. Comput. Eng.*, 21(2), 67–74, 2021, doi:10.4316/AECE.2021.02008

[24] M. Y. Sai, S. Kavya, S. R. Bhimavarapu, M. Mudaliar, and S. Sharma, CPW fed microstrip patch antenna for dedicated short-range communication, *Wireless Personal Commun.*, 122, 3859–3873, 2021, doi:10.1007/s11277-021-09114-7.

[25] S. Sharma and W. Yoon, Multiobjective reinforcement learning based energy consumption in C-RAN enabled massive MIMO, *Radio Eng. J.*, 31 (1), 155–163, 2022.

[26] S. Sharma and W. Yoon, Energy efficient user association with multi-objective optimization for full-duplex C-RAN enabled massive MIMO system, *Adv.Electr. Comput. Eng.*, in press.

[27] S. Sharma and W. Yoon, Multiobjective optimization for energy efficiency in cloud radio access networks, *Int. J. Eng. Res. Technol.*, 12, no. 5, 607–610, 2019.

[28] S. Sharma and W. Yoon, Multi-objective optimization of user capacity for WPCN, *Int. J. Eng. Res. Technol.*, 12(4), 482–485, 2018.

[29] S. Sharma and W. Yoon, Multi-objective energy efficient resource allocation for WPCN, *Int. J. Eng. Res. Technol.*, 11(12), 2035–2043, 2018.

[30] S. Sharma, K. Thakur, and A. Sharma, A DNN Based Method to Approximate Resource Allocation in a Cell-Free Massive MIMO Network, Australian Patent, Application No. 2021105950.

[31] S. Sharma, K. Thakur, and G. R. Sinha, Method and System for Clock Synchronization In IOT Networks Using Bayesian Estimation Australian Patent, Application No. 2021105878.

[32] S. Sharma, K. Thakur, G. R. Sinha, A. Sharma, P. Vyas, and S. Mandal, A Downlink NOMA System Using Interference Cancellation in Cognitive Radio Network to Enhance Spectral Efficiency, German Patent, Application No. 202022101502.3

[33] S. Sharma, G. S. Verma, and K. Thakur, Integration of ML and IoT for healthcare systems, In: *Computational Intelligence and Applications for*

Pandemics and Healthcare, pp. 115–138, 2022, doi:10.4018/978-1-7998-9831-3.ch006.
[34] A. Sharma and S. Sharma, Co-electrodeposition of nanocomposites, In: H.S. Nalwa, Ed., *Encyclopaedia of Nanoscience and Nanotechnology*, American Scientific Publishers, published on 22.06.2017.
[35] J. Liu, J. Shi, and S. J. Hu, Quality assured setup planning based on the stream of variation model for multi-stage machining processes, *IIE Trans.: Quality Relia. Eng.*, 41, 323–334, 2009.
[36] S. S. Manvi and G. Krishna Shyam, Resource management for Infrastructure as a Service (IaaS) in cloud computing: a survey, *J. Network Comput. Appl.*, 41, 424–440, 2014.
[37] R. D. Meller, B. Montreuil, C. Thivierge, and Z. Montreuil, Functional design of physical internet facilities: a road-based transit center, In: *Proceedings of the 12th IMHRC*, 2016, Gardanne, France, 2016, p. 26. Available: https://digitalcommons.georgiasouthern.edu/pmhr_2012/26.
[38] L. Monostori, B. Kádár, T. Bauernhansl, *et al.*, Cyber-physical systems in manufacturing, *CIRP Ann. – Manufact. Technol.*, 65(2), 621–641, 2016.
[39] A. A. Nazari Shirehjini and A. Semsar, Human interaction with IoT-based smart environments, *Multimedia Tools Appl.*, 76(11), 13343–13365, 2017.
[40] A. Y. C. Nee, S. K. Ong, G. Chryssolouris, and D. Mourtzis, Augmented reality applications in design and manufacturing, *CIRP Ann. – Manufact. Technol.*, 61(2), 657–679, 2012.
[41] C. Ok, S. Lee, P. Mitra, and S. Kumara, Distributed energy balanced routing for wireless sensor networks, *Comput. Ind. Eng.*, 57(1), 125–135, 2009.
[42] A. Pantelopoulos and N. G. Bourbakis, A survey on wearable sensor-based systems for health monitoring and prognosis, *IEEE Trans. Syst. Man Cybern.*, Part C (Applications and Reviews), 40(1), 1–12, 2010.
[43] T. Qu, M. Thürer, J. Wang, *et al.*, System dynamics analysis for an internet-of-things-enabled production logistics system, *Int. J. Prod. Res.*, 55(9), 2622–2649, 2017.
[44] Reaidy, P.J., Gunasekaran, A., and Spalanzani, A. Bottom-up approach based on internet of things for order fulfillment in a collaborative warehousing environment, *Int. J. Prod. Econ.*, 159, 29–40, 2015.
[45] Rymaszewska, A., Helo, P., and Gunasekaran, A. IoT powered servitization of manufacturing – an exploratory case study, *Int. J. Prod. Econ.*, in press.
[46] A. Saeed, A. Ahmadinia, A. Javed, and H. Larijani, Random neural network based intelligent intrusion detection for wireless sensor networks, *Proc. Comput.Sci.*, 80, 2372–2376, 2016.
[47] F. K. Shaikh, S. Zeadally, and E. Exposito, Enabling technologies for green internet of things, *IEEE Systems J.*, 11(2), 983–994, 2017.
[48] E. Sun, X. Zhang, and Z. Li, The internet of things (IOT) and cloud computing (CC) based tailings dam monitoring and pre-alarm system in mines, *Safety Sci.*, 50(4), 811–815, 2012.

[49] Y. S. Tan, Y. T. Ng, and J. S. C. Low, Internet-of-things enabled real-time monitoring of energy efficiency on manufacturing shop floors, *Proc. CIRP*, 61, 376–381, 2017.
[50] F. Tao, J. Cheng, and Q. Qi, IIHub: an industrial internet-of-things hub towards smart manufacturing based on cyber-physical system, *IEEE Trans. Ind. Inf.*, 99, 1–1, 2017.
[51] F. Tao, Y. Cheng, L. D. Xu, L. Zhang, and B. H. Li, CCIoT-CMfg: Cloud computing and internet of things-based cloud manufacturing service system, *IEEE Trans. Ind. Inf.*, 10(2), 1435–1442, 2014.
[52] Tao, F. and Qi, Q. (2017) New IT driven service-oriented smart manufacturing: Framework and characteristics. *IEEE Transactions on Systems, Man, and Cybernetics: Systems*, 99, 1–11. [Google Scholar]
[53] L. Thames and D. Schaefer, Cybersecurity for industry 4.0 and advanced manufacturing environments with ensemble intelligence, In: *Cybersecurity for Industry 4.0: Analysis for Design and Manufacturing*, Cham: Springer, 2017, pp.243–265.
[54] K. Thramboulidis and F. Christoulakis, UML4IoT—a UML-based approach to exploit IoT in cyber-physical manufacturing systems, *Comput. Ind.*, 82, 259–272, 2016.
[55] S. Sharma and W. Yoon, Energy efficient user association with multi-objective optimization for full-duplex C-RAN enabled massive MIMO system, *J. Commun. Technol. Electron.*, in press.
[56] C. A. Tokognon, B. Gao, G. Y. Tian, and Y. Yan, Structural health monitoring framework based on internet of things: a survey. *IEEE Internet of Things J.*, 4(3), 619–635, 2017.
[57] H. Vincent, L. Wells, P. Tarazaga, and J. Camelio, Trojan detection and side-channel analyses for cyber-security in cyber-physical manufacturing systems, *Proc. Manufact.*, 1, 77–85, 2015.
[58] X. Wang, H. Chen, and D. Le, A novel IPv6 address configuration for a 6LoWPAN-based WBAN, *J. Netw. Comput. Appl.*, 61, 33–45, 2016.
[59] H. Yang, S. T. S. Bukkapatnam, and L. G. Barajas, Local recurrence based performance prediction and prognostics in the nonlinear and nonstationary systems, *Pattern Recogn.*, 44(8), 1834–1840, 2011.
[60] H. Yang and Y. Chen, Heterogeneous recurrence monitoring and control of nonlinear stochastic processes, *Chaos*, 24(1), 013138, 2014.
[61] H. Yang and G. Liu, Self-organized topology of recurrence-based complex networks, *Chaos*, 23(4), 043116, 2013.
[62] X. Yang and L. Chen, Using multi-temporal remote sensor imagery to detect earthquake-triggered landslides. *Int. J. Appl. Earth Observ. Geoinform.*, 12 (6), 487–495, 2010.
[63] X. Zhang and Q. Huang, Analysis of interaction structure among multiple functional process variables for process control in semiconductor manufacturing. *IEEE Trans. Semicond. Manuf.*, 23(2), 263–272, 2010.
[64] Y. Zhang, Y. Xiang, and X. Huang, Password-authenticated group key exchange: a cross-layer design. *ACM Trans. Internet Technol.*, 16(4), 1–24, 2016.

[65] Y. Zhang, G. Zhang, J. Wang, S. Sun, S. Si, and T. Yang, Real-time information capturing and integration framework of the internet of manufacturing things. *Int. J. Comput. Integr. Manufact.*, 28(8), 811–822, 2015.

[66] R. Y. Zhong, L. Wang, and X. Xu, An IoT-enabled real-time machine status monitoring approach for cloud manufacturing, *Proc. CIRP*, 63, 709–714, 2017.

[67] S. Wang, C. Zhang, and J. Wan, A smart factory solution to hybrid production of multi-type products with reduced intelligence, in *Proc. IEEE Inf. Technol., Netw., Electron. Autom. Control Conf.*, Chongqing, China, May 2016, pp. 848–853.

[68] S. Wang, J. Wan, D. Zhang, D. Li, and C. Zhang, Towards smart factory for Industry 4.0: a self-organized multi-agent system with big data based feedback and coordination, *Comput. Netw.*, 101, 158–168, 2016.

[69] W. Dai, V. Vyatkin, J. H. Christensen, and V. N. Dubinin, Bridging service-oriented architecture and IEC 61499 for flexibility and interoperability, *IEEE Trans. Ind. Informat.*, 11(3), 771–781, 2015.

[70] M. G. Valls, I. R. Lopez, and L. F. Villar, ILAND: an enhanced middleware for real-time reconfiguration of service oriented distributed real-time systems, *IEEE Trans. Ind. Informat.*, 9(1), 228–236, 2013.

[71] M. Short and K. Burn, A generic controller architecture for intelligent robotic systems, *Robot. Comput.-Integr. Manuf.*, 27(2), 292–305, 2011.

[72] L. Morales-Velazquez, R. de J. Romero-Troncoso, R. A. Osornio-Rios, G. Herrera-Ruiz, and E. Cabal-Yepez, Open-architecture system based on a reconfigurable hardware–software multi-agent platform for CNC machines, *J. Syst. Archit.*, 56(9), 407–418, 2010.

[73] W. Lepuschitz, A. Zoitl, M. Vallée, and M. Merdan, Toward self-reconfiguration of manufacturing systems using automation agents, *IEEE Trans. Syst., Man, Cybern. C, Appl. Rev.*, 41(1), 52–69, 2011.

[74] M. Adamski and J. Tkacz, Formal reasoning in logic design of reconfigurable controllers, In: *IFAC Proceedings Volumes*, 45(7), 1–6, 2012.

[75] Chen, J. Wan, L. Shu, *et al.*, Smart factory of industry 4.0: key technologies, application case, and challenges, *IEEE Access*, 6, 6505–6519, 2018, doi:10.1109/ACCESS.2017.2783682.

[76] H. J. Kim, Y. K. Lee, Y. A. Jeon, and J. Y. Son, ICT convergencebased application service development to support the re-configurability of door trim assembly line, In: *Proceeding of the International Conference on Information and Communication Technology Convergence (ICTC)*, Jeju, South Korea, 2016, pp. 1126–1128.

[77] E. Järvenpää, N. Siltala, and M. Lanz, Formal resource and capability descriptions supporting rapid reconfiguration of assembly systems, In: *Proceeding of the IEEE International Symposium on Assembly and Manufacturing (ISAM)*, Fort Worth, TX, 2016, pp. 120–125.

[78] D. Gyulai, B. Kádár, A. Kovács, and L. Monostori, Capacity management for assembly systems with dedicated and reconfigurable resources, *CIRP Ann.*, 63 (1), 457–460, 2014.

[79] D. Zhang, Y. Qian, J. Wan, and S. Zhao, An efficient RFID search protocol based on clouds, *Mobile Netw. Appl.*, 20(3), 356–362, 2015.

[80] R. Y. Zhong, Q. Y. Dai, T. Qu, G. J. Hu, and G. Q. Huang, RFIDenabled real-time manufacturing execution system for mass-customization production, *Robot. Comput.-Integr. Manuf.*, 29 (2), 283–292, 2013.
[81] D. Zhang, Z. He, Y. Qian, J. Wan, D. Li, and S. Zhao, Revisiting unknown RFID tag identification in large-scale Internet of Things, *IEEE Wireless Commun.*, 23(5), 24–29, 2016.
[82] V. Coskun, B. Ozdenizci, and K. Ok, A survey on near field communication (NFC) technology, *Wireless Pers. Commun.*, 71 (3), 2259–2294, 2013.
[83] S. Choudhury, P. Kuchhal, R. Singh, and Anita, ZigBee and Bluetooth network based sensory data acquisition system, *Procedia Comput. Sci.*, 48, 367–372, 2015.
[84] X. Li, D. Li, J. Wan, A. V. Vasilakos, C.-F. Lai, and S. Wang, A review of industrial wireless networks in the context of industry 4.0, *Wireless Netw.*, 23 (1), 23–41, 2017.
[85] Y. Gao, W. Dong, L. Deng, C. Chen, and J. Bu, COPE: improving energy efficiency with coded preambles in low-power sensor networks, *IEEE Trans. Ind. Informat.*, 11(6), 1621–1630, 2015.
[86] J. Lee, L. Kim, and T. Kwon, Flexicast: energy-efficient software integrity checks to build secure industrial wireless active sensor networks, *IEEE Trans. Ind. Informat.*, 12(1), 6–14, 2016.
[87] J. Luo, J. Hu, D. Wu, and R. Li, Opportunistic routing algorithm for relay node selection in wireless sensor networks, *IEEE Trans. Ind. Informat.*, 11 (1), 112–121, 2015.
[88] A. Sharma, S. Das, and K. Das, Electrochemical corrosion behavior of CeO2 nanoparticle reinforced Sn–Ag based lead free nanocomposite solders in 3.5 wt.% NaCl bath, *Surf. Coat. Technol.* 261, 235–243, 2015.
[89] A. Sharma, K. Das, H. J. Fecht, and S. Das, Effect of various additives on morphological and structural characteristics of pulse electrodeposited tin coatings from stannous sulfate electrolyte, *Appl. Surf. Sci.* 314, 516–522, 2014.
[90] A. Sharma, D. U. Lim, and J. P. Jung, Microstructure and brazeability of SiC nanoparticles reinforced Al–9Si–20Cu produced by induction melting, *Mater. Sci. Technol.* 32(8), 773–779, 2016.
[91] A. Sharma, D. E. Xu, J. Chow, M. Mayer, H. R. Sohn, and J. P. Jung, Electromigration of composite Sn-Ag-Cu solder bumps, *Electron. Mater. Lett.* 11(6), 1072–1077, 2015.
[92] A. Sharma, Y. S. Shin, and J. P. Jung, Influence of various additional elements in Al based filler alloys for automotive and brazing industry, *J. Welding Join.* 33(5), 1–8, 2015.
[93] A. Sharma, S. Bhattacharya, S. Das, H. J. Fecht, and K. Das, Development of lead free pulse electrodeposited tin based composite solder coating reinforced with ex situ cerium oxide nanoparticles, *J. Alloys Comp.* 574, 609–616, 2013.
[94] A. Sharma, S. Das, and K. Das, Pulse electroplating of ultrafine grained tin coating, In: *Electroplating of Nanostructures*, 2, Croatia: InTech, 2015.

[95] A. Sharma, S. Das, and K. Das, Pulse electrodeposition of lead-free tin-based composites for microelectronic packaging, *Electrodeposition Compos. Mater.* 3, 253–274, 2016.

[96] S. Mishra, A. Sharma, D. H. Jung, and J. P. Jung, Recent advances in active metal brazing of ceramics and process, *Metals Mater. Int.* 26(8), 1087–1098, 2020.

[97] A. Sharma, S. Bhattacharya, S. Das, and K. Das, Fabrication of Sn nanostructures by template assisted pulse electrodeposition, *Surf. Eng.* 32(5), 378–384, 2016.

[98] S. Bhattacharya, A. Sharma, S. Das, and K. Das, Synthesis and properties of pulse electrodeposited lead-free tin-based Sn/ZrSiO4 nanocomposite coatings, *Metal. Mater. Trans. A* 47(3), 1292–1312, 2016.

[99] A. Sharma, C. S. Cheon, and J. P. Jung, Recent progress in electroless plating of copper, *J. Microelectron. Pack. Soc.* 23(4), 1–6, 2016.

[100] M. J. Chae, A. Sharma, M. C. Oh, and B. Ahn, Lightweight AlCuFeMnMgTi high entropy alloy with high strength-to-density ratio processed by powder metallurgy, *Metals Mater. Int.* 27(4), 629–638, 2021.

[101] D. H. Jung, A. Sharma, D. U. Lim, J. H. Yun, and J. P. Jung, Effects of AlN nanoparticles on the microstructure, solder ability, and mechanical properties of Sn-Ag-Cu solder, *Metal. Mater. Trans. A* 48(9), 4372–4384, 2017.

[102] W. S. Jeon, A. Sharma, and J. P. Jung, Liquid metal embrittlement of galvanized TRIP steels in resistance spot welding, *Metals* 10(6), 787, 2020.

[103] S. K. Dewangan, M. Mohan, V. Kumar, A. Sharma, and B. Ahn, A comprehensive review of the prospects for future hydrogen storage in materials-application and outstanding issues, *Int. J. Energy Res.*, 5, 100033, 2022.

[104] B. Sharma, M. Kumar, V. Kumar, and A. Sharma, Boron nitride nanotubes modified on a lead-free solder alloy for microelectromechanical packaging, *ACS Appl. Nanomater.*, 5, 13626–13636, 2022.

[105] H. Lee, A. Sharma, and B. Ahn, Microstructural evolution and compressive properties of nanocrystalline Ti–Fe alloy fabricated via cryomilling and spark plasma sintering, *J. Mater. Sci.*, 57(38), 1–12, 2022.

[106] J. Lee, H. Kao, and S. Yang, Service innovation and smart analytics for Industry 4.0 and big data environment, *Proc. CIRP*, 16, 3–8, 2014.

[107] P. Lei, L. Zheng, C. Li, and X. Li, MT connect enabled interoperable monitoring system for finish machining assembly interfaces of large-scale components, *Procedia CIRP*, 56, 378–383, 2016.

[108] S. L. Levin and S. Schmidt, IPv4 to IPv6: challenges, solutions, and lessons, *Telecommun. Policy*, 38(11), 1059–1068, 2014.

[109] D. Li, M. Ren, and G. Meng, Application of internet of things technology on predictive maintenance system of coal equipment, *Proc. Eng.*, 174, 885–889, 2017.

Chapter 11

Intelligent automation using IoT and machine learning

B. Padmaja[1], Thota Anjusree[1], E. Krishna Rao Patro[2], Tilottama Goswami[3] and M. Nagaraju[1]

Abstract

Internet of Things (IoT) is an ecosystem of vast connected devices known as objects that continuously generate data. Very large data come from these devices through various sensors. Machine learning (ML) is data-driven and generates meaningful insights from it. ML tries to find patterns from past data and build predictive models that are useful in predicting future behavior and events. There are various machine-learning techniques and approaches to analyze this huge data such as K-Nearest neighbors, Naïve Bayes, Support Vector Machine, Classification trees, Random forests, Bagging, Feedforward neural network, Linear regression, Support vector regression, Regression trees, and Random forests. The ML algorithms examine the data to derive valuable information for a better user experience. With the advancement of ML in IoT, the devices can inspect the data collected, derive meaningful information, create smart applications, and be autonomous. These devices can make decisions and learn from the data collected. Further enhancements in the domains of IoT and ML are making the applications smarter and smarter. As it is said, "Work smarter not harder." ML and IoT are playing a major role in making applications smarter. IoT helps collect data, and ML analyses this data to make smarter decisions. The underlying philosophy of IoT and ML is that smart equipment is not only better than humans at data capturing and analyzing in real-time, but they are also better at communicating critical information that can be used to fast driving and more accurate business decisions. IoT with ML will change everything without the need for human-to-human collaboration making human life much simpler and easy by automating everything. All of the devices in an IoT system would collect data from our living patterns in real-time, adjusting their settings or features to optimize our living experience and ensure the products

[1]Department of Computer Science and Engineering (AI & ML), Institute of Aeronautical Engineering, India
[2]Department of Computer Science and Engineering, Institute of Aeronautical Engineering, India
[3]Department of IT, Vasavi College of Engineering, India

work best for our life needs. All of the devices in an IoT system would collect data from our living patterns, modify their settings or features to improve our lifestyle, and ensure the best commodity for our life. When these applications are integrated into the housing, industries, and financial applications, they provide significant value to end users while providing new revenue opportunities to service and system providers. The futuristic applications of IoT and ML are smart homes, smart cities, etc. These IoT and ML applications are going to be epic in the field of innovation.

Keywords: Big data analytics; Internet of Things; Machine learning; Smart cities; Smart homes

"When the Internet of Things (IoT) and Machine Learning (ML) are used together, they are regarded as the pinnacle of technological innovation."

11.1 Introduction

Our gadgets have become smarter and more autonomous in recent years. Everything is connected to the Internet, from smart cities to self-driving cars to kitchen appliances. We can control machines in our homes using phones from our workplaces. The Internet of Things (IoT) is a new world wide web innovation that is quickly attaining attraction due to advances in many fields like sensor networks, mobile devices, wireless communications, networking, and cloud technologies [1].

11.1.1 IoT

IoT stands for Internet of Things, is an interface of Internet-connected appliances uniquely identified by their IP addresses. The IoT is an innovation in the field of intelligent applications. This material device (embedded with sensors, software, and machine learning (ML) technologies) connects and transforms data with gadgets. There are three paradigms associated with the IoT: Internet-oriented (middleware), things-oriented (sensors), and semantic-oriented (knowledge). The IoT refers to natural and computer-generated gadgets with distinct identities that enable wise implementation that makes energy, logistics, industrial control, retail, agriculture, and other sectors "smarter." Figure 11.1 depicts the various applications of IoT.

11.1.2 History behind IoT

One of the earliest and most well-known IoT experiments occurred in 1982 when C.S. graduate students at Carnegie Melon University connected a Coca-Cola vending machine to the Internet. The program code returns the drinks' temperature and determines their availability. IoT was developed by computer scientist Kevin Ashton in the year 1999. In the year 2000, LG electronics launched the first world refrigerator embedded with an IoT interface. This was followed by a small rabbit-shaped robot to

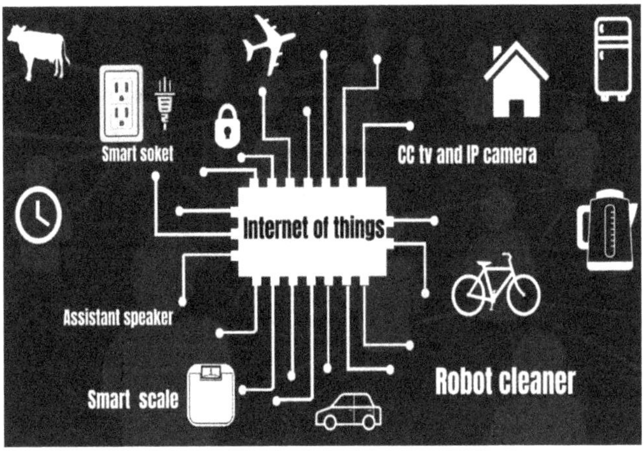

Figure 11.1 Applications of IoT

report news, weather forecast, and the stock market in the year 2005. In Switzerland, the first international conference on IoT was held in the year 2008. There are 27 billion devices connected with IoT in today's world and that may rise to 100 billion IoT embedded with ML devices by 2030.

11.1.3 ML

ML is a part of Artificial Intelligence (AI) that gives computers the capacity to learn without being explicitly programmed. ML utilizes the patterns found in existing data to predict the outcomes of future unknown data. ML uses statistical models to develop predictions. There are three types of ML.

11.1.3.1 Supervised learning

The machine is trained with labeled data and gives the output based on the input data; this type of ML is known as supervised learning. Continuous training is performed on the training data sets until the desired level of output is achieved [2]. Supervised learning is classified into two types, they are regression and classification. Figure 11.2 shows the flow diagram of the supervised learning model.

$y = f(x)$, where y is the dependent variable and x is the independent variable.

11.1.3.1.1 Regression

Continuous variables are commonly predicted using regression algorithms. Linear regression, regression trees, non-linear regression, Bayesian linear regression, and polynomial regression are some of the popular regression algorithms.

11.1.3.1.2 Classification

Classification algorithms are used when the output is absolute. The most common algorithms are Random Forest, Decision trees, Logistic Regression, Support vector machines, etc.

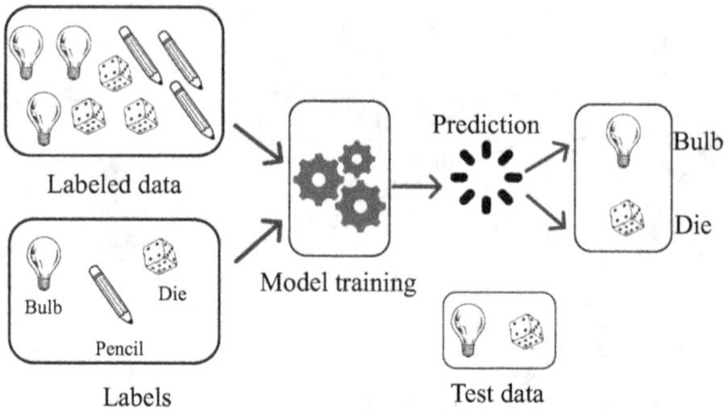

Figure 11.2 Supervised learning

11.1.3.2 Unsupervised learning

In this type of ML, the machine is trained with unlabeled data and predicts the output without any supervision. This algorithm aims to divide and group the data into clusters of similar properties and features [3]. Some approved unsupervised algorithms are K-means clustering, KNN (k-nearest neighbors), hierarchal clustering, anomaly detection, neural networks, principal component analysis, independent component analysis, Apriori algorithm, singular value decomposition, etc.

11.1.3.3 Reinforcement learning

In this type of ML, the machine is instructed based on trial and error to make decisions. In this type of learning, the machine learns from mistakes. For every good decision, the machine is rewarded and for every mistake, it is punished. Some of the popular reinforcement algorithms are Q-learning, SARSA (State-Action-Reward-State-Action), DQN (Deep Q Network), DDPG (Deep deterministic policy gradient), etc.

Data is meaningless until it is considered and prepared into useful information. ML allows the implementation of IoT systems to draw and generate facts from lower-level data by filtering, processing, categorizing, condensing, and considering details.

11.1.4 Data analysis in IoT using ML algorithms

Millions of devices create large amounts of data as part of the IoT. Data powers ML, which generates insight from it. ML identifies patterns in past behavior and builds models to help predict future behavior and events. The various algorithms that are used for classification in ML are K-nearest neighbors, Naïve Bayes, Support Vector Machine, Classification trees, Random forests, Bagging, Feedforward neural network, etc. The ML algorithms that are used for regression are linear regression, support vector regression, regression trees, random forests, etc. The algorithms used for feature extraction include principal component analysis, canonical correlation analysis, etc. The algorithms used for anomaly detection are one-class support vector machines, etc. Table 11.1 shows the existing different ML algorithms and their use cases in IoT [4].

Table 11.1 Existing ML algorithms and their use cases

ML algorithms	IoT use cases
Classification	Smart traffic
Clustering	Smart traffic, smart health
Anomaly detection	Smart traffic, smart environment
Support vector regression	Smart weather prediction
Linear regression	Economics, market analysis, energy usage
Classification and regression trees	Smart cities
Support vector machines	All applications
K-nearest neighbors	Smart homes, smart applications
Naïve Bayes	Smart agriculture, smart homes, smart applications
K-means	Smart city, smart home, smart applications, controlling air pollution, traffic control
Density-based clustering	Smart applications
Feed forward neural network	Smart health
Principal component analysis	Monitoring public places
Canonical correlation analysis	Monitoring public places
One-class support vector machines	Smart human activity control

11.1.5 Existing ML algorithms and their use cases in IoT

11.2 Physical design of IoT

The IoT device has an assemblage for connecting to wired and wireless devices. These include sensor input–output units, Internet-associated units, memory and storage units, and audio–video units. An IoT device can collect a large number of data sensors such as temperature, humidity, and light intensity. The observed data can be sent to devices, cloud-based servers, or storage. IoT devices can be linked to actuators, allowing them to communicate with other entities, including non-IoT devices and systems in the devices nearby. IoT devices are of varied types such as home appliances, smartphones and computers, wearable electronics, energy systems, retail payment systems, printers, industrial machines, healthcare systems, surveillance cameras, wearable objects, smart watches, LED lights, motor vehicles, and automated machinery [5]. Figure 11.3 depicts the design components of IoT. IoT applications generate data when dealing with data analytic systems such as ML models that provide valuable information to guide actions nearby or far off.

11.2.1 IoT protocols

IoT protocols are rules that keep an eye on the communication and exchange of data sent to the network. IoT protocols ensure that details from one device or sensor get discovered and interpreted by another device, network, or service [6]. Figure 11.4 visualizes the different IoT protocols.

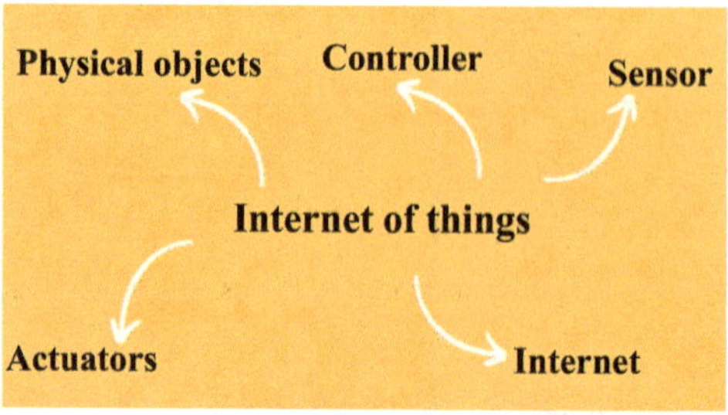

Figure 11.3 Physical components in IoT

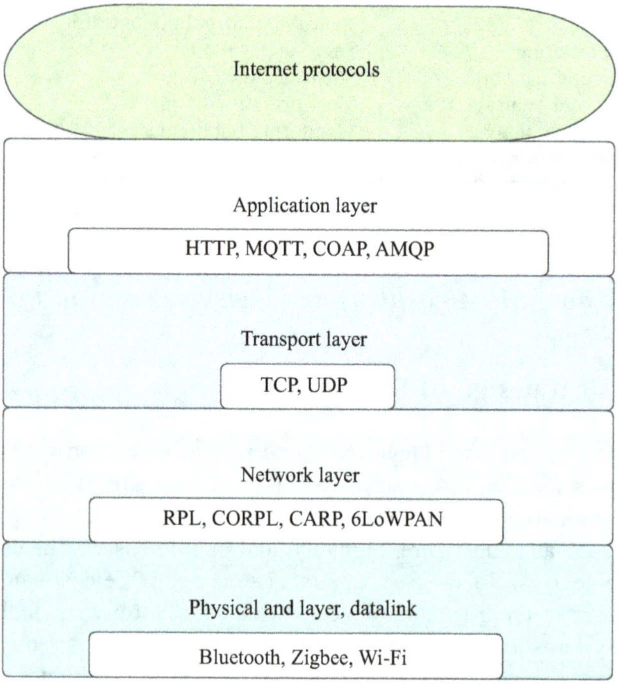

Figure 11.4 IoT protocols

11.2.1.1 Link layer

The link layer determines the information that is sent over the network's material frameworks, such as copper wire, coaxial cable, or a radio wave. The Internet operates by dividing data into small pieces known as packets. The underlying software and hardware divide all large data packets sent between two users into smaller packets. The technology governs and conveys instructions through the physical device across to which they are connected, such as coaxial cables. Some popular link layer protocols used in IoT include 802.3-Ethernet, 802.11-Wi-Fi

802.16-Wimax, 802.15.4-LR-WPAN, and 2G/3G/4G-Mobile communication. Figure 11.5 visualizes the different components of a IoT device [7].

11.2.1.2 Network/Internet layer

The transmission of the IP segment from the content, origin, and destination network is the responsibility of the network layers. This layer performs the person addressing and packet routing. Person verification is done by using IPv4, IPv6, etc. [8].

11.2.1.3 Transport layer

The transport layer protocols allow start-to-finish message transfer of the underlying network. Some of the transport layer protocols are transmission control protocol (TCP), user datagram protocol (UDP), etc.

11.2.1.4 Application layer

The application layer ensures that data from the sender is received in an acceptable format by the receiver. This protocol specifies how applications interact with protocols to send data across Wi-Fi. Using ports, application layer protocols make programs process connections. Hypertext Transfer Protocol (HTTP) is an application layer protocol used in IoT that follows a request-response model. The HTTP protocol employs Universal Resource Identifiers (URIs) to identify HTTP resources, whereas constrained application protocol (CoAP) is used for machine-to-machine applications in environments with constrained devices and networks [9], WebSocket, MQTT (Message Queue Telemetry transport), XMPP (Extensible Messaging and Presence Protocol), DDS (Data distribution service) provide QoS (Quality of Service) control and configurable reliability, AMQP (Advanced Message Queuing protocol), etc.

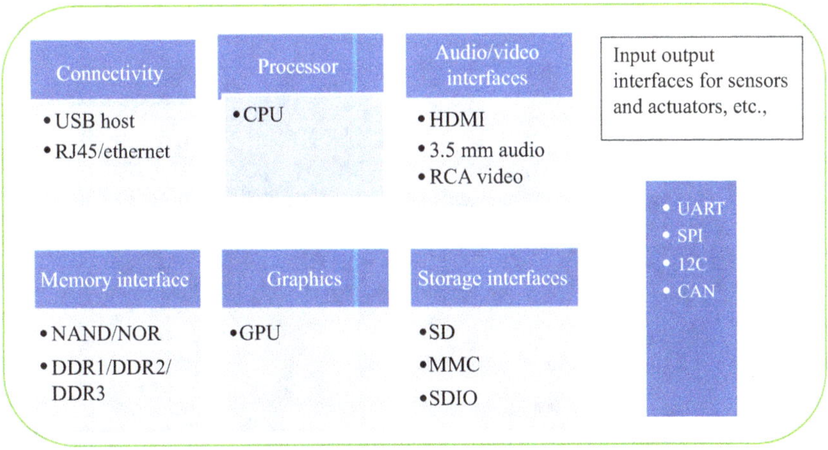

Figure 11.5 Components of a IoT device

11.3 IoT-enabling technologies

IoT is embedded with various innovations such as wireless sensor networks, cloud computing, big data analytics, security rules, architectures, communication rules, web services, mobile Internet, and semantic search engines [10]. Figure 11.6 depicts IoT-enabling technologies.

11.3.1 Cloud computing

Cloud-based computing is data processing services like servers, storage, databases, networking, software, analytics, and intelligence over the Internet called "cloud" to offer faster innovation, flexible resources, and economies of scale. Cloud computing helps in reducing the overall cost of the computation, helping minimize operating costs and running equipment quality by increasing performance [11]. Cloud computing involves the process of preparing and equipping a network of data processing, service, and storage resources on demand and providing these resources as charging services to the users, in a "pay as you go" model. The various types of cloud computing services provide to the end users are Infrastructure-as-a-Service (Iaas), Platform-as-a-service (Paas), and Software-as-a-Service (SaaS).

11.3.2 Big data analytics

Big data is the storage of a database quantity and speed in terms of temporal variation, which are so large that existing data processing tools struggle to store, manage, process, and inspect the details. It involves a certain process, beginning with data cleansing, data wrangling, data processing, and visualization.

11.3.3 Embedded systems

An embedded system is a computer system that includes an embedded unit and program to execute a fixed job. The key elements of an embedded system comprise a microprocessor or microcontroller, memory such as RAM (Random Access

Figure 11.6 IoT-enabling technologies

Memory), ROM (Read only memory), cache, networking units such as Ethernet, Wi-Fi adapters, input/output units, and storage such as flash memory [12].

11.4 Physical components used in IoT

To be considered an IoT system, four components must be properly integrated: sensors/devices, connectivity, data processing, and user interface. The term IoT refers to real-time objects that are installed with sensors, processing power, programs, and automation that connect and transfer information with other gadgets and systems via communication networks [13].

11.4.1 Relay switch

Relays are electromechanical devices that use an electromagnet to move a pair of contacts from open to closed. A relay switch linked to an instrument can switch on or off an appliance depending on the order of the IoT device network.

11.4.2 Sensors

A sensor is an instrument that measures physical information from its surroundings and converts it into useful data that can be interpreted by humans or machines. There are many sensors in the IoT that generate a large quantity of data for analysis by ML. A few sensors are a temperature sensor, motion detector, etc.

11.4.3 Microcontrollers

A microcontroller is a small and low-cost processor that is designed to perform particular tasks such as displaying microwave information, and receiving remote signals. The processor, memory (RAM, ROM, EPROM), communication interface, timers, counters, and other components comprise the microcontroller. The microcontroller will be able to control various sensors with a Wi-Fi module.

11.4.4 Radio frequency identification (RFID) tags

RFID tags are a variety of inquiry system which uses intelligent barcodes to identify items. RFID stands for "radio frequency identification," and RFID tags make use of radio frequency technology. These radio waves send information from the tag to a reader, which then sends the data to an RFID.

11.5 Smart applications of IoT embedded with ML

IoT embedded with ML has a broad range of applications and it influences various concerns. These applications of IoT and ML make the job of a human being much simpler by automating daily tasks.

11.5.1 Smart homes

Smart homes are an IoT application that incorporates ML. In-home automation contains a single interconnection that remote control lights and applications in a residence. This system's appliance communicates with the lights via electronic relay switches. A local database stores the status information for each light or appliance. Locally deployed representational state transfer (REST) services enable recovering and notifying the status of each light in the database. The controller service regularly notices the status of the appliance by retrieving the status from the database and activating the relay switches as needed. The locally deployed application includes user interaction for monitoring lights or appliances. Because the appliance is linked to the network that can be controlled remotely. [14]

11.5.1.1 Smart lighting

A mobile application can connect wirelessly enabled Internet-connected lights. Sensors are embedded in smart lights. LED lighting systems are used in these lights. They have IP capabilities. The lights can adapt their lighting to the environment. These aid in energy conservation by adjusting to lighting conditions and turning on or off or dimming lights as needed. LED lights and IP-enabled lights are the technologies that enable smart lighting. Smart home lighting saves energy by controlling the lights accordingly. These contribute to a pleasant atmosphere. A ML-enabled LED lighting system uses existing information gathered to enhance and control the lighting system, making it more convenient. It detects changes in lighting and dynamically adjusts luminary brightness based on user preferences.

11.5.1.2 Smart appliances

TV'S, refrigerators, washing machines, music systems, AC, etc., are controlled and managed by remote controllers. Figure 11.7 visualizes the different smart home appliances. These applications can be remotely monitored and controlled. By using these smart appliances, human labor can be reduced and work can be efficiently done. Smart appliance management is easy and also provides information to the users remotely. These can notify when the work is complete and can be controlled remotely.

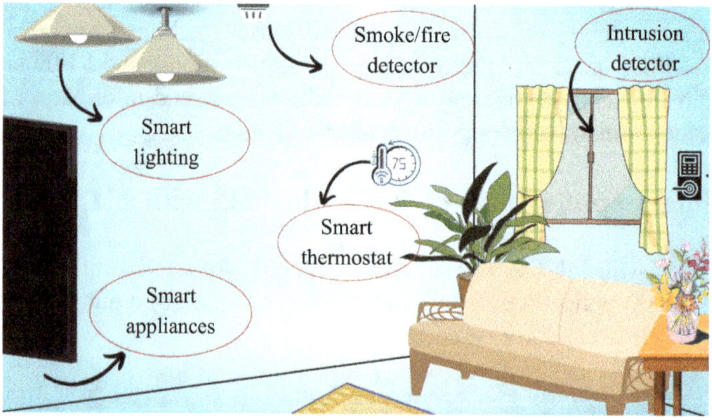

Figure 11.7 Smart homes enabled with IoT and ML

11.5.1.2.1 Smart washing machines

Smart washers and dryers can be monitored far away and notified. To help you multitask, smart washers and dryers use an Internet connection. Typically, you can control your laundry appliances with a smartphone app. In addition to remote control and monitoring, the app typically includes instructions for doing laundry and maintaining your appliances. Some smart laundry appliances can assist you in deciding by providing downloadable cycles for special items such as bathing suits, sleeping bags, and more. Some models will even recommend a dryer cycle based on your washer cycle.

11.5.1.2.2 Smart thermostats

Smart thermostats permit monitoring the temperature far off. Smart thermostats are one of the most effective ways to reduce energy costs while also keeping our homes comfortable. Thermostats detect the temperature of the air that surrounds them. The thermostat keeps track of which parts of the house use more energy than others and at what times of the day.

11.5.1.2.3 Smart refrigerators

Smart refrigerators can keep track of the items stored using RFID and sends updates to users when a commodity is unavailable. Smart refrigerators have touchscreen controls and link to the network to provide a variety of features. Smart refrigerators have internal cameras, more flexible cooling options, and the ability to interact while away from home using your smartphone or tablet. Smart refrigerators can communicate with other appliances in residences. IoT enabled with machine learning helps in detecting the optimum conditions such as temperature for storing food and can help in determining the quality of food and can help in optimum food wastage solutions.

11.5.1.2.4 Smart alarm clocks

The smart alarm clock using IoT will have features such as a weather reporting system using a temperature sensor and a humidity sensor. It will also accurately display the traffic status of roads on that specific day. A smart alarm clock can set alarms, play soothing music while you sleep, and answer questions about the day ahead in the morning.

11.5.1.2.5 Smart lock

The smart lock enables and controls the security of the entire house. This can control the electricity supply to the house. And can alert police and nearby neighbors in case of any intrusions.

11.5.1.2.6 Smart bed

A smart bed can collect information about sleep cycles and the quality of sleep. Based on the user it can optimize and can help increase the quality and duration of sleep. A smart bed can automatically warm up the bed in winter.

11.5.1.2.7 Smart vacuum cleaner

A smart vacuum cleaner analyses the entire house and cleans and mops the house whenever it finds dust automatically without any human intervention. Smart robotic vacuums use filters in conjunction with powerful suction to filter allergens and keep them from invading your breathing space. A smart robotic vacuum will detect different types of flooring and will adjust accordingly. They can even detect the edges of a room or obstacles ahead, allowing for truly hands-free floor cleaning.

11.5.1.2.8 Smart doorbells

A smart doorbell can alert the people inside the house about the person standing outside. This can give appropriate messages when the owner is not present in the house. Even when you are not at home, a smart doorbell camera lets you see who is on the other side of your door. In the event of a home invasion, the smart doorbell saves the videos it captures. These videos can then be sent to the police or other authorities in charge of the situation.

11.5.1.3 Intrusion detection

Intrusion detection systems detect and alert trespass using security cameras and motion sensors. It can notify us via SMS or e-mail with a short video clip. It can even send an alarm to cyber police and our neighbors. Advanced systems can even send detailed alerts with images or video clips. A cloud-controlled intrusion detection system makes use of location-aware services, which detect and store the geo-tag control in a home automation system independently. The system recognizes the interruption, extracts the subject, and generates alerts using image processing.

11.5.1.4 Smoke gas detectors

Smoke is a common prior warning sign of a fire. Smoke sensors are fixed in homes and buildings. These alerts send out signals when they detect smoke, harmful gases, LPG leaks, and so on. Smoke detectors use optical detection, ionization, or air sampling methods to detect smoke. These send signals to a fire alarm system, generating alerts. A smart smoke or gas detector can provide visual feedback on its status, such as health and battery level, as well as raise alerts in human voice notifying the location of the problem and send a message or an e-mail to a user and the local fire safety department. This system identifies chemical leakage and provides a visual perception.

11.5.1.5 Indoor air quality monitoring

Indoor air quality monitoring systems use wireless sensor networks to collect data on the inside air quality and density of various gases. Poor inside air quality is difficult to detect. Smart indoor air quality monitor, we gain insights into the air we breathe at home. It examines the indoor factors that contribute to bad air quality such as particulate matter, volatile organic compounds, carbon monoxide, humidity, and temperature levels. When poor air quality is detected, an air purifier is automatically turned on to help improve your indoor air quality.

11.5.2 Smart cities

11.5.2.1 Smart parking

Smart parking makes parking spaces more accessible for drivers. It detects the available parking spaces and sends data via the Internet. These provide status

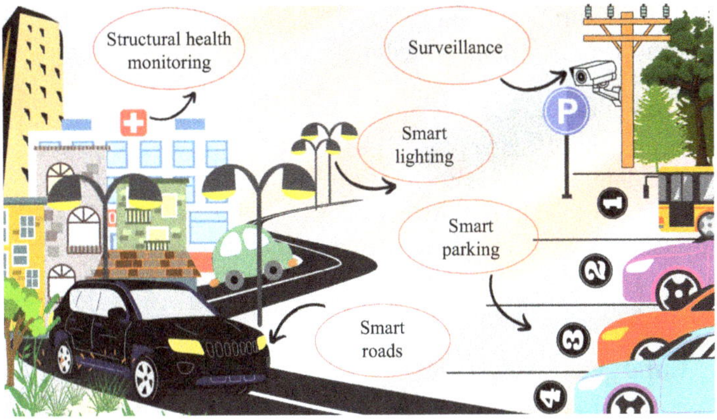

Figure 11.8 Smart cities enabled with IoT and ML

updates and information on available slots. Smart parking is powered by IoT systems embedded with machine learning that detect the available parking spaces and convey the details to the smart parking equipment. Drivers with smartphones and in-car navigation systems can access these applications. Each parking slot has smart parking sensors that detect whether it is empty or occupied. IoT embedded by machine learning can help in parking availability monitoring, parking place, dynamic pricing schemes, remote parking monitoring, and automatic guidance. Figure 11.8 shows the different IoT and ML-enabled smart systems.

11.5.2.2 Smart street lighting

It is used to help in saving energy. Existing street light systems must be manually turned on and off. Existing street lighting systems consume a lot of energy and are expensive to maintain. Smart street lighting enables automatically controlled of changing environmental conditions. Smart streetlights linked to the network can be remotely operated. Smart lights embedded with sensors can interact with other lights and exchange data on the prevailing condition of lighting.

11.5.2.3 Smart roads

Smart roads with sensors that alert traffic cops, improving driving safety on public roads and reducing traffic congestion and accidents. These provide information, travel time, and updates in the event of bad driving conditions, traffic congestion, or accidents. The system gives a clear view of the road ahead, allowing them to react to potential hazards as soon as possible.

11.5.2.4 Structural health monitoring

Structural health system aims to detect cracks and mechanical breakdowns in infrastructure assets such as buildings and bridges. Wireless sensor networks (WSNs) are used to monitor the structures such as buildings and bridges by collecting information from sensors deployed at different points. This system aids in calculating the structure's remaining life. These systems can provide warnings in the event of a structure's imminent failure.

11.5.2.5 Surveillance

Police command control centers are linked to CCTV to monitor public safety and security. Depending on the weather, day or night, the surveillance cameras can switch from normal to infrared night modes and vice versa. WSNs are used by monitoring systems to collect surveillance data such as motion detection data. It is possible to build city-wide close observation of video clips that are combined with cloud storage solutions and video analytics applications.

11.5.2.6 Emergency response

Constructions, water and gas pipelines, transportation, and power transformers are examples of infrastructure that can be monitored using IoT systems. Fire stations, water pipelines, public transportation, LPG leaks, and power outages are all monitored, and alerts are sent to the appropriate departments to minimize loss. It adapts dynamically by operating traffic lights, changing driving policies, changing driver behavior, and implementing critical security controls. This system reduces the latency of emergency vehicles such as ambulances and police cars while minimizing traffic disruption. Figure 11.9 shows the emergency services enabled with IoT and ML. A large number of data are collected and shared through sensors and analyzed through machine learning which helps in the rerouting of traffic, and evacuation alerts to the public in case of emergencies.

11.5.3 Environment

11.5.3.1 Weather monitoring

IoT-based monitoring of weather systems tracks rainfall, floods, cyclones, droughts, global warming, temperature, humidity, and UV rays. WSNs are used in weather monitoring systems to forecast temperature, humidity, and atmospheric pressure, which is then recorded and analyzed. The system consists of multiple

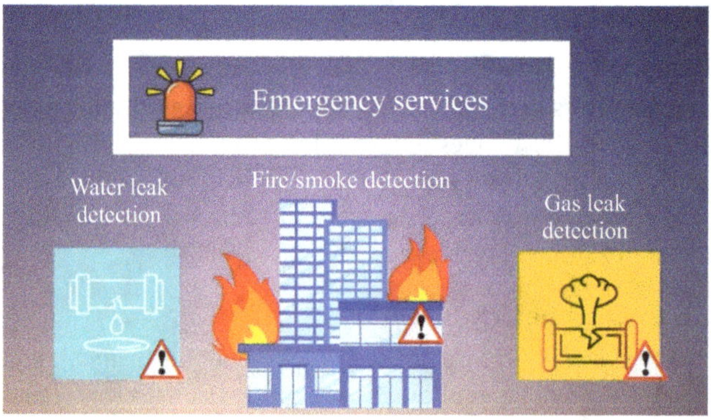

Figure 11.9 Intelligent emergency management

controls placed in various locations to monitor the temperature, humidity, and pressure of an area. Temperature, pressure, and humidity sensors are installed in the end nodes. Weather and air quality in the atmosphere can be monitored using IoT embedded with ML.

11.5.3.2 Air pollution and noise pollution monitoring

Because of urban development, air and noise pollution in cities has increased to alarming levels. Air and noise pollution cause health problems in humans causing sleeplessness and stress. Monitoring air and noise pollution in cities can assist policymakers to reduce pollution in residential areas, schools, hospitals, and parks. IoT-based air and noise pollution monitoring systems make use of air and noise monitoring stations located throughout the city. Data is collected and analyzed using ML to create a map of air and noise pollution. The area is installed with sound and air quality detectors. IoT-based air quality along with a noise pollution monitoring system that enables monitoring and verification of live air quality and noise pollution monitoring of specific areas by IoT. The system detects or senses the presence of harmful gases and compounds in the air using air sensors, which constantly send data to a microcontroller. The system also measures the level of noise and transmits it to an online server via IoT. The system technology is so user-friendly and simple to use that it can be installed in homes, schools, and small squares. Harmful gases from factories, automobiles, and industries are measured by the sensors and the information is conveyed to the respective pollution controlling board to take necessary action.

11.5.3.3 Fire detection

Fire alarms play a very important role in industries, stores, malls, apartment complexes, and parking lots. Detection of fires and smoke early helps us save lives. So, an IoT-based fire detection system with temperature and smoke sensors gets the presence of a fire in the building and also transmits relevant information via IoT and is analyzed through ML.

11.5.3.4 Forest fire detection

Fires in forests can affect reserves, assets, people, and wildlife. Forest fires are caused by a variety of factors, including human error in lighting, volcanic eruptions, and sparks from rock fall. Fire hazards are common in many areas, and IoT systems can help a community address this issue by using wireless IoT sensors, especially when the location of a wildfire is inaccessible by five-gigabit cellular broadband services or even wired broadband internet. Smoke, temperature, and other relevant data could be reported by IoT sensors. The forest fire noticing system consists of multiple nodes placed in various places to monitor the carbon dioxide and atmospheric levels in a forest. Temperature, humidity, and carbon dioxide sensors are installed in the end nodes. The forest fire monitoring system is based on wireless sensor networks that employ multi-criteria detection using an artificial neural network (ANN). To detect forest fires, the ANN combines sensing data with multiple attributes like temperature, humidity, visible light, and so on.

11.5.3.5 River floods detection

River floods have a significant impact on biodiversity. Floods occur as a result of constant rainfall, raise river levels and increased water overflows the bank and hits the road, causing it to swell. It can occur at any time of year. IoT sensors detect the possibility of flooding and warn us by sending alerts to our smartphones. River flood monitoring systems based on IoT employ several sensor nodes that use ultrasonic sensors to monitor the water level. Data from multiple sensor nodes are combined in the cloud. This equipment sends notifications when there is an increase in water level. It is described as a river flood monitoring system that uses wireless sensor nodes outfitted with various sensors to measure river and weather conditions. Figure 11.10 shows the different applications of IoT.

11.5.4 Energy

11.5.4.1 Smart grids

WSNs are used in smart grids to observe the grid at various points. A smart grid is a data communication interface that is connected to the electrical grid that reviews data transmission, distribution, and utilization of electricity. Smart grid technology provides suppliers and customers with predictive information and recommendations for administering power. The equipment's durability and the grid's integrity can be assessed using IoT-based sensing and measurement technologies. Smart meters can control electricity consumption in-turn turn off the power supply and helps to prevent power theft [15].

11.5.4.2 Renewable energy systems

The output of renewable energy resources varies, and merging can enhance grid practical performance issues. At the point of interconnection, an IoT-based interface connected with transformers measures electrical units and inflow power to the grid. Closed-loop controls can be used in wind energy systems to minimize the voltage at the link and to enhance continuous power generation.

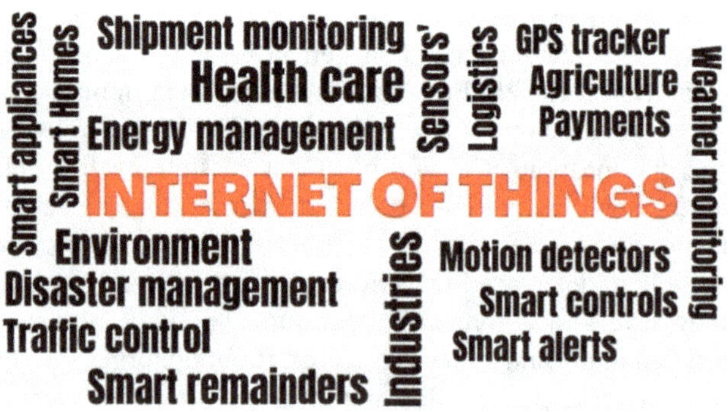

Figure 11.10 IoT and ML applications

11.5.4.3 Prognostics

Smart grids, power plants, wind turbines, and farms have a large number of pivotal parts that work to function properly. Machines must be carefully monitored because wear and tear of critical components, as well as sudden changes in operating conditions, can cause failures. Thousands of sensors in energy systems collect real-time maintenance data to monitor conditions and predict failures. Prognostic real-time health management systems based on IoT can forecast machine or energy system performance by interpreting how far a system deviates from its normal operating profiles. ML can provide predictions for impending failures in energy systems and equipment.

11.5.5 Retail

11.5.5.1 Inventory management

Inventory management is important nowadays, overstocking can result in excess storage costs, spoilage risks, and result in revenue loss. IoT systems employ RFID tags to aid in management and maintenance, RFID tags linked to the products allow them to be located, allowing inventory levels to be accurately determined and products to be restored. RFID readers attached to retail warehouses can be used to track items. Using the data collected by RFID readers, IoT systems enable remote inventory monitoring. ML has given us the power of smart algorithms that can help us track and manage inventory much better than a typical human can. Careful analysis of IoT data by ML systems will aid in identifying patterns that a human cannot see. This procedure will undoubtedly improve inventory management decision-making.

11.5.5.2 Smart payments

Contactless payments, for example, are enabled by technologies such as near field communication (NFC) and Bluetooth. NFC instructions allow smartphones and other devices to communicate with one another. End users can use their NFC-enabled smartphones to store credit card information and make payments by bringing the smartphones close to terminals.

11.5.5.3 Smart vending machines

The smart vending machine allows inventory monitoring, elastic pricing products, promotions, and contactless payments via NFC. User preferences can be remembered and learned over time using smart mobile applications that interact with smart vending machines. These smart vending machines provide a user-specific interface. Smart vending machines can decrease the rate of perishable items as the date of expiry approaches. Customers' purchase choices and preferences can be used to recommend new products to them.

11.5.6 Smart logistics

11.5.6.1 Route generation and scheduling

Details are collected from different sources such as IoT, and data-driven transportation systems can provide route guidance, and vehicle routing, and predict customer demands by processing the data into useful information through machine learning. The number of possible route combinations grows exponentially based on vehicle availability and different modes of transportation. IoT-based systems powered by the cloud can respond quickly to customer queries and serve a large transportation interface.

11.5.6.2 Vehicle fleet tracking

Vehicle fleet tracking devices use GPS technology to locate vehicles on demand. In event of deviations from planned routes, alerts can be generated. Vehicle location and route data can be aggregated and analyzed using ML to detect traffic congestion on routes, assignments, and alternative route generation. The system can analyze vehicle messages to detect unexpected incidents and handle data, allowing corrective actions to be followed.

11.5.6.3 Smart shipment monitoring

The conditions inside containers can be monitored by the shipment monitoring system. It contains a single node for tracking shipments. These devices monitor vibration levels using accelerometers and gyroscope sensors. The temperature, pressure, and humidity are used in IoT-based shipment monitoring using ML which allows effective decision-making.

11.5.6.4 Smart remote vehicle diagnostics

These vehicle diagnostic systems can identify and warn faults in vehicles. These diagnostic systems collect data such as the overall status of vehicle subsystems using onboard IoT devices. Data like this is collected by joining systems with IoT device interfaces. Vehicle diagnostic systems based on IoT can send alerts and recommend corrective actions.

11.5.7 Agriculture

11.5.7.1 Smart irrigation

IoT soil moisture monitoring devices in the field of irrigation can aid in determining the best watering methods. WSNs are used in soil monitoring systems to keep an eye on soil moisture in different areas. A single node in the smart irrigation system observes soil moisture levels and controls the irrigation system. This system's device collects soil moisture data from sensors. The moisture levels are constantly monitored by the controller service. If the soil water retention level falls below a certain level, the irrigation system is activated. Actuators such as solenoid valves can be used to control the irrigation system. The information is sent to cloud computing by the controller, for storing, and recovering moisture data stored in the cloud database. A cloud-based representational state transfer (REST) web service is used. This application uses to check soil moisture levels over time, that aid in irrigation scheduling choices. Smart irrigation systems have the potential to increase crop yields while conserving water.

11.5.7.2 Smart greenhouse control

Greenhouse house control is set with glass or plastic roofs that give the required conditions for plant growth. Sensors measure the requirements of the plant and climatological conditions which can be automatically controlled by actuation devices such as valves that release water and switches that control air circulation. IoT systems in greenhouse control play an important role in aiming for productivity improvement. Data from various sensors is collected to control procedures. This system employs a wireless sensor network to monitor and control plants' temperature and humidity for better agricultural production.

11.5.8 Smart industries

11.5.8.1 Smart machine diagnosis and prognosis

Machine prognosis is an indicator of performance based on information on current operating conditions and variations in normal operating conditions. IoT plays a significant role in industrial machine prognosis and diagnosis. Machine sensors can check operating conditions in temperature levels. Sensor data measurements are performed on intervals ranging from a fraction of a second, resulting in large amounts of data being generated. IoT-based systems combined with cloud-based storage and analytics can aid in the storage, collection, and analysis of such large amounts of information.

11.5.8.2 Indoor air quality monitoring

In industries, maintaining air quality is pivotal to workers' health and safety since the air inside the industries contains poisonous gases that are hazardous to health. Using various gas sensors, IoT-based gas monitoring systems can assist in keeping an eye on the inside quality of air in an organization. Indoor air quality varies depending on the location. WSNs based on IoT devices can detect dangerous regions and take corrective action to ensure proper ventilation. The air quality monitoring includes both stationary sensors for accurate readings and adjustment and mobile sensors for coverage. IoT-based AI air quality monitoring devices determine inside air quality [16].

11.5.9 Health and lifestyle monitoring

11.5.9.1 Smart health and fitness monitoring devices

IoT devices that can be worn enable undisturbed and continuous monitoring of emotional criteria and can aid in the monitoring of health and fitness. These gadgets can be used in the form of belts or wristbands. These devices form a body area network that informs us about the health condition of an individual.

11.5.9.2 Smart wearable electronics

Wearable electronic gadgets, smart watches, smart glasses, and wristbands incorporate electronics into clothing and accessories provide a variety of features to help us with our day-to-day activities. That encourages us to healthy way of living. Smartwatches that run mobile operating systems offer users to access mobile applications. Smart glasses allow users to use voice commands to take pictures and video clips, GPS, and flight updates. Smart shoes use embedded sensors to monitor walking and running speed, and the number of jumps, with smartphones to view the data. Smart wristbands can monitor activity and in reducing their calories.

11.5.10 Smart transportation

11.5.10.1 Navigation control

IoT embedded with ML helps in smart navigation and GPS such as optimizing the route based on traffic and weather conditions. With advancements in wireless communication, the demand for devices for safe mobility. For maximum probability data estimation with IoT embedded with artificial intelligent, devices with multiple input and output methods are used for smart navigation control and GPS. IoT embedded with ML helps in optimizing the maps for locating stores, hotels, malls, transportation, educational institutions, hospitals, etc. [17].

11.5.11 Smart maintenance

11.5.11.1 Water supply/water meters

IoT embedded with ML helps in monitoring water usage and helps in planning water thus saving water. The IoT enables new water management solutions, allowing for more efficient use of this valuable resource. Comprehensive water management strategies can save up to 20% on water costs, making a significant difference in cities. Water projects can be especially complicated because many cities rely on aging infrastructure. The IoT also allows municipalities to reduce operational costs associated with construction, maintenance, and other activities.

11.5.11.2 Supply of electricity

IoT embedded with ML helps in monitoring electricity usage and helps in planning and consuming less electricity. IoT technology will aid in the efficient distribution of electricity. It will also result in abrupt changes in its manufacturing and delivery to end users. Because of IoT, renewable electricity now accounts for 8.4% of global electricity production. Mobile applications for smart electricity metering. Implementing IoT allows electric power companies to improve efficiency.

11.5.12 Disaster management

Disaster management aims to mitigate potential disaster damage and provide the injured person with immediate assistance. An efficient rescue operation is required after the occurrence of a disaster. To plan an effective relief operation, various types of information about the impact play an important part in disaster management needs. IoT embedded with machine learning aids in disaster prediction based on sensor data collected in the past.

11.5.13 Health care

E-health control and prevention are the two primary goals of future health care. People can now choose to be monitored by doctors even if they do not live in the same country or location. IoT-assisted e-health is extremely useful for tracing and monitoring people's health histories. IoT healthcare solutions may also benefit specialists by allowing them to collect data to advance their medical calculations.

11.5.14 Military

IoT and ML in the military field can help in detecting enemy encroachment and enable better security of the country. The IoT powered by ML helps with a variety of benefits in areas of monitoring and focusing critical points in the war zone. Military IoT has a wide range of "things," including battlefield sensors, soldier health, armaments, identifying equipment, outfit, drones, ships, planes, tanks, and body sensors. They work together to deliver important information to the battlefield [18].

11.5.15 Business intelligence

IoT with ML can help in providing better solutions for improving business by analyzing customer reviews. Business intelligence can assist in analyzing patterns that are constantly changing in a portion of the overall industry; changes in client behavior and spending patterns; client inclinations; organizational capacities and trade market. On its assembly line, a company can automate quality inspection and defect tracking. IoT enabled with ML can help in forecast demand patterns and consumption of the customers. It helps in improving business by improving customer user experience. IoT with ML helps in creating smart offices. IoT enabled with ML can help in improving employees' experience in the office based on their past data and can help in assisting with their queries.

11.5.16 Aviation industry

11.5.16.1 Smart airports

IoT and ML can make smart airports real automated and sensor-based systems that collect information from passengers. To guide the crowd at various airport checkpoints such as passport verification, security checks, and even luggage submission. This will make the entire process more convenient and automated, resulting in fewer chances of overcrowding. Thereby reducing the issues relating to airport management authorities and passengers who face the hectic and time-consuming procedural formalities. IoT enabled with ML applications can help in improving the experience of the passengers on the flight ranging from aircraft safety to comfortable travel, cabin climate control, and even emergency alerts.

11.5.16.2 Smart travel management

IoT devices can record and identify customer interest when booking a seat through their online website and food choices while traveling. IoT-enabled solutions such as barcodes or chips can assist passengers in finding their luggage as they prepare to board a flight. Passengers can use their smartphones to check the location and condition of their luggage using IoT devices and features. These devices combined with AI and inspection cameras at various stages of automated handling detect the transport of banned items via illegal means.

11.5.16.3 Effective aircraft maintenance

IoT can be a good alternative for getting the most out of airplane maintenance practices. IoT systems can monitor the current and on-air conditions of arriving planes and the condition of their various parts. This information can be sent to the concerned engineers ahead of time, allowing them to classify which maintenance function to perform first and on which aircraft. This will help to reduce plane runway time and ensures proper maintenance is carried out.

11.5.17 Childcare and elderly care

Child care is an important and rewarding job in the world. With many other responsibilities, parents face difficulty to ensure their child's well-being and safety throughout the day. Taking care of infants, looking after kids, noticing nutrition absorption, continuous care of their health, and assigning an interactive playmate to the child in the form of toys Internet of Toys are the areas of applications in child care. Toys can be embedded with sensors that can give notifications to parents about their children. When IoT and ML are used they can alert people in case of emergencies.

According to a World Health Organization report, an aging population is becoming increasingly common in both developed and developing countries. IoT with ML can help in monitoring elderly people at ease. An acoustic sensor in IoT is capable of automatically detecting falls and reporting them to caregivers. Wearable devices can help in monitoring health of the old people and automatically consult doctors. These devices can help in assisting old people and can take care of their well-being.

11.6 Conclusion

We are living in a smart world. IoT embedded with ML is a revolution which made life easier and more comfortable. It is a landmark development in technology. It has enhanced a lot of change in the lifestyle of people. In every sector, IoT-based tools integrate us for better living like smart homes, smart payments, smart lighting, smart weather monitoring, detection of forest fires, flood control, smart grids, health management, and smart irrigation systems to save resources by increasing crop yield and greenhouse management. Industrial IoT applications include machine diagnosis and prognosis systems that aid in fault prediction and determination of indoor air quality systems. IoT interfaces in health and fitness monitoring systems and wearable gadgets. When these applications are integrated into the rating, billing, and financial applications, they provide significant value to end users while providing new revenue opportunities to service and system providers. The IoT is distinguished by three core elements: asset digitization, asset data collection, and computational algorithms. ML helps in taking smart decisions. Various IoT-based sensors are used in self-driving cars. We can conclude it will help people find a better solution to their problems.

The IoT deals with all hardware components connected to the Wi-Fi that are uniquely identified by their IP numbers and can transfer data over the network without human intervention. Physical devices in an IoT network are typically things embedded with sensors that exchange information. IoT is an ecosystem of vast connected devices known as objects that continuously learn from data. ML is data-driven and generates meaningful insights from it. ML tries to find patterns from past data and build predictive models that are useful in predicting future behavior and events. Very large data come from these devices through various sensors. The ML algorithms examine the data to derive valuable information for a better user experience. With the advancement of machine learning in IoT, the devices can inspect the data collected, derive meaningful information, create smart applications, and be autonomous. These devices can make decisions and learn from the data collected. Further enhancements in the domains of IoT and ML are making the applications smarter and smarter. As it is said, "Work smarter not harder." ML and IoT are playing a major role in making applications smarter. IoT helps collect data, and ML analyses this data to make smarter decisions. The futuristic applications of IoT and ML are smart homes, smart cities, etc. These IoT and ML applications are going to be epic in the field of innovation.

Although the Internet is ubiquitous, it still serves as a means of communication between individuals. After the Internet revolution, it was all Internet of people, all the information is being shared between individuals. In today's generation, now it is all about IoT that are collecting information whether it is offline or online and analyzing that data to automate our daily tasks. In the industry 4.0 revolution, industrial IoT (IIoT) is changing and enhancing manufacturing and industrial processes. The underlying philosophy of IIoT is that smart equipment is not only better than humans at data capturing and analyzing in real-time but they are also better at communicating critical information that can be used to fast driving and more accurate business decisions. IoT with ML will change everything without the need for human-to-human collaboration making human life much simpler and easy by automating everything. All of the devices in an IoT system would collect data from your living patterns in real-time, adjusting their settings or features to optimize your living experience and ensure the products work best for our life needs. All of the devices in an IoT system would collect data from our living patterns, modify their settings or features to improve our lifestyle, and ensure the best commodity for our life. IoT refers to Internet-connected gadgets with distinct identities. ML can be used to connect and transform data with gadgets to make them smarter. When these applications are integrated into the housing, industries, and financial applications, they provide significant value to end users while providing new revenue opportunities to service and system providers. The IoT is distinguished by three core elements: asset digitization, asset data collection, and computational algorithms. ML helps in taking smart decisions. Various IoT-based sensors are used in self-driving cars. We can conclude it will help people solve their problems better.

11.7 Further scope and limitations

Advances in technologies such as data analytics, cloud computing, embedded systems, and many key enabling technologies have reduced the cost of smart solutions for smart cities. This has enabled many industries to upgrade with IoT technology. With 5G connectivity, IoT will become crucial for automation. The increasing research and advancement in deep learning also have helped in the increasing the accuracy of the prediction. The advancement in IoT and ML has its limitations. The IoT devices can be hacked and there are chances of data breaches if the IoT devices are not properly secured. The data security is the most challenging task in IoT. The future research on major security threats to IoT devices in terms of architectural, data, communication security, and malware analysis can be an added advantage.

References

[1] A. McEwen and H. Cassimally. *Designing the Internet of Things.* Chichester, UK: Wiley, 2013.
[2] J. Gubbi, R. Buyya, S. Marusic, and M. Palaniswami. "Internet of Things (IoT): A vision, architectural elements, and future directions", *Future Generation Computer Systems*, 29(7), 1645–1660, 2013.
[3] M. Jindal, J. Gupta, and B. Bhushan. "Machine learning methods for IoT and their Future Applications," in *2019 International Conference on Computing, Communication, and Intelligent Systems.*
[4] M.S. Mahdavinejad, M. Rezvan, M. Barekatain, P. Adibi, P. Barnaghi, and A.P. Sheth. "Machine learning for internet of things data analysis: a survey", *Digital Communications and Networks*, 4(3), 161–175, 2018.
[5] M. Schofield. *Managing Crisis Using Interconnected Devices Powered by the Internet of Things (IoT).* Hershey, PA: IGI Global, 2022 (Chapter 2).
[6] https://azure.microsoft.com/en-us/solutions/iot/iot-technology-protocols/
[7] S.N. Swamy, D. Jadhav, and N. Kulkarni. "Security threats in the application layer in IoT applications", in *2017 International Conference on I-SMAC (IoT in Social, Mobile, Analytics, and Cloud) (I-SMAC)*, 2017.
[8] P.P. Ray. "A survey on Internet of Things architectures", *Journal of King Saud University – Computer and Information Sciences*, 30, 291–319, 2016.
[9] https://azure.microsoft.com/en-in/resources/cloud-computing-dictionary/what-is-cloud-computing/
[10] "Internet of Things: enabling technologies, security and social implications," New York, NY: Springer Science and Business Media LLC, 2021.
[11] https://militaryembedded.com/ai/deep-learning/iot-ai-and-the-future-battlefield
[12] N.M. Htaik, N.A. Maung, and W. Zaw. "Enhanced IoT-based interoperable and configurable middleware using semantic web techniques," in *2018 15th International Conference on Electrical Engineering/Electronics, Computer, Telecommunications and Information Technology (ECTI-CON)*, 2018.
[13] S. Tanwar, *Blockchain for 5G-enabled IoT.* New York, NY: Springer Science and Business Media LLC, 2021.
[14] M. Jindal, J. Gupta, and B. Bhushan. "Machine learning methods for IoT and their future applications," in *2019 International Conference on Computing, Communication, and Intelligent Systems (ACCESS)*, 2019.

[15] E. Hossain, I. Khan, F. Un-Noor, S. Shazali Sikander, and Md. S.H. Sunny. "Application of big data and machine learning in smart grid, and associated security concerns: a review", *IEEE Access*, 99, 1, 2019.

[16] T. Manglani, A. Srivastava, A. Kumar, and R. Sharma. "IoT-based air and sound pollution monitoring system for smart environment", in *2022 International Conference on Electronics and Renewable Systems (ICEARS)*, 2022.

[17] F. Zantalis, G. Koulouras, S. Karabetsos, and D. Kandris. "A review of machine learning and IoT in smart transportation", *Future Internet*, 11, 94, 2019.

[18] K.S. Niraja, Murugan, and C.S.R. Prabhu. "Security risks in Internet of Things: a survey", in *2017 IEEE International Conference on Computational Intelligence and Computing Research (ICCIC)*, 2017.

Chapter 12

Recent trends in applications of cognitive sensors for smart manufacturing and control

Amitabh Mishra[1]

Abstract

Industrial Internet of Things (IIoT) is focused on optimization of industrial working in several fields including assembly lines, manufacturing processes and production of any kind, including agriculture. As the equipment and system are data intensive and largely dependent on wireless transmission of data from the interface point of the primary sensing element to where it is going to be utilised, adding cognitive capabilities to the radio system involved can add benefits of optimal utilisation of the frequency spectrum and faster transmission of critical process data. IIoT also offers advanced functionalities such as preventive maintenance, asset tracking and better optimized resources in the operation and control of industrial processes for manufacturing, production, or processing. This chapter focuses on the evolution of the sensors and process control equipment from traditional versions to those being used today for IIoT.

Keywords: Industrial Internet of Things (IIoT); Smart sensors; Cognitive radio; Cyber physical systems; Sensor networks; Industry 4.0

12.1 Introduction

The Internet of Things (IoT) can be considered as the latest revolution in digital technology that creates a vast interconnected network of digital devices, embedded sensors, and other physical objects in our lives through the Internet. The IoT setup lets data be exchanged between the devices, collected, and analysed for the purpose of information collection and automation or remote control. While, on the one hand, home devices such as security cameras can be accessed and thermostats, sprinklers and lights can be controlled from thousands of miles away, the other aspect could involve the remote control of water quality, agriculture, traffic lights or even vehicle fleets, electric grids, industrial robots and manufacturing processes in industries.

[1]Department of Computer Science, The University of West Florida, USA

Industrial IoT (IIoT) is that prong of IoT that improves production and manufacturing processes by employing intelligent sensors and actuators connected by the Internet [1]. More rigorous standards and robust designs are required for IIoT sensors because they need to operate for long periods in harsh environments. Linking of these automation assets through IIoT makes real-time analysis, smarter monitoring and better-informed decision making possible, thus letting the operators run the processes more intelligently. Inefficiencies in supply chains and processes can be identified quickly due to better asset traceability and at times even predicted, reducing production and maintenance costs while maintaining or improving product quality. Smart algorithms and availability of a variety of real-time data related to temperature, vibrations, and other parameters to monitor wear and tear and make predictive maintenance possible, without waiting for an equipment or machinery breakdown to happen, thus saving on production losses, and improving on facility management. IIoT is also geared towards better energy management for sustainable and green manufacturing methods.

Smart IIoT sensors can keep a real-time check on the functioning and regime of various processes and machine operation involved, sensing parameters of interest, aggregating data from the process and machinery and reporting it, also helping create operational benchmarks. The benchmarks can then be used to detect minor to major deviations that could impact process performance, thus enabling smart manufacture. Thus, the IIoT systems can also create a system of predictive maintenance to avoid breakdowns and production losses [2].

IIoT offers an important advantage of asset tracking and management by letting vendors, manufacturers, and buyers track the entire supply chain about the location, condition, and status of Internet-connected goods through real-time data and arrive at informed business decisions. Corrective actions can be taken in case of material damage and precautions initiated by concerned parties as they get immediate notifications. Data from IIoT sensors can further be fed to big data analytics for the next level of business intelligence [3]. All these features of IIoT help in improving customer satisfaction due to customer-centric manufacture, supply, and feedback systems that it enables.

While the basic sensing mechanisms and control strategies have not changed despite the arrival of IIoT, several additional requirements such as miniaturization, local processing, and advanced networking capabilities have risen to make IIoT possible. We take a look at the basic sensing and control systems that are being enhanced and modified for IIoT in the next section.

12.1.1 Instrumentation and control systems in manufacturing and control

Human senses cannot fathom the exact quantitative information about the parameters connected to events occurring around us. Measurement is the result of an opinion formed about the relative size or intensity of the physical quantity of interest after comparing the sensed value with a chosen unit, considered a standard. Sensors let us measure the aspects of the physical quantities of interest that are beyond the range and precision of our unaided human senses. A sensor is a device that recognizes the advances in physical or other parameters through measurement for the purpose of acquiring qualitative and quantitative information about the quantity of interest.

As shown in Figure 12.1, the sensor starts with a Primary Sensing Element for sensing the value, or state of the process parameter of interest, and creating a corresponding output signal that is fed to a ***Transducer*** for converting the output into another form of energy that is better suited for measurement and control. The next stage has a ***Manipulation Element*** that applies mathematical processing to the signal and conditions it for the subsequent stages. Subsequently, the system has a ***Data Transmission Element*** that sends the signal from one location to another, usually a base station or a control room, keeping the information contained well preserved. The next stage has a ***Data Processing Element*** that scales, linearizes, adds bias, compensates for any errors, and prepares the data for display, storage and for use in a control loop. The last stage in the system has a ***Data Presentation Element*** that provides the information to the human operator as well as saves and uses the data signal for computation of any control signal that should be sent back to the field to operate any final control elements that control the value of the process parameter [4].

Sensing elements of a certain type can perform measurements only if they have a contact with the medium whose characteristics are being measured while there are another variety of sensors that do not need to connect physically with the medium.

The information from a sensor can then be used for knowledge, generation of alarms, or for discrete and continuous control of the quantity. Sensors used for Industrial Automation are devices that take the quantity of interest as input and produce a corresponding output. Often, the input quantity is converted into an output quantity through one or more steps of conversion, commonly known as transduction which essentially means change in the form of energy.

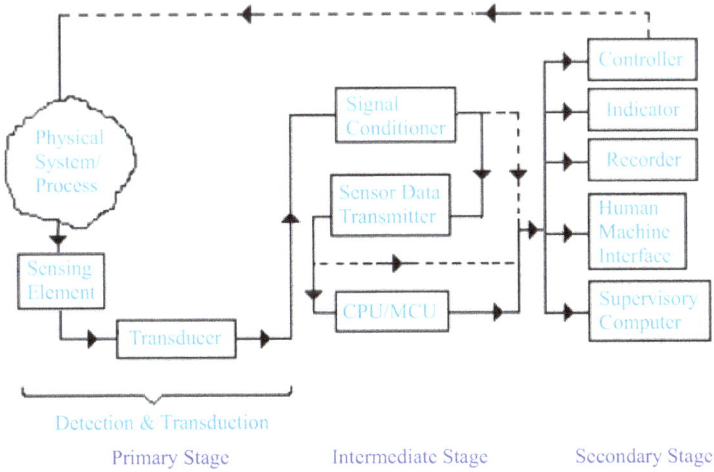

Figure 12.1 Generalised sensing and control automation loop

We see sensors and the information provided by them all around us starting with fitness devices that measure human body parameters, home automation sensors for home security, lighting control, temperature control, smart sprinkler control, in kitchen devices for time or temperature control; in vehicles for the measurement of speed, miles travelled, fuel remaining, smart headlights, and so many other sensors with newer smarter security and communication features [5]. The biggest application of these sensors is evidenced in the industries for providing information about the various aspects of the industrial process in focus and for utilising this information for process control.

12.1.2 Sensors used in industrial automation

In the field of industrial automation, sensors are deployed to play a dynamic role for making the process used for making of the products automatic, smart, and self-reliant. These sensors enable the process engineers and operators to identify, sense, measure, analyse, and process a multitude of alterations like variations in pressure, flow, temperature, location, and in other vital parameters that occur in any industrial manufacturing process.

Sensors also perform a major task of forecasting and averting a plenty of potential errors and losses resulting from them, as well as in generating process alarms and enabling control through the manipulation of final control elements and actuators.

12.2 Types of industrial sensors

There exist a variety of sensors used for manufacturing processes and associated controls, every one of these designed to track and gather data related to specific process parameters and apparatus performance standards. These sensor types include the following classes of sensors based on the quantity of interest. We discuss some of the common ones from amongst a vast number of such sensors available.

12.2.1 Temperature sensors

Temperature monitoring is important for industries such as petroleum processing in which several products separate through fractional distillation within a single degree change in temperature, to greenhouse farms or food industry in which prolonged exposure to incorrect temperature can cause a wastage worth millions of dollars, in data centre HVAC systems which need to be running to keep the centre cool, in power plants to monitor boiler temperature or in pharmaceuticals which require freezers to be maintained at ultra-low temperature for their vaccines. A typical IIoT temperature sensor is shown in Figure 12.2.

The degree of hotness or coldness of an object, region, or material depends upon the molecular activity in the object which gets converted into kinetic energy, measuring which provides with the information about the degree of hotness or coldness of the object. Temperature sensors use the energy of the object or system being monitored to change the kinetic energy of the sensor.

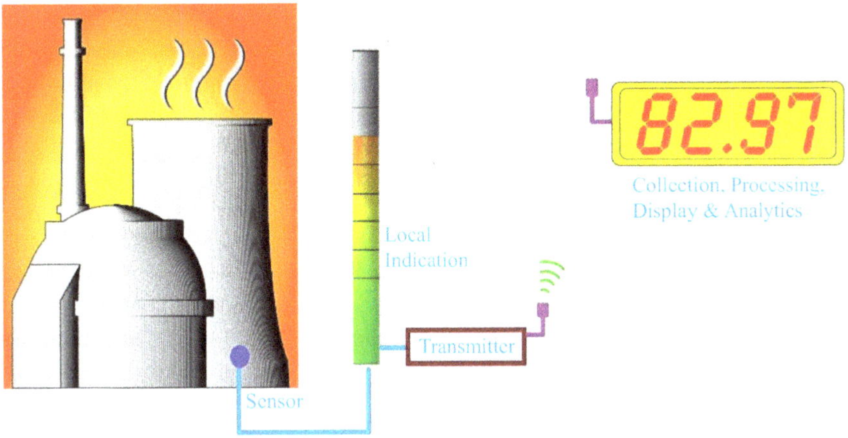

Figure 12.2 IIoT temperature sensor

In the case of a liquid-filled thermometer, the change in kinetic energy causes an expansion in the liquid inside, which rises in the tube by a proportional amount giving the temperature of the object.

Bimetallic thermometers use strips made with two dissimilar metals that have different coefficients of thermal expansion. The strips are bonded together to create the thermometer. When the thermometer is subjected to a change in temperature, the metals bend differently causing the bonded thermometer to bend to a side by the amount proportional to the temperature causing the deformation.

The Bourdon tube temperature sensor consists of a gas-tight container associated with a small-cross section, spiral capillary tube that delivers its output to a secondary, a local indicator. The sensor enlarges or contracts with variation in temperature. This causes the Bourdon gauge to displace, thus converting pressure to a mechanical displacement that can be used to read temperature on the scale in the indicator.

Thermistors are semiconductors made from metal oxides, sintered at high temperature and mixed to have a high thermal coefficient of electrical resistance. They undergo change in resistance with change in temperature. This sensor offers extremely high sensitivity and high-resolution measurements over limited temperature ranges in applications such as measurement of motor winding temperature and in medical thermometry.

A thermocouple measures temperature using two thermoelectric principles called Seebeck effect and Peltier effect. To create a thermocouple, two wires that have disparate electrical properties are soldered and made to join only at their ends creating two junctions. Of the two junctions, one junction is placed at a higher temperature and the other at a lower temperature, and this produces an equivalent thermal electromagnetic force that appears as a voltage between the two wires. This voltage generates an analogous electric current in the loop formed by the wires.

No external energy, power, or excitation is additionally required for the working of thermocouple sensors. The hot junction that normally measures the temperature is in contact with the point in the process, to measure the temperature at the point. The colder junction, used for providing reference, is open and is connected to the sensitive, and well calibrated indicating or recording instrument.

Resistance temperature detector (RTD) sensors use the principle of change of resistance of a wire in response to the temperature to which the sensor is exposed. The wire material is frequently a noble metal such as platinum that has a positive temperature coefficient of resistance. RTDs are passive sensors and need external energy to be supplied to them to make them work.

In case of a radiation pyrometer sensor, thermal radiation emitted from a hot object is collected and focused on the sensor. The pyrometer sensor element could be a thermocouple, or a thermopile created out of multiple serially connected thermocouples that infers the temperature based on the Seebeck and Peltier effects that make the thermocouple produce a small voltage in proportion to the temperature being measured. The sensing element could also be created making the radiant energy strike a photon detector that causes a release of electrical charges and produces an output proportional to the radiant energy. Radiation pyrometers are non-contact type temperature instruments that find use in applications where a direct contact of the region of interest is not possible or feasible, for example, for monitoring the temperature of a continuously moving hot sheet in steel and metal product industries.

Silicon-based temperature-sensing ICs are now being used as digital temperature sensors. These sensors are designed with the required analog to digital conversion and obviate the need for additional signal conditioning, thus simplifying the control system design [6].

Health of a piece of equipment can be very reliably ascertained by measuring its temperature, especially if the equipment involves movement of some kind and friction could possibly be involved. Normal operating temperature of such equipment is known by prior measurement and observation. At times, maintaining this temperature might need additional cooling, and a measure of inlet and outlet temperatures of the flowing coolant can reveal critical information about the equipment related to its health. Any sudden spikes or uncontrolled fluctuations in the measured temperature can speak volumes about the equipment's performance and health. Such spikes and changes might be indicative of equipment wear, loose mechanical connections or rise in friction. If cooling is involved, the temperature rise might indicate a failure of the coolant flow. Unchecked, this could lead to equipment damage due to overheating and interruptions in production.

12.2.2 Force sensors

The force sensor uses change in shape and dimension of the component under test as a measure of the force that caused it. By extension, such sensors can also be used to measure pressure and torque. Strain gauge sensors are the most common type of force sensors used. They work on the principle of electrical resistance change to be a linear function of the mechanical strain caused by the force and could be made from semiconductors too. These gauges, available as metal wires or foils, are bonded or cemented directly onto the surface of the structural member which is being examined so that any changes in strain in the member are transmitted directly to the gauge material, thereby changing its resistance.

12.2.3 Flow sensors

Flow has been traditionally measured by creating an artificial obstruction in the path of the fluid, measuring the difference in pressure before and after the obstruction and using the pressure differential to calculate the flow as it is directly related to the flow. A variety of obstructions such as orifices, plates and weirs can be used for flow measurement. Flow of any fluid is generally measured inferentially through the measurement of differential pressure or the velocity of the fluid through a known area as it is a direct indication of the fluid volume rate.

Flowmeters can be broadly categorized as velocity flowmeters, mass flowmeters, head flowmeters, and positive displacement flowmeters. Head flowmeters work by reading the pressure difference proportional to the square of flow created by placing a restriction in the flow path. The restriction forms the primary device part, and the differential pressure measurement system forms the secondary part of these sensors. They are simple, dependable, and provide more flexibility in comparison with other flow measurement mechanisms. Positive displacement flowmeters sense and find out the volume of the flow by trapping samples of the fluid at a fixed rate which provides the flow by totalizing through computing the multiplication of the sample volume with the figure of samples collected. These sensors are best suited for flow applications that require automatic batching and accounting.

Venturi tubes are prepared by gradually narrowing the flow pipe diameter and gauging the resultant difference of pressure across the two ends of the cone. This constriction is followed by a cylindrical section. An expanding section follows the cylindrical section and is designed to return the flow back almost to the original fluid pressure.

These sensors are used in applications using large diameter pipes that need a low pressure drop but high measurement accuracy, such as in waste treatment plants where choking of the pipe needs to be avoided while the measurement is important. These sensors are most accurate but the largest and most expensive.

Flow nozzles are designed on similar lines as the venturi tube. The outlet of the flow nozzle is a constriction with an elliptical cross-section designed to hinder the flow but does not have a specific area beyond the outlet for pressure recovery.

A pitot tube flow sensor is a differential pressure-based flow detector made up from two hollow tubes to gauge the pressure within a pipe used for flow by connecting the pipe tappings at different points. One of the two tubes measures the velocity head plus potential head, also known as impact or stagnation pressure at a point in the flow. The speed of the fluid reduces to zero at the opening of the tube and forms the input with high-pressure for a differential pressure sensor element.

There is a second pressure tap further ahead along the flow that provides input for the low-pressure value.

A turbine-metre has a bearings-supported multi-blade rotor fixed inside the pipe through which the fluid is flowing, placed perpendicular to the flow. The velocity of the rotor varies in proportion to the velocity of the fluid that is related to the overall volume flow rate through the pipe.

In ultrasonic flow sensors, ultrasonic sound pulses created by an oscillating piezoelectric crystal are made to travel through moving fluids to determine the flow rate of fluids as their velocity gets modified in accordance with the velocity of the fluid. At times two of such transducers can be used to send sound waves in opposite directions with one of them measuring the fluid velocity and the other measuring the flow rate.

Another variant of these sensors employing two transducer probes works on the principle of Doppler effect with both probes fitted in the same casing on one side of the pipe. One of these probes transmits an ultrasonic sound wave of constant frequency into the fluid, the speed of which can provide the indication of flow.

If the fluid has a certain minimum electrical conductivity, flow can also be measured using electromagnetic flowmeters that create a moving conductor across a magnetic field by placing an electromagnet around a non-conducting section of the pipe through which the fluid is flowing, as shown in Figure 12.3. The voltage induced between the measuring terminals is directly proportional to the velocity of the fluid, from which the flow can be ascertained.

Yet another mechanism of flow measurement uses the hot wire anemometer sensor shown in Figure 12.4. This sensor uses the principle of change in resistance of a piece of heated resistive material in the path of the fluid and measuring the change of resistance caused by the amount of cooling of the sensor depending on

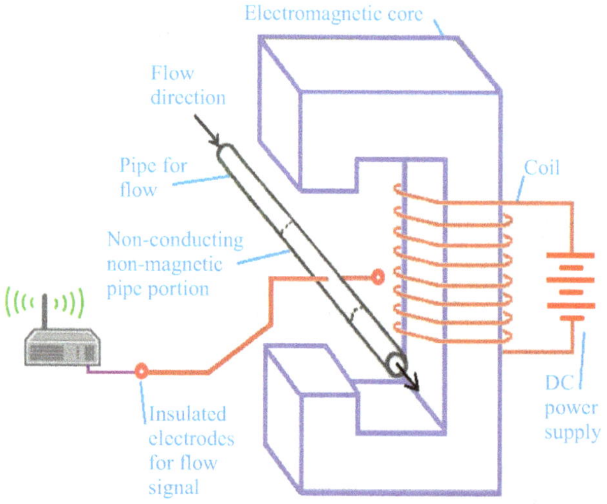

Figure 12.3 Electromagnetic flowmeter

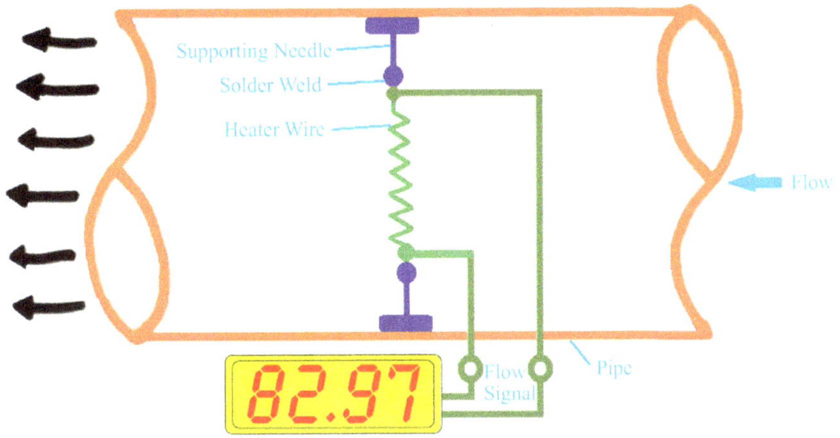

Figure 12.4 Hot wire anemometer sensor

the rate at which the fluid flows through the pipe. The rate of reduction of temperature of the element is related to the flow rate.

12.2.4 Pressure sensors

Pressure sensors find use in industrial hydraulic systems and in climate control systems. The following section briefly discusses some of the common kinds of pressure sensors used since ages.

A Bourdon tube pressure gauge sensor comprises a twisted, elastic tube with an elliptical cross-section that is bent in circular shape to create almost three-fourth of a circle and closed at one end. The open end of the tube allows pressure to enter, causing a tendency of the tube to unwind and straighten, attempting to change to circular cross-section, deforming in accordance with the amount of pressure causing it. The free end of the tube is connected to a pointer that can move on a calibrated scale by an amount proportional to the pressure causing the deformation, thus indicating the pressure locally. The sensor can additionally be fed to an air nozzle to for air pressure control or modeled to convert the pointer movement into an electrical signal to be transmitted to the control room. These sensors are simple to construct at nominal cost, and accurate in operation over a large range of pressure measurement. However, their performance is affected by shocks and vibrations and can degrade over time due to hysteresis effect

The bellows sensor has a circular, metallic but thin-walled cross-section with multiple folds that make it collapsible or expandable axially in accordance with the pressure applied from the free end of the sensor. The axial expansion can be used to directly read the pressure on a scale, or used to generate an analogous electric current with the output being very linear. The expansion of the bellows tube can be limited by adding a restraining spring to the sensor. These sensors are small in size and not very expensive but the range of these sensors is limited. They can also be used for measuring pressure for slurries.

Vacuum sensors are used for measurement of pressure less than the atmospheric pressure and could be difficult to measure. Vacuum can be measured using the principle of change of electrical resistance of a wire with temperature. In high vacuum conditions, the convection lessens, and the wire temperature rises, causing a change in the current in the circuit due to change in wire resistance. These sensors find use in aircrafts, vehicles, weather instrumentation, and pneumatic machinery systems as well as for secondary measurements of speed, altitude, fluid level, and flow.

12.2.5 Strain sensors

Strain is the expanse of distortion of a body (bending, warping, elongation, or expansion) in accordance with an applied force and is indirectly used to measure force or pressure. It is mostly measured using a strain gauge sensor created using a material with electrical properties such that its electrical resistance changes in amount and direction in proportion to the extent of strain experienced by the sensor. If the strain gauge sensor is neatly bonded to a structural member subjected to force or pressure, the sensor's strain is dependent on this force or pressure. The electrical resistance of some semiconductor materials can also be altered in response to strain due to this property known as piezoresistivity. Strain gauges can be created as fine wires, as shown in Figure 12.5 or thin foils as shown in Figure 12.6 that are then bonded to the structural member.

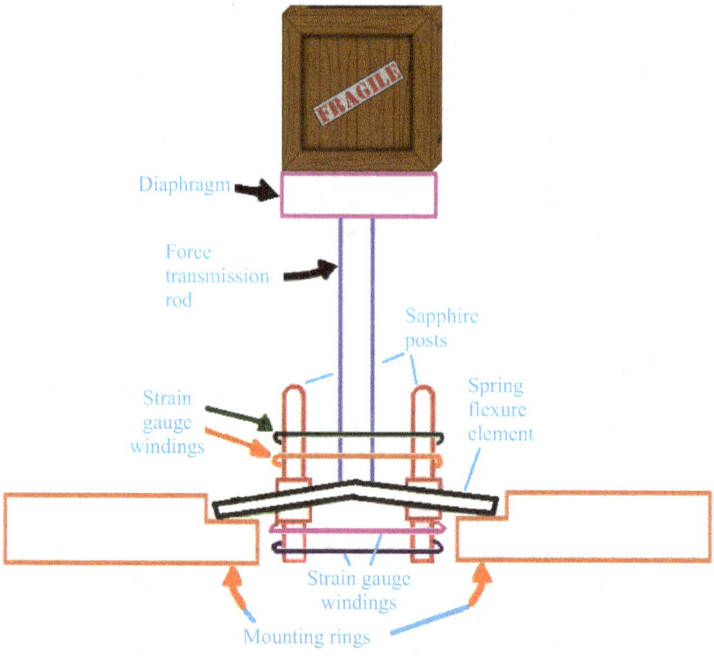

Figure 12.5 Balanced wire strain gauges

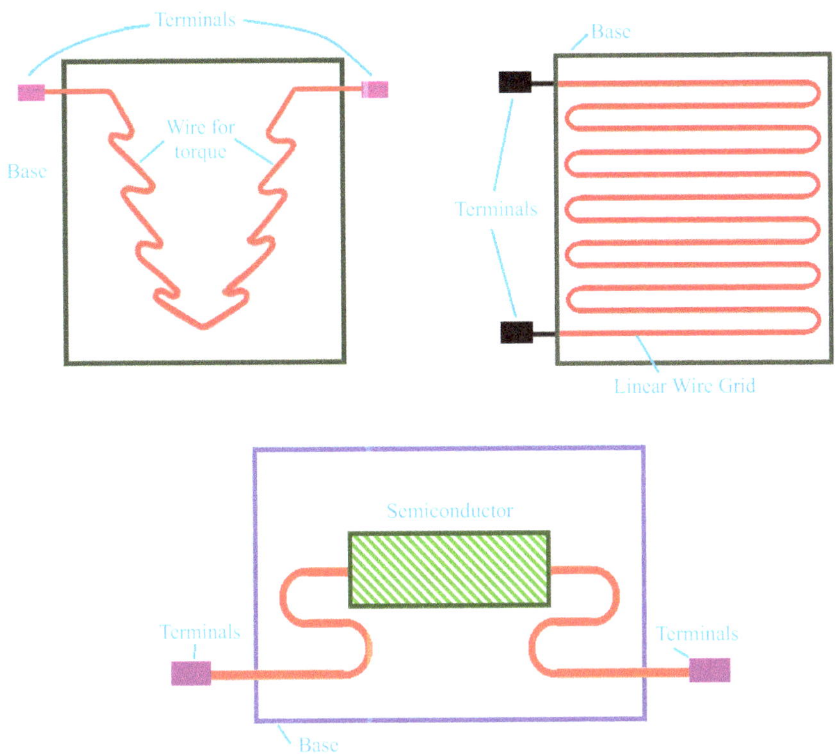

Figure 12.6 Strip strain gauges

12.2.6 Torque sensors

Torque sensors are frequently used in rotating machinery such as motors, generators, and prime movers for measuring the speed of rotation, oscillating torque, mass, and moment of inertia. Multiple strain gauge transducers can be connected in an electrical bridge arrangement for torque measurement in these sensors. These sensors are frequently used for motor condition monitoring for measuring strain and torque for ascertaining the total well-being of equipment and recognising the early cautionary symbols of imminent breakdowns, thus making it easier to identify problems before they have a chance to influence the productivity adversely.

12.2.7 Vibration or seismic sensors

In seismic accelerometers, the movement of a piece of mass in response to an applied force is gauged and linked with acceleration. Multi-axial accelerometer sensors screen rotating machinery for vibration and take measurements for velocity, movement, and acceleration, thus establishing the extremities of vibration. If the vibration is within norms, it indicates normal machinery operation. Any fluctuations or deviations from normal values of vibration are indicative of abnormal

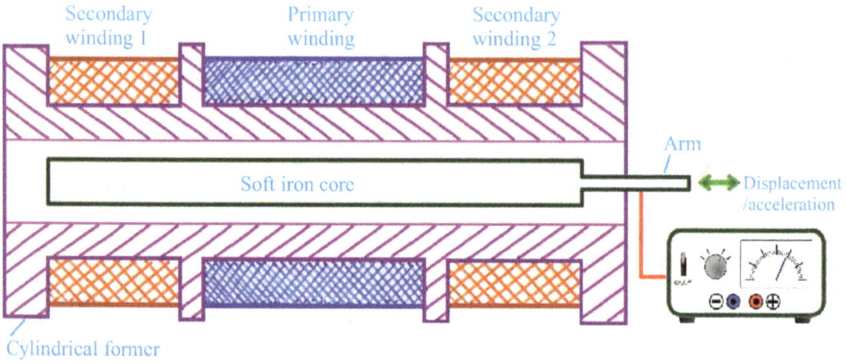

Figure 12.7 Linear variable differential transformer sensor for displacement, velocity, and acceleration

operating conditions that might call for additional checks and inspections if some maintenance is required. Thus, these sensors help with predictive maintenance.

This measurement can utilize a sensor employing a linear or rotary variable differential transformer that measures the changes in displacement or its rate with time. The transformer measures the quantity through a change in magnetic flux by converting the change to a change in the voltage induced in the transformer secondaries caused by the movement of a magnetic core between the transformer coils. The magnetic core is connected to the element generating the quantity that is to be measured, as depicted in Figure 12.7.

Vibrations can also be measured in real time using interferometric holography through wave front reconstruction utilising interference between coherent laser beams. This method works best for small amplitude vibrations.

12.2.8 Angular velocity sensors

Angular velocity is a quantity of interest in rotating machines such as motors and generators. It can be measured digitally by using optical sensors that involve the creation of light pulses by disrupting the path of a light beam to an optical sensor by a perforated disk fitted to the rotating equipment and then counting the pulses, as shown in Figure 12.8 [7].

Another mechanism shown in Figure 12.9 could use the change of magnetic field caused by a toothed rotor in a magnetic pickup, thereby varying the reluctance of the air gap between the rotor and the pickup to cause electrical pulses that can be counted. In the stroboscopic method shown in Figure 12.10, the angular velocity is measured by adjusting the flashing frequency of a light beam focused on a mark in a disc attached to the rotating shaft such that the mark is visible at specific instants of time.

12.2.9 Level measurement sensors

Levels of liquids or solids in tanks, vessels, drums, or silos require measurement from automobile and airplane fuel tanks to several industries such as petroleum

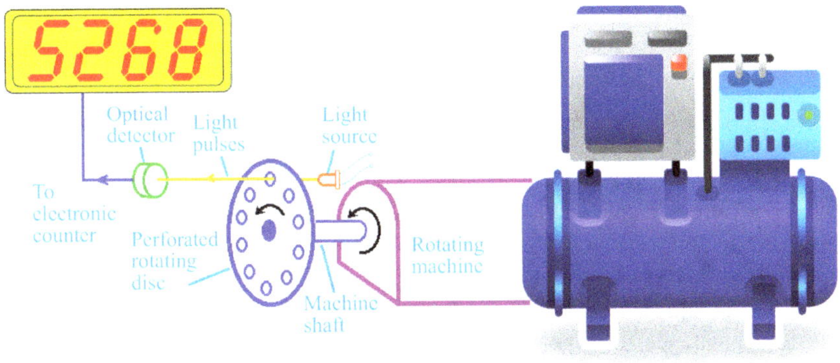

a. Disc-based optical angular velocity measurement

Figure 12.8 Perforated disc angular velocity sensor

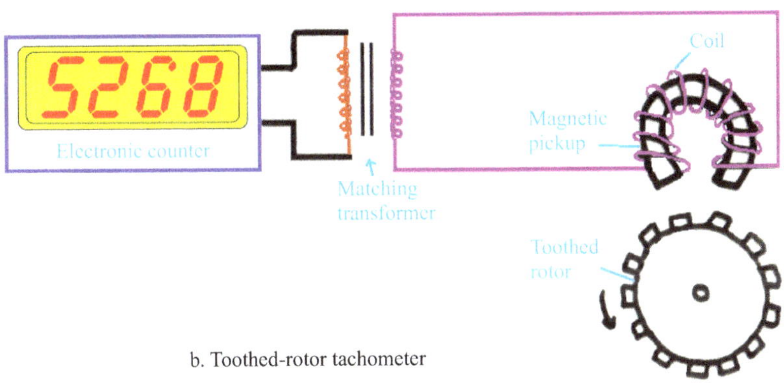

b. Toothed-rotor tachometer

Figure 12.9 Toothed rotor angular velocity sensors

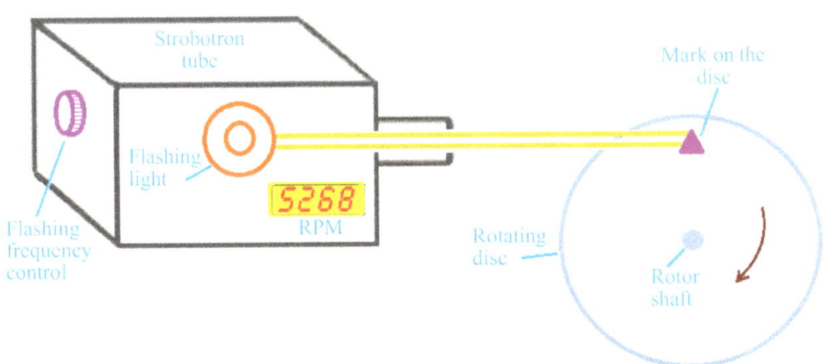

c. Stroboscopic angular velocity measurement

Figure 12.10 Stroboscopic angular velocity sensors

processing, food processing and in power plants to keep track of the amount of material in the tank. Power plants cannot have dangerously low levels of water in their drums that produce steam for running their turbines. Silos and containers for any solid or liquid materials in industries cannot be allowed to overflow. The gauge glass method provides the simplest technique by which liquid level is measured in a vessel. A transparent tube is taken out of the bottom of the vessel and the other end is open to the atmosphere. The liquid in the vessel will rise in the transparent tube by an amount equal to the height of liquid in the vessel.

Other types of level measurement sensors can use raising or lowering of a floatable probe, a pressure head between the fluid level and bottom of the vessel, change in capacitance, change in conductivity of a conductive probe, change in tuning fork frequency, amount of interference in optical/infrared radiation, scattering of nucleonic particles, and reflections in case of radar-based and ultrasonic sensing.

12.2.10 Position or linear velocity sensors

Displacement, change in position or change of linear velocity, can be measured using several types of sensors. One common mechanism is to use an electromagnetic sensor that works on the principle of change in magnetic reluctance caused by a change in flux linkages due to changes in the air gap as shown in Figure 12.11. This change in reluctance produces a proportional change in measured voltage output of the electromagnetic sensor. Depending on the sensing

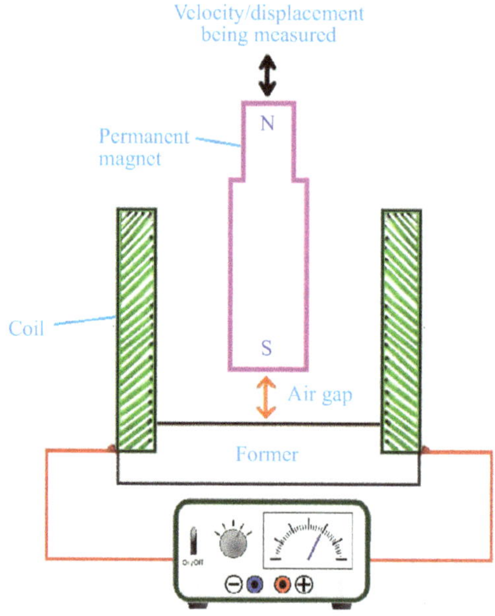

Figure 12.11 Magnetic displacement/velocity sensor

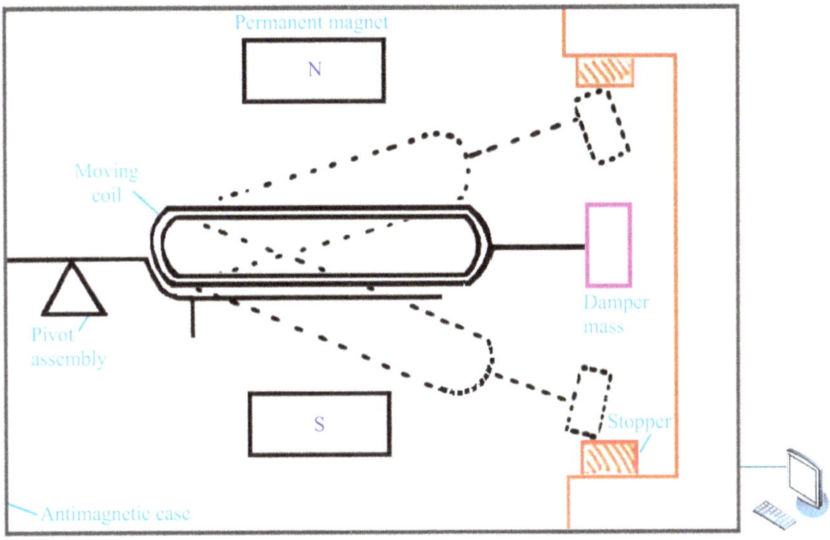

Figure 12.12 Moving coil-based displacement/velocity sensor

element directly connected to the displacement, variable reluctance type electromagnetic displacement transducers could be constructed in moving magnet and moving coil varieties. Such sensing elements are robust, inexpensive, and economical to manufacture in bulk. They face negligible wear and tear and require hardly any maintenance while providing a very linear, highly stable, and consistent output correlated to the velocity.

The moving magnet type uses a permanent magnet free to move in the region around an electrical coil. The magnet provides a constant magnetic field and is in physical contact with the displacement or velocity to be measured. The movement of the magnet causes a change in reluctance because of the variation in the air gap between the moving coil and the permanent magnet. The electrical coil gets a voltage induced in it that is proportional to the velocity being measured. The polarity of this voltage is dependent on the direction of movement. In contrast, in a moving coil type sensor, the magnet is fixed but the coil is free to move with velocity. The moving coil type velocity sensor shown in Figure 12.12 is good for the measurement of velocities that could either be linear, sinusoidal, or in any random development. The only limitations of such sensors are their limited frequency response and sensitivity to vibrations and stray magnetic fields.

12.2.11 Radiation sensors

Measurement of nuclear radiation is important in and around places where radioactive materials are found and mined, stored, and used. The radiation consists of a variety of particles and rays that could be dangerous for human and other life form cells beyond a specific intensity. The radiation from elements with high atomic numbers could be a mix of alpha particles (Helium nucleus with a positive charge), beta particles (electrons with a negative charge), gamma particles or rays – all of which travel with a very high velocity comparable to that of light rays and have variable amounts of penetration power.

The most common types of radiation sensors use a Geiger Muller tube shown in Figure 12.13 that is an airtight sealed glass envelope filled with an inert gas containing a fine wire anode surrounded by a cylindrical cathode. There is a window transparent to radiation on one side of the tube that lets the radiation enter inside the tube. This entering radiation then ionises some gas atoms through collisions with gas molecules producing a number of pulses in proportion to the number of alpha, beta or gamma radiation. The pulses are fed to a counter connected to a readout that informs the operator about the amount of radiation.

A variant of the Geiger Muller counter is an ionization chamber that is most effective in detecting alpha particles. The difference between the two is that in case of the ionisation chamber, the cylindrical outer wall of the tube itself acts as the cathode for a fine wire anode as shown in Figure 12.14. The cylinder is also filled with an inert gas that ionizes instantaneously because of the radiation entering the chamber through a window in the chamber. The radiation particles cause a proportional number of pulses that are then fed to a counter.

More sensitive than the previous two sensors are the Scintillation Counter sensors that can detect very low levels of radiations and X-rays as well. This sensor exposes crystals of material such as zinc sulphide, sodium iodide, or naphthalene that have a unique property of producing brief flashes of light in response to radiation. These flashes can then be amplified by a photomultiplier tube and fed to a counter to indicate the intensity of radiation as shown in Figure 12.15.

Photons can be detected by solid state ionisation sensors that use radiation to ionise atoms of extrinsic semiconductor material producing a pulsed output. These sensors work at liquid nitrogen temperatures but provide very high sensitivity to

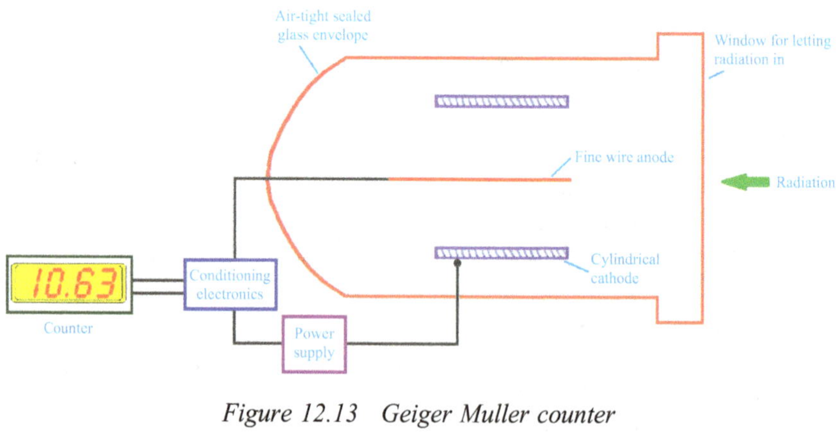

Figure 12.13 Geiger Muller counter

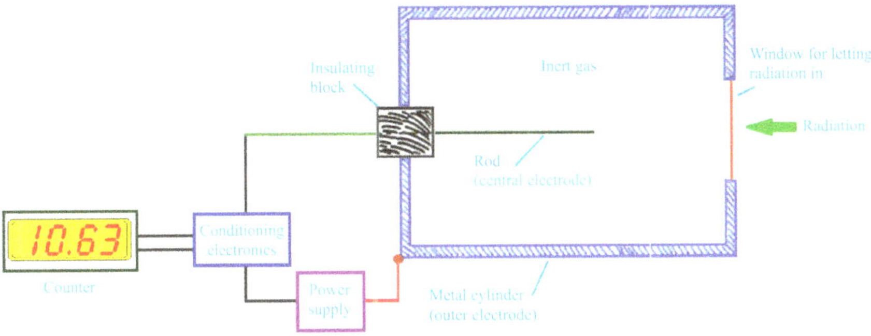

Figure 12.14 Ionization tube counter

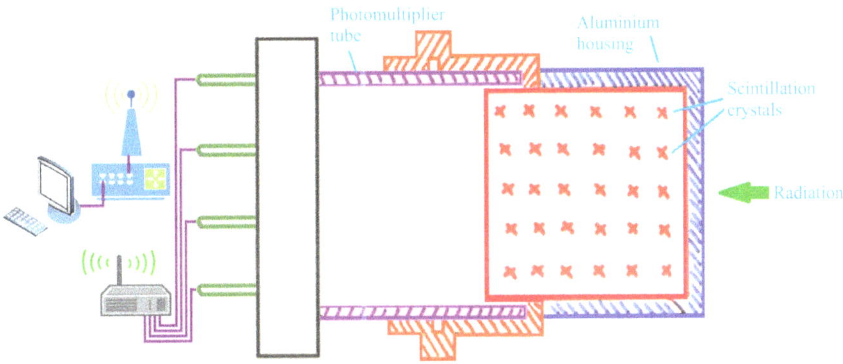

Figure 12.15 Scintillation photomultiplier

photon radiation. Neutrons can be detected by creating fission in uranium isotopes to cause alpha, beta, and gamma radiations that indirectly give a count of the number of neutrons causing them.

12.2.12 Gyroscopes

A gyroscope is a device for measuring or maintaining orientation or gyroscopic inertia, based on the principle of conservation of angular momentum. It is highly used in aircrafts and robotics for providing inertial navigation through motion measurement. This sensor has a spinning wheel on an axle. As a result of the spinning, the sensor tends to resist changes to its orientation because of the angular momentum of the spinning wheel. Any required damping in the movement can be provided by springs.

12.2.13 Limit switches

Limit switches are mechanical devices that can be used to operate when a movement of a mechanical equipment part reaches a certain limit and perform simple, on–off type control. They can be operated by a physical displacement or change in fluid level. Also, any physical quantity such a pressure, flow, or temperature that can be converted to displacement can serve as a limit switch for that quantity. Physical quantities such as an attachment on a valve shaft can move with the opening or closure of the valve and trip a limit switch for signalling and further control purposes based on the valve position.

12.3 Adapting standard commercial sensors for IIoT in cyber physical systems

The IIoT used for cyber physical systems need to have a very robust overall design, one that guards against system failures as failures would be catastrophic, resulting in situations ranging from downtime and wastage to threats to lives of personnel involved.

Spearheaded by AT&T, the Industrial IoT Consortium (IIC) is a group formed by technology companies such as Intel, IBM, Cisco, and General Electric that aims at speeding up the use of IIoT in industries. One of the major focuses of the group is design of IIoT sensor systems with security, which is one of the biggest risks and hurdles in the path of IIoT growth. Managing a vast number of IIoT devices, their identification, and authentication with a view of eliminating rogue devices is another connected challenge connected to the scalability of the IIoT systems. As the systems are still under evolution, patch management for firmware changes is another prime issue.

Traditional sensor systems are simple devices that translate physical quantities into equivalent electrical signals that travel from the point of sensing to the point of usage by humans over wires. While creating sensors for IIoT, it is important to enable them with wireless transmission capabilities to get rid of cabling hassles and improve the flexibility of installation and usage [8]. The designers are also required to keep their cost low to enable large scale use of these sensors. If they are physically small and consume low electrical power, they can be added unobtrusively to the process being monitored, and without much power drain. As they are mostly battery operated, and batteries have limited lifetime, self-sustained power generation and energy harvesting are also desirable features for such sensors, to reduce or eliminate frequent battery replacements. Each of these sensors needs to be uniquely identified and they are used in very high numbers, hence self-identification, self-calibration, and validation, with in-built diagnostics and self-repair are also desirable. If the sensor electronics includes an embedded design involving a processor with memory and storage, it can also perform some data pre-processing to ease the burden of processing on the HMI and supervisory computer systems and the cloud.

Traditional sensor systems now need to have connectivity all the way to the cloud, which should capture and neatly segregate the humongous amount of IIoT data and make it available for processing for business analytics. The sensor connectivity also needs to implement machine to machine communication.

12.3.1 Cognitive sensing requirements for smart IIoT sensors

Cognitive radio (CR) techniques allow better utilization of frequency spectrum for wireless communication by actively sensing the spectrum for current usage, finding out frequency bands that are not being currently used (called white spaces) and utilising any frequencies not being used at that instant in real-time [9]. CR makes the radio system dynamically reconfigurable and lets it change its operating spectral parameters as per the current spectral availability using smart antennas. This new cognitive concept for radios has been made possible by a new development called software-defined radio (SDR) that enhances the radio functionality and makes the wireless connectivity configurable, adaptable, and flexible. Additionally, it improves the usage of frequency spectrum, addressing the problem of under-utilization or non-utilization of spectral resources. CR not only ensures that critical data will be transmitted immediately using the currently available wireless resources but also helps in addressing the challenge of spectral scarcity issues being faced by wireless system designers and engineers.

Licensed frequency spectrum bands have been shown to be under-optimally used, based on several spectral usage studies as the bands are reserved for exclusive use by the licensee party, also known as the primary or legacy user. Hence, when the primary licensee is not using the bands, they are still not available for use by other parties, called secondary users that could have been benefited by temporary sharing of the frequency bands due to exclusive allocation policies that have been traditionally followed by the world.

Dynamic spectrum access (DSA) or spectral agility is possible if CR is used, and any unused frequencies can be made available for opportunistic transmission of information or sensor data by other, unlicensed parties that may be willing to split and share the cost of frequency band usage and solve the problem of non-availability of a transmission frequency at the instant of transmission. The system would work just fine, if the primary user to whom the frequency band is licensed does not face any denial of service or frequency interference issues with respect to its transmissions or faces any disruptions in communications. Mandates have also been put in place to ensure that military communications and radars are not impacted because of such opportunistic spectrum sharing. If the presence of the primary user's transmission is not detected by spectral sensing, secondary users could be allowed to use their frequencies, but the primary user is free to reclaim its frequencies for its use at any point of time it needs them. The secondary users could try switching to an alternate unused frequency when this happens. The secondary users might not be required to obtain express permissions from the primary users in the frequency bands such as the 5.4 GHz band (5.470–5.725 GHz) as already legalized by the authorities such as the FCC.

Implementing cognitive radios for industrial sensors has hurdles in its way because the key to making them work is the ability to sense the spectrum usage reliably [10]. The thresholds of energy from a transmission band in the vicinity of the secondary user could be low due to shadowing or multipath fading, making the task very difficult. The reliability in sensing the presence or absence of the primary or legacy user is of paramount importance.

12.4 Design considerations for IIoT sensors used in manufacture and control

The sensor fabrication technology also needs to address the issue of creation of sensor arrays containing multiple sensors of same or multiple types in the same package by allowing for higher integration and changes in fabrication techniques and adding intelligence to sensor arrays. Subsequent sections try to focus on the different aspects of design of IIoT sensors for cyber physical systems [11]. MEMS technologies are immensely useful for creating the sensors used for cyber physical systems.

12.4.1 MEMS (micro-electro-mechanical systems) sensors

The MEMS inertial sensors are frequently used for the measurement of motion, acceleration, tilt, and orientation in industrial automation. These sensors use ZigBee wireless technology for communication. They are dynamic and have a big range of sensing capabilities. They also find use in entertainment or sports. They could be used for detecting motion or impact, activating vehicle airbags and automotive ABS, and for sensing dynamic acceleration in vehicles, boats, and aircrafts. They also find use in computers, mobile phones, washing machines and even in monitoring of nuclear reactors [12].

Multiple MEMS motion sensors can now be integrated with machines for custom interactions, and the cost of such sensors would still be very low [13]. Advanced signal processing systems in these sensors resolve noise cancellation snags, condition the signal making it smooth, and make the system customizable.

12.4.2 Smart sensors for advanced uses

Using sensor fusion, information created using data from numerous sensors can be united and connected to create inferences about hidden complications. For example, data from vibration and temperature sensors can be used in the prediction of mechanical breakdown in case of moving machines.

It is also possible to combine the functions of both these sensors in one sensing device; or create a soft sensor by combining the functionality in software.

Smart IIoT network involves using sensors to sense and measure the real-world variable and transduce the sensed quantity into a stream of digital data that is sent to a gateway using a combination of wireless and wired links.

Every sensor is designed with a built-in microprocessor unit (MPU) that runs application algorithms for filtering, amplification, signal conversion, compensation, and other signal conditioning jobs specific to the process. The MPU's abilities can be used for advanced functions such as sensor self-diagnostics. As the MPU is highly capable, it can also run quick recalibration for changes in production pattern, thus reducing the load on the IIoT networks' central resources. What is even better is that the system is remotely programmable or can be calibrated, frequently by using wireless terminals. This swift recalibration procedure allows for creation of personalised products that are unique in terms of manufacture, packaging, inspection, and dispatch even if the production aims at a single-unit batch while the costs are still low, and typically at mass-production price [14].

The MPU also monitors the production sequence to catch any unacceptable drift in production parameters, generating alarms and warnings if required; thus, enabling preventive action by process operators and avoiding catastrophic failures that cause downtime, production losses and a hit on the brand value.

Reactive reporting by exception allows the sensor to only transmit data if the value of the changes appreciably from previous sensed sample values, and the threshold can be set and altered quickly as the system is digital. This decreases the burden of computing on the base station's computing resources and cuts down on the power required by the sensor, which could be coming from a battery. Most IIoT sensors are powered by batteries and may or may not use additional energy harvesting after total battery drain, something very beneficial for an IIoT system.

Smart sensors could also be built with two or more sensing elements in the same probe package, to provide the sensor with reliability by redundancy and improved monitoring feedback. Such design also works well for sensor self-diagnostics, in the event that one of the two elements drifts or fails.

New IIoT applications require that the weight, size, form factor, power requirement, sensor packaging and cost of IIoT sensors must keep diminishing.

Smart sensors are fabricated by multi-chip integration that interconnects microelectromechanical system (MEMS) sensor elements with CMOS integrated circuits (ICs) through the board substrate or the IC package. The techniques used today do not require the use of any routing layers creating a complete, intelligent sensor system capable of wireless communications on one chip. The benefits of this integration are lower package costs, smaller footprints, reduced parasitics, and higher interconnect densities. The combined package has the provisions for providing device power supply as well as all required signal conditioning [15].

Industrial control systems today can use edge programmable controllers that include the functionality of programmable logic controllers (PLCs) and continuous control systems that have scalable I/O and communication options. PLCs offer a rugged, purpose-built design with many typically expandable options for I/O and device communication. Added functionality includes processing of vast hordes of sensor data for general-purpose and supervisory purposes, storage, and networking for a wide variety of applications such as database management or cybersecurity that are highly demanding [16]. All this functionality, along with support for writing custom programs in high-level languages, is available today in a single package. The systems also support standard industrial communication protocols such as ISA100 that could be used for continuous control applications such as wireless valve control.

12.4.3 Energy economy and harvesting for sensor operation

Batteries are used by default to power up smart IIoT sensors. They need to be rechargeable, have a long life and provide high-energy density. In addition, the sensor designs should make smarter use of the battery power in sensing, processing, and transmission of data. Smarter transmission algorithms like data compression, use of statistical data, and reactive data transmission can be used to reduce the drain on the battery and increase its lifetime.

Smart IIoT sensors used for industrial applications need to have long life and the fact that they are battery powered is the biggest hurdle for this design. MEMS systems have enabled the use of miniature mechanical energy generators which can be used in these sensors if feasible. Where they are not an option, robust and effective energy harvesting needs to be added to sensor design to achieve this goal. Energy in any form from the ambience or from movement of some sensor components can be utilized for the generation of power as well as for recharging any sensor batteries. A moving magnet or moving coil used in the sensor can act as generators and create sufficient electrical energy to transmit the data samples through wireless signals to the receiving unit. Ambient optical energy, temperature, solar power or fluid flow can also be used to create this electrical power through transduction. Thus, these generators can support or help eliminate battery power.

The sensors need to be designed with transceiver electronics that consume low power for transmission and reception. Communication protocols need to be modified with an objective of reducing the amount of data transmission. Reducing the distance of transmission through routing could also help achieve this objective.

The physical layer of the communications suite should also be designed to use noise-immune wireless frequencies such as 915 MHz, 1,800–1,900 MHz, 2.4 GHz, and 5 GHz spectral ranges that are globally accepted as standards for industrial control applications.

12.4.4 Potential benefits of IIoT-based manufacturing and control

As the use of cables is eliminated because of wireless transmissions, these IIoT sensor and automation systems do not incur costs for cabling, cable protection, junctions, and terminations with respect to material cost and labour required, saving a lot of money, and reducing installation and labour costs. Hence, they find use in difficult to access locations such as for hatch covers, access ports and in hopper flaps or diverters.

Most of the troubleshooting and maintenance in conventional sensor and automation systems is attributed to repairing or replacing damaged cables and associated paraphernalia. Going wireless also cuts down on maintenance costs and personnel required and increases the system productivity and uptime. Typical applications could include valve position monitoring,

The IIoT devices are excellent candidates for harsh and hazardous working environments as their low energy, spark-free versions are also available today due to intrinsically safe design strategies. They could be used for monitoring of tank, vessel, and drum levels containing corrosive or hazardous liquids, and fire vent positions.

As the IIoT sensors and switches are now available in their wireless avatars, they can be placed in remote locations that are difficult or infeasible to reach through cable enabling flexible control system topology, for example, in rotating and moving machinery and equipment. Small sizes of the new IIoT devices allow ease of installation and adding them to equipment and machinery such as in control of cranes, with much reduced design constraints. Such devices are also perfect for distant locations where the availability of mains electrical power is infeasible.

12.5 Some IIOT use cases

12.5.1 Smart fault indicator

A new variety of fault indicators can now be used for indications of faults in overhead electrical power transmission lines. They have special wireless sensors that combine current and temperature measurement managed by an ultra-low power microcontroller and are installed on the network junctions. The base station concentrators for these sensors are mounted on the electrical poles. The current (measured and analysed in frequency domain) and temperature values are sensed and transmitted over wireless links to the concentrators. The concentrators further relay the data to the main station using the cellular network with the help of a GSM modem in real-time, making continuous remote monitoring of faults possible. This data is then processed at the main station and used for real-time display, trends, and recording, alarm generations, control, data pattern analysis, and diagnostics. Faults and their locations are identified fast, action is taken immediately and any downtime in the power distribution network is greatly reduced. The system also has built in diagnostics to report any sensor problems or errors. The sensor and microcontroller unit works on low power which allows for long battery life and is very beneficial as the unit is mounted on power lines and is not easily accessible.

12.5.2 Smart position sensors

Smart IIoT position sensors can be created by combining multiple magnetoresistive (MR) sensors for non-contact sensing of position and are finding increasing use in medical, aerospace, and industrial applications for measurement and control [17]. They are lightweight and are used in machinery, valves, or elevators, for alignment and positioning by ascertaining the location of a magnet fixed to any moving objects, with the sensors affixed in pairs along the direction of movement of this fixed magnet. The outputs of the MR sensor pairs are compared, to find out which sensor pair is the closest to the fixed magnet, and is producing the highest output, thereby helping ascertain the position of the object very accurately.

The sensor array is also interfaced with an application-specific integrated circuit chip to add features of highly accurate direct digital output, fast calibration, and self-diagnostics. The durability and lifetime of this sensor system is high as it is non-contact and there is no wear and tear or mechanical failure due to friction.

Because of their small size, wireless capabilities, and well-sealed packaging, they can be deployed quickly in harsh environments with no wiring required.

12.5.3 Smart sensors with self-diagnostics and repair abilities

Smart IIoT sensors are being increasingly used in areas with harsh conditions and potential gas hazards such as industrial manufacturing and metallurgical complexes, hydrocarbon industries, warehouses, harbours, and sewage treatment plants. These sensors are critical for human safety, while accessing or maintaining the sensors for battery replacement itself or in case of failure might be problematic, and yet reliability needs to be high. They could also be used for detection of fire or intruders. Efforts are being made to create self-configurable smart wireless sensors for gas detection and analysis with multiple probes for improved reliability, self-diagnostics, and repair capability [18]. The sensor system shown in Figure 12.16 is managed by an additional microcontroller for performance enhancement, in addition to the main processor, both of which are created using FPGA that consumes very low power. The sensor node uses energy harvesting to circumvent battery power issues and can locate operational problems and potential operational failure sites. Redundancy sensor design guards against sensor failure by switching to the hot standby sensor in case the one actively sensing and collecting data fails abruptly or degrades in performance. When a degradation is detected, the responsible sensor can be cut off from the power supply, thus saving energy and the standby sensor used, for consistently reliable measurement [19]. The second sensor can also be used to validate the data being sensed by the active sensor. The diagnostic system's design considers all possible types of component level failures in the sensor system and self-actuated remedies in accordance. This also allows the sensor system to also act promptly to report and repair the system, making it highly reliable, very low maintenance and energy efficient.

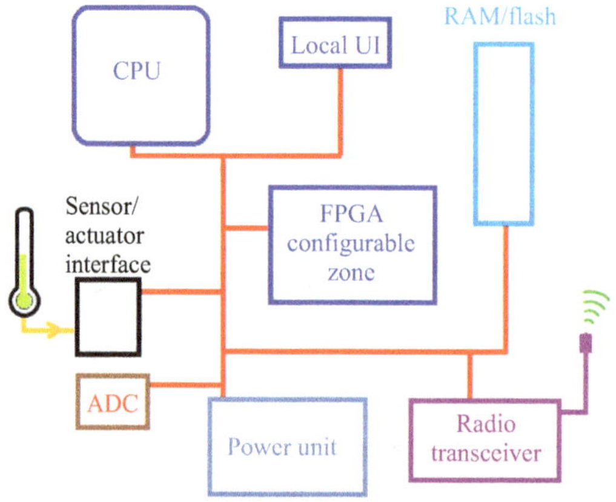

Figure 12.16 Hardware configuration for a self-reconfigurable sensor node

12.5.4 Dynamic alarms in plant operations and management

Alarm systems for supporting large modern industrial facilities provide invaluable help to the operators by signalling them in several ways about situations that require the operator's attention, acknowledgement and possible corrective action. The alarm systems alert the operators using multiple ways such as audio and visual, and involve buzzers, indicating lamp and annunciation panels, light boxes or are available in Data Acquisition Systems (DAS), Programmable Logic Controller (PLC), Distributed Control Systems (DCS), Human Machine Interface (HMI), or Supervisory Control and Data Acquisition (SCADA) systems. In the latter, more advanced systems, multiple alarms can be flexibly configured through software and could be set for warning or corrective actions for failsafe operations. The alarms are critical in the plant operation for human and equipment safety, prevention of abnormal operations, hazards, and possible shutdowns. In case of applications involving alarms related to very high-speed parameters, the alarm design is hardware based for superfast response.

Frequent changes in process conditions and modes of operation in large plants managing extremely complicated production or processing processes need interoperability between multiple connected subsystems. The sensing and control automation equipment used in such plants require a comprehensive alarm system that can be dynamically modified to affect the changes in multiple connected or related alarms. Such configurability is also connected to the context of conditions of plant operation or changes in the product or process and involves maintenance of several disparate databases to be applied under different contexts and treating them as a single database as per context.

In smart control sensor systems, the operators can configure the alarms and HMI interface screens based on the requirements like priority, operator roles, and operation location using computers, handheld terminals, or even cell phones. Supervisory programs run on top of the control programs take in all events and alarm information and use it for downtime or episode analysis about any problems faced and corrected, to create a knowledge base that improves plant equipment utilization as well as productivity. The supervisory programs also help in rationalization, creating alarm performance benchmarks and generating an alarm philosophy and auditing document.

12.5.5 IIoT for oil and gas sector

IIoT helps configure, safely operate, and manage the fleet of unmanned aircraft that use thermal and visual imaging sensors to scan pipelines and contain critical data for oil companies [20].

In natural gas pumping stations, gas supply valves used to be traditionally operated manually and their open or closed positions required to be monitored by the control room. The positions would be ascertained manually. Two miniature, spark-less limit switches attached to the valve top can be activated electromechanically by a shaft attached to convey open and closed states of the valve now over wireless links to the control room. The switches are inexpensive and need no battery or cabling for power or for sending the signal to the control room. The system enhances the overall operation by placing the control stations far away from the zone where there is a risk of explosions. Such switches can also be used very safely, without any cable-related hassles to ascertain that while filling the tanker trucks with gas, the filling pipe is correctly placed and then send this information to the control room over wireless link.

12.5.6 Tunable diode laser-based gas analysers

The traditional gas analysers used infrared or ultraviolet spectroscopy for detecting gas concentrations and used to be low on accuracy, precision, and hence reliability for low gas concentrations, difficult to maintain, had to be frequently calibrated, and subject to wear and tear. Newer gas analyser cells using tunable diode laser shown in Figure 12.17 work at high resolutions, overcome these shortcomings and are drift-free and accurate even for measuring trace concentrations of gases. In these gas analysers, tunable diode lasers emit rays in the near-infrared region of the spectrum [21]. These rays are made to pass through the gas sample to be analysed before reaching a detector. The gas absorbs the energy in the rays according to its composition and reduces the amount of radiation that reaches the detector.

These analysers are robust, have a quick response with high sensitivity, and require hardly any maintenance or recalibration. They can be used for metrology and climate research, in transportation and in harsh industrial environments such as petrochemicals, petroleum refineries, and natural and other gas processing plants.

12.5.7 Smart agriculture

Agriculture as an industry focuses on maximizing the produce along with ensuring its quality, and its timely delivery to the consumers with best pricing. It can also involve monitoring of the livestock. Despite the best efforts from farmers who

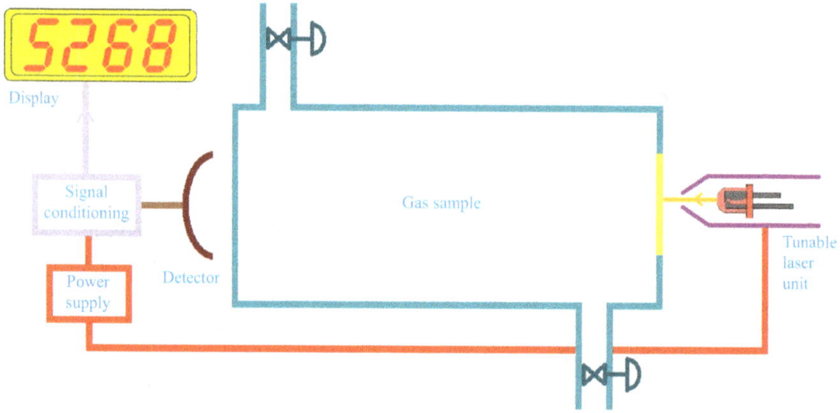

Figure 12.17 A typical sample cell with source and detector

follow conventional farming methods, close to 50% of the produce is lost due to poor pricing and wastage. Agriculture methods have been in serious need of innovation for quite some time, and with the growing world population, the challenges are becoming bigger with time. IIoT applied to agriculture helps with the objectives by real-time asset tracking, irrigation and fertilizer management, smart pest control, monitoring weather, drought conditions and crop health to get better yields by optimising on several factors for a well-informed and reduced risk farm management. IIoT sensors used in the agriculture sector assimilate information on soil moisture, nutrients, and other factors to help farmers control these factors and grow the most productive crop possible, along with help in harvesting automation [22]. They can also be used to track the location of the livestock through RFID tags and their health as well, so that any sick animals in the herd can be isolated from the healthy ones.

12.5.8 Machine health monitoring

Physical quantities such as force, torque, displacement, velocity, acceleration, vibrations, and temperature can be sensed on machines such as robotic arms, motors, prime movers, and CNC to continuously monitor and find out if the operational parameters of the machines are in order and in accordance with the set norms. Any unusual operation or overloads can be proactively and very accurately notified with the help of alarms and serious deviations can be prevented with fail safes [23]. Unusual temperature rise could be a result of friction which can cause serious damage to the machinery if left unchecked. Such a monitoring arrangement with IIoT sensors can also provide for predictive maintenance, longer health of equipment and save against accidents, human hazards, and machinery damage [24]. It would also help maintain the production quality while reducing production and maintenance costs.

12.5.9 Optimization of material handling automated guided vehicles

Automated guided vehicles (AGV) are used on the factory floors of many industries for material handling and are now reducing in size and weight with better guidance mechanisms as trackless AGVs using laser guidance due to smart IIoT sensors. These new AGVs are robustly built to last long in harsh factory environments with improved wireless motion control with safety devices. The AGVs have better integration with shop floor management systems in the control room through industrial grade wireless devices so that there is no interruption in communication [25].

12.5.10 Managing automotive assembly line inventory

Weight-based limit switches can now be used for the management of automotive assembly line inventory. This was previously done manually by informing the inventory personnel to place a stack of materials when required and could involve delays and lost production. Multiple stacks of the inventory part can be placed in a conveyor and moved as required. When a stack of specific automotive parts empties, a limit switch placed at the bottom activates the sliding in of the next full stack of automotive parts as the previous stack's replacement in the same place. This results in faster replenishment of inventory parts and reduced downtime on production.

12.6 The future of IIoT

As wireless communication forms the core of IIoT systems, they continue to face issues due to slow response, signal attenuation, interference from system noise that hampers the feedback from sensors as well as commands to final control elements. Frequent tuning and recalibration of individual sensors to overcome these problems is a requirement in IIoT systems and should be best handled by the sensors themselves.

Industry the world over has seen some major revolutions starting with water or steam propelled machinery, followed by assembly lines and equipment driven by electricity, leading to the next phase involving the use of computers and automation in the industry for manufacturing and process control [26]. The IIoT revolution is currently in a phase that has been termed as Industry 4.0 or the fourth industrial revolution by several experts. The phase is marked by added flexibility and responsiveness to system parameters used to enhance the manufacturing or processing functionality in favour of customer requirements, that could vary from one customer to another. The design of such systems today involves inputs, feedback, and interactive communication from all the concerned parties across the supply and process chain, including the suppliers, material handlers, operators and managers, as well as customers – for setting up production and management specifications.

The production process becomes asynchronous by adding IIoT radio sensors for identification and processing in all the parts involved as these sensors also keep storing information as they move through the manufacturing process. Any variations in the final product or processing required for a specific part/s or subassembly are conveyed to the part or subassembly and maintained in its memory for the process operators to get the specifications and instructions from and act accordingly. All these added functions for smart manufacture point to the need for increased embedded computing at the sensor level.

New wireless technologies such as 5G have already started being used by mobile networks for superfast data transfer of 20 Gbps or more along with low data latency of less than a millisecond. These two features mean that data will be available at the recipient's end in real time and several new applications at remote sites that were previously deemed not viable will now become a reality.

As several industries such as oil and gas are asset intensive, IIoT will pay rich dividends for all such industries and processes by adding features such as predictive maintenance and smart asset management for their speedy growth and expansion.

Combining big data analytics and artificial intelligence (AI) with IIoT paves way for advanced functionality in process management systems [27,28]. In product-based industries, several ambient or atmospheric conditions might be responsible for the final outcome on the product and its quality, but it would not normally be possible to ascertain the amount of individual contribution from each of these conditions or factors. The treasure trove of data collected from IIoT sensors could now be able to train an AI model that would unravel the mystery and help the production managers analyse the process so as to find out such details that connect the conditions with product quality. The operators would then be able to fine tune the process for a better control on the quality of the final product, optimizing on raw material, minimizing wastage, and ensuring consistency in quality.

Finally, planning up an IIoT system with all security and privacy concerns in place is of utmost importance [29]. Ensuring the security triad of confidentiality, integrity, and authenticity with the abundance of IIoT data needs to be ensured at all times, and any unauthorised access or attacks avoided.

Acknowledgments

The author acknowledges the use of Visio drawing software, Libre Office, and some free to use images from wpclipart.com for creating the figures.

References

[1] E. Sisinni, A. Saifullah, S. Han, U. Jennehag, and M. Gidlund, "Industrial Internet of Things: challenges, opportunities, and directions," *IEEE Transactions on Industrial Informatics*, 2018;14(11):4724–4734, doi:10.1109/TII.2018.2852491.

[2] C. P. Kruger and G. P. Hancke, "Implementing the Internet of Things vision in industrial wireless sensor networks," In *2014 12th IEEE International Conference on Industrial Informatics (INDIN)*, Porto Alegre, Brazil, 2014, pp. 627–632, doi:10.1109/INDIN.2014.6945586.

[3] B. Caiazzo, T. Murino, A. Petrillo, G. Piccirillo, and S. Santini, "An IoT-based and cloud-assisted AI-driven monitoring platform for smart manufacturing: design architecture and experimental validation", *Journal of Manufacturing Technology Management*, 2022;34(4):507–534. https://doi.org/10.1108/JMTM-02-2022-0092.

[4] T. J. Marshall and Y. A. Brady, "Process control and automation solutions", In *Biopharmaceutical Processing*, Elsevier, 2018, pp. 579–621, ISBN 9780081006238, https://doi.org/10.1016/B978-0-08-100623-8.00030-X.

[5] C. Pu, "Spam DIS attack against routing protocol in the Internet of Things," In *2019 International Conference on Computing*, Networking and Communications (ICNC), Honolulu, HI, 2019, pp. 73–77, doi:10.1109/ICCNC.2019.8685628.

[6] J. Park, "Signal conditioning", In *Practical Data Acquisition for Instrumentation and Control Systems*, Newnes Press – Elsevier, ISBN 0-7506-5796-0.

[7] L. C. Souza, E. R. Neto, E. S. Lima, and A. C. S. Junior, "Optically-powered wireless sensor nodes towards Industrial Internet of Things", *Sensors (Basel)*, 2021;22(1):57, doi: 10.3390/s22010057. PMID: 35009600; PMCID: PMC8747621.

[8] R. Herrero, "Introduction", In *Fundamentals of IoT Communication Technologies*, New York, NY: Springer, 2022, pp. 3–19, doi:10.1007/978-3-030-70080-5_1

[9] J. Shreyas, S. Ajmani, P. K. Udayaprasad, D. Chouhon, and S. M. Dilip Kumar, "Dynamic routing scheme for linking wireless sensor network towards Internet of Things," In *2021 5th International Conference on Electrical Information and Communication Technology (EICT)*, Khulna, Bangladesh, 2021, pp. 1–4, doi:10.1109/EICT54103.2021.9733501.

[10] M. Ge, "Recognition and detection methods of artificial intelligence in computer network faults under the background of big data", *Wireless Communications and Mobile Computing*, 2022;2022:Article ID 5332876, 13 pp.. https://doi.org/10.1155/2022/5332876

[11] G. Schuh, M. Jarke, A. Gützlaff, I. Koren, T. Janke, and H. Neumann, "Review of commercial and open technologies available for Industrial Internet of Things", In *Design and Operation of Production Networks for Mass Personalization in the Era of CloudTechnology*, Elsevier, 2022, pp. 209–241 (Chapter 8), ISBN 9780128236574, https://doi.org/10.1016/B978-0-12-823657-4.00005-1.

[12] P. Singh, and L. K. Singh, "Instrumentation and control systems design for nuclear power plant: an interview study with industry practitioners", *Nuclear Engineering and Technology*, 2021;53(11):3694–3703, ISSN 1738-5733, https://doi.org/10.1016/j.net.2021.05.025.

[13] M. A. Huff, "MEMS: an enabling technology for the Internet of Things (IoT)", In *Internet of Things and Data Analytics Handbook*, Wiley, 2016, https://doi.org/10.1002/9781119173601.ch9

[14] B. Wang, P. Zheng, Y. Yin, A. Shih, and L. Wang, "Toward human-centric smart manufacturing: a human-cyber-physical systems (HCPS) perspective", *Journal of Manufacturing Systems*, 2022;63:471–490, ISSN 0278-6125, https://doi.org/10.1016/j.jmsy.2022.05.005.

[15] J. Park, "Specific techniques", In *Practical Data Acquisition for Instrumentation and Control Systems*, Newnes Press – Elsevier (Chapter 11), ISBN 0-7506-5796-0

[16] B. Mostefa and G. Abdelkader, "A survey of wireless sensor network security in the context of Internet of Things," In *2017 4th International Conference on Information and Communication Technologies for Disaster Management (ICT-DM)*, Münster, Germany, 2017, pp. 1–8, doi:10.1109/ICT-DM.2017.8275691.

[17] M. A. Khan, J. Sun, B. Li, A. Przybysz, and J. Kose, "Magnetic sensors – a review and recent technologies", *Engineering Research Express*, 2021;3 (2):022005, doi:10.1088/2631-8695/ac0838.

[18] V. -T. Hoang, N. Julien, and P. Berruet, "On-line self-diagnosis based on power – measurement for a wireless sensor node", In *First IEEE Workshop on Highly-Reliable Power-Efficient Embedded Designs*, February 2013, Shenzhen, China.

[19] V. -T. Hoang, N. Julien, and P. Berruet, "Increasing the autonomy of wireless sensor node by effective use of both DPM and DVFS methods," In *2013 IEEE Faible Tension Faible Consommation*, Paris, France, 2013, pp. 1–4, doi:10.1109/FTFC.2013.6577766.

[20] R. Alguliyev, T. Fataliyev, and S. Mehdiyev, "The industrial internet of things: the evolution of automation in the oil and gas complex", In *SOCAR Proceedings*, 2019, pp. 66–71, doi: 10.5510/OGP20190200391.

[21] Z. Zhang, M. Li, J. Guo, B. Du, and R. Zheng, "A portable tunable diode laser absorption spectroscopy system for dissolved CO_2 detection using a high-efficiency headspace equilibrator", *Sensors* 2021;21(5):1723, https://doi.org/10.3390/s21051723

[22] A. Rejeb, K. Rejeb, A. Abdollahi, F. Al-Turjman, and H. Treiblmaier, "The interplay between the Internet of Things and agriculture: a bibliometric analysis and research agenda", *Internet of Things*, 2022;19:100580, ISSN 2542-6605, https://doi.org/10.1016/j.iot.2022.100580.

[23] C.-W. Hsu, J.-H. Lu, T.-C. Wang, J.-C. Huang, and M.-H. Shu, "Industrial Internet of Things on integrated preventive maintenance and enterprise-resource-planning systems: a case study of fastener forming manufacturing processes", *Proceedings of the Institution of Mechanical Engineers, Part B: Journal of Engineering Manufacture*. 2022;237(6–7):1100–1114. doi:10.1177/09544054221116703.

[24] J. Wang, and R. X. Gao, "Innovative smart scheduling and predictive maintenance techniques", In *Design and Operation of Production Networks for Mass Personalization in the Era of CloudTechnology*, Elsevier, 2022, pp. 181–207 (Chapter 7), ISBN 9780128236574, https://doi.org/10.1016/B978-0-12-823657-4.00007-5.

[25] M. Iqbal and R. Omar, "Automatic guided vehicle (AGV) design using an IoT-based RFID for location determination," In *2020 International Conference on Applied Science and Technology (iCAST)*, Padang, Indonesia, 2020, pp. 489–494, doi:10.1109/iCAST51016.2020.9557700.

[26] K. Xia, C. Sacco, M. Kirkpatrick, *et al.*, "A digital twin to train deep reinforcement learning agent for smart manufacturing plants: environment, interfaces and intelligence", *Journal of Manufacturing Systems*, 2021;58 (Part B):210–230, ISSN 0278-6125, https://doi.org/10.1016/j.jmsy.2020.06.012.

[27] A. Nassehi, R. Y. Zhong, X. Li, and B. I. Epureanu, "Review of machine learning technologies and artificial intelligence in modern manufacturing systems", In *Design and Operation of Production Networks for Mass Personalization in the Era of Cloud Technology*, Elsevier, 2022, pp. 317–348 (Chapter 11), ISBN 9780128236574, https://doi.org/10.1016/B978-0-12-823657-4.00002-6.

[28] S. Latif, M. Driss, W. Boulila, *et al.*, "Deep learning for the industrial Internet of Things (IIoT): a comprehensive survey of techniques, implementation frameworks, potential applications, and future directions", *Sensors* 2021;21(22):7518, https://doi.org/10.3390/s21227518.

[29] C. Pu, "Sybil attack in RPL-based Internet of Things: analysis and defenses," *IEEE Internet of Things Journal*, 2020;7(6):4937–4949, doi:10.1109/JIOT.2020.2971463.

Chapter 13

A systematic study on cognitive sensors in robotics, UAVs, and drones

Jay Sanghavi[1], *Abhi Vakil*[1], *Devshree Jadeja*[1] *and Manan Shah*[2]

Abstract

Cognitive sensor technologies are a vital basis for a robust, powerful economy in terms of digital transformation. Cooperative processing of complicated tasks comprising multiple agents, such as humans, robots, or both, can aid cognitive Internet of Things (IoT) control. Aerial robots (UAVs, RPVs, drones) and nano-satellites are fast gaining in popularity, and why accelerating in adoption is explained in the study. By aligning reserves to the scenario at hand, sensors as robots will enable the proper progressive application of remote sensing assets. Cognitive robots and drones with the ability to interpret the analog output from non-digitized industrial machinery employed in industry 4.0 are being discussed in the study. The study sheds some light on how concepts like aerial robotics, cognitive architecture, object-oriented design and cognitive deep learning algorithms, and autonomous drones use cognitive sensors. We have discussed how cognitive sensor technology enables us to cover the broad spectrum of healthcare, industry, and agriculture applications. We addressed the factors in this examination, which are the number one factors that may be advanced with using cognitive sensors when transforming traditional drones, robotics, and UAVs with the help of cognitive sensors.

Keywords: Drones; UAV; Cognitive

13.1 Introduction

Intelligent unmanned autonomous systems that can manage or finish activities without human involvement and use cutting-edge technology are built by humans.

[1]Department of Computer Engineering, School of Technology, Pandit Deendayal Energy University, India
[2]Department of Chemical Engineering, School of Energy Technology, Pandit Deendayal Energy University, India

These systems are unmanned and do not require human supervision. Because these apparatuses are unmanned, there is no requirement for human supervision. Since the dawn of recorded human history, unmanned systems have been developed in what appears to be an endless variety of combinations. At the same time that unmanned systems have become more technologically advanced, the human knowledge base has become more diverse. During World War I, the United States and Great Britain collaborated to develop the world's first airplane to take off and land without a human pilot. In March of 1917, the British Aerial Target, a miniature radio-controlled aircraft, participated in combat for the first time. This event took place. In October of 1918, the Kettering Bug, an airborne torpedo developed by the United States, completed its maiden voyage into the air. Although both appeared to do well in the flight tests, neither was actually in the conflict. In the early 1900s, drones were the focus of a significant amount of study, the primary objective of which was to provide military men with practice targets. It is generally agreed that A.M. Lows' "Aerial Target" project, which began in 1916 and was named after him, was the precursor to powered unmanned aerial vehicles (UAVs). Conventional controls have been replaced by microcontrollers, system-on-chip (SOC), and single-board computers due to developments in both the speed of UAV processors and computer technology. Without these advancements, the development of modern drones would not have been conceivable (SBC). Recent significant advances in artificial intelligence are directly connected to the development of unmanned autonomous systems. The Internet of Things (IoT) has rapidly become a pervasive technology that is implemented in a variety of contexts, such as in the provision of intelligent services and applications. This progression is inextricably tied to the progress that has been made in contemporary wireless communication technologies. This is a direct consequence of the IoT becoming a pervasive technology. The proliferation of disparate products that are more interconnected with one another on a wider scale has led to the accumulation of enormous volumes of data over time.

Because of advancements in online robust control algorithms, sensor technologies with affordable prices and form factors, and power-efficient, high-performance embedded computers, autonomous agents can now detect, plan, and react to their environments in a matter of milliseconds. This is made possible by recent technological advances. The utilization of drones in consumer activities as well as general aviation came about concurrently with the development of electrical power systems and cognitive computers. Throughout the past few years, the IoT has developed from a general concept into a fully-fledged corporate enterprise. This evolution has occurred at a rapid pace. Predictions made about the IoT in its early phases that were previously considered to be ludicrous are starting to feel more feasible as more evidence comes in to back them. We continue to see a need to incorporate technology from the IoT into applications that are crucial to the industry. This will allow us to save both money and time, as well as reduce the complexity of operational processes. Take, for example, the autonomous drone, which, at the touch of a button, can now collect data from structures such as cell towers. This not only saves time and resources but also eliminates the dangers that are associated with human inspection. When I was in my early 20s, the idea of drones being used to check buildings looked like something that came straight out of a science fiction movie.

Because AI is now capable of performing tasks that previously required the intelligence and labor of humans as well as imitating human behavior, the digital economy has experienced a significant transformation as a result of these capabilities. In order to generate codes, machine learning software requires huge volumes of data in a variety of formats, such as structured, unstructured, and semi-structured data. Researchers are putting a lot of effort into developing cognitive intelligence in AI-based robots and drones in order to cut down on the number of accidents that could be caused by misunderstandings. It is anticipated that cognitive robots will find applications in a diverse array of fields, such as artificial intelligence, robotics, cognitive science, neurology, and a great number of other fields. In order to accomplish this objective, cognitive intelligence will be merged with artificially intelligent robots and drones, which will be used in conjunction with one another. The development of cognitive intelligence has made it possible for businesspeople to access more intelligent talents, such as planning, highly developed motor coordination, attentional allocation, and perception. These are just some of the benefits that have resulted from this development. The incorporation of cognitive intelligence enables machines like robots and drones that employ artificial intelligence to become significantly more capable of independently taking in relevant data and coming to decisions. Systems that have been programmed with artificial intelligence are able to carry out cognitive tasks such as acquiring and storing memories, learning new information, interpreting data, deciding what to do, and solving problems. According to the forecasts provided by the international technology market, the application of these tools will increase the responsiveness, efficiency, worker happiness, and flexibility of the workplaces associated with Industry 4.0. Robots and drones that have been programmed with AI are able to perform improved forms of automated data processing, interpretation, and decision-making that incorporate cognitive intelligence. According to the forecasts provided by the international technology market, the application of these tools will result in an increase in the responsiveness, efficiency, worker happiness, and flexibility of the workplaces associated with Industry 4.0. If given the appropriate instruction, humanoid robots equipped with cognitive science and AI are capable of performing many of the tasks that have traditionally been carried out by people in a wide variety of industries. This is the case provided that the robots are given the appropriate amount of training. The most successful information technology firms in the world have an obligation to devise creative answers to the challenge of integrating cognitive intelligence into their products without jeopardizing existing computer systems or the employment of humans.

Unmanned aerial vehicles, also known as UAVs, are a type of military aircraft that are capable of flying on their own and employing a wide variety of technologies, such as sensors, target designators, offensive weapons, and other tools, to either destroy or otherwise interfere with their targets. These vehicles are also referred to as UAVs. Unmanned aerial vehicles (also known as UAVs) have an edge over their crewed counterparts in terms of both their range and their durability. The reason for this is that UAVs are not required to have a crew, to have life support equipment, or to meet any of the other safety regulations that are often included in crewed aircraft. As a result of this, UAVs are able to fly for longer periods than their human equivalents. After the end of World War II, the Israeli military forces were able to begin the development of UAVs by deconstructing modern aircraft and incorporating IR sensors and trainable televisions into their designs. This process took place after the war ended. When it came to determining goals and keeping track of progress, many people felt that it was a tremendously fruitful endeavor.

Saleem in addition to the other individuals (2015) stated that UAVs, often known as UAVs, were initially and largely utilized for military objectives; but, as technology evolved, new uses also began to develop. UAVs can now be employed for a variety of purposes. UAVs of the current generation are often employed in people's day-to-day lives for a variety of reasons, including their remarkable performance and incredibly low costs. Applications range from streaming videos in real-time and taking photographs to managing traffic, restoring damaged infrastructure, and monitoring environmental conditions. UAV, more commonly referred to as UAVs, are essential to the development of 5G networks because of their ability to improve the performance of the IoT and provide variety to 5G communications [1].

13.2 Brain processing and the IoT

The IoT, a technological revolution, is ushering in a new era of pervasive connectivity, processing, and communication (IoT). Over the past 10 years, efforts have been made globally to enhance IoT research and development from the academic community, service, superior organizations, network and service providers, etc. Technically, the connectivity, control, computation, and communication domains have received the bulk of emphasis. In actuality, these are really essential subjects. IoT, however, is all muscle and no brain, similar to an uncomfortable stegosaurus, as it lacks comprehensive cognitive faculties.

13.3 A view of cognitive censoring in robotics

In the technological field of cerebral robotics, robots that can learn from their mistakes, from human teachers, and even on their own are utilized, which helps them get better at interacting with their surroundings. The use of robots and other machines that can carry out a variety of tasks in our daily lives will only grow. Persistent cognitive robots still need to be adaptable enough to deal with ambiguous situations and changing environments in order to be really revolutionary. The goal of the group is to provide digital or robotic organisms higher-level cognitive capacities, such as the ability to think about goals, vision, actions, other agents' mental states, and teamwork. Cognitive modules enable the robot to interact socially with people in a way that is both complicated and simple and sensible. Their human-like appearance is extremely fascinating to children and those who are less used to direct engagement with robots. Most robotics specialists base their hypotheses about the development of robot mobility and intelligence on our understanding of how animals think. The major goal is to provide the robot the ability to act and react in ways that are representative of situations in real life. The objective of the cognitive robotics group is to provide robotic or digital animals with higher-order cognitive abilities. The ability to think about goals, a vision, activities, the thoughts of other agents, teamwork, etc. could be among these abilities. Because it is outfitted with cognitive modules, the robot can interact with people in social situations in a way that is intelligent, intelligible, and reasonable. Children and people with limited prior experience to robots are particularly drawn to the device's humanoid features.

The idea of a conscious machine is appealing, but creating one is still debatable and challenging. Here, we talk about how this issue pertains to our work on creating a humanoid cognitive robot that learns to do tasks via imitation learning. Our discussion is divided into three parts. In order to deepen our understanding of phenomenal awareness, we must first analyze the technique we used. The finding of computational correlates of the brain is the foundation of this approach. In the second section, we present a newly developed cognitive robotic system that imitates human demonstrations to learn new abilities. Instead of just copying the demonstrator's movements exactly, this humanoid robot uses cause-and-effect reasoning to determine its objectives. Its cognitive components, in particular, put a strong emphasis on top-down control of a working memory that houses the explanatory interpretations the robot develops throughout learning. In conclusion, we describe our continuing study into the cognitive system that our robot employs to learn imitation, including its low-level cognitive neuromotor components, working memory use, and causal reasoning procedures. In light of our first results, we propose that top-down cognitive control of selective attention, in particular its gating processes, is a critical possible computational component of consciousness in humanoid robots. In light of this, we conclude that constructing advanced neurocognitive control systems for cognitive robots and using them to look for computational correlates of consciousness is an important step toward improving our knowledge of consciousness. It also provides a possible and practical approach for developing a fantastically aware machine in the next few years.

13.4 Why should cognitive robotics get fussed with IoT?

Robotics is gaining wide acceptance to perform over their performance thresholds, whether it is when transmitting enormous amounts of sensor data over a shaky mobile network or when connections to robot controllers are prolonged. This is certainly relevant when there are problems with the mobile network. The IoT streamlines 5G-enabled obtaining solutions. To provide simple and approachable 5G for testing, do IoT Fieldlab's mobile installation will be employed. It consists of just a flight case with a 5G antenna and parallelization for connecting your robot applications to a lightning-fast, secure private network wherever you are. This installation will provide easy-to-use 5G for use during the testing process. You can test out 5G and ask for assistance from our business partners, including Statistical Analysis System (SAS), a worldwide leader in advanced analytics, to get a configuration that works properly. Each sensor required for mobile and wandering robots must be monitored cognitively in order for them to adequately carry out the required work. To handle low-level (touch) and low-level (distance) detectors, accordingly, from the standpoint of an embedded system, a cognitive architecture must have a high-level (site and auditory) and low-level (posture, payload, strain, and distance) detector system. It must be a pretty efficient and rapid system. It has been found that insects may respond to various sensory inputs by using quick reflexes that entail a precise sensory-motor connection [2].

13.5 Cognitive IoT and cognitive robotics as a plinth of smart manufacturing and sustainability

The field of robotics known as "social robotics" focuses on how robots and humans (and, to a lesser extent, robots and other robots) interact in real-world settings. By giving robots the capacity to comprehend, perform thoughtful thought, and respond to inquiries in challenging contexts, cognitive robotics seeks to imitate human intelligence.

As we have seen, social robots can come in a variety of shapes and sizes, from humanoid machines to creatures controlled by advanced AI. The design of the robotic hardware required for social interaction, the ability to gain understanding, and the ability to mimic human touch are just a few of the many topics that are the subject of studies in this field.

The purpose of seeking cognitive approaches to information gathered by sensing devices is what we refer to as the "cognitive Internet of Things." The IoT is dedicated to providing back-end services for ubiquitous sensing, monitoring, and mapping, while the robotic communities are specialists at delivering action, engagement, and self-directed mobility.

The two might be combined to make a much more useful Internet of Robotic Things. Even the simplest and oldest known animals have a diversity of sensors crammed into every cell. All of these behaviors contribute to the acquisition of time–space knowledge and the emergence of complex behaviors. Simple actions like movement can pave the way for more complex behaviors like cognition and the processing of perceptual input.

The robot is designed in a cloud environment with IoT capabilities, and it can gather information from various sources, including the human body and its near surroundings. As a result, the robot is better able to create the simulated environment required to execute the numerical model of robot control principles. The IoT has been using cognitive techniques to manage its decentralized structure in recent years. At RobotHouse, an experimental workplace created to resemble the office of the future, you may future-proof your career. Startups, SMEs, and chance events can create and test robotic solutions in an industrial context. This layout is meant to promote effective teamwork. A highly fast Wi-Fi backbone has been created for this environment as a result of the collaboration between RobotHouse and Do IoT Fieldlab. While working with our startup ecosystem, business partners, and reputable organizations like SAS, the undeniable global leader in analysis tools, you will get the opportunity to test out 5G technology. The RoboHouse community is dedicated to cutting your workload, improving your job satisfaction, and showcasing your company's promising future [3].

13.6 Contentment for the industry in which cognitive robots offer applications of 4.0 and sustainability

Technology advancements that enable autonomous operation for mobile robots are what are driving Industry 4.0. On the shop floor, this is causing a revolution. Robotics was used by loudspeaker and subwoofer manufacturer Visionary Electronics to enable the metal cleaning procedure. This made it possible for the firm to make goods of greater quality. In a nutshell, the industries that manufacture goods have the highest demand for the technology under discussion. The network of welding, dispensing, and painting robots are technologically capable of determining when maintenance has to be done in order to prevent failures. The goal is to improve the human–robot connection and make it feel more natural through the emulation of living beings' mental processes and the creation of a thorough grasp of navigation. The recurrence of these steps will enable the achievement of this goal. This special issue focuses on the fusion of social applications and cognitive robots as its main area of research and main area of study. Robotics are frequently used in the product creation, finishing, transport, and assembly stages in the automotive sector. [As an instance:] Robotics is being used more and more in the production of cars. (The process of creating, finishing, transferring, and assembling a product.) Robotics are utilized in the material handling industries where manufactured products are prepared for distribution to choose, sort, package, load, and unload products. The processes of selecting, sorting, loading, and unloading products are sped up in the aftermath.

In order to standardize them and lessen their harmful effects on the environment, products that have reached the end of their useful lives (EOL) must be disassembled. Despite efforts made throughout, manual disassembly continues to be carried out. Manual disassembly is still done because it is interesting to determine how much and what kind of end-of-life (EOL) items will be recovered. This chapter presents a conflict resolution approach with cognitive robotics as its guiding premise. Reasoning, execution monitoring, learning, and revising are four separate categories of cognitive processes that are included in the system. One of the cognitive processes is reasoning. Putting the suggested methods to the test involved using LCD panels. After the data has been collected, the system can be changed such that it can be customized to any product model without requiring prior product knowledge [4,5].

13.7 The advancement of mutualistic human–robot synergy

The growing interest is shown by a recent announcement in the new publication *International Journal of Humanoid Robotics* (*IJHR*). Since 1995, the Cognitive Robotics Laboratory at Vanderbilt University has been working on the creation of a humanoid robot that they have dubbed the Intelligent Soft-Arm Control (ISAC). A versatile humanoid robot that may work beside a person as a companion or assistance at home or in a factory, ISAC was originally created to help the physically impaired. For concurrent, distributed robot control, we have created a multi-agent architecture based on a different design philosophy and a reliable human–robot interface. Our group focuses a larger emphasis on the cognitive aspects of the humanoid than many other humanoid research organizations across the globe that pay more attention to human-like motion control and efficient walking pattern construction. According to the work given here, ISAC can acquire new abilities using memory systems that are currently being developed together with two agents, the self-agent and the human agent. The humanoid's internal representation of an individual is called the human agent. It contains details on the person's whereabouts, what they are doing, and how they are feeling, according to observations and interactions. The self-agent is the internal representation of the humanoid. It gives the system a sense of how it is doing in terms of its hardware, routines, and responsibilities. We discuss the Sensory EgoSphere (SES) and Procedural Memory (PM), two types of short- and long-term memories, respectively, while talking about robot memory systems. A geodesic hemisphere serves as the SES's interface with the outside world and serves as a database for the humanoid's short-term memory. It makes it easier for ISAC to follow a spatially indexed map of relative sensory input in its environment. Incorporating both fundamental and meta behaviors into a single data structure, PM serves as a basis for the acquisition of new skills [6].

13.8 An amateur drone surveillance system powered by the cognitive IoT

The cognitive challenges that can be completed in Dragnet can be accomplished through the utilization of a wide number of different methods. Figure 13.1 shows that the major basic enabling technologies for Dragnet are detection, localization, tracking, and control. These are also the technologies that are shown in Figure 13.1. These technologies are just some of the many others that are now available. In this section of the paper, we will present a synopsis of each of the significant enabling techniques, as well as a discussion of the potential technical obstacles and unresolved concerns that lie ahead.

Although it is common knowledge that UAVs used by the military have been used on battlefields for a considerable amount of time, interest in the numerous applications of commercial or civilian drones has only lately grown around the

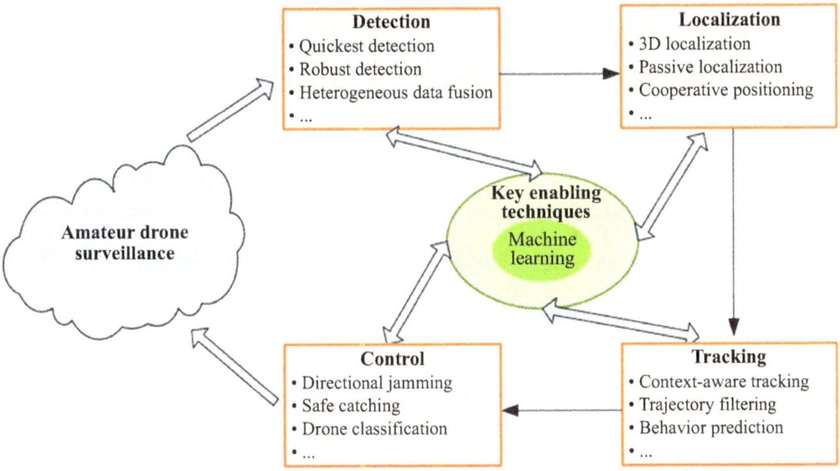

Figure 13.1 Critical empowering approaches for drone surveillance by amateurs

world. As a direct consequence of this, amateur drone surveillance is a relatively new subfield that is yet extensively unexplored. It is possible to derive useful information from military anti-drone technology.

13.9 Cognitive sensors in drone summary

Mini-UAVs, often referred to as drones, have received immense prominence as they have a diverse range of effects on transmission, videography, agribusiness, surveillance, and a multitude plethora of public functions [7]. However, many dangers to privacy, security, and safety are associated with using amateur drones. Amateur drone surveillance becomes a crucial, largely untapped area to address to meet these challenges. In this paper, we first give a succinct overview of the most recent research on drone surveillance. Then, we showcase Dragnet, a phrase used for enthusiast drone observation using the recently created cognitive IoT devices framework. Following that, we go into painstaking depth into the key Dragnet enabling techniques and the underlying hurdles and unresolved issues. An excellent simulation of the identification of unauthorized amateur drones was also offered. We are confident that this required field will yield lucrative research results, and we have only begun to scratch the surface. Table 13.1 shows framework, features, and benefits of revolution of drones.

Aftab *et al.* [8] mentioned that adopting a cognitive methodology for network management in an IoT environment driven by drones provides IoT users with various cutting-edge benefits. In this research, we propose a conscious cluster-based networking strategy for drone-based IoT to alleviate networking-related challenges. For drone-based cognitive IoT, we provide a self-organized clustering scheme (HSCS) that is a mix of glowworm swarm optimization (GSO) and the

Table 13.1 Framework, features, and benefits of revolution of drones

Framework	Features	Benefits and constraints
Anti-UAV defense system (AUDS)	Users may remotely distinguish and track small UAVs, classify them, and decide whether to obstruct their activities using a smart-sensor and effector program called AUDS	1. It features a software-defined intelligent directional RF inhibitor, an electro-optical video for tracking and categorization, and a digital radar for target tracking 2. It works around-the-clock in all kinds of weather 3. It has a passing time minimum of 0.01 ms^2 4. It has a 10 km sensing range
Autonomous drone defense with dedrone	Dedrone is a platform for airspace protection that recognizes, characterizes, and eliminates all drone threats	1. Continuous airspace surveillance 2. Automatic notification and alarm 3. Evaluation of drone DNA, RF and visual sensor data, and pattern recognition for drone categorization
Compact laser weapons system (CLWS) from Boeing Tracks and Deactivates UAVs	A target can be acquired, tracked, and identified with a laser weapon system called a CLWS, or it can even be destroyed	1. It is portable and can be put up in 15 min 2. It can destroy objects such as UAVs from 22 miles away and 20 pounds in 10 s with a 10 kW energy beam
The U.S. Army's Enhanced Area Protection and Survivability (EAPS) Technology	In addition to rocket, artillery, and mortar fire, the EAPS is a missile-based system that combats threats posed by drones, unmanned aircraft systems, and mortars	1. It employs a 50-mm cannon to fire system interceptors, a precise tracking radar interferometer as a sensor, a fire control desktop, and an RF transmitter and receiver to discharge the projectile into an engagement "basket" 2. The computations are completed on the surface, and the RF then sends the information up to the round

dragonfly method (DA). In addition, based on the findings of the behavioral research of DA, we have developed a system that allows for the effective monitoring of members of the cluster, so assuring the success of clustered management. Cluster maintenance, which improves the stability of the network, makes use of a process that identifies dead cluster members so that it may be carried out. The results of our proposed HSCS simulation are compared to the ones obtained by the existing hybrid bio-inspired clustering method BICSF. The results indicate that the proposed HSCS performs better than the BICSF in terms of the numerous assessment criteria used during performance evaluation.

Recent years have seen a rise in consumer curiosity about fully autonomous drones, which has resulted in the development of a fresh market segment for highly efficient drones [9]. However, in the professional drone business, the primary concerns that need to be addressed in order to achieve high levels of reliability, safety, and performance are the flight stack and the architecture that lies behind it. In this work, we analyze ArduCopter, a well-known open-source flying stack, to get a better understanding of the performance criteria used by the research community. We also study the effect that flight stack has on domain-specific applications. Our characterizations and benchmarks indicate that the flying range of the drone may be greatly increased if the underlying software for the flight controller is significantly improved. We want to utilize this research as the basis for developing a fundamental model for a multifunctional drone that is capable of updating its software and switching missions while it is in flight.

According to Dell'agnola (n.d.), using drones in search and rescue (SAR) missions may be highly emotionally exhausting. Large amounts of mental labor may impair human performance, which in turn raises the risk of mission compromise and failure, which can have disastrous consequences. Monitoring the cognitive load of rescue workers is essential in order to prevent them from making dangerous choices. The search and rescue drone strikes in this study contain three different levels of cognitive strain simulation. We provide a recursive feature elimination approach that makes use of the eXtreme Gradient Boosting (XGBoost) algorithm and the Shapley Additive explanations (SHAP) score. The objective of this method is to choose the features that are the most reflective of the whole. To do this, we investigate the usage of many machine learning techniques, including XGBoost, K-Nearest Neighbors, Logistic Regression, Gaussian Naive Bayes, and Support Vector Machine. According to the results of our research, an XGBoost equipped with 24 differentiating characteristics has the ability to properly assess two and three degrees of cognitive strain on an unseen test set gathered from 24 different volunteers, with an accuracy of 80.2% and 62.9%, respectively.

13.10 5G-cognitive drone system for electric grid preventive maintenance

Institute of Electrical and Electronics Engineers (n.d.) said that as part of the 5G-PPP NRG-5 project, a power grid Preventive Monitoring as a Service (PMaaS) program is being advanced, targeting the vertical domain of smart energy. The operation of this service's complex, bandwidth-intensive, computationally intensive, and time-sensitive application depends on swarms of semi-autonomous drones while still meeting all logistical, communication, and mission criteria. This research investigates 5G advancements inventing VNFs to enable low-latency 5G-enabled cognitive monitoring using drones, resulting in more effectively operating energy networks at the edge network, reducing maintenance costs, and enhancing the QoE provided by the utilities to the citizens. The automatic mode with a pre-selected flight path will be used for the initial tests. This will include the regular planned predictive maintenance tests together with the research gas plant's particular tasks (e.g. activity alarm, intruder detection, changes in the terrain, etc.). A second flight plan might govern the areas around each fundamental matter.

When producing panoramas from drone-based aerial photographs, the uneven shooting angle typically degrades the final image quality, according to SEO (2022). According to the findings of this research, a generative adversarial network (GAN). A deep convolutional network is meant to lower the wrongly estimated hypothesis in conjunction with a transform established by a specific baseline approach, such as the randomly sampled consensus method, in order to mitigate these polluting effects (RANSAC). The suggested discriminator of the GAN used in the proposed technique decides whether or not the target hypothesis survey determined by RANSAC is true. The correctness of the discriminator's inference is then confirmed by recalculating the generator to determine the correlation between the produced samples and the target sample. We put the recommended tactic to the test on a variety of aerial photographs obtained by both random sampling and drones. Even in challenging receiver-operated scenarios, the method that has been suggested has been shown to be stable and works quite well. In addition, the layout of the training samples may be improved, and the procedures must be used to photographs with various file types, resolutions, volumes, etc.

According to [10], "New aerial projects are enabled by the advances in aviation and transportation technology in the early twenty-first century," and "a variety of activities can use unmanned aircraft systems (UAS) that are generally integrated into the National Airspace System." (New aerial projects are enabled by the advances in aviation and transportation technology in the early twenty-first century) (NAS). The primary emphasis of this paper is an in-depth investigation of operations involving several UAS. In the beginning of the year 2020, NASA and Uber Technologies will begin collaborating on the design and development of operational ideas, roles and responsibilities, and ground control station (GCS) concepts for the purpose of enabling food delivery operations using a large number of incredibly small UAS (sUAS). A cognitive tour was used as the method for collecting the data. Under close observation, the subject matter experts (SMEs) presented data that outlined the predicted risks, workloads, and informational concerns associated with the contingency scenario.

For instance, while enhancing a future remote HITL pilot testing, the visual and educational elements of the TSD and Timeline displays were taken into account. Research is also undertaken on various HAT tools and technologies and guidelines for inter-operator vehicle transfers.

Santana *et al.* [11] concluded that technology is needed for crewless aerial vehicles (UAVs) to converse with bases, directors, equipment, computers, and even similar crewless aerial vehicles, in addition to the ability to fly autonomously. However, UAVs typically operate in unlicensed frequency bands where they face competition from other wireless networks and the growing number of mobile devices. By using software transmission rather than a transmission frequency set by the hardware, respect to efficient communication is made possible by cognitive radio (CR). CR makes strategic use of free transmissions and chooses which ones to use according to the needs of the application. It is projected that CR, which is regarded as an essential enabler, would also significantly promote the development of smart cities, 5G, the IoT, and the Internet of Future Things (IoFT). This study provides an overview of the existing testbed possibilities for real-world data trials in the field of computer-aided communication (CR), as well as the building blocks for an easy-to-use and low-cost testbed. Additionally, the study identifies the key components that are required for a testbed. It also sheds light on the considerable potential as well as the forthcoming challenges that the industry will face.

Gótzy *et al.* (n.d.) mentioned that the development of airborne surveillance drones became conceivable due to the increasing computational power of embedded systems. Robots working alone are, however, less effective than those creating a collective. Area surveillance systems are frequently deployed in many industries, including securing warehouses or other industrial, commercial, or military sites. Furthermore, these devices are deployed to frighten or catch intruders. These gadgets are employed in combination with human guards. Aerial drones that can find and locate attackers may replace these sensors and guards. In order to address the issue, we provide a distributed, communication-enhanced system that uses a sizable army of mobile, autonomous robots that patrol a region and intercept targets they discover. The robots may communicate with one another inside their limited environment by employing local sensing. We used a square grid to implement the multi-orbit-surrounding (MOS)-based aerial surveillance idea, demonstrating its effectiveness in promptly locating and surrounding an intruder in a little time frame.

Basak *et al.* concluded that, unfortunately, despite its several beneficial uses, drones are also being used for malicious operations like drug smuggling or to pose a threat to locations that require high levels of security along with nuclear power plants and airports. According to the existing localization and neutralization technology, the drone has already been located and classified. This research focuses on recognizing and classifying drones utilizing radio frequency (RF) technology and the broadcast signal's frequency pattern. We created a one-of-a-kind drone RF dataset with the help of commercial drones and made an effort to completely evaluate two different detection and classification frameworks: the two-stage and the combined frameworks. We demonstrate via thorough study that the You Only Look Once (YOLO) framework has superior detection performance compared to Goodness-of-Fit (GoF) spectrum sensing in a simultaneous multi-signal scenario, and classification performance equivalent to that of the Deep Residual Neural Network (DRNN) architecture.

According to Fadel *et al.* [12], some people think that CR sensor networks (CRSNs) may operate as a dependable, resilient, and effective communications infrastructure that can cope with the current and future energy management demands of an intelligent grid. Some of the current and future applications of CRSN-based smart grids include substation automation, overhead transmission line monitoring, home energy management, advanced metering infrastructure, wide-area situational awareness, demand response, outage management, distribution automation, and asset management. Other applications include wide-area situational awareness, demand response, and outage management. The purpose of this research is to address these applications by proposing honey bee mating optimization-based routing algorithms as well as cooperative channel assignment techniques. The proposed architecture not only significantly reduces the likelihood of data packet loss in the intricate smart grid spectrum settings but it also guarantees outstanding post quality across all sensor nodes. In terms of packet delivery ratio, latency, and power consumption, the validity of the model has been tested, and the results demonstrate that it has successfully fulfilled the QoS criteria of the majority of the SG applications that were mentioned.

Because of the unpredictability of natural disasters, Klaine *et al.* [13] came to the conclusion that there has to be a strong communication network structure in place before a catastrophe may occur. This is in addition to emergency response teams that have been established in advance. In this regard, the deployment of a network of drones (crewless aerial vehicles) is being researched as a possible solution for crises since it is intelligent, mobile, and adaptable. In light of the findings of this study, an intelligent solution based on reinforcement learning is proposed as a means of determining the locations at which a large number of drone small cells (DSCs) should be deployed in the event of an emergency. Although there are constraints on backhaul and radio access networks, the primary goal of the proposed technique is to expand the number of users that the system is capable of supporting. This is the case even if there are such limitations on drones. The Q-learning approach outperforms all other methods in all measures, according to the findings reported in current literature. Therefore, intelligent DSCs are a good solution for the purpose of making it possible for an emergency communication network to be created rapidly and efficiently.

Uhlmann and Schulte (2011) discuss various aspects of automation when integrating multiple discrete unmanned sensor platforms into a military helicopter situation. Artificial cognitive units implement some of the complex task executions aboard highly automated vehicles. The strategy used, artificial cognition, uses arguments relevant to the scenario at hand to support goal-directed behavior throughout task execution, instead of supporting task execution beyond a given predefined set of instructions. Tasks assigned by human operators are developed at a level of abstraction that can also be generically assigned to human subordinates within a mission. Much like human subordinates, UAVs use their psychological abilities to adapt their task execution to the current context of attention and data about teammate task assignments. Experiments have shown that artificial perception units can control multiple UAVs with a modest effort during military missions heavier than aircraft. Artificial noise mounted on UAVs has been shown to effectively support human operators' task of accelerating mission safety by providing reconnaissance information about flight paths and operational areas. Furthermore, given the level of control required over the UAV, control subjects said that they taught the UAV at a relative or global level, that is, a cognitive skill.

Almasoud and Kamal [14] consider cognitive crewless aerial vehicles (UAVs) while discussing data distribution to groups of IoT devices. When a suitable channel is available, cognitive UAVs share the wireless spectrum with primary users (PUs) whenever they enter the environment. The authors logically formulated this issue to optimize the fewest bits received by an IoT device. The issue is a diverse collection of nonlinear programs (MINLP). After approximately the cone problem, the authors suggested an iterative convex approximation-based algorithmic program to reveal this flaw. Karush–Kuhn–Tucker (KKT) points are where the method converges, as demonstrated by the authors. The authors often simulate the implications of the PU's OFF-amount model, maximum transmit slot type, and maximum tolerated interference threshold on the relatively few bits received from IoT devices.

Uhrmann and Schulte [15] evaluate the situation included in the operation through an unknown, perhaps hazardous terrain. This study shows how operational data may be fed into an artificial cognitive system to allow the commander of a human helicopter to control several UAVs. The finding showed that task-based guided crewless aerial vehicles boost overall mission effectiveness and offer tactical advantages. The outcome of the behavioral experiments demonstrates that task-based directing only requires a small portion of the operator's mental capacity, allowing him to continue performing his duties as the commander of a crewed helicopter. Additionally, it has been discovered that operators instinctively use the new elastic tightness of UAV control to balance the authority between humans and the UAV. The scans revealed that they reduced the pilot's burden and improved his attentiveness and alertness during the flight.

Studying a CR network in which an UAV is a secondary user is what [16] set out to do (SU). Based on the findings of this research, a low-cost energy management technique for improving the performance of UAVs is proposed. In the configuration that we provide, the UAV receives data transmissions over the BS while also picking up the periodic spectrum. The authors proposed an algorithmic program that would be supported by alternating improvement and classification methodologies in order to maximize the scientific energy of the UAV. This would be accomplished by optimizing the amount of time spent sensing as well as the amount of secondary transmission power. When used in combination with the PSO method, the algorithm and comparison between them were developed because it provides the optimal allocations for secondary transmission power and sensing time. Despite having a completely distinct set of fundamental presumptions, the suggested method outperforms the PSO algorithm and has a more regular look than the latter. When the optimal solution to a problem is discovered within the constraints of power and collision restrictions, as well as when spectrum scarcity is relieved, energy efficiency is typically enhanced. The process that is used to include UAVs into the CR network is one that is both practical and effective.

13.11 The system's simulated cooperative spectrum sensing has been optimized for UAV usage

Wu *et al.* [17] addressed an energy-efficient UAV-based twine CR system is being developed in order to overcome the issues raised by the UAV usage of the spectrum and the power constraint. The authors suggest utilizing an energy-efficient VCSS with the SZOF rule and a VCSS inside the periodic spectrum sensing frame structure in order to reduce the standard number of possibilities while maintaining detection performance. To fulfil this need, we have designed the VCSS optimization problem for the UAV-based interweaves CR system. On the basis of the precision and dependability of our theoretical calculations, the simulation results demonstrate that the UAV-based interweaving approach not only dramatically decreases the number of alternatives but also achieves the ideal detection performance and throughput for $K = N$.

Saleem *et al.* [18] focused on integrating UAVs with CR technologies. The main objectives were to discuss problems, opportunities, and research challenges in integrating UAVs and CR technology. The author of this chapter explored the issue of the lack of spectrum that UAV systems confront. The inherent properties of CR technology and crewless aerial vehicles require a solution to the aforementioned layer-wise integration problems. The authors also looked at the option to adopt CR technology for UAV systems.

Liu *et al.* [19] evaluate the expeditious knowledge of wireless sensor network (WSN), collecting using UAV support. Sensing data from ground sensors were gathered by UAVs flying overhead. Online air-ground cooperation offered a solution due to the unknowns of active sensor dispersion and the UAV's constrained coverage capabilities. Ground sensors created terrestrial data transfer clusters to increase transmission effectiveness. The author created a coalition formation game model for a group of ground sensors and looked into the stability of the coalition structure to support the transmission correlation and Lo's dependability. A data upload method for data collection was created to eliminate data transfer issues. Investigating the theme's flying speed enhanced the UAV's various security levels. The simulation findings demonstrated that UAV flight planning would increase the effectiveness of data uploading and that the effect of air-ground combined online optimization is substantially more significant than the UAV's unilateral data collection.

Hu *et al.* [20] investigated numerous spectrum sharing options between terrestrial wireless networks and UAV networks. Both spectral efficiency (SE) and energy efficiency (EE) optimization, two key UAV network issues, were examined by the authors. To protect the PU, a hybrid transmission strategy that concurrently supports UAV position and modifies spectrum detection and power allocation is provided. In this paper, an efficient iterative approach for obtaining system parameters in order to accomplish the trade-off between SE and EE is proposed (sensing time, sensing threshold, and UAV transmit power). When the position of the PT remains unchanged for a number of frames, a multi-frame mixed sensing theme is likely to be seen. SE–EE trade-off curves and computer simulations validate the suggested optimization technique. In computer science and engineering, the hybrid mode performs more effectively than the overlay and underlay alternatives. The performance of the EE will be improved by making use of the multi-frame linked continuous testing.

Liu *et al.* [21] describe a UAV-based approach for enhancing signal intensity in the LOS channel to improve spectrum sensing performance. The UAV captures and transmits periodic spectrum data while circling the PU. In addition to this, the properties of spectrum sensing performed by straight-line UAVs are investigated. In order to improve the UAVs turnout while simultaneously reducing the amount of interference that may get through, improved spectrum sensing has been supplied. In order to improve spectrum sensitivity and selectivity in deeply fading channels, the use of virtual CSS has been suggested. A method for joint optimization is presented as a means of determining the most appropriate response to the improvement problem. The simulations reveal that there is a compromise between sensing and throughput during a UAV-based CR and that the virtual CSS is superior in this regard. There is a sensor radian that is perfect in every way, and it increases effective throughput.

Daniel *et al.* [22] recommend that small-scale UAVs be powered by public wireless infrastructure. It has been demonstrated to be a workable solution with many merits, including high coverage, pre-existing infrastructures, and frequencies that may be paired with various failsafe and redundancy systems to handle latency and reliability difficulties. In this study, we demonstrated that communication awareness and autonomous UAVs might improve the rescue tactic at a fair price based on CBRN reconnaissance CONOPS.

Wan *et al.* [23] said cognitive automation employs an assortment of machine learning-based technologies, notably speech recognition and natural language processing. Due to Cognitive Process Automation (CPA) and AI, healthcare data management and compliance are proliferating at astonishing speeds. All industrial sectors begin this digital shift together. Based on the fundamental needs of jobs over network use, we classify the capabilities of blending the edge functions and the sophisticated core characteristics of a conventional healthcare robot. In order to understand the psychology and movement of actual, artificial, and animal beings, the approach relies on computer models.

Elevated collaboration between humans and robots is required due to the rapid growth of tailored services facilitated by cutting-edge AI technology. Perception has the potential to record passive inputs, which enhances data collection. This edge function should first understand the perceived data since that information will subsequently serve as the basis for cloud analysis. Data center-oriented connection, which also enables edge activities, is necessary for cloud-based core services.

Beetz *et al.* [24] said robotic competency is acceptable in certain circumstances. However, the necessity for considerable knowledge of the world to consider unpredictable conditions, constellations, and workloads makes real-world implementation critical. Human computation has demonstrated its ability to close this information asymmetry and gather data, but it is highly reliant on the quality of the supplied data. Human–robot interaction (HRI), at the crossroads of several current AI subdomains, presents a challenge for AI. A solver robot abstract machine is utilized to conceive, create, and use autonomous robots that have cognition that carries out typical movement patterns, which is presented in this study.

CRAM grants autonomous robots with simple decision-making inference in real-time instead of just mandating them to be pre-programmed. Autonomous robots with this kind of CRAM programming are much more generic, adaptable, and reliable than control systems lacking such cognitive abilities. According to CRAM, the comprehensive domain need not be explicitly defined in a database of abstract expertise. Instead, it incorporates the adoption of symbolic representations to generalize the findings into perception and act pathways, as well as the basic data structures of the control algorithms.

The paper by [25] has a core idea behind this study: to seriously consider scientific results, especially when considering the preconditions and causes of how and why cognition and intelligence arose in nature. We want to be clear that we do not think that constructing humanoid robots is a feasible aim and that we do not think that robots whose behavior mimics that of ants or other similar critters are a reasonable starting point. The hardest part may be finding and assembling a solid set of assumptions. The report highlights two potential strategies: Sewers as a natural environment for artificial intelligence and self-organizing behavior and sensors in robots. Dealing with a range of resource constraints and accumulating broadly applicable behavior patterns are necessary when using real robots to test AI theories or models. Several of its theories and methodological approach are diverse, despite being a part of the behavior-oriented AI paradigm, formalizing and automating knowledge, common sense, and sophisticated expertise across all spheres of human existence.

Buiu [26] found a revolutionary cognitive robotics lab course using the most recent developments in mobile robots, communication networks, software agents, and instructional tools. The merged education strategy uses simulation and trials on real robots, as well as in-class trials and remote control through the Web. At the same time, actual teachers and artificial chatbots supervise the students.

The outcomes of the evaluation demonstrate the validity and popularity of this tactic. Most remote robotics applications outlined thus far in the literature have relied on a primary robot control leveraging servo motor genera. The implementations covered in this study are concentrated on rigorous degrees in control relevant to cognitive robotics. This work presents several cognitive robotics experiments as part of a hybrid instructional approach supported by Distributed Laboratory (DLAB).

Various educational and technical views on the experiments have been provided. They establish a connection with main Braitenberg vehicles assessed through client–server simulation on both local and distant robots. In addition to supporting human users in doing research, interactive software agents are responsible for the implementation of a number of security elements. The laboratory has robot simulators that were built using a variety of programming languages. The learner may implement Braitenberg vehicles by making use of either MATLAB® or Visual Basic or Visual C. KIKS, which is a simulation environment for Khepera robots, is given in the same manner.

Lv and Qiao [5] brought up the point that the cognitive computing paradigm that is used for the control system does not mirror the safety device that is used by the collaborative robot. When compared to both the DBN and MLP algorithms, the error rate of the DBNLP algorithm model is much lower. When there are a substantial number of less postings, resource distribution is both simpler and faster.

The feasibility of an interactive robot control system built on the foundation of cognitive computing is investigated in this research. *Methods*: This study combines cognitive computing, deep belief network algorithms, and collaborative robots in order to develop a model for a cognition computing system that illustrates deep belief networks and is extended to the control loop of industrial automation. Specifically, the goal of this research is to accomplish this by creating a model that can be applied to industrial automation. In addition, simulation is used in this investigation to analyze, contrast, and evaluate the performance of deep belief networks (DBNs) and multilayer perceptrons (MLPs). Methods were provided by the cognitive computing system model of deep belief network and linear perceptron (DBNLP).

Fang *et al.* [27] discovered that today's commercial robots perform effectively in relatively static environments with set landmarks, meaning that environmental variations are minor. Human movements have little impact on them. Robots are becoming increasingly vital for inventing new applications as the IoT expands rapidly. Extensive and cognitive perception skills are needed to identify humanity in a challenging situation. To accomplish this, a radar-integrated CMOS-based robot sensing platform is provided. Geolocation and human identity are surprisingly accurate.

The SLAM action potentials procedure incorporates localization, state, estimated speed, and anticipated route in a customizable manner while synthesizing coordinates using an amplified Kalman filter mechanism. Since such a robotic work environment is multifaceted and time-sensitive, multidisciplinary sensing, and real-time computing are crucial for ensuring seamless interactions with high levels of safety and efficiency.

Diversified sensing and real-time computation are necessary for facilitating interactions with high levels of safety and efficacy in a robotic work environment that is both complicated and time-sensitive.

The phase processing approach's accuracy in subject localization and vital sign recognition was shown when it was used for the identification of real persons based on the characteristics of their vital signs.

In addition, the robotic radar sensing platform demonstrates its ability to differentiate between people in different lighting and climate conditions, laying the groundwork for the development of adaptable robots in the field of supervision.

IoT and AI, according to Hu *et al.* [20], have been crucial in driving modernization in intelligent manufacturing, supporting economic growth, and boosting people's quality of life. Edge computing can assist an intelligent factory in increasing its computing power, network bandwidth, and memory space to the edge of the IoT. This is a response to resource allocation as well as data downlink and uplink processing across production and production workflows. Edge computing can also help an intelligent factory connect to the IoT. The Affective Interaction Intelligence Robot (iRobot), which is powered by AI and the IoT cloud platform, may be better able to handle the customer's emotional concerns. As a direct result of this, this has become a well-liked topic for study in the field of agile manufacturing. In the field of intelligent manufacturing, one of the most important study topics is the iRobot-Factory, which is an intelligent robot factory. This chapter describes a highly networked and deeply integrated intelligent production line and provides an overview of the overall composition, design, and advantages of such a factory. In order to boost user engagement and emotion detection, it discusses various issues that still need to be overcome with load balancing, customized robots, and cloud-end fusion. The results of the experiments showed that by implementing our method, chip assembly and operating efficiency may be significantly boosted while simultaneously lowering the amount of system instructions.

Umar and Mosisa [28] discovered the autonomous Internet of Mobile Things (IoMT), the autonomous IoT, and the Internet of Everything (IoE) are three facets that are developing as part of the IoT.

This chapter's goal is to present a high-level overview of the IoRT's guiding principles, technology, possible pitfalls, future innovations, and current applications. The intensification of cross-cultural tensions is resulting in the emergence of new cross-regional issues. IoT networks will soon be joined by collaborative robots, which will help them adapt to change and rapidly expand their service offerings. Another forward-thinking and cutting-edge notion in robotics is the use of sensors and several other network technologies. Robots that previously had limited communication abilities are gaining the ability to be more personable. One of the following advancements that is predicted for IoT applications in different intelligent devices is the blending of IoT, cognitive computing, and AI technologies.

A vital ingredient of the potential implementation of the IoT and AI is placing robotics in a prominent position in the company's service. These traits frequently make it possible for IoT to be applied across a variety of domains, leading to benefits such as improved cognitive capabilities that expand the application of machine learning, improved device dependability, responsiveness, and reliability, a slight decrease in consumer threats, and assistance with human and machine–robot interaction.

Arena and Patanè [29] said that cognitive robotics research aims for a robot to design new, efficient behavior patterns that are specifically fitted to the features of the natural world by leveraging its fundamental knowledge, which can be built beforehand or learned. This robot should be able to modify how it uses sensors to get information from the surrounding environment to complete a task. These skills are innate to all organic things, but implementing them effectively inside a robotic system is a considerable barrier.

The finding suggests how sensory events cause cognitive schemes to form in nonlinear spatial–temporal brain lattices. To execute the task correctly, a cognitive system is intended to control all the sensors required for walking and roaming robots. In terms of hardware implementation, a rapid and successful multi-sensory system that can communicate with both high- and low-level sensors should be able to differentiate a cognitive framework from others.

Rusu *et al.* [30] found that ubiquitous computing, sensors, effectors, and computing hardware are disseminated among daily things and covertly incorporated. Since objects are labeled with RFID tags and the boards include tag readers, we have cabinets inside them in sensor-equipped surroundings. The following describes a typical setting. A service robot connects to ubiquitous computing, sensing, and actuation infrastructure and blends itself into it.

After connecting, the robot perceives what is inside the cabinet in the same fashion it perceives in its hand: by receiving and interpreting pertinent sensor data while not physically connected to the sensor. Regarding the prospects of service robotics, service robots will not be independent, solitary entities but skilled, adaptable agents in operating settings with sensors and effectors. The goal of the robotics community is to make it possible for robots to carry out all sense, thought, and action selection just like people and animals do.

Amato *et al.* [2] found that robotic ecologies are networks of various robotic devices that work together to execute complex tasks and are well-integrated into everyday surroundings. Even if they have the potential to become more frequent, one of the most challenging challenges is understanding how to diminish the amount of planning, pre-programming, and human oversight required in practical implementations by enabling them both unsupervised and adaptively. The RUBICON project is developing techniques for learning that will result in more efficient, versatile, and cost-effective robotic ecosystem interaction. In contrast to self-organizing that is impacted by inputs from its activities and the user, RUBICON gradually and progressively learns how to use the available sensors, actuators, and other functional components. Our approach incorporates a broad range of techniques from fields as diverse as cognitive robotics, machine learning, planning, agent-based control, and wireless sensor networks.

13.12 Outline of the assimilation of cognitive radio with UAVs

The IoT emerged as a result of the increasing networkability of consumer gadgets. The popular practise of connecting computer equipment, monitors, and many other "things" (sometimes referred to as "appliances") to the Internet in order to transmit data and increase capabilities and efficiency is referred to as the "Internet of Things" (IoT).

IoFT is plagued with problems relating to the law, software, and hardware. Users might be put in danger by weaknesses in the integrated security of the UAV. Every layer of IoFT—from the hardware to the networks to the sessions and applications themselves—is vulnerable to attack. UAVs are need to be linked as the IoT continues to develop. On the other hand, flying ad hoc networks and these aircraft were not the same, which led to the development of the Internet of Flying Things (FANETS). The Internet of Future Things (IoFT) is a relatively young academic area, despite the fact that its applications are widespread. Due to the unpredictability of the wireless medium and the great mobility of UAVs, this topic is generally seen as very challenging. As a direct consequence of this, the functionality of network protocols designed for use in networks that also have a ground component may not be adequate for use with UAVs. It is essential that the high mobility of UAVs be taken into consideration throughout the process of creating transport procedures for CR-based UAVs.

The experiment we ran in this paper comprised attacking the prototype CR-based UAV with an outside jamming assault. To counteract the impacts of the jamming assault, the CR should identify the jammed frequency range as being occupied and divert to a standby frequency. CR-based UAV hardware and software. We also examined CR for UAV communications and the first steps to integrate it with UAVs. This work also presented issues like how CR impacts UAV energy use and how UAV mobility affects CR spectrum mobility and spectrum sharing. We also described a cheap CR-based UAV testbed. UAV CR algorithm study uses machine learning to predict PU arrivals using UAV data.

13.13 Scope and future work

Although we have only begun to scratch the surface of this incredibly important topic, I am confident that further research will yield discoveries that will be extremely useful. It is projected that as these disciplines continue to make technological advancements, regulating algorithms will grow more challenging. In order to ensure that judgments are made quickly and accurately, it will be important to use both human cognition and digital models. Numerous studies have demonstrated that testing a novel cognitive drone in a real-world setting where previously recommended preventative maintenance checks are still necessary presents one of the biggest obstacles in the design process. One of the main challenges in the design process is this. Thinking about how the drone fits into the economy, security, and basic needs of humans is another difficult topic. It is crucial that the training techniques are enhanced in order to take use of both the neural network performance and ad hoc topographical deformation procedures while attempting to infer acknowledgements. Many researchers are having difficulty getting beyond this important obstacle. It is essential to finish this phase of picture processing to get the greatest outcomes. Up until now, theoretical research and management of issues like synchronization among various cognitive swarms of drones have been the only approach. This still holds true even if study is being done to determine whether or not it can be used in events that occur in real life. You might encounter similar difficulties when using RF technologies to describe and identify drones. The performance of the classification process is liable to alter when there are unidentified or more recent drone signals present, as was extensively detailed in the discussion of the limits in the preceding section. The main goal of the work that will be done in the near future is to establish a dynamic, cross-layer transmission architecture for WSN-based smart grid applications. The future holds a lot of potential thanks to the oncoming wave of developments in this field. It will be able to create algorithms that are quicker, more accurate, and more efficient as a result of improvements in processing power and the widespread application of machine learning, artificial intelligence, and deep learning. This will enable the resolution of several now intractable issues. This might open the door to more investigation into cognitive UAVs. The social and behavioral effects of the widespread usage of robots and drones should receive major financing, similar to how the medical industry devotes significant amounts to research. However, when it comes to image processing, the training samples' organization might also be enhanced, and we need to employ methods with images that have a range of formats, resolutions, sizes, and other characteristics. Additionally, we must employ strategies with photographs that feature a variety of subjects. The ability to plan a flight path for UAVs that is based on the amount of spectrum that is being used allows for the resolution of one of the most intriguing problems in unmanned aerial path planning. Channel occupation mapping can be used to achieve this. This approach is one choice for resolving the current problem. The method of forecasting PU arrivals using factual data received by UAVs in conjunction with machine learning approaches is included in the investigation into CR algorithms for UAVs. This is a crucial aspect of the inquiry. Additionally, it is imperative to have CR hardware and software that complies with the rules and laws that govern UAVs. While studies on unrestricted drone use may be able to address the issue of inadequate real-time analysis, our goal is to develop a strong system that can recognize and categorize all drone emissions. This is true even though the challenge brought on by the absence of real-time analysis can be overcome.

The deployment of emergency cellular networks using drone reinforcement learning will give rise to a brand-new area of the study known as 3D placement. In the event of an emergency, this will consequently result in improved problem-solving skills inside DSCs. It can be used to obtain an even higher degree of precision thanks to the large range of add-ons that are available. One of these improvements could involve the placement of drones using additional machine learning algorithms, such as SARSA. Additionally, value function approximation, which combines supervised learning with reinforcement learning, can be used to derive a reasonable estimate of the value associated with each state-action pair. The machine learning algorithm's performance can be enhanced using this estimate. The algorithm would then be able to generalize, evaluate the worth of each state-action combination in light of the system's features, choose the best course of action for as-yet-undiscovered states, and prevent the environment from becoming discretized. The upcoming optimization study will be fascinating because it will look at each user's connections as well as how satisfied they are with other metrics like robustness and latency. This gives one the impression that the study will be both intriguing and strange. In conclusion, it has been suggested that additional restrictions may be placed on DSCs in the distant future. Both the period of time the DSC could stay in the air and the amount of electricity it could use to propel itself would have a limit. As a result, it would be possible for the software to draw in the maximum number of users while also enabling: to increase the number of users and flight duration for each DSC, the ideal balance between mobility and coverage must be found.

It is becoming more and more important to have robots that can interact with people in public while still keeping themselves safe. These robots' success depends on their capacity to anticipate not only how people will respond to the tasks they perform but also how other robots will respond to those tasks. It is therefore important to merge the two research areas of "control architectures that purposefully take into account the necessity for experience acquisition and usage" and "physical systems developed expressly to interact with an endless cosmos." In both of these domains, building physical systems with the express intent of interacting with an infinite cosmos is the focus. Because they can sense the world in which they operate, concentrate on its most important features, make plans, predict the outcomes of their own actions as well as the actions of other agents, and learn from past experiences, cognitive robots are able to achieve their goals. In order to deal with the inherent ambiguity of their natural surroundings, they do this by continuously learning, researching, and exchanging knowledge with one another.

The creation of self-sufficient robots currently makes substantial use of principles and methods drawn from cognitive research. Cognitive science can be incorporated into the study of robotics in a variety of ways. The creation and management of human–robot interaction as well as the application of computational cognitive models as the robots' reasoning apparatus are a few of these methods. The goal of this conference was to bring together experts in the domains of cognitive science, robotics, and human–machine interfaces in order to debate and research this emerging subject. The bulk of the time, cognitively realistic (also known as "person-in-the-loop") situations cannot be handled by logically strong C++ systems. In this context, embodied systems that are more complex than simple software agents are referred to as "robotics." Autonomous driving cars and mobile robots are two examples of such systems [27].

13.14 Conclusion

The IoT has become actual from the conceptual era. Autonomous robots, drones, and cameras may now examine machinery, perform preventive maintenance, and give the operations crew analogue data. AI is currently being used to support many sentient solutions, including healthcare, autonomous transportation, and other areas. As a next-generation application based on AI, cognitive computing offers individualized interactions and services that mimic human behavior for human contact. On the other hand, a lot of data is produced by applications for smart cities like healthcare, intelligent transportation, the retail sector, and firefighting. How to effectively manage the massive amount of data is a constant worry. The fundamental focus of Industry 4.0 and the ultra-smart society is adaptive anticipatory interactions, and cognitive robotics is a sort of embodied cognition that make use of the physical morphology, kinematics, and dynamics of the robot as well as its surroundings to achieve this. We revealed just how complex (slower rate) devices may be utilized to teach more advanced (better level) ones with data that links "meanings" to visual information employing bio-inspired computational methods that are transmitted by sensory events. UAVs may communicate more effectively in general and perform activities that are difficult for conventional UAVs when partnered with CR. CR offers intelligent wireless communication by employing software transmission rather than a transmission frequency set by the hardware. Through the use of artificial and computational intelligence technologies, cognitive robots offer a novel way of implementing and mimicking natural intelligence. With the aid of these technologies, cognitive robots that can reason, learn, and seamlessly connect to cognitive systems can be made.

References

[1] Dias Santana, G. M., de Cristo, R. S., and Lucas Jaquie Castelo Branco, K. R. (2021). Integrating cognitive radio with unmanned aerial vehicles: an overview. *Sensors*, 21(3), 830.

[2] Amato, G., Bacciu, D., Broxvall, M., *et al.* (2015). Robotic ubiquitous cognitive ecology for smart homes. *Journal of Intelligent & Robotic Systems*, 80(1), 57–81.

[3] Vongbunyong, S., Kara, S., and Pagnucco, M. (2013). Application of cognitive robotics in disassembly of products. *CIRP Annals*, 62(1), 31–34.

[4] Javaid, M., Haleem, A., Singh, R. P., and Suman, R. (2021). Substantial capabilities of robotics in enhancing industry 4.0 implementation. *Cognitive Robotics*, 1, 58–75.

[5] Lv, Z. and Qiao, L. (2020). Deep belief network and linear perceptron based cognitive computing for collaborative robots. *Applied Soft Computing*, 92, 106300.

[6] Xu, Y., Qian, H., and Wu, X. (2014). *Household Service Robotics*. London: Academic Press.

[7] Ding, G., Wu, Q., Zhang, L., Lin, Y., Tsiftsis, T. A., and Yao, Y.-D. (2017). *An Amateur Drone Surveillance System Based on Cognitive Internet of Things*. http://arxiv.org/abs/1711.10738

[8] Aftab, F., Khan, A., and Zhang, Z. (2019). Hybrid self-organized clustering scheme for drone based cognitive Internet of Things. *IEEE Access*, 7, 56217–56227. https://doi.org/10.1109/ACCESS.2019.2913912

[9] Jijina, S., Amyette, A., Shoghi, N., Hadidi, R., and Kim, H. (2020). Understanding the software and hardware stacks of a general-purpose cognitive drone. In *Proceedings – 2020 IEEE International Symposium on Performance Analysis of Systems and Software, ISPASS 2020*, pp. 212–214. https://doi.org/10.1109/ISPASS48437.2020.00036

[10] Smith, C. L., Sadler, G. G., Tyson, T. L., *et al.* (2021). A cognitive walk-through of multiple drone delivery operations. In *AIAA Aviation and Aeronautics Forum and Exposition, AIAA AVIATION Forum 2021*. https://doi.org/10.2514/6.2021-2330

[11] Santana, G. M. D., de Cristo, R. S., and Branco, K. R. L. J. C. (2021). Integrating cognitive radio with unmanned aerial vehicles: an overview. In *Sensors (Switzerland)* (Vol. 21, Issue 3, pp. 1–27). Basel: MDPI AG. https://doi.org/10.3390/s21030830

[12] Fadel, E., Faheem, M., Gungor, V. C., *et al.* (2017). Spectrum-aware bio-inspired routing in cognitive radio sensor networks for smart grid applications. *Computer Communications*, 101, 106–120. https://doi.org/10.1016/j.comcom.2016.12.020

[13] Klaine, P. v., Nadas, J. P. B., Souza, R. D., and Imran, M. A. (2018). Distributed drone base station positioning for emergency cellular networks

using reinforcement learning. *Cognitive Computation*, 10(5), 790–804. https://doi.org/10.1007/s12559-018-9559-8

[14] Almasoud, A. M. and Kamal, A. E. (2019). Data dissemination in IoT using a cognitive UAV. *IEEE Transactions on Cognitive Communications and Networking*, 5(4), 849–862.

[15] Uhrmann, J. and Schulte, A. (2012). Concept, design and evaluation of cognitive task-based UAV guidance. *Journal of Advanced Intelligent Systems*, 5(1).

[16] Pan, Y., Da, X., Hu, H., Zhu, Z., Xu, R., and Ni, L. (2019). Energy-efficiency optimization of UAV-based cognitive radio system. *IEEE Access*, 7, 155381–155391.

[17] Wu, J., Chen, Y., Li, P., *et al.* (2021). Optimisation of virtual cooperative spectrum sensing for UAV-based interweave cognitive radio system. *IET Communications*, 15(10), 1368–1379.

[18] Saleem, Y., Rehmani, M. H., and Zeadally, S. (2015b). Integration of cognitive radio technology with unmanned aerial vehicles: issues, opportunities, and future research challenges. *Journal of Network and Computer Applications*, 50, 15–31.

[19] Liu, D., Xu, Y., Xu, Y., *et al.* (2020). Opportunistic data collection in cognitive wireless sensor networks: air–ground collaborative online planning. *IEEE Internet of Things Journal*, 7(9), 8837–8851.

[20] Hu, L., Miao, Y., Wu, G., Hassan, M. M., and Humar, I. (2019). iRobot-Factory: an intelligent robot factory based on cognitive manufacturing and edge computing. *Future Generation Computer Systems*, 90, 569–577.

[21] Liu, X., Guan, M., Zhang, X., and Ding, H. (2018). Spectrum sensing optimization in an UAV-based cognitive radio. *IEEE Access*, 6, 44002–44009.

[22] Daniel, K., Rohde, S., and Wietfeld, C. (2010). Leveraging public wireless communication infrastructures for UAV-based sensor networks. In *2010 IEEE International Conference on Technologies for Homeland Security (HST)*, pp. 179–184.

[23] Wan, S., Gu, Z., and Ni, Q. (2020). Cognitive computing and wireless communications on the edge for healthcare service robots. *Computer Communications*, 149, 99–106.

[24] Beetz, M., Mösenlechner, L., and Tenorth, M. (2010). CRAM—a cognitive robot abstract machine for everyday manipulation in human environments. In *2010 IEEE/RSJ International Conference on Intelligent Robots and Systems*, pp. 1012–1017.

[25] Christaller, T. (1999). Cognitive robotics: a new approach to artificial intelligence. *Artificial Life and Robotics*, 3(4), 221–224.

[26] Buiu, C. (2008). Hybrid educational strategy for a laboratory course on cognitive robotics. *IEEE Transactions on Education*, 51(1), 100–107.

[27] Fang, Z., Lou, L., Tang, K., *et al.* (2021). A CMOS-integrated radar-assisted cognitive sensing platform for seamless human-robot interactions. In *2021 IEEE International Symposium on Circuits and Systems (ISCAS)*, pp. 1–4.

[28] Umar, S. and Mosisa, E. K. (2021). A survey on evolution of cognitive robotics with Internet of Things. *International Journal of Scientific Research in Science, Engineering and Technology*, 8(2), 337–344.
[29] Arena, P. and Patanè, L. (2009). Simple sensors provide inputs for cognitive robots. *IEEE Instrumentation & Measurement Magazine*, 12(3), 13–20.
[30] Rusu, R. B., Maldonado, A., Beetz, M., and Gerkey, B. (2007). Extending Player/Stage/Gazebo towards cognitive robots acting in ubiquitous sensor-equipped environments. In *ICRA Workshop for Networked Robot Systems*.

Chapter 14

Sensor data fusion and processing in smart agriculture: crop quality assessment, crop damage, smart planning

Ghulam Mustafa[1,2#], Qurban Ali[3#], Hengbiao Zheng[2], Meng Zhou[2], Tao Cheng[2], Yan Zhu[2], Xia Yao[2], Yuhong Liu[1] and Sarfraz Hussain[1]

Abstract

Agricultural lands are generally vast, remote, and sensitive to weather, making data collection difficult. Despite these limitations, additional data are being captured as technologies evolve and prices reduce. Proximal, aerial, satellite, and ancillary data are obtained to study the crop for various objectives of smart agriculture. Geographically unequal areas lack vital data to address their challenges. Even in regions with adequate infrastructure and resources, data being obtained for agricultural environments produce knowledge gaps. Many aspects must be considered and measured for a region's whole description or examination of the targeted problem. Data from a single type of sensor is not very effective for algorithms, and results are imprecise. However, fusing sensor data is one of the key approaches for crops' quality assessment, damage, and smart planning for their management. The data fusion explores complementarities and synergies of extracting trustworthy and relevant information from data. Exploring the correspondence and interactions between various types of data is the core concept behind data fusion, which aims to derive more trustworthy and applicable information regarding the topics being investigated. Despite some success, several obstacles still limit a more general adoption of this type of strategy. This is especially true for the highly complicated

[1]Key Laboratory of Integrated Regulation and Resource Development on Shallow Lakes of Ministry of Education, College of Environment, Hohai University, China
[2]National Engineering and Technology Center for Information Agriculture, Key Laboratory for Crop System Analysis and Decision Making, Ministry of Agriculture, Jiangsu Key Laboratory for Information Agriculture, Jiangsu Collaborative Innovation Center for Modern Crop Production, Nanjing Agricultural University, China
[3]Key Laboratory of Integrated Management of Crop Diseases and Pests, Department of Plant Pathology, College of Plant Protection, Nanjing Agricultural University, Ministry of Education, China
[#]These authors contributed equally to this chapter.

ecosystems that can be found in rural and agricultural settings. In this part of book, we provide a comprehensive review on data fusion to agricultural problems and study of plant disease using different sensor data for smart agriculture. Conclusively, the study contributes on emphasizing the data acquisition scales, data types, and their fusion for decision making in smart agriculture.

Keywords: Data scale; Data fusion; Data processing; Sensors

14.1 Introduction

The quantity (and caliber) of sensors being used to gather data in various contexts has been continually increasing. Today, a wide range of technology is being used to "sense" complicated environments, such as agricultural fields, producing enormous amounts of data that is examined to provide meaningful information about the area or plants being monitored. More research has been conducted to uncover the information in the remotely sensed data [1,2,3]. However, it is commonly difficult to adapt the improvements made in experiments for the conditions being practiced. There are primarily two causes for this. First, the data utilized in these trials typically do not cover all the variables related to the issue at hand. The investigations presented in scientific books are usually limited in scope. The findings in publications may appear promising, they frequently do not include information regarding how well the suggested strategy works in actual or controlled situations. Second, the accepted sensing technology might not be able to gather enough data to resolve and provide enough information, even if the data fully represent the varied conditions present in practice. The diverse conditions can exhibit identical visual signals, even sophisticated artificial intelligence algorithms fed with RGB digital photos sometimes fail to identify plant diseases from their symptoms [4].

Using data fusion (DF) techniques can help to reduce gaps caused by data limitations. "Data fusion" terminology describes combining data from more than one source of information so that more accurate, consistent, and concise information can be produced rather than obtained from any individual data source [5]. This type of approach has increased since the mid-1990s [6], largely due to its increased use in agricultural experiments. Arguably, DF techniques may face their greatest challenge when maximizing the synergies and complementarities between different types of data and data sources. Resultantly, the variety of data sources, approaches, and DF process cannot be formalized that is remedy to all agricultural applications. Despite falling into a slightly different contexts, Bleiholder and Naumann [7] applied a three-step process to most applications for formalization of DF. Importantly, the first step is to identify the most relevant sources of information or attributes. In the second step, identifying a correspondence between data sources can be quite easy if they are similar, but it can be challenging if the data sources are different. This issue is addressed in the following paragraph by describing three types of DF. Image data sources, for example, must be aligned properly, as misalignments can cause inconsistent representations that are unreliable. The third step involves applying the actual DF, once the data have been properly identified and consistent. However, it is often neglected to deal with existing data inconsistencies [7]. By extracting and exploring the metadata associated with the data being fused, auxiliary tools can reduce inconsistencies and (at least partly) alleviate this situation [8].

Three groups of DF techniques are most commonly divided [9]: (a) raw data (also called "low-level" or "early integration"), combining raw and preprocessed data into a matrix, using data that is of the same type and is normalized properly to produce a single matrix. (b) This method involves extracting features from a variety of data types and concatenating them into a matrix, which is commonly called "mid-level" integration or "intermediate integration," and is typically applied when pieces of information can be treated in a way that generates complementary and compatible features. (c) It is more appropriate to apply classification and regression algorithms separately to each type of data at a decision level (also referred to as a high level or a late integration), when data sources are too distinct to combine earlier, and then combine the outputs of each model. Based on Ouhami *et al.* [10], there are three types of DF methods: knowledge-based, evidence-based, and probability-based. Despite being useful in different contexts, the first classification appears to be more applicable to this project (Figure 14.1). Various scales can be used for collecting agricultural data, including proximal, aerial, and orbital (satellites). Where proximal scale represents the direct organ (leaf, spike, stem, flower etc.) testing, aerial scale highlights the aerial examination (drones, fixed-wing aircraft, helicopters, balloons, blimps and dirigibles, rockets, pigeons, and kites) and orbital scale refers the all kinds of satellites. Figure 14.1 illustrates these scales.

In addition to navigation systems for autonomous vehicles [11–13], plant disease detections [14–16], fruit detections [17–19], the definition of heterogeneous management zones [20–22], plant phenotyping [23], and soil analysis [24–26]. The proximal data can also be used in a variety of other applications. It is mainly used for the detection of certain objects and for the estimation of agricultural variables

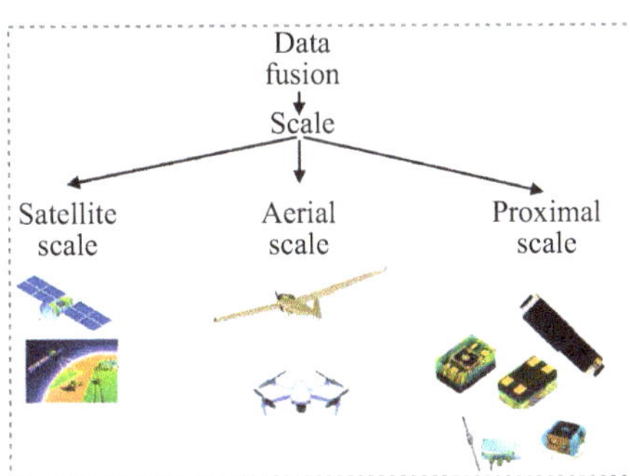

Figure 14.1 Data scales for agricultural observations. Figures were created with Biorender (https://biorender.com/).

(such as soil moisture and nitrogen content) when aerial data are collected with unmanned aerial vehicles (UAVs) (UAVs are used mainly for the detection of certain plant species and fruits for aerial view data collection) [27,28]. In addition to mapping soil moisture, satellite data are used to determine crop type [29,30], evapotranspiration [31–34], biomass [35,36], phenological states [37,38], and nitrogen status [39,40]. There are several studies that use data that originate from different scales instead of using the same scale as in most DF approaches [10,36,41,42].

14.1.1 Explain sensors

With sensors, agricultural control and supervisory systems can be developed, and management practices can be evaluated. Sensors also provide data that can be used to develop agricultural management systems. Furthermore, sensors are involved in the development of algorithms to automate the prescription of inputs based on data collected by sensors [43]. Many of the problems faced by farmers can be solved through smart farming or precision agriculture. Seeds, fertilizers, and water can increase yield more efficiently. The sensors in your field will allow you to map your field so that you will be able to know how much soil there is on every inch [44]. Different kinds of sensors are employed to acquire relevant data from the field and based on this data, we make DF and conclude our objectives more precisely and accurately.

1. Location sensors
   These sensors measure longitude and latitude in the field with high accuracy. A global positioning system (GPS) unit is also included with these sensors in order to calculate the location using different satellites [45]. Example NJG1157PCD-TE1, NJR NJG1157PCD-TE1.
2. Optical sensors
   These light-based sensors can analyze soil composition and structure. These sensors use different frequencies of light reflected by the soil to determine factors like soil texture, organic matter, and moisture levels. They can also determine a plant's overall health by examining its leaves [45].
3. Electrochemical sensors
   There has been a long history of employing such sensors. Soil nutrients and acidity levels can be measured with these sensors. These sensors' electrodes pick up on ion concentrations in the ground, providing information about the soil's structure [46].
4. Mechanical sensors
   Soil compaction and mechanical resistance can be detected with the help of these sensors. These sensors can measure how hard a plant is working to take up water, which will allow us to better control our irrigation strategies [47]. An example is Honeywell FSG15N1A.
5. Soil electrical resistivity sensors
   This is a tool for determining how much a given soil's resistance to electrical current has changed because of a change in moisture levels. They can also be

used to identify the friction angle of soil and the physical features of soil as well [48].

6. Airflow sensors
   This sensor works by forcing a calibrated volume of air into the ground at a predetermined depth. Soil permeability can be evaluated with its help. Parameters in the soil can be determined, including the soil's moisture, compaction, and structure [49]. Since each property results in a unique "signature."
7. Agricultural weather station
   These are the discrete units spread out around the area that collect data about the environment. Depending on your needs, they may measure things like temperature, position, humidity, solar radiation, wind speed, and more [50].
8. Yield monitoring
   Food grains like wheat and corn can have their weight measured using a yield monitoring equipment that is often mounted on the harvester. You can get results based on time, distance, and even GPS, with up to 30 cm accuracy [51].
9. Variable rate fertilizer
   These devices may spread liquid, granular, or gaseous fertilizers in the field based on a map of the area and data from optical sensors. Based on the GPS map, they can automatically distribute the appropriate amount of fertilizers [52].
10. Weed mapping
    To accomplish this, image processing equipment can be mounted on top of a vehicle or a drone to capture footage or stills from the field. The yield can be analyzed using this technique as well. The GPS in the gadget generates a weed map that may be used for either human or automated weed collection [53,54].
11. Variable spraying
    The automated spraying system will spray herbicides once the weed map is ready, and it can also adjust the concentration of herbicides based on the needs of the land [55].
12. Salinity mapping
    A salinity meter can be used for this purpose. The salinity meter allows us to track the field's salinity over time [49].
13. Mobile as a tool
    A smartphone's camera and color-identifying features, as well as its GPS and accelerometer, make it useful for field applications, including soil and crop observation [56].

14.2 Data types

There is often a considerable discrepancy between the informational value of various data types. However, the value of a single data source does not guarantee its efficacy when combined with other datasets. In reality, the degree of a variable's complementarity with respect to the other variables is the most crucial signal for including or not including it [57]. If the factors in question have a high degree of complementarity, even if they perform poorly individually, they can considerably boost the performance of the other variables [17]. However, even if all of the factors are underperforming on their own, they may perform well when combined [36]. On the other hand, variables that do well when evaluated individually may not perform well when combined with other variables if they have a great deal of redundancy in common. There can be a decline in quality as a result of poor synergy [58] and significant degrees of redundancy [59] across various data sources. The introduction of new data types may not always lead to better outcomes, but it can make models more resistant to outliers and other factors that weren't taken into account when they were first built [24]. Moreover, data complementarity can be improved using redundancy reduction methods [e.g., principal component analysis (PCA)]. Determining whether or not a certain piece of data should be included is not a simple task and typically needs extensive experimentation, investigating various combinations of data [25,60]. To optimize the efficiency of the DF process, it may be necessary to account for the unique characteristics of various types of data. For instance, thermal pictures are significantly affected by factors like time of day and weather conditions during collection [17,61]. In addition, data obtained by low-cost sensors might be particularly noisy and prone to outliers [12], which could lead to inaccurate conclusions rather than better ones. Applying noise-reduction and outlier-removal methods is suggested in such scenarios [23,26]. Torres *et al.* (2020) thoroughly examined several outlier elimination methods in the discussion of water management. Many researchers overlook the problem of false data, which reduces the depth and quality of the combined data [62], despite its prevalence.

Broad-band multispectral or narrow-band hyperspectral data collection is the primary use for sensors, while they can also be used for image purposes. Satellites, manned, and unmanned aircraft also carry sensors; the ground serves as a base for mobile and stationary sensors used in the field and laboratory settings. Continually, objects on Earth send out, bounce back, and gather electromagnetic radiations [63]. Generally, remote sensing technology can tell the difference between two things by analyzing the electromagnetic waves that are sent, received, and absorbed by them. The visible (0.4–0.7 mm), infrared (0.7–15 mm), and microwave (0.75–100 cm) portions of the electromagnetic spectrum are usually used for remote sensing. Remote sensing big data is characterized by its extreme volume, complexity, and spread over space and time, because of these elements and their relationship to the rate and nature of data capture. Higher spatial, spectral, and temporal resolutions are now possible because of advancements in remote sensing technologies [64]. The generation of pictures and spectral data in large volumes and extreme complexity is the focus of agricultural remote sensing, a specialized subject that drives decisions for agricultural development. For making educated decisions about fertilization, irrigation, and pest management, remote sensing is used in agriculture to keep tabs on soil parameters and crop stress.

Visible-NIR (near infrared) (0.4–1.5 mm) sensors for plant vegetation studies; SWIR (short wavelength infrared) (1.5–3 mm) sensors for plant moisture studies; TI (thermal infrared) (3–15 mm) sensors for analyzing the temperature of crop field surfaces or crop canopies, as well as microwave sensors for analyzing soil moisture are used [65]. Vegetation structure over agricultural lands can now be measured using light detection and ranging (LiDAR) and synthetic aperture radar (SAR) [65,66]. Remote sensing has emerged as a crucial technology in modern precision agriculture to actualize site-specific agricultural field management that considers within-field variability in soil, plant stress, and treatment efficacy. The complexity and volume of remote sensing data are expected to grow rapidly with the rapid advancement of remote sensing technology, notably the usage of new sensors with greater resolutions. Finding ways to efficiently extract relevant information from such large datasets is now of paramount importance to users to better analyze the data, find answers to queries, and find solutions to issues. Big data include information gathered via remote sensing [67]. Storage, information extraction, fast processing, information fusion, and using a lot of remote sensing data are all areas that are getting much research attention right now [68,69]. Here, we will discuss some of the research regarding plant disease using agricultural remote sensing with feature selection and data utilization.

14.3 Imaging and non-imaging spectroscopy for disease detection

Detailed and highly conducive information on crop systems and individual plants can be obtained using new sensor systems. Different sensor types can assess the target object's characteristics/parameters based on signal-object interactions. Plant physiological parameters can be measured using a variety of sensor types, as detailed by numerous scientists [70,71]. It is possible to measure the thermal properties of plants [72]; the fluorescence of chlorophyll [73], and the reflectance of plants [74]. Plant disease detection requires more sophisticated sensor systems due to the lack of diagnostic potential for stress response and photosynthesis changes in plants. Non-imaging and imaging hyperspectral sensors can be used to detect, identify, and quantify plant disease in this study. However, there are a variety of approaches to fuse optical sensors for plant science that have been documented in the literature. As the technology of remote sensing has progressed, sensors have upgraded from multispectral to hyperspectral to ultraspectral [75]. The spectral range covered by these devices provides a plethora of data. However, only a few spectral regions are relevant, depending on the measured object and the intended use. Hyperspectral sensors with a spectral resolution of up to 1 nm have a high correlation between their narrow spectral bands, and out of this redundant information is measured. An awareness of object spectral properties and signal-object interaction is crucial for the development of disease diagnosis techniques using remote sensing devices. Reflectance sensors are currently divided into imaging and non-imaging sensors based on their spatial scale, spectral resolution, and data assessment method [76]. For example, airborne or spaceborne far-range systems with a smaller spatial resolution, or near-range sensing systems with maximal spatial resolution, are all examples. The smallest detectable symptom or structure defines the maximum spatial resolution, which is defined by the minimum pixel size. Since the advent of hyperspectral narrowband sensors and the subsequent transition from multispectral broadband sensors, the amount and quality of available data has increased dramatically, among this we need to focus on most relevant information.

14.3.1 Disease detection with non-imaging hyperspectral data

Non-imaging multi- and hyperspectral remote sensing technology is one of the most advanced methods to monitor the changes in the spectral characteristics of crops in the world. Its application is much earlier than imaging hyperspectral technology and used for the plants' different aspects [66]. Scientific literature has shown that the physio-chemical damages caused by different pathogens of unlike diseases are various. There is not any single monitoring method that works well for a variety of diseases [77]. Similar to imaging spectroscopy, non-imaging spectroscopy is also well applied to classify diseases [78]. Huang *et al.* [79] found that hyperspectral reflectance has the potential to detect rice leaf folder damage severity in rice. Similarly, Lu *et al.* [80] used a non-imaging spectrometer to compute spectral vegetation indices and discriminate various tomato leaf diseases using a K-nearest neighbor classifier. Mahlein *et al.* [77] developed distinct disease indices by employing hyperspectral signatures of healthy and sick sugar beet leaf samples obtained in a laboratory setting using a spectrometer that does not involve imaging. Delalieux *et al.* [81] gathered hyperspectral data using a non-imaging spectrometer from healthy and *Venturia inaequalis* diseased apple leaves and evaluated multiple multivariate statistical techniques. Resultantly, concluded that the SWIR range contains relevant information for the early discrimination of healthy and infected leaves. The great concern of non-imaging over imaging research is the loss of spatial information, which helps to give information characteristic to diseased areas.

14.3.2 Disease detection with imaging hyperspectral data

The Airborne Imaging Spectrometer, developed at the Jet Propulsion Laboratory of the National Aeronautics and Space Administration in the early 1980s, marked the beginning of hyperspectral images (HSI) [82]. In HSI, spectroscopy and imaging are brought together to capture spatial and spectral data about a scene or object at the same time [83]. This results in a ternary picture with a spatial "*X*" and "*Y*" axis and a spectral "*Z*" axis, commonly referred to as a "data-cube" or a "hypercube" [84]. Whereas, in order to check for yellow rust disease in wheat and citrus canker in grapefruits, a push-broom type scanner has been utilized [85]. Diseased leaf tissue could be detected and precisely described thanks to a pixel-by-pixel mapping of spectral reflectance in the visible and near-infrared region [86]. However, observing spectral reflectance patterns of foliar illnesses in detail is facilitated by pixel-wise attribution of disease-specific symptoms and healthy tissue [15]. HSI with high spatial resolution is necessary for distinguishing between some disease symptoms and those of other diseases and stresses [87]. As a result, numerous feature extraction techniques have been established for tracking various plant states. Physical and morphological plant traits are just two examples of the types of qualities that have been used in feature extraction. A tool for identifying leaf diseases in cucumber plants was demonstrated by Zhang *et al.* [88]. The application initially used *k*-means clustering to identify the contaminated area of the leaf, and then used color and shape extraction to determine its location with an accuracy of 85.7%. In a similar vein, Guo *et al.* [89] used a Bayesian technique with texture and color cues to identify downy mildew, anthracnose, powdery, and gray mold infection with accuracies of 94.0%, 86.7%, 88.8%, and 84.4%, respectively. To counter this, Vianna *et al.* [90] created a neural network based pattern recognition approach to identifying the most widespread form of tomato late blight.

HSI technology has two main applications in crop pest monitoring: the most important is the use of imaging spectroscopy for diseased area identification through spectral mapping, PCA, vegetation index method, Partial least squares, and machine vision methods [91]. The second aspect is the inversion of disease severity based on spectral imaging techniques. The selection of sensitive bands developed the vegetation index to establish a disease severity model [92]. Such applications are relatively simple and have a single method, mainly used for determining disease severity and assessing losses [92].

14.3.3 Disease detection with fluorescence imaging

When measuring plant photosynthesis, various chlorophyll fluorescence parameters are used. This type of fluorescence imaging instrument uses an LED or laser light source to measure electron transfer from the photosynthesis of chlorophyll [91,93]. Biotic and abiotic stresses on the leaf area have been studied using this method to determine the effects on photosynthetic activity. For the detection and quantification of fungal infections, fluorescence imaging and image analysis techniques have been shown to be useful [94]. The existing chlorophyll fluorescence imaging systems in conventional agricultural greenhouses or field conditions because these systems require plants to be prepared following a strict protocol. The extraction of fluorescence properties has become the primary focus of the researchers that can improve the examination of the plants.

14.4 Data fusion techniques

Literature's wide range of DF methods suggests there is no one approach well suited to all (or even most) of the fields that could benefit from such a process. To prevent biasness, Tavares *et al.* [95] suggested using independent sample subsets (independent validations) to determine the best DF approach for a given application and set of attributes. Experiments sometimes show that DF is ineffective simply because it is not appropriate for the problem that is being tackled [27], which is an important point to dress. DF may not always be the most effective method for avoiding the potentially misleading behavior of trying to make your findings suit your preconceived notions. Many researchers still utilize regression models (at the proximal and aerial scale) and spatiotemporal approaches (at the orbital scale). These strategies have been tried repeatedly, and each time they've proven successful. There is no way to compare apples to apples because different studies utilize different datasets, but there does appear to be little advancement and some degree of redundancy across studies that take this method. While machine learning approaches have been used for some time, their full potential has not yet been fully explored. Some explanations for this are as follows. First, it is not easy to put into practice DF using machine learning and artificial intelligence models. Various data formats must be carefully managed to ensure they work together inside the framework of the model because so many machine learning approaches have poor interpretability [96]. Compatibility is important, yet it may be difficult to find a way to achieve it in most situations. Second, there is an absence of databases that comprehensively document all the complexities of a specific application's variability. Data gaps are a concern for any method [97], but the impacts are magnified when machine learning methods are used. Third, a lot of computing power is needed for model training in machine learning approaches (deep learning in particular). Recently, this crisis has become substantially less severe as more powerful computers have become widely available [1], along with deep learning's exceptional capacity for integrating and extracting information from different types of images [19,98].

14.4.1 Data fusion impediments

The data that needs to be combined, the diversity and heterogeneity of the technologies, and the nature of the application itself are the primary causes of problems. The data brought into the framework may be inconsistent, poorly correlated, or presented in multiple ways. It is possible to classify each of these obstacles as: (1) data imperfection, (2) data correlation, (3) data inconsistency, and (4) data disparateness [99].

1. *Data imperfection*: It is essential to keep in mind that sensor data is not always reliable and accurate. Effectively dealing with the flaw requires the application of data fusion algorithms. (a) Imprecision, (b) uncertainty, and (c) granularity are the three flaws through which to examine the data. (a) The imperfection of data leads to experimental results that vary from one attempt to the next; this is known as imprecision. (b) A low degree of trust in the data is called uncertainty. It leads to inaccuracies and errors and imprecision occurs when there is a focus on multiple items instead of just one. Incomplete data is used to describe vague information that is lacking certain necessary details. Hence, controversy invades as ambiguity when the data is well-defined and exact but imprecise. Imprecision lies when multiple pieces of information and a lack of detail make this data hard to pin down. Example: in the sentence "the river is longest," the information "longest" is not well-defined, therefore, it might be interpreted differently depending on who

you ask. (c) Granularity is the capability to choose between things described by data based on the specified set of quality or condition.

2. *Data inconsistency*: Over or under confidence in a data fusion technique is possible due to this issue, which is especially prevalent in wireless sensor networks. It needs to be handled appropriately, and input data inconsistencies can arise from false positives, data collisions, or information that was not adequately coordinated. In the first fusion stage, sensor data must be changed from each sensor's local frame to a common frame, which introduces an alignment challenge known as a sensor registration problem.
3. *Inconsistency in the data*: The sensor's working environment is a major contributor to data discrepancies, along with the sensor's own inherent imprecision. The conflicting nature, disorder and outlier of data are also crucial factors for data inconsistency.
4. *Data disparateness*: Data that are significantly different from one another in terms of kind, quality, or nature are said to be "disparate." They are inefficient and incompatible with one another, making it impossible to meet the information needs of objectives. These data are of a poor quality; they are inaccurate, inconsistent, unclear, and inconsistent.

14.5 Low, mid, and high-level data fusion

The degree of fusion to be implemented is also not a simple decision. Low-level fusion has been used in most research projects since it is easier to implement and requires fewer computer resources [100]. Some research seems to suggest that higher fusion levels tend to provide better outcomes [101], with the authors claiming that the poorer results found when lower level fusion is performed are likely due to data redundancies originating from complementing information from various sensors [24]. However, contrary findings have been found in other research, with the low-level method emerging victorious [57]. Few research has compared the various fusion levels, making it unable to make firm conclusions. More studies comparing the effects of various degrees of fusion are required to shed light on the optimal approach to data in various settings. To directly merge different data sources is the goal of low-level fusion. Since they can have varying value ranges and presentation formats, normalization is typically required to ensure compatibility between databases [102]. Normalization is typically a simple process when the sole distinction is the range of values. However, the normalizing procedure can be complex or even impracticable if the data utilized are of different natures (e.g., pictures and meteorological data), in which usage of higher DF levels may be needed. For information extraction purposes, mid-level fusion can be broken down into two stages: the first is the processing of data to generate significant features, and the second is the fusing of these features to arrive at a single response [103]. Various features are sometimes retrieved from the same dataset, even though in most situations, the features are extracted from different sources of data (the features themselves are often also different). Combining features from the same dataset isn't always "data fusion," but there are occasions when the strategies used are

Table 14.1 Data types and data fusion approaches

S. no.	Data types		Data fusion approaches
1	Elevation models and topographic records	1	Decision rules
2	Fluorescence images	2	Deep learning
3	Historical data	3	Geostatistical tools
4	Hyperspectral images	4	Kalman filter
5	Inertial measurements	5	Machine learning
6	Laser scanning	6	Majority rules
7	Multispectral images	7	Model output averaging
8	Position measurements	8	PCA and derivatives
9	RGB images	9	Regression methods
10	SAR images	10	STARFM-like statistical approaches
11	Soil measurements	11	Others
12	Spectroscopy		
13	Thermal images		
14	Weather/environmental measurements		
15	Others		

similar to those used in true DF since the features represent such distinct parts of the data. Although high-level fusion is the most effective of the three types of fusion, it is also the least common. The two most obvious explanations for this are: To begin, high-level fusion might be resource intensive, particularly during the learning phase [100]. Using deep neural networks in any of the classifiers will increase this effect significantly. Second, picking the proper variables to feed a single model can be difficult, so selecting variables for numerous models at once might be cumbersome. However, as the research on DF matures and the properties of each type of model/classifier are more understood, the use of high-level fusion is expected to increase (Table 14.1).

14.6 Data acquisition and fusion at different scales

14.6.1 Data acquisition at proximal scale

Prediction of soil parameters, demarcation of homogenous zones, and robotic navigation and control are where proximal-scale research has focused most closely. Disease and fruit detection, water content and stress prediction, phonological state assessment, and yield prediction are just some of the applications that may be found here. The cited works examined further into satellite data, and the research integrated ground-level information with satellite imagery. Spectrometers, Cameras (thermal, RGB, multispectral, hyperspectral), resistivity, and conductance sensors, GPS, inertial measurement unit (IMU), and meteorological data were just some of the data sources used. There has been increasing work to investigate sensor complementarities in the field (Table 14.2), but most problems still lack solid answers [115].

Table 14.2 Studies considered most relevant for proximal scale (1st column), objective of the study (2nd column), the data fusion methodology (3rd column) and different types of data – numbered in Table 14.1 – that is fused (4th column)

Conducted study	Purpose of study	Fusion methodology	Data type
[37]	Crop phenological states estimation	Particle filter scheme	7, 10, 14
[61]	Detection of fruit	LPT and fuzzy logic	9, 13
[24]	Examination of soil properties	RK, PLSR	4, 11
[102]	Estimation of crop nutrition status	PCA	2, 12
[26]	Predicting soil properties	PLSR	1, 11, 12
[104]	Determination of soil nutrients	PLSR	1, 11
[101]	Prediction of toxic elements	SPA, PLSR, PCA	2, 12
[105]	Assessments for agricultural vulnerability	RF, XGBoost, binary relevance	3, 14
[106]	Predicting the diverse types of soil properties	PCA, PLSR, SMLR	11, 12
[36]	Biomass estimation in grasslands	Simple quadratic combination	7, 15
[15]	Disease detection in plant	Kohonen self-organizing maps	2, 4
[107]	Detection of water stress	Least squares support vectors machine	2, 4
[108]	Determination of water holding capacity in different zones	ANN, MLR	11, 12
[58]	Site specific seeding potential in potato	PLSR	7, 12
[109]	Protein and yield prediction in wheat	Canonical powered partial least-squares	12, 14
[110]	Yield prediction in wheat	XY-fused networks, SKN, CP-ANN	7, 12
[111]	Topsoil mapping	KNN, PLSR	1, 11, 12
[19]	Detection of fruits	CNN, scoring system	7, 9
[23]	3D phenotyping for agriculture	Linear interpolation	9, 14
[112]	Orchard mapping through mobile robot localization	Projection on the RGB images using laser data	9, 6
[95]	Fertility analysis of tropical soil	Least squares modelling, SVM, PLS	2, 7
[60]	Assessment of health soil	PLSR	11, 12
[113]	Soil texture prediction	PCA, SMLR, PLSR	12, 2
[28]	Cotton plants detection	Discriminant analysis	7, 12
[114]	Wheat spikes recognition	Gram–Schmidt fusion algorithm	7, 9

14.6.2 Data acquisition at aerial scale

A growing body of research is being done on the use of UAVs to address agricultural issues, however these studies remain far outnumbered by proximal and

Table 14.3 Studies considered most relevant for proximal scale (1st column), objective of the study (2nd column), the data fusion methodology (3rd column) and different types of data – numbered in Table 14.1 – that is fused (4th column)

Conducted study	Purpose of study	Fusion methodology	Data type
[116]	Moisture estimation in root zone	GLM, GBM, DRF, NN	5, 7
[117]	Weed detection in rice field	Haar wavelet transformation	7, 9
[118]	Spectral and temporal response of surfaces	Bayesian data imputation	4, 7
[59]	Soya bean phenotyping	RF, SVR, ELR	7, 9, 13
[119]	Yield prediction for soya bean	RF, SVR, PLSR, DNN of two types	7, 9, 13
[38]	Crop monitoring	PLSR, RF, SVR, ELR	7, 9
[32]	Evapotranspiration estimation	MSDF-ET	7, 9, 13
[120]	Vegetation monitoring in arid and semi-arid lands	Decision tree	4, 6
[27]	Determination of biomass and leaf nitrogen estimation in sugarcane	Linear regression and PCA	6, 7
[121]	Biomass estimation in wheat	Linear regression and normalized difference indices	4, 7

orbital strategies. Most of the research is concentrated on phenotyping and water management, but there is also a significant amount of work done on crop monitoring and object detection i.e. weeds, crops, etc. (Table 14.3). All the methods rely on some sort of digital image as their foundation.

14.6.3 Data acquisition at satellite scale

A lot of the articles that used satellite images wanted to either fill in data gaps in a primary data source by combining it with another source of data (like SAR images and combining optical) [37,98], or improve the spatial resolution of the relatively blurry images taken by satellites with high revisit frequencies [29,30]. In the second case, the details of the high spatial resolution images and the revisit frequencies of their counterparts are usually carried over into the fused results. However, the quality of the fused data usually does not match the quality of the data from real missions, especially when surface changes happen quickly and subtly [122]. Tao *et al.* [123] found that different sensors and image processing algorithms lead to data that isn't always consistent. This can make it hard to notice changes that happen quickly.

Images and products from Landsat and MODIS are still the most popular, but other satellites like Sentinel, GeoEye, Worldview, and others are becoming more popular. DF has been used on satellite images for a long time, and well-known methods like STARFM and its variations are still used often. However, interest in machine learning methods, especially deep learning models, has been steadily growing. By far, the most common use is water management in its many forms (estimating evapotranspiration, mapping irrigated areas, detecting drought, etc.). A few other common uses are figuring out the yield, keeping an eye on crops, classifying the land cover, and predicting the properties of the soil.

Extremely dispersed and diverse regions are a big obstacle to overcome on the orbital scale [124,125]. Masiza *et al.* [57] argue that "effective mapping of a fragmented agricultural landscape is a function of objectively produced datasets, tailored to geographic context, and an informed optimization of mapping

Table 14.4 Studies considered most relevant for proximal scale (1st column), objective of the study (2nd column), the data fusion methodology (3rd column) and different types of data – numbered in Table 14.1 – that is fused (4th column)

Conducted study	Purpose of study	Fusion methodology	Data type
[98]	Crop classification mapping	SegNet, RF, 2D and 3D U-Net	7, 10
[37]	Determination of crop phenological stages	Particle filter scheme	7, 10, 14
[127]	In-field examination of soil properties	PLSR, RK	4, 11
[128]	Estimation of nitrogen uptake for wheat grains	BK	4, 7
[129]	Moisture monitoring in soil surface	Kriging and linear regression analysis	7, 15
[128]	Crop classification and discrimination	Voting system	7, 10
[130]	Crop mapping	Pixelwise matching	7, 10
[118]	Spectral response on temporal basis	Bayesian data imputation	4, 7
[31]	Evapotranspiration mapping	Empirical exploration for different band relationships	7, 13
[38]	Crop monitoring	ELR, SVR, PLSR, RF	7, 9, 13
[57]	Mapping for smallholder crop farming	NB, ANN, SVM, RF, XGBoost	7, 10
[36]	Evapotranspiration estimation	MSDF-ET	7, 9, 13
[110]	Yield prediction for wheat	SKN, XY-fused networks, CP-ANN	7, 12
[124]	Drought monitoring	RF	7, 15
[97]	Crop type mapping for classification	RF	7, 10
[131]	Crop classification	Committee of MLPs	7, 10
[6]	Classification using multisource remotely sensed data	Bayesian formulation	7, 10
[132]	Monitoring of land cover	FARMA	7, 10
[133]	Estimation of soil moisture content	ANN	7, 10
[134]	Winter wheat mapping in urban region	RF, SVM	7, 10

algorithms," indicating that there is no simple solution to this issue. Target locations, however, may sometimes be too small to be captured by satellite photos due to their pixelated nature [126]. It may be necessary in such circumstances to combine the photos with other images or data types received at higher resolutions (either from above or up close) (Table 14.4).

The difficulty of locating and gathering reference data for the purposes of validating the procedures used is another significant problem. This issue may be more challenging if the reference data is collected on-site. It is also crucial to keep in mind that, even if reference data can be gathered, variations in granularity and the locations of the sample sites may make comparisons with the fused data challenging or impossible [124].

14.7 Conclusions

The vast amount of data currently being produced in the agricultural environment is being dealt efficiently, with considerable effort being put into this endeavor. Although significant progress has been made, numerous obstacles continue to hinder strategies based on data fusion from being used more frequently in real-world settings. These three elements seem to be very important. First, even though a lot of data is being produced, the complexity of the agricultural environment prevents the databases from being large enough to account for all the variations observed in actual practice. Because of this, ideas and techniques put forth in the literature frequently fall short in actual, practical settings. Second, the findings stated in scientific journals need to be judged against such constraints because the datasets utilized for study almost always contain some gaps. Miserably, this is rarely the case, and frequently, those findings serve as the foundation for developing technologies that are very certain to fail. Third, several technological, economic, political, social, and environmental constraints hinder the adoption of the technology even if it is sufficiently reliable for use in real-world applications [135,136].

The potential for growth remains despite these obstacles. In addition to technical developments, improvements have been made in the quantity and caliber of data being gathered. Numerous research teams are making datasets available under the findable, accessible, interoperable, and reusable principles, which is a practice that is steadily expanding [1]. The representativeness of datasets typically improves significantly when data are gathered from a range of sources. Additionally, "citizen science principles" [137,138] are being used across disciplines with encouraging results. These principles call for the participation of people outside the research community in an effort to generate databases. Adoption barriers frequently erode once a technology really offers advantages to potential consumers. All of these initiatives will likely result in a continued narrowing of the gap between academic research and actual adoption for crop quality assessment, crop damage and smart planning.

List of abbreviations

ANN	artificial neural network
BK	block kriging
CNN	convolutional neural network
CP-ANN	counter-propagation artificial neural networks
DEM	digital elevation model
DNN	deep neural network
DRF	distributed random forest
FARMA	fusion approach for remotely-sensed mapping synthetic aperture radar (SAR) of agriculture
GBM	gradient boosting machine
GLM	generalized linear model
KNN	k-nearest neighbors
MLPs	multilayer perceptron
MLR	multiple linear regression
MSDF-ET	multi-sensor data fusion model for actual-multi-sensor data fusion model for actual
NB	Naïve Bayes
NN	neural net
PCA	principal component analysis
PLS	partial least square
PLSR	partial least square regression
RF	random forest
RK	regression kriging
SKN	supervised Kohonen networks
SMLR	stepwise multiple linear regression
SPA	successive projections algorithm
SVR	support vector regression
XGBoost	extreme gradient boosting

References

[1] Barbedo, J.G., 2021. Deep learning applied to plant pathology: the problem of data representativeness. *Tropical Plant Pathology* 1–10, 85–94.

[2] Coble, K.H., Mishra, A.K., Ferrell, S., and Griffin, T., 2018. Big data in agriculture: a challenge for the future. *Applied Economic Perspectives and Policy* 40, 79–96.

[3] Kamilaris, A., Kartakoullis, A., and Prenafeta-Boldú, F.X., 2017. A review on the practice of big data analysis in agriculture. *Computers and Electronics in Agriculture* 143, 23–37.

[4] Barbedo, J.G., 2018. Factors influencing the use of deep learning for plant disease recognition. *Biosystems Engineering* 172, 84–91.

[5] Munir, A., Blasch, E., Kwon, J., Kong, J., and Aved, A., 2021. Artificial intelligence and data fusion at the edge. *IEEE Aerospace and Electronic Systems Magazine* 36, 62–78.

[6] Solberg, A.H.S., Jain, A.K., and Taxt, T., 1994. Multisource classification of remotely sensed data: fusion of Landsat TM and SAR images. *IEEE transactions on Geoscience and Remote Sensing* 32, 768–778.

[7] Bleiholder, J. and Naumann, F., 2009. Data fusion. *ACM Computing Surveys (CSUR)* 41, 1–41.

[8] Caruccio, L. and Cirillo, S., 2020. Incremental discovery of imprecise functional dependencies. *Journal of Data and Information Quality (JDIQ)* 12, 1–25.

[9] Dalla Mura, M., Prasad, S., Pacifici, F., Gamba, P., Chanussot, J., and Benediktsson, J.A., 2015. Challenges and opportunities of multimodality and data fusion in remote sensing. *Proceedings of the IEEE* 103, 1585–1601.

[10] Ouhami, M., Hafiane, A., Es-Saady, Y., El Hajji, M., and Canals, R., 2021. Computer vision, IoT and data fusion for crop disease detection using machine learning: a survey and ongoing research. *Remote Sensing* 13, 2486.

[11] Li, Y., Jia, H., Qi, J., *et al.*, 2020. An acquisition method of agricultural equipment roll angle based on multi-source information fusion. *Sensors* 20, 2082.

[12] Zaidner, G. and Shapiro, A., 2016. A novel data fusion algorithm for low-cost localisation and navigation of autonomous vineyard sprayer robots. *Biosystems Engineering* 146, 133–148.

[13] Zhang, Q., Chen, Q., Xu, Z., Zhang, T., and Niu, X., 2021. Evaluating the navigation performance of multi-information integration based on low-end inertial sensors for precision agriculture. *Precision Agriculture* 22, 627–646.

[14] Cruz, A.C., Luvisi, A., De Bellis, L., and Ampatzidis, Y., 2017. X-FIDO: an effective application for detecting olive quick decline syndrome with deep learning and data fusion. *Frontiers in Plant Science* 8, 1741.

[15] Moshou, D., Bravo, C., Oberti, R., *et al.*, 2005. Plant disease detection based on data fusion of hyper-spectral and multi-spectral fluorescence imaging using Kohonen maps. *Real-Time Imaging* 11, 75–83.

[16] Shankar, P., Johnen, A., and Liwicki, M., 2020. Data fusion and artificial neural networks for modelling crop disease severity. *Presented at the 2020 IEEE 23rd International Conference on Information Fusion (FUSION)*, IEEE, pp. 1–8.

[17] Gan, H., Lee, W.S., Alchanatis, V., Ehsani, R., and Schueller, J.K., 2018. Immature green citrus fruit detection using color and thermal images. *Computers and Electronics in Agriculture* 152, 117–125.

[18] Li, P., Lee, S.-H., Hsu, H.-Y., and Park, J.-S., 2017. Nonlinear fusion of multispectral citrus fruit image data with information contents. *Sensors* 17, 142.

[19] Sa, I., Ge, Z., Dayoub, F., Upcroft, B., Perez, T., and McCool, C., 2016. Deepfruits: a fruit detection system using deep neural networks. *Sensors* 16, 1222.

[20] Anastasiou, E., Castrignanò, A., Arvanitis, K., and Fountas, S., 2019. A multi-source data fusion approach to assess spatial-temporal variability and delineate homogeneous zones: a use case in a table grape vineyard in Greece. *Science of the Total Environment* 684, 155–163.

[21] Castrignanò, A., Buttafuoco, G., Quarto, R., *et al.*, 2018. A geostatistical sensor data fusion approach for delineating homogeneous management zones in precision agriculture. *Catena* 167, 293–304.

[22] Guerrero, A., De Neve, S., and Mouazen, A.M., 2021. Data fusion approach for map-based variable-rate nitrogen fertilization in barley and wheat. *Soil and Tillage Research* 205, 104789.

[23] Sampaio, G.S., Silva, L.A., and Marengoni, M., 2021. 3D reconstruction of non-rigid plants and sensor data fusion for agriculture phenotyping. *Sensors* 21, 4115.

[24] Afriyie, E., Verdoodt, A., and Mouazen, A.M., 2021. Data fusion of visible near-infrared and mid-infrared spectroscopy for rapid estimation of soil aggregate stability indices. *Computers and Electronics in Agriculture* 187, 106229.

[25] Huo, Z., Tian, J., Wu, Y., and Ma, F., 2020. A soil environmental quality assessment model based on data fusion and its application in Hebei Province. *Sustainability* 12, 6804.

[26] Ji, W., Adamchuk, V.I., Chen, S., *et al.*, 2019. Simultaneous measurement of multiple soil properties through proximal sensor data fusion: a case study. *Geoderma* 341, 111–128.

[27] Shendryk, Y., Sofonia, J., Garrard, R., Rist, Y., Skocaj, D., and Thorburn, P., 2020. Fine-scale prediction of biomass and leaf nitrogen content in sugarcane using UAV LiDAR and multispectral imaging. *International Journal of Applied Earth Observation and Geoinformation* 92, 102177.

[28] Zhang, H., Lan, Y., Suh, C.P.-C., *et al.*, 2013. Fusion of remotely sensed data from airborne and ground-based sensors to enhance detection of cotton plants. *Computers and Electronics in Agriculture* 93, 55–59.

[29] Abowarda, A.S., Bai, L., Zhang, C., *et al.*, 2021. Generating surface soil moisture at 30 m spatial resolution using both data fusion and machine learning toward better water resources management at the field scale. *Remote Sensing of Environment* 255, 112301.

[30] Bai, L., Long, D., and Yan, L., 2019. Estimation of surface soil moisture with downscaled land surface temperatures using a data fusion approach for heterogeneous agricultural land. *Water Resources Research* 55, 1105–1128.

[31] Guzinski, R., Nieto, H., Sandholt, I., and Karamitilios, G., 2020. Modelling high-resolution actual evapotranspiration through Sentinel-2 and Sentinel-3 data fusion. *Remote Sensing* 12, 1433.

[32] Mokhtari, A., Ahmadi, A., Daccache, A., and Drechsler, K., 2021. Actual evapotranspiration from UAV images: a multi-sensor data fusion approach. *Remote Sensing* 13, 2315.

[33] Semmens, K.A., Anderson, M.C., Kustas, W.P., *et al.*, 2016. Monitoring daily evapotranspiration over two California vineyards using Landsat 8 in a multi-sensor data fusion approach. *Remote Sensing of Environment* 185, 155–170.

[34] Wang, T., Tang, R., Li, Z.-L., Jiang, Y., Liu, M., and Niu, L., 2019. An improved spatio-temporal adaptive data fusion algorithm for evapotranspiration mapping. *Remote Sensing* 11, 761.

[35] Jimenez-Sierra, D.A., Benítez-Restrepo, H.D., Vargas-Cardona, H.D., and Chanussot, J., 2020. Graph-based data fusion applied to: change detection and biomass estimation in rice crops. *Remote Sensing* 12, 2683.

[36] Moeckel, T., Safari, H., Reddersen, B., Fricke, T., and Wachendorf, M., 2017. Fusion of ultrasonic and spectral sensor data for improving the estimation of biomass in grasslands with heterogeneous sward structure. *Remote Sensing* 9, 98.

[37] De Bernardis, C., Vicente-Guijalba, F., Martinez-Marin, T., and Lopez-Sanchez, J.M., 2016. Contribution to real-time estimation of crop phenological states in a dynamical framework based on NDVI time series: data fusion with SAR and temperature. *IEEE Journal of Selected Topics in Applied Earth Observations and Remote Sensing* 9, 3512–3523.

[38] Maimaitijiang, M., Sagan, V., Sidike, P., Daloye, A.M., Erkbol, H., and Fritschi, F.B., 2020. Crop monitoring using satellite/UAV data fusion and machine learning. *Remote Sensing* 12, 1357.

[39] Brinkhoff, J., Dunn, B.W., Robson, A.J., Dunn, T.S., and Dehaan, R.L., 2019. Modeling mid-season rice nitrogen uptake using multispectral satellite data. *Remote Sensing* 11, 1837.

[40] Chlingaryan, A., Sukkarieh, S., and Whelan, B., 2018. Machine learning approaches for crop yield prediction and nitrogen status estimation in precision agriculture: a review. *Computers and Electronics in Agriculture* 151, 61–69.

[41] Cucchiaro, S., Fallu, D.J., Zhang, H., *et al.*, 2020. Multiplatform-SfM and TLS data fusion for monitoring agricultural terraces in complex topographic and landcover conditions. *Remote Sensing* 12, 1946.

[42] Hu, S., Mo, X., and Huang, F., 2019. Retrieval of photosynthetic capability for yield gap attribution in maize via model-data fusion. *Agricultural Water Management* 226, 105783.

[43] Han, J., Park, C., Kwon, J.H., Lee, J., Kim, T.S., and Jang, Y.Y., 2020. Performance evaluation of autonomous driving control algorithm for a crawler-type agricultural vehicle based on low-cost multi-sensor fusion positioning. *Applied Sciences* 10, 4667.

[44] Finn, M.P., Lewis, M., Bosch, D.D., *et al.*, 2011. Remote sensing of soil moisture using airborne hyperspectral data. *GIScience & Remote Sensing* 48, 522–540.

[45] Hu, P., Chen, Y., and Sonkusale, S., 2015. Low cost spectrometer accessory for cell phone based optical sensor. *Presented at the 2015 IEEE Virtual Conference on Applications of Commercial Sensors (VCACS)*, IEEE, pp. 1–5.

[46] Yew, T.K., Yusoff, Y., Sieng, L.K., Lah, H.C., Majid, H., and Shelida, N., 2014. An electrochemical sensor ASIC for agriculture applications. *Presented at the 2014 37th International Convention on Information and Communication Technology, Electronics and Microelectronics (MIPRO)*, IEEE, pp. 85–90.

[47] Yurui, S., Qingmeng, Z., Zhaolong, Z., and Lammers, P.S., 2007. Measuring soil physical properties by sensor fusion technique. *Presented at the 2007 2nd IEEE Conference on Industrial Electronics and Applications*, IEEE, pp. 142–146.

[48] Sai, V.V. and Hemalatha, T., 2017. Computational methods for simulating soil parameters using electrical resistivity technique. *Presented at the 2017 8th International Conference on Computing, Communication and Networking Technologies (ICCCNT)*, IEEE, pp. 1–7.

[49] Schriber, S., 2019. Smart agriculture sensors: helping small farmers and positively impacting global issues, Too. Mouser Electron: http://www. sci-encedirect. com/science/article/pii/S016719871500029X4.

[50] Math, R.K.M. and Dharwadkar, N.V., 2018. IoT Based low-cost weather station and monitoring system for precision agriculture in India. *Presented at the 2018 2nd International Conference on I-SMAC (IoT in Social, Mobile, Analytics and Cloud)*, IEEE, pp. 81–86.

[51] Sun, L. and Zhu, Z., 2011. A RS/GIS-based system for monitoring crop yield. *Presented at the 2011 International Conference on Computer Distributed Control and Intelligent Environmental Monitoring*, IEEE, pp. 17–20.

[52] Xu, T., Su, N., Wang, R., and Song, L., 2015. A novel variable rate fertilization system based on the Android platform. *Presented at the 2015 IEEE International Conference on Progress in Informatics and Computing (PIC)*, IEEE, pp. 395–398.

[53] Šeatović, D., Kutterer, H., and Anken, T., 2010. Automatic weed detection and treatment in grasslands. *Presented at the Proceedings ELMAR-2010*, IEEE, pp. 65–68.

[54] Xiu, W., Zhijun, M., Wei, M., and Jianbin, J., 2010. The application of GPS for weed investigation in winter wheat field. *Presented at the 2010 World Automation Congress.*

[55] Yang, J. and Gu, M., 2018. Design of the auto-variable spraying system based on ARM9&Linux. *Presented at the 2018 2nd IEEE Advanced Information Management, Communicates, Electronic and Automation Control Conference (IMCEC)*, IEEE, pp. 1–2487.

[56] Petrellis, N., 2017. A smart phone image processing application for plant disease diagnosis. *Presented at the 2017 6th International Conference on Modern Circuits and Systems Technologies (MOCAST)*, IEEE, pp. 1–4.

[57] Masiza, W., Chirima, J.G., Hamandawana, H., and Pillay, R., 2020. Enhanced mapping of a smallholder crop farming landscape through image fusion and model stacking. *International Journal of Remote Sensing* 41, 8739–8756.
[58] Munnaf, M.A., Haesaert, G., Van Meirvenne, M., and Mouazen, A.M., 2020. Map-based site-specific seeding of consumption potato production using high-resolution soil and crop data fusion. *Computers and Electronics in Agriculture* 178, 105752.
[59] Maimaitijiang, M., Ghulam, A., Sidike, P., *et al.*, 2017. Unmanned Aerial System (UAS)-based phenotyping of soybean using multi-sensor data fusion and extreme learning machine. *ISPRS Journal of Photogrammetry and Remote Sensing* 134, 43–58.
[60] Veum, K.S., Sudduth, K.A., Kremer, R.J., and Kitchen, N.R., 2017. Sensor data fusion for soil health assessment. *Geoderma* 305, 53–61.
[61] Bulanon, D., Burks, T., and Alchanatis, V., 2009. Image fusion of visible and thermal images for fruit detection. *Biosystems Engineering* 103, 12–22.
[62] Li, D., Song, Z., Quan, C., Xu, X., and Liu, C., 2021. Recent advances in image fusion technology in agriculture. *Computers and Electronics in Agriculture* 191, 106491.
[63] Lee, K.-S., Cohen, W.B., Kennedy, R.E., Maiersperger, T.K., and Gower, S. T., 2004. Hyperspectral versus multispectral data for estimating leaf area index in four different biomes. *Remote Sensing of Environment* 91, 508–520.
[64] Huang, Y., Chen, Z., Tao, Y., Huang, X., and Gu, X., 2018. Agricultural remote sensing big data: management and applications. *Journal of Integrative Agriculture* 17, 1915–1931.
[65] Mulla, D.J., 2013. Twenty five years of remote sensing in precision agriculture: key advances and remaining knowledge gaps. *Biosystems Engineering* 114, 358–371.
[66] Zhang, C. and Kovacs, J.M., 2012. The application of small unmanned aerial systems for precision agriculture: a review. *Precision Agriculture* 13, 693–712.
[67] Ma, Y., Wu, H., Wang, L., *et al.*, 2015. Remote sensing big data computing: challenges and opportunities. *Future Generation Computer Systems* 51, 47–60.
[68] Jagannathan, S., 2016. Real-time big data analytics architecture for remote sensing application. *Presented at the 2016 International Conference on Signal Processing, Communication, Power and Embedded System (SCOPES)*, IEEE, pp. 1912–1916.
[69] Rathore, M.M.U., Paul, A., Ahmad, A., Chen, B.-W., Huang, B., and Ji, W., 2015. Real-time big data analytical architecture for remote sensing application. *IEEE Journal of Selected Topics in Applied Earth Observations and Remote Sensing* 8, 4610–4621.
[70] Mishra, P., Asaari, M.S.M., Herrero-Langreo, A., Lohumi, S., Diezma, B., and Scheunders, P., 2017. Close range hyperspectral imaging of plants: a review. *Biosystems Engineering* 164, 49–67.

[71] Mustafa, G., Zheng, H., Khan, I.H., *et al.*, 2022. Hyperspectral reflectance proxies to diagnose in-field Fusarium head blight in wheat with machine learning. *Remote Sensing* 14, 2784.

[72] Lenthe, J.H., Oerke, E.C., and Dehne, H.W., 2007. Digital infrared thermography for monitoring canopy health of wheat. *Precision Agriculture* 8, 15–26.

[73] Buschmann, C. and Lichtenthaler, H.K., 1998. Principles and characteristics of multi-colour fluorescence imaging of plants. *Journal of Plant Physiology* 152, 297–314.

[74] Tian, L., Xue, B., Wang, Z., *et al.*, 2021. Spectroscopic detection of rice leaf blast infection from asymptomatic to mild stages with integrated machine learning and feature selection. *Remote Sensing of Environment* 257, 112350–112350.

[75] Meigs, A.D., Otten, L.J., and Cherezova, T.Y., 1998. Ultraspectral imaging: a new contribution to global virtual presence, in: *1998 IEEE Aerospace Conference Proceedings (Cat. No. 98TH8339)*. IEEE, pp. 5–12.

[76] Melesse, A.M., Weng, Q., Thenkabail, P.S., and Senay, G.B., 2007. Remote sensing sensors and applications in environmental resources mapping and modelling. *Sensors* 7, 3209–3241.

[77] Mahlein, A.K., Rumpf, T., Welke, P., *et al.*, 2013. Development of spectral indices for detecting and identifying plant diseases. *Remote Sensing of Environment* 128, 21–30.

[78] Rumpf, T., Mahlein, A.K., Steiner, U., Oerke, E.C., Dehne, H.W., and Plümer, L., 2010. Early detection and classification of plant diseases with support vector machines based on hyperspectral reflectance. *Computers and Electronics in Agriculture* 74, 91–99.

[79] Huang, J., Liao, H., Zhu, Y., Sun, J., Sun, Q., and Liu, X., 2012. Hyperspectral detection of rice damaged by rice leaf folder (*Cnaphalocrocis medinalis*). *Computers and Electronics in Agriculture* 82, 100–107.

[80] Lu, J., Ehsani, R., Shi, Y., de Castro, A.I., and Wang, S., 2018. Detection of multi-tomato leaf diseases (late blight, target and bacterial spots) in different stages by using a spectral-based sensor. *Scientific Reports* 8, 1–11.

[81] Delalieux, S., Van Aardt, J.A.N., Keulemans, W., Schrevens, E., and Coppin, P., 2007. Detection of biotic stress (*Venturia inaequalis*) in apple trees using hyperspectral data: non-parametric statistical approaches and physiological implications. *European Journal of Agronomy* 27, 130–143.

[82] Campbell, P.K.E., Middleton, E.M., McMurtrey, J.E., Corp, L.A., and Chappelle, E.W., 2007. Assessment of vegetation stress using reflectance or fluorescence measurements. *Journal of Environmental Quality* 36, 832–845. https://doi.org/10.2134/jeq2005.0396

[83] Green, A.A., Berman, M., Switzer, P., and Craig, M.D., 1988. A transformation for ordering multispectral data in terms of image quality with implications for noise removal. *IEEE Transactions on Geoscience and Remote Sensing* 26, 65–74.

[84] Goetz, A.F. and Rowan, L.C., 1981. Geologic remote sensing. *Science* 211, 781–791.

[85] Qin, J., Burks, T.F., Kim, M.S., Chao, K., and Ritenour, M.A., 2008. Citrus canker detection using hyperspectral reflectance imaging and PCA-based image classification method. *Sensing and Instrumentation for Food Quality and Safety* 2, 168–177.

[86] Mahlein, A.-K., Steiner, U., Hillnhütter, C., Dehne, H.-W., and Oerke, E.-C., 2012. Hyperspectral imaging for small-scale analysis of symptoms caused by different sugar beet diseases. *Plant Methods* 8, 3–3.

[87] Bravo, C., Moshou, D., West, J., McCartney, A., and Ramon, H., 2003. Early disease detection in wheat fields using spectral reflectance. *Biosystems Engineering* 84, 137–145.

[88] Zhang, S., Wu, X., You, Z., and Zhang, L., 2017. Leaf image based cucumber disease recognition using sparse representation classification. *Computers and Electronics in Agriculture* 134, 135–141.

[89] Guo, P., Liu, T., and Li, N., 2014. Design of automatic recognition of cucumber disease image. *Information Technology Journal* 13, 2129.

[90] Vianna, G.K., Oliveira, G.S., and Cunha, G.V., 2017. A neuro-automata decision support system for the control of late blight in tomato crops. *International Journal of Computer and Information Engineering* 11, 455–462.

[91] Bauriegel, E. and Herppich, W.B., 2014. Hyperspectral and chlorophyll fluorescence imaging for early detection of plant diseases, with special reference to Fusarium spec. infections on wheat. *Agriculture* 4, 32–57.

[92] Bauriegel, E., Giebel, A., Geyer, M., Schmidt, U., and Herppich, W.B., 2011. Early detection of Fusarium infection in wheat using hyper-spectral imaging. *Computers and Electronics in Agriculture* 75, 304–312. https://doi.org/10.1016/j.compag.2010.12.006

[93] Murchie, E.H. and Lawson, T., 2013. Chlorophyll fluorescence analysis: a guide to good practice and understanding some new applications. *Journal of Experimental Botany* 64, 3983–3998.

[94] Konanz, S., Kocsányi, L., and Buschmann, C., 2014. Advanced multi-color fluorescence imaging system for detection of biotic and abiotic stresses in leaves. *Agriculture* 4, 79–95.

[95] Tavares, T.R., Molin, J.P., Javadi, S.H., Carvalho, H.W.P., and de, Mouazen, A.M., 2020. Combined use of vis-NIR and XRF sensors for tropical soil fertility analysis: assessing different data fusion approaches. *Sensors* 21, 148.

[96] Mancipe-Castro, L. and Gutiérrez-Carvajal, R., 2022. Prediction of environment variables in precision agriculture using a sparse model as data fusion strategy. *Information Processing in Agriculture* 9, 171–183.

[97] Pott, L.P., Amado, T.J.C., Schwalbert, R.A., Corassa, G.M., and Ciampitti, I. A., 2021. Satellite-based data fusion crop type classification and mapping in Rio Grande do Sul, Brazil. *ISPRS Journal of Photogrammetry and Remote Sensing* 176, 196–210.

[98] Adrian, J., Sagan, V., and Maimaitijiang, M., 2021. Sentinel SAR-optical fusion for crop type mapping using deep learning and Google Earth Engine. *ISPRS Journal of Photogrammetry and Remote Sensing* 175, 215–235.

[99] Barde, A. and Jain, S., 2018. A survey of multi-sensor data fusion in wireless sensor networks. *Presented at the Proceedings of 3rd International Conference on Internet of Things and Connected Technologies (ICIoTCT)*, pp. 26–27.

[100] Useya, J. and Chen, S., 2018. Comparative performance evaluation of pixel-level and decision-level data fusion of Landsat 8 OLI, Landsat 7 ETM+ and Sentinel-2 MSI for crop ensemble classification. *IEEE Journal of Selected Topics in Applied Earth Observations and Remote Sensing* 11, 4441–4451.

[101] Fang, L., Li, X.A., Ty, B., and Al, A., 2021. Measurement of potentially toxic elements in the soil through NIR, MIR, and XRF spectral data fusion. *Computers and Electronics in Agriculture* 187.

[102] Comino, F., Ayora-Cañada, M., Aranda, V., Díaz, A., and Domínguez-Vidal, A., 2018. Near-infrared spectroscopy and X-ray fluorescence data fusion for olive leaf analysis and crop nutritional status determination. *Talanta* 188, 676–684.

[103] Guijarro, M., Riomoros, I., Pajares, G., and Zitinski, P., 2015. Discrete wavelets transform for improving greenness image segmentation in agricultural images. *Computers and Electronics in Agriculture* 118, 396–407.

[104] La, W.-J., Sudduth, K.A., Kim, H.-J., and Chung, S.-O., 2016. Fusion of spectral and electrochemical sensor data for estimating soil macronutrients. *Transactions of the ASABE* 59, 787–794.

[105] López, I.D., Figueroa, A., and Corrales, J.C., 2021. Multi-label data fusion to support agricultural vulnerability assessments. *IEEE Access* 9, 88313–88326.

[106] Mahmood, H.S., Hoogmoed, W.B., and van Henten, E.J., 2012. Sensor data fusion to predict multiple soil properties. *Precision Agriculture* 13, 628–645.

[107] Moshou, D., Pantazi, X.-E., Kateris, D., and Gravalos, I., 2014. Water stress detection based on optical multisensor fusion with a least squares support vector machine classifier. *Biosystems Engineering* 117, 15–22.

[108] Mouazen, A.M., Alhwaimel, S.A., Kuang, B., and Waine, T., 2014. Multiple on-line soil sensors and data fusion approach for delineation of water holding capacity zones for site specific irrigation. *Soil and Tillage Research* 143, 95–105.

[109] Øvergaard, S.I., Isaksson, T., and Korsaeth, A., 2013. Prediction of wheat yield and protein using remote sensors on plots—Part I: assessing near infrared model robustness for year and site variations. *Journal of Near Infrared Spectroscopy* 21, 117–131.

[110] Pantazi, X.E., Moshou, D., Alexandridis, T., Whetton, R.L., and Mouazen, A.M., 2016. Wheat yield prediction using machine learning and advanced sensing techniques. *Computers and Electronics in Agriculture* 121, 57–65.

[111] Piikki, K., Söderström, M., and Stenberg, B., 2013. Sensor data fusion for topsoil clay mapping. *Geoderma* 199, 106–116.

[112] Shalal, N., Low, T., McCarthy, C., and Hancock, N., 2015. Orchard mapping and mobile robot localisation using on-board camera and laser scanner data fusion – Part B: mapping and localisation. *Computers and Electronics in Agriculture* 119, 267–278.

[113] Wang, S., Li, W., Li, J., and Liu, X., 2013. Prediction of soil texture using FT-NIR spectroscopy and PXRF spectrometry with data fusion. *Soil Science* 178, 626–638.

[114] Zhou, C., Liang, D., Yang, X., Xu, B., and Yang, G., 2018. Recognition of wheat spike from field based phenotype platform using multi-sensor fusion and improved maximum entropy segmentation algorithms. *Remote Sensing* 10, 246.

[115] Anagnostis, A., Benos, L., Tsaopoulos, D., Tagarakis, A., Tsolakis, N., and Bochtis, D., 2021. Human activity recognition through recurrent neural networks for human–robot interaction in agriculture. *Applied Sciences* 11, 2188.

[116] Babaeian, E., Paheding, S., Siddique, N., Devabhaktuni, V.K., and Tuller, M., 2021. Estimation of root zone soil moisture from ground and remotely sensed soil information with multisensor data fusion and automated machine learning. *Remote Sensing of Environment* 260, 112434.

[117] Barrero, O. and Perdomo, S.A., 2018. RGB and multispectral UAV image fusion for Gramineae weed detection in rice fields. *Precision Agriculture* 19, 809–822.

[118] Gevaert, C.M., Suomalainen, J., Tang, J., and Kooistra, L., 2015. Generation of spectral–temporal response surfaces by combining multispectral satellite and hyperspectral UAV imagery for precision agriculture applications. *IEEE Journal of Selected Topics in Applied Earth Observations and Remote Sensing* 8, 3140–3146.

[119] Maimaitijiang, M., Sagan, V., Sidike, P., Hartling, S., Esposito, F., and Fritschi, F.B., 2020. Soybean yield prediction from UAV using multimodal data fusion and deep learning. *Remote Sensing of Environment* 237, 111599.

[120] Sankey, T.T., McVay, J., Swetnam, T.L., McClaran, M.P., Heilman, P., and Nichols, M., 2018. UAV hyperspectral and lidar data and their fusion for arid and semi-arid land vegetation monitoring. *Remote Sensing in Ecology and Conservation* 4, 20–33.

[121] Zheng, H., Cheng, T., Zhou, M., *et al.*, 2019. Improved estimation of rice aboveground biomass combining textural and spectral analysis of UAV imagery. *Precision Agriculture* 20, 611–629. https://doi.org/10.1007/s11119-018-9600-7

[122] Gao, F., Hilker, T., Zhu, X., *et al.*, 2015. Fusing Landsat and MODIS data for vegetation monitoring. *IEEE Geoscience and Remote Sensing Magazine* 3, 47–60.

[123] Tao, G., Jia, K., Wei, X., *et al.*, 2021. Improving the spatiotemporal fusion accuracy of fractional vegetation cover in agricultural regions by combining vegetation growth models. *International Journal of Applied Earth Observation and Geoinformation* 101, 102362.

[124] Park, S., Im, J., Park, S., and Rhee, J., 2017. Drought monitoring using high resolution soil moisture through multi-sensor satellite data fusion over the Korean Peninsula. *Agricultural and Forest Meteorology* 237, 257–269.

[125] Shen, Y., Shen, G., Zhai, H., Yang, C., and Qi, K., 2021. A Gaussian kernel-based spatiotemporal fusion model for agricultural remote sensing monitoring. *IEEE Journal of Selected Topics in Applied Earth Observations and Remote Sensing* 14, 3533–3545.

[126] Cammalleri, C., Anderson, M., Gao, F., Hain, C., and Kustas, W., 2014. Mapping daily evapotranspiration at field scales over rainfed and irrigated agricultural areas using remote sensing data fusion. *Agricultural and Forest Meteorology* 186, 1–11.

[127] Casa, R., Castaldi, F., Pascucci, S., Basso, B., and Pignatti, S., 2013. Geophysical and hyperspectral data fusion techniques for in-field estimation of soil properties. *Vadose Zone Journal* 12, vzj2012.0201.

[128] Castaldi, F., Castrignanò, A., and Casa, R., 2016. A data fusion and spatial data analysis approach for the estimation of wheat grain nitrogen uptake from satellite data. *International Journal of Remote Sensing* 37, 4317–4336.

[129] Chen, C.-F., Valdez, M.C., Chang, N.-B., Chang, L.-Y., and Yuan, P.-Y., 2014. Monitoring spatiotemporal surface soil moisture variations during dry seasons in central America with multisensor cascade data fusion. *IEEE Journal of Selected Topics in Applied Earth Observations and Remote Sensing* 7, 4340–4355.

[130] Forkuor, G., Conrad, C., Thiel, M., Ullmann, T., and Zoungrana, E., 2014. Integration of optical and Synthetic Aperture Radar imagery for improving crop mapping in Northwestern Benin, West Africa. *Remote Sensing* 6, 6472–6499.

[131] Skakun, S., Kussul, N., Shelestov, A.Y., Lavreniuk, M., and Kussul, O., 2015. Efficiency assessment of multitemporal C-band Radarsat-2 intensity and Landsat-8 surface reflectance satellite imagery for crop classification in Ukraine. *IEEE Journal of Selected Topics in Applied Earth Observations and Remote Sensing* 9, 3712–3719.

[132] Thomas, N., Neigh, C.S., Carroll, M.L., McCarty, J.L., and Bunting, P., 2020. Fusion approach for remotely-sensed mapping of agriculture (FARMA): a scalable open source method for land cover monitoring using data fusion. *Remote Sensing* 12, 3459.

[133] Yahia, O., Guida, R., and Iervolino, P., 2021. Novel weight-based approach for soil moisture content estimation via synthetic aperture radar, multispectral and thermal infrared data fusion. *Sensors* 21, 3457.

[134] Zhou, T., Pan, J., Zhang, P., Wei, S., and Han, T., 2017. Mapping winter wheat with multi-temporal SAR and optical images in an urban agricultural region. *Sensors* 17, 1210.

[135] Barbedo, J.G.A., 2022. Data fusion in agriculture: resolving ambiguities and closing data gaps. *Sensors* 22, 2285.

[136] da Silveira, F., Lermen, F.H., and Amaral, F.G., 2021. An overview of agriculture 4.0 development: systematic review of descriptions, technologies, barriers, advantages, and disadvantages. *Computers and Electronics in Agriculture* 189, 106405.

[137] Irwin, A., 2002. *Citizen Science: A Study of People, Expertise and Sustainable Development.* London: Routledge.

[138] Silvertown, J., 2009. A new dawn for citizen science. *Trends in Ecology & Evolution* 24, 467–471.

Chapter 15

Incremental learning of plant diseases and new plant types: moving towards a smart agriculture system

Yash Mittal[1] *and Uttam Kumar*[1]

Abstract

In machine learning (ML) and deep learning, once a model has been trained and deployed, learning new classes is difficult. In this context, class incremental learning has traditionally been accomplished by retraining an already existing deep neural network (DNN) across newly added classes using either fine tuning techniques, partially freezing the DNN, or by joint training. The drawback of traditional approaches is that either accuracy dramatically declines when adding new classes one at a time or compute power gradually increases. In this work, three conventional and ten continuous ML algorithms have been compared based on online learning techniques for class incremental learning in the plant domain. The basic premise is to initially learn healthy plants leaf data from publicly available Plant Village dataset and subsequently incrementally learn new plants leaf diseases in the first set of experiments. In a separate second series of experiments, the models were trained with plants leaf data from Plant Seedling dataset available in the public domain. Then, the models incrementally learned new plants on the go. The current comparison is motivated by the aforementioned caveats. Task agnostic accuracy and forgetting metrics were utilized to evaluate and compare results of all the algorithms used in the experiments.

Experimental results revealed that incremental ML models may learn new classes of plant diseases and plant types (as classes) incrementally and continually with only an insignificant forgetting, without retraining the model from scratch. These models tend to save the compute resources and training time without any downtime in the current system. Models' performance in terms of overall average accuracy did not degrade even after adding multiple plants-related diseases and new plants. This study highlighted the value of incremental class learning as a significant advancement in the field of agricultural plants that can pave the way for smart agriculture.

[1]Spatial Computing Laboratory, Center for Data Sciences, International Institute of Information Technology (IIIT), Bangalore, India

Keywords: Incremental class learning; Plant classification; Plant disease; Catastrophic forgetting; Smart agriculture

15.1 Introduction

According to the United Nations report, the current world population of 7.6 billion is expected to reach 8.6 billion in 2030, 9.8 billion in 2050, and 11.2 billion in 2100. As a result of the continuously growing population, the agriculture sector is likely to face many challenges, such as reduction in total cultivable land and need for more intensive crop production. With the ongoing research for new technological advancements and interventions during the third wave of agricultural revolution, precision farming is expected to aid in mitigating some of these issues [1]. Nonetheless, disease occurrence and spread in plants/crops impact the overall yield and their quality. Therefore, identification and treatment of plant disease(s) have been recognized as one of the important aspects of precision agriculture. For instance, convolution neural network (CNN) models have been built to classify fixed number of plant diseases [2]. However, when the number of plant diseases are variable in the spatio-temporal domain, the models require re-training from scratch to identify these new disease types, which in turn requires more compute resources and re-training time.

Aiming to create artificially intelligent systems that can constantly learn to address new tasks from new input while maintaining knowledge acquired from previously taught tasks [3,4], incremental learning (IL) attempts to build these capabilities. In most IL scenarios, learners are given tasks in a series of clearly defined training sessions, each of which has only one task worth of data accessible for that learning. After each training session, the learner should be capable of performing all previously seen tasks on unseen data. This type of learning model's biological influence is obvious because it accurately captures how humans learn new information and apply it to past activities [5]. When faced with new tasks to learn, we draw on prior knowledge, and subsequently apply the newly learned information to those tasks. This is completely opposed to the prevalent supervised learning paradigm, where labelled data for all the tasks are available concurrently during a deep network training session. Incremental learners are graded on all the previously learned tasks, but they only have access to data from one task at a time. *In other words, learning from input from the present task without forgetting previously learned tasks is the main challenge in incremental learning.* Lack of data from earlier tasks makes it difficult for the classifier created by the naive approach of fine tuning to categorize the data, which has been successfully used in the domain transfer problems. Catastrophic forgetting [6], or the propensity of a neural network to under fit old data when new ones are investigated, is the problem that must be solved in every situation. The effect of catastrophic forgetting is alleviated either by increasing the model's capacity to accommodate new knowledge or by storing exemplars of past classes in a bounded memory and replaying them in each new state.

Previous work to detect plant diseases using CNN-based deep learning systems starting from simple 9-layer deep CNN to identify plant diseases have shown classification accuracies of 96.46–99.48% [7,8]. However, none of the existing works have attempted to address the issue of catastrophic forgetting and near real-time model adaptation when new disease classes are introduced after the model has been trained. The earliest reported study in 1989 by McCloskey and Cohen [2] concluded that the algorithms trained with backpropagation suffer from catastrophic forgetting which was later confirmed by Ratcliff [9] on a wider range of tasks trained with backpropagation. With the resurgence of deep learning from around 2011 [10], the problem of catastrophic forgetting quickly gained renewed attention [11,12]. This led to a surge of work in IL, continual learning and lifelong learning. Class incremental learning has now become an interesting field of research recently due to its high memory efficiency and adaptive modeling. In the recent study of different class IL algorithms for visual tasks [13], the authors compared different class incremental algorithms on standard datasets like ILSVRC, VGGFACE2, Google Landmarks, and CIFAR100 and found that after multiple stages of addition of new classes, accuracy drops to significant levels. The above review indicated three key challenges as listed below:

(i) **Retraining from scratch** – As discussed earlier, and also reported in [6], authors have shown how CNN models can be built to classify fixed number of plant diseases. However, when the number of plant diseases is variable, the model requires re-training from scratch with overhead on compute resources and time. We attempt to address this issue by model tuning at the time when new classes are added. Retraining will not be required, thus saving compute resources and training time.

(ii) **Retention of complete data for retraining** – To add new classes to an existing trained model, new classes along with all the historical data in memory are required to re-train the model. Access and storage of historical data may not be feasible in several scenarios due to the reasons of data privacy, data purging laws, etc. We try to overcome this issue by utilizing exemplars and non-exemplars-based class IL methods where few or no historical data samples are required to re-train the model respectively.

(iii) **Significant drop in accuracy with high catastrophic forgetting** – The challenge of class IL algorithms [13] with significant drops in accuracy is addressed here. We show that even after the addition of multiple classes over multiple tasks, accuracy does not drop drastically, with minimum forgetting on new classes with the plant dataset. Moreover, class IL [14] showed promising results for plant classification that may have larger domain shifts.

In this study, we attempt to address the above challenges and show the efficacy of class IL [15] to identify new plant-related diseases and new plant classes, where the existing model can be adapted in near real time to the new plant disease and plant data without forgetting the old ones or re-training the model from scratch. Another important aspect that came out of the analysis was that class IL which works well only when the domain shift between tasks is small, actually worked quite well for plant disease and plant identification that generally exhibits larger domain shifts.

15.2 Methodology

In this section, we discuss class IL algorithms which are broadly divided into three categories as described below.

15.2.1 Regularization approaches

To reduce catastrophic forgetting, several methods combine the classification loss and regularization terms. Some regularize on the weights and estimate an importance metric for each parameter in the network [16–21], while others focus on the importance of remembering feature representations [22–27]. Majority of these methods were formed inside the task-IL context and have been examined [28]. We briefly review them because of their significance for class-IL as well. In particular, regularization of feature representations is frequently utilized in class-IL. Finally, we will go through a few regularization methods that have recently been developed just for class-IL.

Weight regularization – The primary goal of the first class of strategies is to stop weight drift that has been found to be important for earlier tasks. They accomplish this by learning each task and then calculating the previous relevance of each parameter in the network (which is presumed to be independent). When training on new tasks, the importance of each parameter is used to penalize changes to them. That is, in addition to the cross-entropy classification loss, an additional loss is introduced as shown by (15.1) below:

$$\mathcal{L}_{\text{reg}}(\theta^t) = \frac{1}{2}\sum_{i=1}^{|\theta^{t-1}|} \Omega_i(\theta_i^{t-1} - \theta_i^t)^2, \tag{15.1}$$

where θ_i^t is the weight i of the network currently being trained, θ_i^{t-1} is the value of this parameter at the end of training on task $t-1$, $|\theta^{t-1}|$ is the number of weights in the network, and Ω_i contains important values for each network weight.

Kirkpatrick *et al.* [12] proposed elastic weight consolidation (EWC) in which Ω_i is calculated as a diagonal approximation of the Fisher Information Matrix (FIM). While ignoring the impact of those parameters along the learning trajectory in weight space, it reflects the model's significance at the bare minimum after each job is learnt. In [21], the parameter space was rotated that offers a better approximation of the FIM to enhance EWC. While training, the model was expanded with fixed parameters; this does not increase the network's capacity but adds to the computational and memory costs. An extension of the Kronecker factorization technique [29] was suggested for block-diagonal approximation of the FIM in an effort to improve its approximation in EWC. They additionally demonstrated accommodation of batch normalization through Kronecker factorizations.

The Path Integral technique (PathInt) [19], in contrast, accumulates the changes in each parameter online over the course of the entire learning trajectory. Starting from pretrained models result in an underestimation of the importance while batch modifications to the weights cause an overestimation. Memory Aware Synapses (MAS) [18] also suggests that this should be addressed via Ω_i online by collecting the sensitivity of the learnt function (the magnitude of the gradient). The Riemanian Walk (RWalk) algorithm [17] fuses the FIM approximation and the online path integral to determine the relative weights of each parameter. RWalk also takes advantage of examples to boost performance even more.

Newer trends in regularization – In recent work on class-IL, a number of regularization strategies have been put forth. For example, Zagoruyko and Komodakis [30] proposed to use the attention of the teacher network to guide the student network. This method was utilized by Learning without Memorizing (LwM) [31] for class-IL. The basic idea is that when training a new task, the network's attention strategy from earlier training should not alter. It is anticipated that the factors used to determine a particular class designation will not change. The attention distillation loss reinforces Eq. (15.2):

$$\mathcal{L}_{AD}(\mathrm{x};\theta^t) = \left\| \frac{Q^{t-1}(\mathrm{x})}{\|Q^{t-1}(\mathrm{x})\|_2} - \frac{Q^t(\mathrm{x})}{\|Q^t(\mathrm{x})\|_2} \right\|_1,$$

where the attention map Q is given by: (15.2)

$$Q^t(\mathrm{x}) = \mathrm{Grad} - \mathrm{CAM}\ (\mathrm{x},\ \theta^t,\ c)$$
$$Q^{t-1}(\mathrm{x}) = \mathrm{Grad} - \mathrm{CAM}\ (\mathrm{x},\ \theta^{t-1},\ c)$$

and is generated with the Grad-CAM algorithm [32]. To create a coarse localization map representing the image regions that contributed to the prediction, Grad-CAM computes the gradient with respect to a target class c. The target class label cannot be used here because it was not included during the training of the preceding model θ^{t-1}. Instead, computing the attention maps is suggested using the prior class predicted with the highest probability (15.3):

$$c = \mathrm{argmaxh}(x;\ominus^{t-1}). \tag{15.3}$$

Deep Model Consolidation (DMC) is another modern technique that builds on LwF [26]. It is based on the observation that there is an asymmetry between previous and new classes throughout training: new classes have explicit, strong supervision, whereas old classes have less-effective supervision that is communicated through knowledge distillation. To remove this asymmetry, double distillation loss on the model θ^{t-1} trained on previous classes was applied and a newly trained model θ^t for the new classes (allowing this model to forget previous tasks) as shown in (15.4) and (15.5):

$$\mathcal{L}_{DD}(\mathrm{u};\theta) = \frac{1}{N^t}\sum_{k=1}^{N^t}(\mathrm{o}_k(\mathrm{u}) - \mathrm{o}_k^{\circ}(\mathrm{u}))^2, \tag{15.4}$$

where $\mathrm{o_k}^{\circ}$ are normalized logits :

$$\mathrm{o}_k^{\circ}(\mathrm{u}) = \begin{cases} \mathrm{o}_k^{t-1}(\mathrm{u}) - \dfrac{1}{N^{t-1}}\displaystyle\sum_{l=1}^{N^{t-1}}\mathrm{o}_l^{t-1}(\mathrm{u}) & \text{if } 1 \le k \le N^{t-1} \\ \mathrm{o}_k^{t}(\mathrm{u}) - \dfrac{1}{N^{t}}\displaystyle\sum_{l=1}^{N^{t}}\mathrm{o}_l^{t}(\mathrm{u}) & \text{if } N^{t-1} < k \le N^t \end{cases} \tag{15.5}$$

Here $o^{t-1}(u)$ refers to the logits from the network trained on previous tasks, and $o^{t}(u)$ the ones trained on the new classes. Because the algorithm does not have access to data of previous tasks, auxiliary data "u" is used, which can be any unlabeled data from a similar domain.

Global Distillation (GD) [27] also suggested using external data to extract information from earlier tasks in a similar manner. The present task was first taught to a teacher, whose confidence is then calibrated using examples and external data. After that, they triple-distill a new model using input from the teacher, the prior model, and their group (plus the cross-entropy loss for current task and exemplars). Data from the present task and external datasets were utilized with the teacher and the prior task, but only the external data was used with the ensemble. Finally, by weighting the loss in accordance with the quantity of data, they carried out fine tuning while avoiding task recency bias. The external dataset sampling method was based on the predictions of the model.

In another related work, current task data was not trained directly [33], but rather an expert teacher was trained first. The method, which is similar to LwF but used saved exemplars rather than new data, further distilled an old model with a small fraction of previous task data reserved to preserve performance on old tasks. Based on an analysis of Incremental Classifier and Representation Learning (iCaR) [34], it was asserted that the nearest exemplar mean classifier was biased and suggested employing a dynamic threshold moving approach to correct it by retaining an up-to-date scaling vector. This classifier bias was trained with distillation loss. Finally, Hou *et al.* [35] technique included the less-forget constraint, which is a variation of LwF. They suggested regularizing on the cosine similarity between the L2-normalized logits of the previous and current network rather than regularizing network predictions (see (15.6)):

$$\mathcal{L}_{lf}(\mathrm{x};\theta) = 1 - \frac{\langle \mathrm{o}^{t-1}(\mathrm{x}), \mathrm{o}^{t}(\mathrm{x}) \rangle}{\|\mathrm{o}^{t-1}(\mathrm{x})\|_2 \|\mathrm{o}^{t}(\mathrm{x})\|_2}, \tag{15.6}$$

where <.,.> is the inner product between vectors. This regularization is less sensitive to task imbalance because the comparison is between normalized vectors; the loss reduces bias against new classifications.

15.2.2 Rehearsal approaches

Rehearsal-based methods keep a limited number of examples [2], such as exemplar rehearsal [36,37], or create artificial images or features [38–41], such as pseudo-rehearsal. Rehearsal techniques try to stop people from forgetting previous tasks by replaying the created or stored data from those tasks. Majority of the rehearsal techniques combine the use of exemplars to address inter-task confusion with strategies that address other factors that contribute to catastrophic forgetting. iCaRL was the first of its kind to suggest the use of exemplar rehearsal for class-IL [2]. Since then, the bulk of class-IL methods have used this method. This section focuses on the decisions that must be made while using exemplars.

Different memory types: After the model has been modified to the new task and training, the exemplar memory must be extended. Some exemplars must be deleted in order to make room for the new ones if the memory has a fixed maximum size across all tasks (fixed memory). By doing this, it is ensured that the memory capacity is fixed and constant. More the activities and classes learned, less is the rehearsal time allotted for each class. The memory could be increased after mastering a predetermined number of jobs in order to better support the new distributions. However, previously deleted samples are lost, making the timing of expansion crucial. The only additional samples required if the memory is permitted to expand are those from the present task. This requires a linear increase of memory, which may not be appropriate in some applications, but it encourages the classes to have a stable representation throughout rehearsal across all jobs. To provide an equal representation of all classes, the number of exemplars per class is required to be same in both the scenarios.

Sampling strategies: Random sampling from the available data is the simplest technique to choose exemplars to add to the memory and has been found to be successful with minute computing cost [2,17]. Inspired by the authors of [42], iCaRL proposed to select exemplars based on their corresponding feature space representations (herding). Exemplars are chosen iteratively for each of the classes using the approach. An exemplar is chosen at each phase so that, when added to the exemplars of its class, the resulting exemplar mean is most comparable to the actual class mean. When certain exemplars need to be removed, the order in which they are added is crucial and is taken into consideration. Despite typically outperforming random, this iterative selection method raises processing costs. Two additional sampling techniques are suggested in RWalk [17]. In the first one, entropy of the softmax output is calculated, and examples with greater entropies are chosen (entropy). This makes it mandatory to choose samples with more evenly distributed scores across all classes. In the second method, exemplars are similarly chosen depending on how close they are to the decision boundary (distance) assuming that the feature space and the decision boundaries do not change much. The pseudo-distance to the decision boundary for a given sample (x_i, y_i) is calculated by $f(x_i; \varphi)^T V_y$, meaning that smaller the distance, more closer to the decision boundary. With the exception of random, the order in which the exemplars are selected for these sampling strategies is recorded after a decreasing order of importance. The exemplars with the lesser importance are the ones deleted first if a fixed memory is used as some memory must be freed to make place for new exemplars.

Task balancing: The trade-off between the two components of the loss as well as the quantity of samples from each class at each training step determine the weights of the new classes in comparison to the old ones when rehearsal is used during the training of a new job. Most methods draw training batch samples from a pool that includes both fresh data and rehearsal exemplars [2,17,35,36]. The trade-off between the cross-entropy loss and the other losses that prevent forgetting implies that batches plainly have an excess of new samples. In contrast, a more equitable training system was suggested [37] in which batches were equally split between new and previous groups. This appears to have some positive impacts in adjusting for the imbalance in the tasks.

15.2.3 Bias correction approaches

The issue of task-recency bias, which is the tendency of an incrementally learnt network to be biased towards classes in the most recently learned task, is addressed through bias-correction approaches. This is mostly due to the network having seen numerous examples of the classes in the most recent task but none (or very few in the case of rehearsal) from older tasks at the conclusion of training. Hou *et al.* [35] observed that the classifier bias favors the most recent classes and that the classifier norm is larger for the new classes than for the old classes which is a direct consequence of this.

The earlier mentioned iCaRL method [2] combined exemplars and Learning without Forgetting, using a classifier layer and cross-entropy during training. They do not use the classifier during inference in order to avoid the task recency bias. Instead, they calculate the class mean of the feature representation's exemplars before applying the nearest exemplar mean to classify the data. The method was shown to be significantly less prone to the task recency bias because this process is independent of the weights and biases of the final layer.

Castro *et al.* [37] end-to-end IL (EEIL) method included one straightforward yet efficient solution to prevent task-recency bias. At the end of each training session, they recommend introducing a new stage called balanced training. For a limited number of iterations during this phase, an equal number of exemplars from each class was used. They introduced a distillation loss on the classification layer, but only for the classes from the current task, to prevent forgetting the new classes. Balanced training results in overfitting the stored exemplars when these exemplars are not accurate.

Wu *et al.* [36] provided another simple and successful solution for reducing task-recency bias – Bias Correction (BiC). To the network, they included an additional layer intended to correct task bias. A training session consisted of two parts. They trained the new task using the cross-entropy loss and the distillation loss during the first stage. As a validation set for the second phase, they split a very small portion of the training data. To address the task-recency bias, learning a linear transformation on top of the logits, O_k was recommend. The converted logits are given in (15.7):

$$q_k = \alpha_s O_k + \beta_S, \quad c_k \in C^S \tag{15.7}$$

where α_s and β_s are the parameters which compensate for the bias in task s. There are only two parameters for each task that are common by all classes in that task (initialized $\alpha_1 = 1$ and $\beta_1 = 0$). With the exception of the parameters for the active task α_t and β_t, all network parameters are locked throughout the second phase. These are optimized with a standard softmax on the transformed logits q_k using the set-aside validation set. Finally, they only apply a weight decay on β parameters and not on the α parameters.

As previously mentioned, Hou *et al.* [35] also noted task-recency bias and proposed to swap the conventional softmax layer in their Learning a Unified Classifier Incrementally through Rebalancing (LUCIR) technique for a cosine normalization layer in accordance with (15.8):

$$\mathcal{L}_{cos}(\mathrm{x};\theta^t) = \sum_{k=1}^{N^t} yk \log \frac{\exp\left(\eta\langle\frac{f(\mathrm{x})}{\|f(\mathrm{x})\|}, \frac{V_k}{\|V_k\|}\rangle\right)}{\sum_{i=1}^{N^t}\exp\left(\eta\langle\frac{f(\mathrm{x})}{\|f(\mathrm{x})\|}, \frac{V_i}{\|V_i\|}\rangle\right)} \tag{15.8}$$

where *f(x)* are the feature extractor output, $<.,.>$ is the inner product, V_k are the classifier weights (also called class embedding) related to class k, and η is a learnable parameter which controls the peakedness of the probability distribution.

The issue of inter-task confusion was also addressed [35] where the margin ranking loss was used to prevent new classes from taking up the same space as classes from earlier tasks. The current embedding is pushed further from the embeddings of the K most similar classes by the following loss (15.9):

$$\mathcal{L}_{mr}(\mathrm{x}) = \sum_{k=1}^{K} \max\left(m - \left\langle \frac{f(\mathrm{x})}{\|f(\mathrm{x})\|}, \frac{V_y}{\|V_y\|}\rangle + \langle\frac{f(\mathrm{x})}{\|f(\mathrm{x})\|}, \frac{V_k}{\|Vk\|} \right\rangle, 0\right) \tag{15.9}$$

where V_y refers to the ground truth class embedding of x, V_k is the embedding of the closest classes, and m is the margin.

Recently, Belouadah and Popescu [15] suggested Class-IL with dual memory that overcomes task-recency bias (IL2M). Their approach is comparable to BiC [36] in which correction of the predictions made by the network was suggested. However, IL2M correction is based on the saved certainty statistics of predictions of classes from prior tasks, unlike BiC, which learns to correct the predictions by adding an additional layer. They compute the rectified predictions of the previous k classes using the definition of m = argmax $\hat{y}$ (x) as given in (15.10):

$$\widehat{y}^r_k(\mathrm{x}) = \begin{cases} \widehat{y}k(\mathrm{x}) \times \frac{\bar{y}^p_k}{y^t_k} \times \frac{\bar{y}^t}{y^p} & \text{if } m \in C^t \\ \widehat{y}k(\mathrm{x}) & \text{otherwise.} \end{cases} \tag{15.10}$$

Here $y^k p$ (superscript p refers to past) is the mean of the predictions $\hat{y}_k$ for all images of class c_k after training the task in which class c_k is first learned ($c_k \in Cp$), and y^p is the mean of the predictions for all classes in that task. Following the corresponding training session, y_k^p and y^p are both immediately saved. After training the new assignment, y_k^t is the mean of the predictions made for all images in class c_k (this is computed based on the exemplars). Similarly, y^t is the new task's mean forecast for every class. The correction is only used when the anticipated class is a new class ($m \in C^t$). The authors contend that if the projected class is an older class, no correction is necessary because the prediction does not suffer from task-imbalance.

15.3 Data and experiments

In this section, we discuss the data, experimental setup, and evaluation metrics to evaluate different approaches as discussed in Methodology section on the Plant Village and Plant Seedling dataset.

15.3.1 Dataset

Two most popular datasets – Plant Village dataset (PVD) and Plant Seedling dataset (PSD) – were used in the experiments. PVD consists of plant disease data with a total of 39 classes including one background class and PSD has approximately 960 unique plants belonging to 12 species at several growth stages.

In PVD data, augmentation methods like scaling, rotation, principal component analysis (PCA), color augmentation, noise introduction, gamma correction and image flipping were used to increase the dataset size from 55,636 images to 61,486 with 1,143 background class images so that a more generalized model could be developed. This dataset contains 13 different plant leaves which were subdivided into *healthy* and one or more type of *infected leaf* for each of the plants. The sample images of the random classes are shown in Figure 15.1, and Figure 15.2 renders the transformations that were applied in the augmentation process of the sampled images in the training data. PSD comprises annotated RGB images with a physical resolution of roughly 10 pixels/mm (Figure 15.3). The database has been recorded at Aarhus University Flakkebjerg Research Station with a collaboration between University of Southern Denmark and Aarhus University. Details of the different classes in both the dataset are available in class order tables.

15.3.2 Class order

In PVD, the class ordering was done in such a way that initial model was trained on most of the healthy leaf images, and then later in each task, new classes of diseases for specific plants were added. Subsequently, in task 1, all the diseases of apple and cherry were added. Similarly, in task 2, all the diseases of corn and peach were added. In task 3, grapes and pepper were supplemented. In task 4, 2 classes of potato, 1 class of strawberry and 1 class of tomato were added. Finally in task 5 and 6, all the left-out classes of tomato were included. Initially in task 1, 15 classes were used to train the model and later, in the next 6 tasks, 4 new classes were added in subsequent steps. Thus, a total of 39 classes were used in the entire experiment. The details of this implementation are given in Table 15.1.

In PSD, class ordering ensured that early model was trained on different visible plants, and then later in each task new plants were added. Initially in task 0, 6 classes were added, namely, Black-grass, Charlock, Common Chickweed, Fat Hen, Maize and Small flowered Cranesbill. From tasks 1 to 3, two new classes were included in each of the tasks. In task 1, Cleavers and Common Wheat, in task 2, Loose Silky-bent and Scentless Mayweed, and similarly in task 3, Shepherd Purse and Sugar Beet were added. The details are listed in Table 15.2.

Figure 15.1 Sample images of the random class in PVD. (a) Apple with black rot, (b) grape with leaf blight, (c) tomato with leaf mold, (d) cherry with powdery mildew, (e) potato with early blight, (f) healthy soybean, (g) peach with bacterial spots, and (h) healthy blueberry.

15.3.3 Implementation details

Resnet-18 was used as a backbone architecture for all the experiments. Identical initial conditions and random seed were set to do a fair comparison across different algorithms. Data from previous tasks except exemplars were not available during training. The results were initially evaluated on baselines where fine tuning, model freezing, and joint training with all the classes were performed. Algorithms without exemplars used for experiments were fine tuning, freezing, joint, MAS, PathInt, EWC, and LWM. Algorithms with exemplars used for the experiments were RWalk, EEIL, iCaRL, IL2M, LUCIR, and BiC. Training with only cross-entropy loss was the default fine tuning baseline which was treated as a "lower baseline" for the experiments. In case of freezing, all the layers except the last one (corresponding to the classification layer or head of the network) were frozen after the first task was learned. Finally, incremental joint training was performed, which used all the seen data at each task, starting from the model learned for the previous one. This baseline rendered an upper bound reference for all the learned tasks. Continual hyperparameter

Figure 15.2 Augmented images in PVD. (a) Original image, and (b)–(h) are augmented images.

Figure 15.3 Sample images of all the species from PSD

Table 15.1 Class order in different tasks for PVD

Task	Classes used
0	Apple_healthy, Background_without_leaves, Blueberry_healthy, Cherry_healthy, Corn_healthy, Corn_healthy, Grape_healthy, Peach_healthy, Pepper_bell_healthy, Potato_healthy, Raspberry_healthy, Soybean_healthy, Strawberry_healthy, Tomato_healthy, Squash_Powdery_mildew
1	Apple_scab, Apple_Black_rot, Apple_Cedar_apple_rust, Cherry_Powdery_mildew
2	Corn_Cercospora_leaf_spot_Gray_leaf_spot, Corn_Common_rust, Corn_Northern_Leaf_Blight, Peach_Bacterial_spot
3	Grape_Black_rot, Grape_Esca_(Black_Measles), Grape_Leaf_blight_(Isariopsis_Leaf_Spot), Pepper_bell_Bacterial_spot
4	Potato_Early_blight, Potato_Late_blight. Strawberry_Leaf_scorch, Tomato_Bacterial_spot
5	Tomato_Early_blight, Tomato_Late_blight, Tomato_Leaf_Mold, Tomato_Septoria_leaf_spot
6	Tomato_Spider_mites_Two-spotted_spider_mite, Tomato_Target_Spot, Tomato_Yellow_Leaf_Curl Virus, Tomato_mosaic_virus

Table 15.2 Class order in different tasks for PSD

Task	Classes used
0	Black-grass, Charlock, CommonChickweed, FatHcn, Maize, Small-floweredCranesbill
1	Cleavers, CommonWheat
2	LooseSilky-bent, ScentlessMayweed
3	ShepherdPurse, SugarBeet

framework [28] was used for hyperparameter tuning to select the best model. This framework ensures that the data for that task is available, as in real scenario. For each task, a maximal plasticity search with fine tuning and a stability decay phase were used with the corresponding method which allowed to establish a reference performance and find the best stability-plasticity trade-off.

15.3.4 Evaluation metrics

To evaluate the performance of the algorithms, suitable metrics were considered for class-IL tasks. Here, two major metrics were used – average accuracy and average forgetting. When the task had variable number of classes, a class frequency weighted version was used. Forgetting estimates how much the model forgot about previous tasks i.e., the tendency of an artificial neural network to completely and abruptly forget previously learned information upon learning new information; used more in task-IL cases.

15.4 Results and discussion

In this section, results for both PVD and PSD are presented and discussed considering all the three approaches, viz. regularization, rehearsal, and bias correction approaches with the baselines. Initially, the model was trained to get three baselines: (i) the lower bound by fine tuning, (ii) upper bound by joint training of all the classes, and (iii) by freezing the head and tuning the last layers. Comparison based on average task agnostic accuracy revealed that LwM performed best in regularization-based methods on both the datasets. In rehearsal-based approaches, RWalk performed worst for PVD and best for PSD while EEIL was superior on PVD. Among the bias-correction techniques, LUCIR performed best on both the datasets. LwM was superior to other non-exemplar-based methods while LUCIR rendered the highest accuracy among exemplar methods with an average task agnostic accuracy of 94.9% and 88.5% which is very close to the upper baseline of 99.6% and 92.8% on PVD and PSD. It is to be noted that the average task forgetting for LUCIR is the lowest on PVD and comparable to other methods on PSD. Detailed analysis of the results are presented in Tables 15.3–15.6 and Figures 15.4 and 15.5 demonstrates combined analysis of all the algorithms on both the datasets.

Table 15.3 Average task agnostic accuracy for different algorithms on PVD

	Algorithms	**T0**	**Tl**	**T2**	**T3**	**T4**	**T5**	**T6**
Without exemplars	Fine-tuning	99.9	48.6	36.4	34.5	29.1	18.1	17.5
	Freezing	99.9	57.7	45.6	33.6	30.5	25.9	26.3
	Joint	99.9	99.6	99.3	99.5	99.6	99.5	99.6
	MAS	99.8	63.8	40.8	31.0	35.1	25.7	22.8
	PathInt	99.9	57.4	37.4	41.8	34.9	22.9	20.4
	EWC	99.8	65.0	42.1	37.8	35.0	19.6	18.0
	LwM	99.9	63.6	51.5	43.2	43.0	36.0	29.9
With exemplars	RWalk	99.9	97.8	85.4	68.4	11.9	32.3	34.7
	EEIL	99.9	98.9	96.5	96.7	95.3	92.3	91.4
	iCaRL	98.6	93.6	89.7	89.8	88.8	89.0	87.1
	IL2M	99.9	99.1	96.8	93.9	95.9	93.7	91.8
	LUCIR	99.8	98.5	97.1	96.2	96.6	94.9	**94.9**
	BiC	99.8	98.8	97.7	97.0	96.4	94.7	94.3

Table 15.4 Average task agnostic accuracy for different algorithms on PSD

	Algorithms	**T0**	**T1**	**T2**	**T3**
Without exemplars	Fine-tuning	95.3	27.1	21.9	21.8
	Freezing	95.3	47.6	33.1	26.5
	Joint	95.3	93.0	93.0	92.8
	MAS	95.4	25.6	21.6	28.6
	PathInt	95.3	26.3	21.6	27.7
	EWC	95.4	27.0	22.0	23.6
	LwM	95.3	45.5	30.6	32.5
With exemplars	RWalk	95.8	87.3	86.6	82.8
	EEIL	95.3	87.8	85.0	80.2
	iCaRL	94.7	81.5	74.9	75.3
	IL2M	95.3	88.5	84.6	84.6
	LUCIR	98.2	87.8	89.4	**88.5**
	BiC	97.2	88.3	80.3	80.6

Table 15.5 Average task agnostic forgetting for different algorithms on PVD

	Algorithms	**Tl**	**T2**	**T3**	**T4**	**T5**	**T6**
Without exemplars	Fine-tuning	25.1	40.3	46.6	57.0	66.0	63.2
	Freezing	0.7	3.0	5.2	4.0	3.2	4.5
	Joint	0.1	−0.0	−0.0	−0.1	0.1	−0.0
	MAS	7.5	30.6	23.6	28.5	27.0	22.1
	Pathlnt	12.8	39.1	33.6	42.0	53.0	50.6
	EWC	33.3	60.0	59.0	59.1	65.8	64.3
	LwM	1.0	16.6	30.4	31.5	37.3	43.6
With exemplars	RWalk	2.0	15.0	31.3	87.1	54.6	52.1
	EEDL	1.2	3.8	3.2	4.9	8.2	9.0
	iCaRL	1.9	2.4	2.5	2.5	0.1	2.8
	IL2M	1.5	3.8	7.0	4.2	6.6	8.7
	LUCIR	1.4	1.8	1.6	1.2	2.3	**2.2**
	BiC	1.1	1.7	1.8	2.5	3.7	3.9

Table 15.6 Average task agnostic forgetting for different algorithms on PSD

	Algorithms	**Tl**	**T2**	**T3**
Without exemplars	Fine-tuning	42.7	15.6	5.3
	Freezing	0	0.2	0.1
	Joint	−0.9	1.1	1.4
	MAS	47.6	18.6	3.8
	Pathlnt	43.6	15.6	3.2
	EWC	43.9	15.9	5.5
	LwM	8.4	10.9	12.4
With exemplars	RWalk	7.6	8.1	13.2
	EEIL	6.9	12.9	17.3
	iCaRL	3.0	3.9	3.0
	IL2M	9.8	15.8	13.6
	LUCIR	4.0	3.1	**4.4**
	BiC	4.4	5.0	2.3

15.5 Conclusions

One of the most important challenges associated with a live model in system is to encounter new plant classes in agricultural domain. It is crucial to develop a system which overcomes this challenge without retraining the model from scratch. There could be scenarios where data with which the model was trained previously may not be available and retraining with all data might be infeasible. Retraining model from scratch is not cost and compute effective. In this chapter, the efficacy of incremental learning in solving some plant-related challenges in agricultural domain was demonstrated with experiments. LUCIR performed better than all other techniques on both the PVD and PSD.

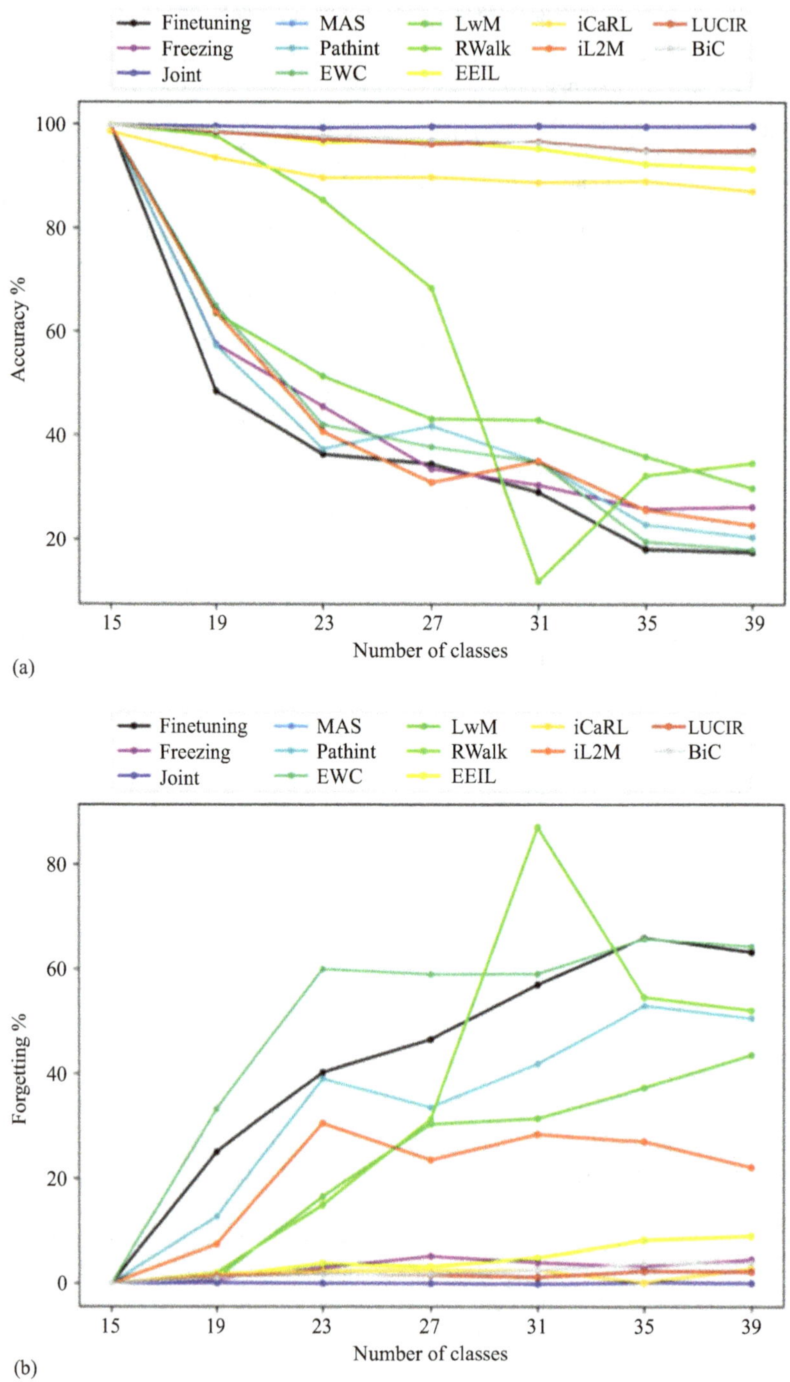

Figure 15.4 Average task agnostic: (a) accuracy and (b) forgetting on different algorithms for PVD

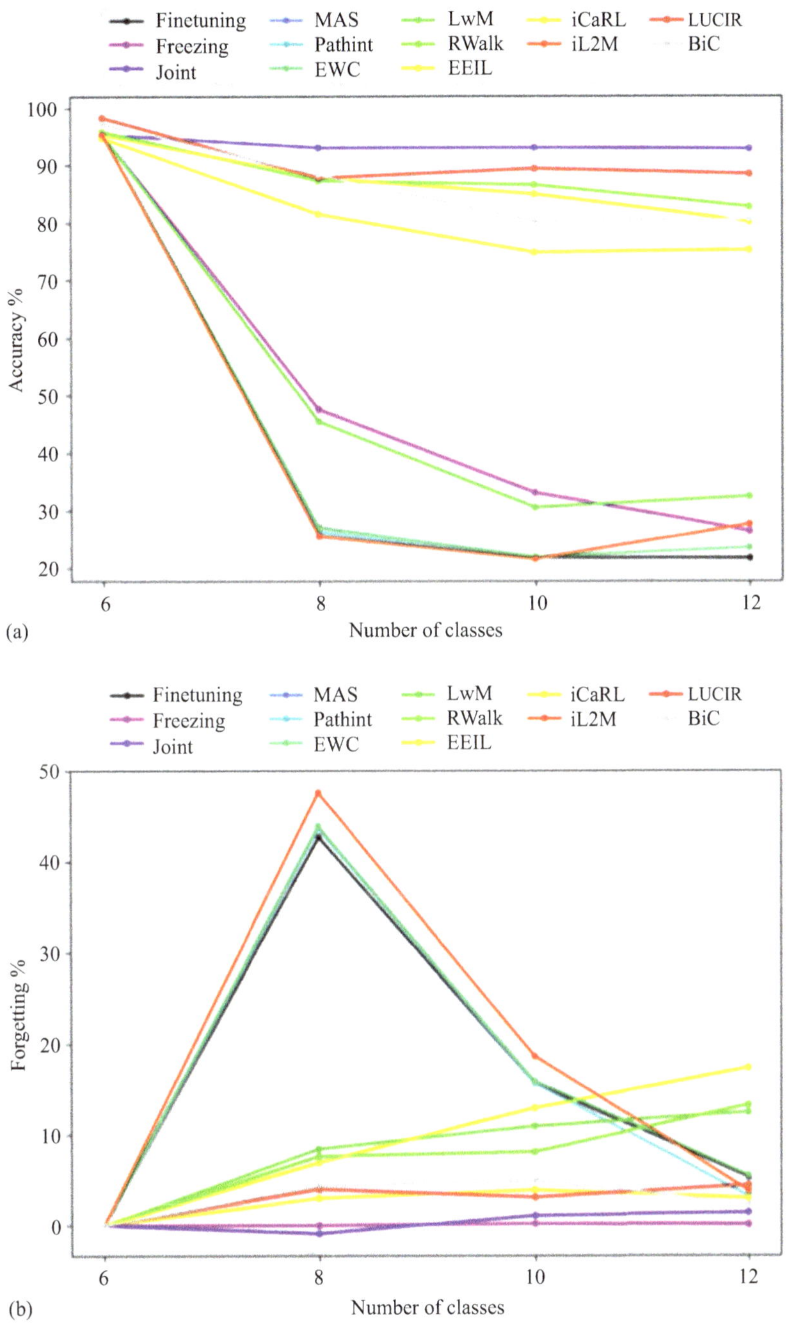

Figure 15.5 *Average task agnostic: (a) accuracy and (b) forgetting on different algorithms for PSD*

Overall, exemplar methods rendered higher accuracy compared to non-exemplar methods. In future, quantitative analysis of the cost and compute power with these class incremental ML systems will be explored and will be compared to traditional systems where models are trained from scratch.

References

[1] V.M. Abdul, A. Joseph, A.J. Ajay, and K. Mufeedha. "Precision farming: the future of Indian agriculture," *J. Appl. Biol. Biotechnol.* 2016;11:68–72. https://doi.org/10.7324/JABB.2016.40609

[2] M. McCloskey and Neal J. Cohen. "Catastrophic interference in connectionist networks: the sequential learning problem," in *Psychology of Learning and Motivation*, vol. 24. Elsevier, pp. 109–165.

[3] S.-A. Rebuffi, A. Kolesnikov, G. Sperl, and C.H. Lampert, "icarl: Incremental classifier and representation learning," in *Proceedings of the IEEE Conference on Computer Vision and Pattern Recognition*, 2017.

[4] S. Thrun, "Is learning the nth thing any easier than learning the first?" in *Proceedings of the Advances in Neural Information Processing Systems*, 1996.

[5] R.M. French, "Catastrophic forgetting in connectionist networks," *Trends Cognit. Sci.*, 1999;3:128–135.

[6] M. Mccloskey and N.J. Cohen. "Catastrophic interference in connectionist networks: the sequential learning problem," *Psychol. Learn. Motivat.* 1989;24:104–169.

[7] G. Geetharamani and J. Arun Pandian. "Identification of plant leaf diseases using a nine-layer deep convolutional neural network. *Comput. Electr. Eng.* 2019;76:323–338. https://doi.org/10.1016/j.compeleceng.2019.04.011

[8] J.P. Schwarz Schuler, S. Romaní, M. Abdel-nasser, H. Rashwan, and D. Puig. "Color-aware two-branch DCNN for efficient plant disease classification," *Mendel* 2022;28:55–62. https://doi.org/10.13164/mendel.2022.1.055

[9] R. Ratcliff. "Connectionist models of recognition memory: constraints imposed by learning and forgetting functions," *Psycholog. Rev.* 1990;97 (2):285.

[10] A. Krizhevsky, I. Sutskever, and G.E. Hinton. "Imagenet classification with deep convolutional neural networks," in *Advances in Neural Information Processing Systems* vol. 25, 2012.

[11] I.J Goodfellow, M. Mirza, D. Xiao, A. Courville, and Y. Bengio. "An empirical investigation of catastrophic forgetting in gradient-based neural networks," 2013, arXiv preprint arXiv:1312.6211.

[12] J. Kirkpatrick, R. Pascanu, N. Rabinowitz, *et al.* "Overcoming catastrophic forgetting in neural networks," *Proc. Natl. Acad. Sci.* 2017;114:3521–3526.

[13] E. Belouadah, A. Popescu, and I. Kanellos. "A comprehensive study of class incremental learning algorithms for visual tasks," CoRR abs/2011.01844, 2020. arXiv:2011.01844. https://arxiv.org/abs/2011.01844

[14] M. De Lange, R. Aljundi, M. Masana, *et al.* "A continual learning survey: defying forgetting in classification tasks," *IEEE Trans. Pattern Anal. Mach. Intell.* 2022;44:3366–3385. https://doi.org/10.1109/TPAMI.2021.3057446

[15] E. Belouadah and A. Popescu, "Il2m: class incremental learning with dual memory," in *Proceedings of the IEEE International Conference on Computer Vision*, 2019.

[16] J. Kirkpatrick, R. Pascanu, N. Rabinowitz, *et al.*, "Overcoming catastrophic forgetting in neural networks," *Proc. Natl. Acad. Sci.*, 2017;114(13):3521–3526.

[17] A. Chaudhry, P. K. Dokania, T. Ajanthan, and P. H. Torr, "Riemannian walk for incremental learning: understanding forgetting and intransigence," in *Proceeding of the European Conference on Computer Vision*, 2018.

[18] R. Aljundi, F. Babiloni, M. Elhoseiny, M. Rohrbach, and T. Tuytelaars, "Memory aware synapses: learning what (not) to forget," in *Proceeding of the European Conference on Computer Vision*, 2018.

[19] F. Zenke, B. Poole, and S. Ganguli, "Continual learning through synaptic intelligence," in *Proceedings of the International Conference on Machine Learning*, 2017.

[20] S.-W. Lee, J.-H. Kim, J. Jun, J.-W. Ha, and B.-T. Zhang, "Overcoming catastrophic forgetting by incremental moment matching," in *Proceedings of the Neural Information Processing Systems*, 2017.

[21] X. Liu, M. Masana, L. Herranz, J. Van de Weijer, A. M. Lopez, and A. D. Bagdanov, "Rotate your networks: better weight consolidation and less catastrophic forgetting," in *International Conference on Pattern Recognition*, 2018.

[22] H. Jung, J. Ju, M. Jung, and J. Kim, Less-forgetting learning in deep neural networks," arXiv, 2016.

[23] Z. Li and D. Hoiem, "Learning without forgetting," *IEEE Trans. Pattern Anal. Mach. Intell.*, 2017;40:2935–2947.

[24] A. Rannen, R. Aljundi, M. B. Blaschko, and T. Tuytelaars, "Encoder based lifelong learning," in *Proceeding of the IEEE Conference on Computer Vision and Pattern Recognition*, 2017.

[25] D. L. Silver and R. E. Mercer, "The task rehearsal method of life- long learning: overcoming impoverished data," in *Conference of the Canadian Society for Computational Studies of Intelligence*, 2002.

[26] J. Zhang, J. Zhang, S. Ghosh, *et al.*, "Class-incremental learning via deep model consolidation," in *Proceeding of the IEEE Winter Conference on Applications of Computer Vision*, 2020.

[27] K. Lee, K. Lee, J. Shin, and H. Lee, "Overcoming catastrophic forgetting with unlabeled data in the wild," in *Proceedings of the IEEE International Conference on Computer Vision*, 2019.

[28] M. Delange, R. Aljundi, M. Masana, *et al.*, "A continual learning survey: defying forgetting in classification tasks," *IEEE Trans. Pattern Anal. Mach. Intell.*, 2021;44:3366–3385.

[29] J. Lee, H. G. Hong, D. Joo, and J. Kim, "Continual learning with extended Kronecker-factored approximate curvature," in *Proceeding of the IEEE Conference on Computer Vision and Pattern Recognition*, 2020.

[30] S. Zagoruyko and N. Komodakis, "Paying more attention to attention: Improving the performance of convolutional neural networks via attention transfer," in *Proceeding of the International Conference on Learning Representations*, 2017.

[31] P. Dhar, R. V. Singh, K.-C. Peng, Z. Wu, and R. Chellappa, "Learning without memorizing," in *Proceeding of the IEEE Conference on Computer Vision and Pattern Recognition*, 2019.

[32] R.R. Selvaraju, M. Cogswell, A. Das, R. Vedantam, D. Parikh, and D. Batra, "Grad-cam: visual explanations from deep networks via gradient-based localization," in *Proceeding of the IEEE Conference on Computer Vision*, 2017.

[33] S. Hou, X. Pan, C. C. Loy, Z. Wang, and D. Lin, "Lifelong learning via progressive distillation and retrospection," in *Proceedings of the European Conference on Computer Vision*, 2018.

[34] K. Javed and F. Shafait, "Revisiting distillation and incremental classifier learning," in *Asian Conference on Computer Vision*, 2018.

[35] S. Hou, X. Pan, C. C. Loy, Z. Wang, and D. Lin, "Learning a unified classifier incrementally via rebalancing," in *Proceeding of the IEEE Conference on Computer Vision*, 2019.

[36] Y. Wu, Y. Chen, L. Wang, *et al.*, "Large scale incremental learning," in *Proceeding of the IEEE Conference on Computer Vision and Pattern Recognition*, 2019.

[37] F. M. Castro, M. J. Marín-Jiménez, N.Guil, C.Schmid, and K.Alahari, "End-to-end incremental learning," in *Proceedings of the European Conference on Computer Vision*, 2018.

[38] H. Shin, J. K. Lee, J. Kim, and J. Kim, "Continual learning with deep generative replay," in *Proceeding of the Advances in Neural Information Processing Systems*, 2017.

[39] O. Ostapenko, M. Puscas, T. Klein, P. Jähnichen, and M. Nabi, "Learning to remember: a synaptic plasticity driven framework for continual learning," in *Proceeding of the IEEE Conference on Computer Vision and Pattern Recognition*, 2019.

[40] Y. Xiang, Y. Fu, P. Ji, and H. Huang, "Incremental learning using conditional adversarial networks," in *Proceeding of the IEEE Conference on Computer Vision*, 2019.

[41] R. Kemker and C. Kanan, "Fearnet: brain-inspired model for incremental learning," in *Proceeding of the International Conference on Learning Representations*, 2018.

[42] M. Welling, "Herding dynamical weights to learn," in *Proceedings of the International Conference on Machine Learning*, 2009.

Chapter 16
Disaster susceptibility analysis in remote sensing

Lea-Tien Tay[1] *and Jaya Sreevalsan-Nair*[2]

Abstract

Natural disasters such as floods, landslides, and forest fires have caused infrastructure destruction, damage to property, injuries, and casualties worldwide. With the advent of sensing technology and data science methodologies, the impact of a disaster can be lessened through in-depth retrospective analysis. Of these, a priori hazard susceptibility analysis, which includes spatial analysis using remotely sensed images of the terrain, is widely adopted. Susceptibility analysis is thus related to spatial aspects of hazard analysis using data captured using both active and passive sensing. In the case of certain disasters, temporal information plays the role of causality owing to which spatiotemporal aspects are relevant. Historical data of occurrences play a vital role in determining the likelihood of disaster occurrence in a new area, or recurrences in an affected site. Such hazard mapping also helps in predicting the degree of the aftermath of a hazardous process in those areas. Overall, state-of-the-art susceptibility analysis uses remote sensing images, land survey data, and other related geospatial and non-spatial data in performing data preprocessing and analysis to calculate the probability of forthcoming hazard events. Data preprocessing involves noise removal, the transformation of raw data to the desired data structure or computed values, dimensionality reduction, feature ranking, and so on. Data analysis includes statistical, machine learning, and deep learning approaches. In this chapter, the widely used preprocessing and data analysis techniques applied in disaster susceptibility analysis are discussed. Additionally, disaster susceptibility analysis-related case studies, e.g. flood and landslide, are presented to elaborate the data science workflows. The resultant maps of the disaster susceptibility analysis are intended to be used by land developers, urban planners, and related authorities in innovating the land use planning and smarter disaster management.

[1]School of Electrical and Electronic Engineering, Universiti Sains Malaysia, Engineering Campus, Malaysia
[2]Graphics-Visualization-Computing Lab, International Institute of Information Technology – Bangalore, India

Keywords: Disaster management; Landslide susceptibility map; Flood susceptibility map; Machine learning; Deep learning; Multi-source data; Image fusion; Artificial neural network; Multilayer perceptron (MLP); ROC analysis

16.1 Introduction

This chapter focuses on the susceptibility mapping of landslides and floods. Landslides are the movement of debris, rocks, or earth material under the influence of gravity, slope gradient, and heavy prolonged rainfall [1,2]. Globally, landslide disasters cause thousands of deaths and injuries, massive damage to properties, and environmental losses every year [3–6]. Flooding ranks first amongst all natural hazards in terms of the consequent economic losses, where they account for one-third of the losses [7]. Flood disasters are also primed to increase owing to indiscriminate urbanization, growing population, global warming, and climate change [8]. Asia is the most affected continent due to floods, as 41% and 61% of the total count of flood events and total economic losses, respectively, occur in Asia [9]. Flash floods are different from rural floods. The former affects the densely populated urban regions, thus causing larger economic losses and more casualties, whereas the latter cause predominantly agricultural losses.

Several research projects have been carried out to investigate disaster areas in various parts of the world. The frequently used approach in these studies is in determining an inventory of historical disaster areas, the effect of geological, topographical, hydrological, and anthropogenic factors on disaster occurrences, and the role of some triggering factors such as rainfall and other climatic changes on landslide or flood occurrences. Landslides occur due to several causative factors, including geomorphology, geology, soil, metrology, land cover, and hydrology [2], which are used for susceptibility analysis. Similarly, the causative or conditioning factors for flood are used as criteria for flood evaluation [10]. These factors may be categorized as (i) hydrologic, (ii) morphometric, (iii) permeability, (iv) land use/land cover (LU/LC) dynamics, and (v) anthropogenic interference [11]. These criteria further have 21 sub-criteria, which include precipitation, river network density, elevation, slope, soil type, topographic wetness index (TWI), LU/LC, normalized differential vegetation index (NDVI), population density, distance from roads, etc. Most methods use a subset of these factors [12,13].

Disaster susceptibility analysis plays a vital role in its prediction and minimization of possible damages in the future. The susceptibility maps are much needed in a timely fashion to alert organizations and people about the possibility of future disaster occurrence. The planners, policymakers, and decision-makers require these maps for emergency mass evacuation and future urban developments. In this chapter, the steps involved in disaster susceptibility analysis are presented. The steps include common preprocessing and data analysis techniques. The results from the models of interest are in the form of susceptibility maps, which provide the quantitative framework for identifying disaster-prone areas. Here, disaster susceptibility analysis for two natural disasters, i.e., landslide and flood are discussed. The case study regions are Malaysia for landslide and India for flood. The discussed approaches can be applied in different case study regions worldwide.

Besides, disaster susceptibility analysis in remote sensing are also useful for various types of natural hazards, such as drought and forest fire. The approaches are repeatable and flexible for other study areas and different natural disaster analysis. The use of multilayer feed-forward artificial neural networks (ANNs), demonstrated for landslides and floods in this chapter, has been extended for susceptibility mapping of other natural hazards, such as forest fires [14], urban heat islands [15], and land subsidence [16]. In such work too, the triggering factors used in the models as independent variables are dependent on the hazard characteristics. For instance, meteorological data, and physical and environmental factors have been used as features for forest fires and urban heat islands, but varied geological factors are alone needed for land subsidence as inputs to the ANNs.

As an extensible model for susceptibility mapping of other natural hazards, there is recent work in the use of deep learning methods of convolutional neural network (CNN) for forest fires [17] and CNN–deep neural network (DNN) model for landslides [18] where the influencing or triggering factors are fed into the models. These triggering factors for forest fires are the same as in landslide susceptibility mapping studies, i.e., geomorphologic, geological, environmental, and human activity related ones. The models can also be simply as an ensemble of machine learning models, where a hybrid ensemble model such as Dagging has been found to be better performing than ANNs and other state-of-the-art methods for flood susceptibility maps [19]. Similarly, multiple models can be applied to the same set of features for comparative analysis, e.g. logit boost outperforms other similar models for land subsidence susceptibility mapping [16], and random forest for forest fire susceptibility mapping [14]. In terms of flexibility of features used, use of multi-scale features in deep learning models is speculated to reduce the uncertainty in the landslide susceptibility map results [20].

16.2 Related work

The use of machine learning and deep learning methods for susceptibility mapping is a relatively recent trend that requires *data fusion* from multiple geospatial data acquisition technologies [21]. The evolution of these methods starts from conventional numerical modeling including geomorphological ones [9,21]. The data-driven methods are ridden with *epistemic* uncertainties [22]. As a trade-off, the state-of-the-art uses *aleatoric* variables for determining probability distributions. Aleatoric uncertainty is that which arises from randomness in the modeling, epistemic uncertainty arise from a lack of information or insufficient historical observations. Given the generality of the governing principle in the susceptibility mapping methods, the literature survey reveals parallels in the analysis of landslides and floods.

16.2.1 Landslides

In recent decades, numerous approaches have been developed to mitigate and reduce the impact of landslides [23,24]. These approaches have identified high-risk areas and recommended the need for appropriate action [25–27]. Because it is almost impossible to predict landslide occurrence in a specific small area, a larger region is taken into account. The focused study region is further divided into various hazard categories [2]. The categorization of the risk level for landslides in a region is known as the zoning of landslides. In order to identify the areas which are highly prone to landslides in the future, it is important that these areas are detected using landslide susceptibility analysis and mapping [28]. The mapping of landslide susceptibility shows the location where local geo-environmental factors are likely to trigger upcoming landslides in the study region. The output maps can be referred to by the authorities to plan future development for the study regions.

The advances in geo-informatics tools like geographic information systems (GIS) and remote sensing have supported the progress of many landslide susceptibility analysis models [28]. Generally, these models are applied in qualitative and quantitative approaches. The qualitative approaches, which were developed extensively from the 1970 to 1980 period, have been applied to measure the degree of hazard based on the experts' knowledge, experiences, and expertise on the topography conditions, using geomorphological investigations, a combination of index map, logical and analytical models [29]. However, the mathematical relationships between landslide causative factors and landslide occurrence are established using geo-informatics tools, statistical analysis, and soft computing models in quantitative approaches used in recent decades [30,31]. There are two types of statistical models, i.e. bi-variate [32,33] and multi-variate [34]. The bi-variate statistical models rely on the landslide density weight values of the respective classes of each causative factor. On the other hand, multi-variate models make use of the sampling of causative factors for the landslide occurrence and non-occurrence dataset. Weight values are statistically computed for each sampling of the causative factors [35]. For the landslide susceptibility analysis, various statistical models such as frequency ratio, information value, certainty factor, Dempster–Shafer evidence theory, Shannon entropy, and weight-of-evidence models have been implemented [36–38].

In recent decades, the machine learning models such as ANNs, decision trees, and support vector machines have also been successfully used for the analysis of landslide susceptibility mapping [39–42]. Machine learning models have been used extensively due to their accurate prediction capability. They have a complex structure and robust performance. One of the widely used machine learning models is ANN. The main advantage of ANN is that it can handle data at any dimension scale ranging from nominal, and ordinal to linear and ratio, and any form of data distribution. In addition, it also deals with qualitative variables that are commonly used in the integrated analysis of spatial data from multiple sources for prediction and classification [43]. It is popular due to its strong capability of nonlinear mapping, high accuracy for learning, and good robustness [44]. A case study of ANN in landslide susceptibility analysis is presented in the next section.

16.2.2 Floods

Different flood-prone areas could be at risk from one or more flood types, namely, pluvial, fluvial, coastal, or groundwater flooding [22], which needs to be accommodated in the flood maps. Flood mapping at the local scale implies several meanings. A systematic classification and ranking of flood maps are done as the following: (i) flood danger map, (ii) flood hazard map, (iii) flood vulnerability map, and (iv) flood damage risk map [45]. The ordering of these maps is sorted by increasing degree of amount and complexity of the information they portray. A flood danger map indicates the spatial distribution of possible spatial distribution of flooding regions, whereas a flood hazard map additionally gives information on flood intensity and the likelihood of its occurrence. A flood vulnerability map also gives information on the exposure or susceptibility of flood-prone elements, e.g. population, built environment, etc. Finally, the flood damage risk map gives the overall impact on the economy. In this parlance, the flood hazard map is synonymous with the flood susceptibility mapping discussed in this chapter. While these flood maps predict future occurrences, they still require *flood extent estimation* for historical events, which is referred to as a flood inundation map and is computed as a segmentation/classification problem [46]. The flood extent map has pixels classified as "flooded" or "non-flooded," whereas the flood susceptibility map marks pixels based on the intensity of flooding, i.e., "very low," "low," "moderate," "high," and "very high" susceptibility [47]. The latter is similar to the zoning of landslides.

The conventional method of using geomorphological and physiographic models and maps was reliant on surveying and in situ measurements [21]. Given its limitations related to fieldwork, domain expertise, and digital data usage, the state-of-the-art gradually shifted to the use of remote sensing and geospatial technologies for multimodal data capture and subsequent modeling. The initial methodology for flood susceptibility mapping relied on geo computation where models are classified as paleo-hydrological, hydro-geomorphological, and hydrological-hydraulic approaches [9]. As an example, the hydrological-hydraulic approach involves the hydrodynamic coupling of 2-dimensional floodplain inundation and 1-dimensional channel flow routing models. Such methods rely on accurate simulation of floods. With the advent of remote sensing using both active and passive sensors, the use of the digital geomorphological model (DGM) has shifted to the use of the digital elevation model (DEM) [21].

Increasingly, data fusion has been adopted for susceptibility mapping as it allows the integration of all possible datasets pertaining to the different conditioning factors in the study areas. Thus, it is ensured that the inputs to any modeling process are rich in information relevant to the hazard. These fused images are now fed into models, such as regression [48], machine learning [49], and deep learning models. An example to the latter is the local spatial sequential long short-term memory (LSS-LSTM) neural networks [50]. The data fusion entails the use of several flood conditioning factors, especially topographic and geo-environmental variables, in these models.

There is also an increasing exclusive focus on flash flood susceptibility to study urban regions. The methods for flood susceptibility prediction (FSP) are similar to those of general floods, where machine and deep learning are extensively used. For the latter, hybrid approaches are used. As an example, a hybrid approach of deep neural network (DNN), analytical hierarchy process (AHP), and frequency ratio (FR) computation has been used for an inference flash-food FSP [51]. Another example involves a novel approach that combines a deep belief network (DBN) with a backpropagation (BP) algorithm, that is optimized by a genetic algorithm (GA). Several machine learning models have been used, and they are compared for their performance. In one such case study, four methods, namely, kernel logistic regression (KLR), radial basis function classifier (RBFC), multinomial naïve Bayes (MNB), and logistic model tree (LMT) are used, of which LMT outperformed the others in a case study [49]. In another study, five machine learning methods are compared, namely, alternating decision tree (ADT), functional tree (FT), kernel logistic regression (KLR), multilayer perceptron (MLP), and quadratic discriminant analysis (QDA), of which ADT performed the best [52].

Overall, the state-of-the-art flash and general flood susceptibility mapping methods are predominantly machine and deep learning ones, that rely on data fusion from disparate sources.

16.3 Case study

The susceptibility analysis for both landslides and floods discussed in this chapter is data-driven and follows similar workflows. The case studies demonstrate the similarities and differences between the processes followed for determining their respective hazard or susceptibility maps. The use of machine learning or deep learning is a key step in these processes. Here, the case studies are chosen to show how ANNs, namely, the multilayer perceptron (MLP), is widely used in susceptibility assessment.

16.3.1 Landslide in Penang, Malaysia

Landslide is one of the major natural disasters in Malaysia due to heavy and prolonged rainfall [53]. Most of the landslides occur in mountainous regions, but a few have occurred in residential areas, resulting in severe fatalities [54]. Landslides have led to serious losses in the hazard-prone areas in Malaysia, including Penang Island. Penang Island is situated off the coast of North-Western Peninsular Malaysia in the Strait of Malacca. In Malaysia, Penang is one of 13 states which is divided into two main parts, i.e., Penang Island and Seberang Perai. Penang Island is the case study for landslides in this chapter because this region has been severely affected by landslides due to heavy prolonged rainfall in recent decades. The geographical coordinates of the Penang Island are 5°15' N to 5°30' N latitudes and 100°10' E to 100°20' E longitudes [41]. It encompasses an area of 285 km^2 with a population size of 700,000. The climate on Penang Island is tropical rainforest, with an average annual precipitation of 2,254–2,903 mm. The average temperatures vary from 29°C to 35°C in daytime, and 26°C to 29°C at night (http://www.met.gov.my). The maximum topographic elevation is 820 m above the sea level, and the slope angle ranges from 0° to 87°. The highest point is Penang Hill, which is situated in the center of the island. At this center area, there are fault lines that start from North to South. The location map of the study area with its historical landslide area is shown in Figure 16.1 [55,56].

Many landslides have often occurred on the island in the past. In 1995, a massive storm caused sixty major landslides in the Penang Hill region. The incident near the Penang Botanical Gardens ruined the pavement and damaged several properties due to uprooting trees. In 2008, there were also 14 landslides in Teluk Bahang and three landslides along Jalan Tun Sardon that caused the power supply cut off and roads to be closed [57]. In 2003, 13 landslides caused havoc in the Penang Hill area. The road was destroyed and rebuild of the road cost around two million Malaysian Ringgit [58]. On 21 October 2017, a landslide struck an affordable housing project construction site in Lengkok Lembah Permai, Tanjung Bungah and this incident claimed the lives of 11 workers after they were buried by the earth from the landslide on a hill side [59]. On 19 October 2018, another landslide incident occurred due to prolonged rain of more than 10 h at the Bukit Kukus project site for the paired road construction site linking Paya Terubong to Relau. Three people have been confirmed dead and 12 people were missing following this incident [60]. On 25 June 2019, there was also a major landslide occurred at a construction site at Jalan Batu Ferringhi, Tanjung Bungah, and four Myanmar workers were buried alive [61].

A few research works have been conducted on Penang Island to assess the landslide susceptibility hazards. Previous works related to the study area used different models such as the frequency ratio model [25,33], analytical hierarchy process [62], Poisson distribution with dominant factors [27], statistical index [63],

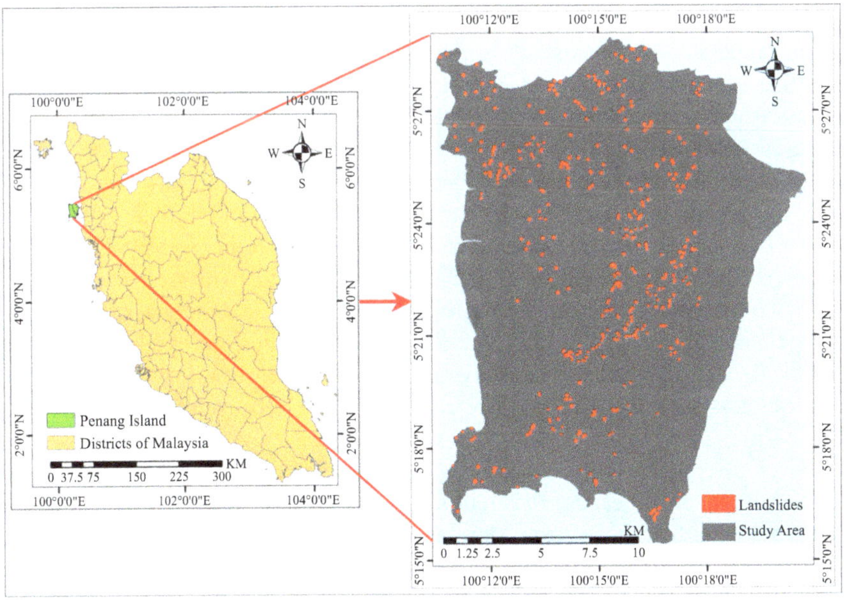

Figure 16.1 The map of Penang Island with landslide inventory

decision tree [39], and artificial bee colony [64]. There are four main steps involved in the landslide susceptibility analysis as follows:

1. Data collection and dataset formation
2. Preprocessing of the spatial dataset
3. Landslide susceptibility analysis
4. Evaluation and assessment of the output

16.3.1.1 Data collection

A landslide inventory map is one of the important elements in generating landslide susceptibility maps and it is used to determine the relationship between landslide occurrence and causative factors. A landslide inventory map is prepared based on historical records, field surveys, and aerial image interpretation. 382 landslide events occurred in different locations on Penang Island (Figure 16.1). The landslides that occurred in the study area can be categorized as debris flows, rotation slides, and rock falls per the landslide classification system [65]. More than 80% of all slides are debris flows and rotation slides. Based on the distribution of landslide and geo-environmental conditions, there are 12 landslide causative factors considered in Penang Island. These factors consist of two types of data, i.e., continuous and categorical. Five continuous factors are elevation, slope angle, distance to road, distance from fault lines, and distance from drainage while the remaining seven categorical factors are curvature, slope aspect, land use, vegetation, precipitation, geology, and soil texture. These factors are collected and converted into a spatial dataset.

First, a digital elevation model (DEM) of 5m resolution is produced by using the topographic map and it represents the elevation map. Other topographic factors such as slope angle, slope aspect, and curvature are computed from the elevation map. Some of the landslides occurred at the roadside along the highways and railways [66]. It is a known fact that the likelihood of a landslide happening decreases with increasing distance from highways [67]. Therefore, line structures such as roads and drainage are also considered significant landslide causative factors. To determine the Euclidean distance from the road and drainage, the buffer zones with intervals of 50 m distance for line features are prepared [68]. The distance to roads and distance from drainage are divided into 10 classes. The fault line is also an important landslide causative factor which generally appears as the potential weaker areas in the hilly region. The fault lines are obtained using the geological map. The buffer zone is divided into 100 m segregations for the distance from the fault lines map. This buffer zone is chosen because the density of fault lines is lower than the density of roads and drainage.

Furthermore, the soil texture of the study area plays an important role in landslide occurrence because different soil types have different landslide susceptibility levels [69]. In hilly areas, slope stability depends on the different physical and mechanical properties of rocks and soils, which can be regarded as a critical factor in slope stability [70]. Therefore, six different types of soil textures such as clay, sand, and urban land, are present in the study area. The geology map contains six types of granites, and more than 70% of the study area is covered by Ferringhi granite, Tanjung Bunga granite, Batu Maung granite, clay, and sand granite. The pertinent research has established the links between land-use and landslides, and the evidence demonstrating the influences of various land-use categories, such as forest and residential areas, on the occurrence of landslides has also been explored [71]. As a result, a total of 17 classes of land use are identified, with forests occupying most of the studied area. In addition, the vegetation factor can help to increase slope stability by lowering the surface ruins and strengthening the mass of the Earth [72]. The vegetation map consists of 14 classes. The forests and fruit plantations represent a significant portion of the study area. Penang, as a tropical island, experiences heavy rainfall caused by storms and the yearly monsoons. Rainfall is a natural triggering factor for landslides because it dampens the soil and causes debris flows, and rock washes away [73]. There are six rain gauge stations installed throughout the island. The data from all stations is collected and the inverse weight distance interpolation technique of ArcGIS software is used to create the spatial distribution of the accumulated average of precipitation. The thematic maps of all causative factors are displayed in Figure 16.2.

16.3.1.2 Preprocessing of spatial database

The spatial database of landslide inventory and causative factors comprises of different range of values and is huge in data size. Therefore, pre-processing of the spatial data is needed before executing the main model. Generally, the preprocessing is implemented in two steps: normalization and dimensionality reduction. Normalization is an important procedure of pre-processing that converts the data from one range to another. The spatial database is normalized so that the range of the whole data set does not differ significantly. There are various normalization techniques such as mean–SD, min–max, and median–IQR. In this case study, mean–SD normalization is applied. The mean–SD normalization depends on the mean and standard deviation of the original data, as given in (16.1). The mean–SD normalized value of sample X_i in data X is given by:

$$X_{Ni} = \frac{X_i - \mu}{\sigma}, \tag{16.1}$$

where μ and σ are the mean and standard deviation of X, respectively.

After the normalization process, the dimensionality reduction technique is carried out. By selecting a smaller dimension in this step, the huge data set is transformed into a dataset with a lower dimension while maintaining the essential attributes of the original dataset. When employing dimension reduction methods, one of the prerequisites is that the relationship between database elements be preserved in the actual space. This is a very crucial criterion for classifying the database in the new space. In this case study, principal component analysis (PCA) is chosen to perform the dimension reduction task. PCA works on both continuous and discrete data types. PCA produces the lower dimensions of the original dataset while maintaining the fundamental information of the features. The low-dimensional features are denoted as principal components (PCs). The first PC

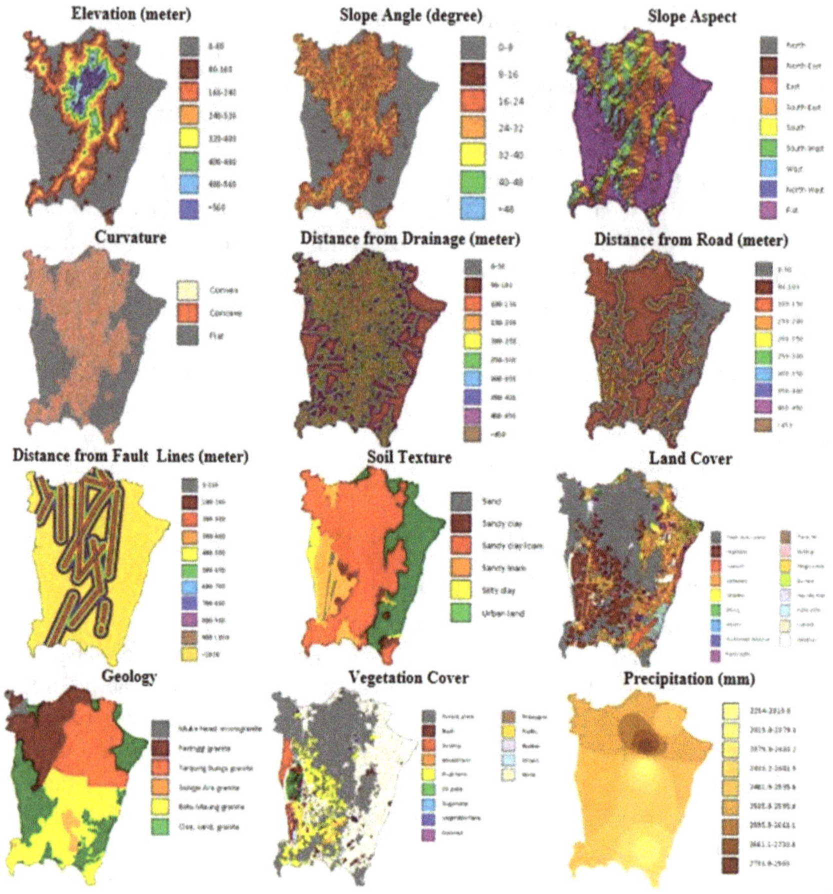

Figure 16.2 The maps of landslide causative factors in Penang Island

reflects the maximum degree of data variability, followed by the second PC with slightly lower variability as compared to the first PC, and so on. The generated PCs are the projections of the original variables. The data is projected in a new space that has minimal correlation and overlapping, which is beneficial for landslide susceptibility analysis.

16.3.1.3 Susceptibility analysis using ANN

An ANN is a powerful computational tool that is inspired by biological neural networks [74]. ANN tends to adapt to the changes in the hidden layer during the training phase. It is applied in many applications due to its convincing ability in solving non-linear engineering problems, high accuracy for learning, and good robustness [44]. ANN is classified into two categories, i.e., feedforward and feedback. The most popular back propagation model, the multilayer perceptron (MLP) is the feedforward type. MLP is one of the widely used tools for classification and prediction problems. The structure of MLP comprises three adjacent layers: input, hidden, and output layers [75] as illustrated in Figure 16.3. Each layer consists of independent processing elements called neurons. These neurons are connected to other layers' neurons through the weights. These weights are modified during the learning process. After the learning process, MLP develops a mapping function between the inputs and outputs [76].

In the learning phase, the 12 landslide causative factors and landslide inventory map are sent to input neurons and labeled as $x_1, x_2, \ldots, x_N$. They are forwarded to hidden neurons through weights denoted by $w_{11}, w_{21}, \ldots, w_{ij}$ The results obtained from hidden neurons are mapped onto the threshold function of each neuron and the outcomes are generated. These outcomes represent the input values to all neurons in the output layer. The final output values are obtained at each output neuron denoted by $y_1, y_2, \ldots, y_k$. In this case study, there is only one output that represents the landslide susceptibility of the trained dataset. Once the learning process is completed at the first iteration, the performance error is calculated by taking the difference between the actual and predicted response variables. The calculated error is propagated backward to update the model. The overall training phase is completed when an acceptable small error is obtained. After completing the training phase, MLP can be used to produce landslide susceptibility output of Penang Island.

The landslide occurrence is labeled as value 1 and landslide non-occurrence is represented by value 0. The dataset is divided into 2 classes in the MLP learning phase, i.e., training dataset and the validation dataset. After the training and validation, the MLP model is built and the whole Penang Island data is sent into the trained MLP model to produce the landslide susceptibility map. The assessment is carried out using the area under the curve (AUC) of the receiver of characteristics (ROC).

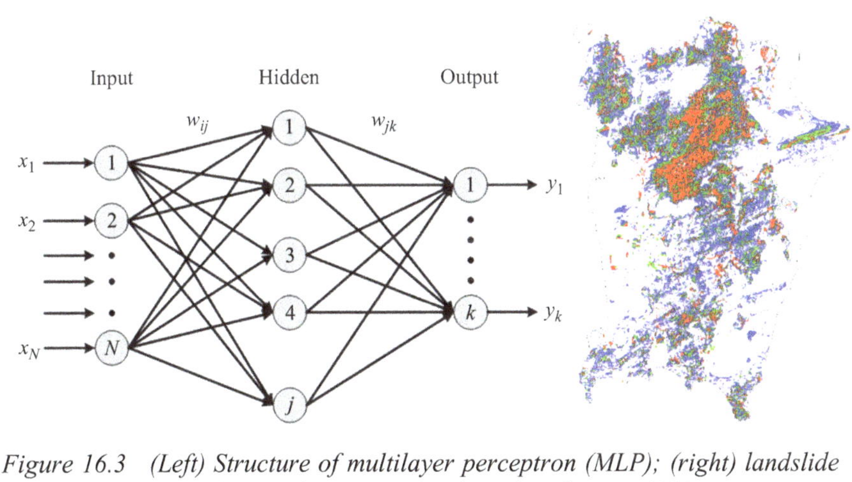

Figure 16.3 (Left) Structure of multilayer perceptron (MLP); (right) landslide hazard map of Penang Island produced using ANN

16.3.1.4 Evaluation and assessment

The accuracy of landslide susceptibility mapping is verified using the ROC. There are two assumptions considered to validate the maps. First, the landslides are related to landslide causative factors such as topography elevation and slope gradient. Second, the triggering factors, i.e., rainfall in this case study, affect the future landslide [77]. In the ROC method, the output indexes of the whole Penang dataset in the trained MLP are arranged in descending sequence. These indexes are named as Landslide Susceptibility Index (LSI). They are divided into one hundred subclasses on the y-axis with 1% separations on the x-axis [26]. The curve presents the accuracy of the trained MLP model in predicting future landslides. The accuracy of this approach is calculated based on the area under the curve [77]. The accuracy of landslide hazard maps of Penang Island produced by ANN is 96.43%.

The landslide hazard map is generated using the LSI obtained from trained MLP of the whole Penang Island dataset. LSI is divided into four major categories. The first category is the 10% of the maximum LSI (90–100%) and it is named a "highly hazardous area" which represents a highly landslide-prone area. The second category is labeled as "hazardous area" and it is the following 10% of LSI (80–90%). The third category denoted as "Moderately hazardous area" corresponds to the next 20% of LSI (60–80%) and the final category named "Non-hazardous area" covers the remaining 60% of LSI (0–60%). The landslide susceptibility map of Penang Island generated using the ANN technique is shown in Figure 16.2. In the figure, red, green, blue, and white color stand for highly hazardous areas, hazardous areas, moderately hazardous areas, and non-hazardous areas, respectively.

16.3.2 Floods in Ganga River Basin, India with a focus on Bihar

Climatological disasters, such as floods, tsunamis, and storms, tend to occur frequently in tropical and sub-tropical regions [13]. Thus, countries such as India, Iran, Bangladesh, etc. are susceptible to such disasters. In particular, floods are affected by climatic-hydrologic factors. The state of Bihar is the 12th largest in geographical area with 94,163 km^2 and third largest by population with that of 1,03,804,637 [78]. Bihar accounts for ~17.2% of the flood-prone areas in the country, where 76% of the state population in its northern parts is under recurring threat. Large-scale losses to lives and the economy occur in the Ganga River Basin (GRB), including regions between Bihar and its neighboring state of Uttar Pradesh [79]. In the case study of the upper and lower Ganga basins [13], the study area includes the Gandak, Ghaghara, and Kosi sub-basins of the Ganga basin in Bihar. The study area spans 83°45'E–85°45'E latitude and 25°15'N–26°15'N longitude, focused on GRB [13]. A large part of the study area is within Bihar, and the remaining is in Uttar Pradesh. Kosi river is known as the *Sorrow of the Bihar* owing to the frequent floods [79]. The study area consists of thick alluvial soil of drift origin, which also contains young loam that is replenished by constant deposition of clay, sand, and silt carried by the rivers [78]. The climate is characterized by four seasons, namely, winter (December–February), pre-monsoon (March–May), monsoon (June–September), and post-monsoon (October–November); and a temperature range from 8.9°C in January to 40.5°C in May [78]. Owing to the monsoon-type tropical climate of Bihar, up to 85% of the rainfall occurs from mid-June to mid-October, i.e., in 4 months, which is largely attributed to the southwest monsoon [11]. A large portion of Bihar is under cultivation, i.e., 72.76% of LU/LC. These regions fall in the vast Indo-Gangetic plain which is at an elevation of 283 m [79]. The location map of the case studies is provided in Figure 16.4.

While GRB experiences frequent and regular flooding, the "2008 Bihar flood" has been a significant flooding event [12]. This has been caused by prolonged rainfall in the upper GRB along with huge water discharge to the Kosi river basin. This led to an embankment breach of about 2 km in length on 18 August 2008, which subsequently caused the Kosi River to change course. The shift in the course further caused flooding in previously non-flooded regions. After 9 days after the breach, the water receded to 1m above the designated level of danger.

Similar to landslides, there are several studies of flood susceptibility mapping based on different machine learning models. The state of the art is in using hybrid models of ensemble base and machine learning ones along with GIS ones [12]. For the study area pertaining to regions in the Ganga basin in the state of Bihar, some of the previous works have used analytical hierarchy process in a GIS framework [11,47], frequency ratio (FR) and Shannon entropy (SE) models [12,80], multilayer perceptron (MLP) with autoencoder models [13], adaptive neuro-fuzzy inference system (ANFIS) [81], and historical flood events along with socio-economic vulnerability mapping [82]. There are four steps in flood susceptibility mapping [12], in line with those of landslide susceptibility mapping:

1. Data collection and dataset formation
   - Curation of flood inventory, a spatial database for flood events, and conditioning factors
   - Selection and preprocessing of satellite images and digital elevation model (DEM)
   - Curation of flood and non-flood sites for training and validation
2. Preprocessing of the spatial dataset
   - Ranking conditioning factors
3. Flood susceptibility analysis
   - Implementation of machine learning or deep learning model
4. Evaluation and assessment of the output

16.3.2.1 Data collection

A flood inventory map is a key element for susceptibility maps, similar to that of landslides. A flood inventory map is generated by determining *n* sites of flooding using historical events and using random sampling of the study area for an equal number of non-flood sites. 300 flood events have been found in GRB during the 2008 flood [13]. The flood sites are obtained from the Landsat 5 multispectral

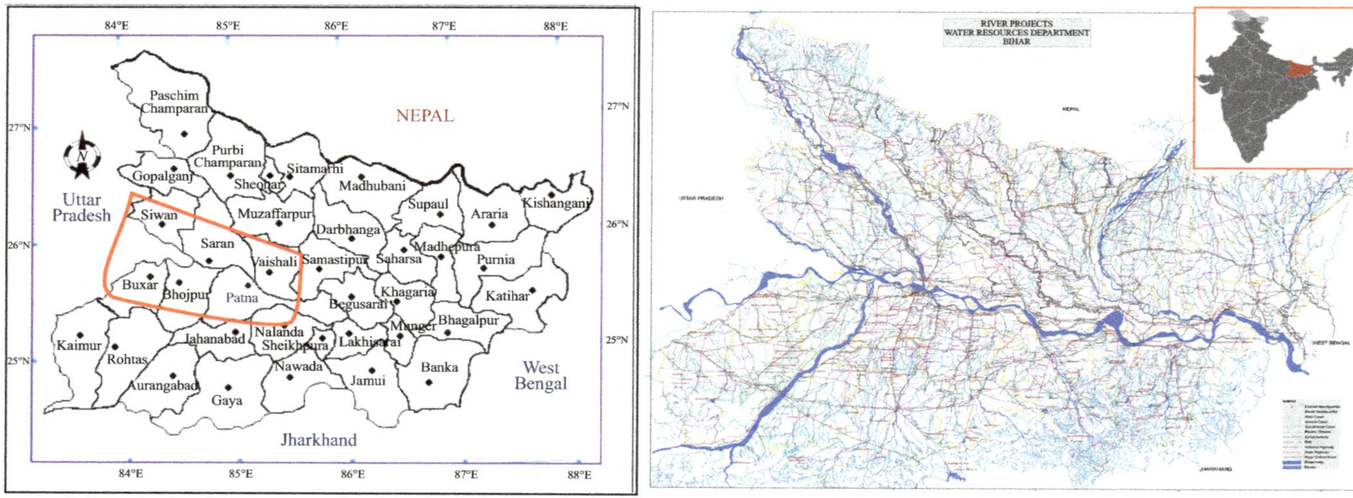

Figure 16.4 *(Left) The map of Bihar showing the region in the study area in the case study of GRB [13] roughly marked by the red boundary. (Right) The river map of Bihar, along with (Inset) its location in India. Maps courtesy from (left) https://state.bihar.gov.in/, (right) https://www.wikiwand.com/en/Geography_of_Bihar, and (inset) https://en.wikipedia.org/wiki/Bihar.*

scanner images from pre- and post-flood dates, i.e., 28 May 2008 and 19 October 2008, respectively, and the digital elevation model (DEM) from shuttle radar topography mission (SRTM) v4 image during February 2000. This DEM has a 90m resolution [83] which is coarser than the DEM used for landslide susceptibility (Section 3.1.1). Since landslides are a geological phenomenon, a higher resolution DEM is required compared to that for floods. The 300 non-flood sites are obtained by random sampling, implemented using the ArcGIS 10.4 software, and pruning using topography maps, surveys, and Google Earth software [13].

The flood inventory map is the target variable, whereas the flood conditioning factors are considered as independent variables. For GRB, 12 factors are used, namely, (i) altitude, (ii) distance to river, (iii) plane curvature, (iv) slope aspect, (v) slope angle, (vi) topographic wetness index (TWI), (vii) distance from road, (viii) rainfall distribution, (ix) LU/LC, (x) soil type, (xi) geomorphology factors, and (xii) river density. These variables are chosen based on an extensive literature survey. Several works evaluate the importance of the larger set of variables using feature ranking, e.g. AHP pairwise comparison matrix [11]. Hydrologic factors have been found to contribute the most to flood conditions, and anthropogenic factors, the least. The hydrologic factors are precipitation, river network density, and stream power index; and the anthropogenic factors are population density, global man-made impervious surface, global human buildup and settlement extent, and distance from roads [11]. It must be noted that the analysis discussed here pertains to largely fluvial, and to a lesser extent, pluvial flooding models, given the focus on GRB and Bihar.

Factors (i)–(vi) are obtained from the DEM, where except for TWI, the others are height-related factors that determine surface runoff. TWI is a water-related factor that reflects the flood occurrence, where, $TWI = \ln\left(\frac{A}{\tan \alpha}\right)$ for catchment area A and slope angle α. All the other factors are obtained from multiple data sources. The distance from the road is obtained from the OpenStreetMap (OSM), and rainfall distribution is procured from the climate forecast system. Anthropogenic factor, such as LU/LC, varies from one land use to another and thus, differentially influences flood occurrences. LU/LC maps are prepared by using the maximum likelihood supervised learning on satellite image data, e.g. dataset from the Climate Change Initiative (CCI) LC dataset [13]. The soil type data is procured from Food and Agricultural Organization (FAO). The geomorphological variables, such as distance to drainage, and river density are available from Google Earth. The final data for training and validation are obtained from Landsat 8 Operational Land Imager (OLI) satellite imagery.

The thematic maps of these flood conditioning factors are prepared, similar to those of causative factors of landslides (Figure 16.1), using ArcGIS and Environment for Visualizing Images (ENVI) software.

16.3.2.2 Preprocessing of spatial database

For preparing the thematic maps, the range of each variable (i.e., the conditioning factor) is converted to a categorical variable. The numerical one is converted to a categorical one using discrete ordered bands. The use of categorical variables leads to the formation of classes, using which one observes contiguous class-wise spatial partitions on the cartographic map. Categorical variables are of two types – *ordinal*, with inherent ordering, and *nominal*, without any ordering. Factors (i), (ii), (v), (vi), (vii), (viii), and (xii) are now ordinal variables, i.e., with ordered classes. The remaining factors become nominal variables. The thematic maps are prepared based on the values of these categorical variables.

The interval values for ordinal variables are determined based on the a priori knowledge of the data. The same process is used for the final hazard zones, which is referred to as the *natural break classification*. The final predicted susceptibility is an ordinal variable that is classified into five values/labels, namely, "very low," "low," "moderate," "high," and "very high" values.

The frequency ratio (FR) helps in determining a strong correlation between different values of the conditioning factors and the occurrence of floods. Thus, FR is computed between each class/value of the conditioning factor and the flood occurrence. The FR for class i of the concerned factor S with n classes is given by:

$$FR(i,S) = \frac{N_{pix}(S_i) / \sum_{i=1}^{n} N_{pix}(S_i)}{N_{pix}(N_i) / \sum_{i=1}^{n} N_{pix}(N_i)}, \tag{16.2}$$

where $N_{pix}(S_i)$ is the number of pixels containing floods in the ith class of S, and $N_{pix}(N_i)$ is the number of all pixels for the selected class. The value of FR is proportional to the correlation between flood occurrence and the respective variable value. In this case study, the altitude of 45–50 m class, above 7° slope angle class, the "flat" slope aspect class, and the "flat" class in the plane curvature are highly correlated with flood occurrence. At the same time, altitude above 65.7 m class, 5–7° slope angle class, the "east" aspect class, and the "convex" class in plane curvature have been found to have the least correlation.

Apart from determining correlations, the sum of FR across all values of condition factors has been referred to as Flood Susceptibility Index (FSI) and can be used for flood susceptibility mapping directly [12].

In some works, multicollinearity analysis is implemented to check the collinearity among conditioning factors [12]. Factors, i.e., variables, with linearity are avoided for determining unbiased outputs. The conditioning factors described in the case study of GRB have been found to have no collinearity in similar studies.

16.3.2.3 Susceptibility analysis using ANN

In the case study [13], two MLP models have been compared for the susceptibility analysis, namely, (i) MLP and (ii) MLP along with an autoencoder neural network. The first model uses the rectified linear unit (ReLU) function [84] as the activation function after processing in the hidden and output layers. The ReLU function is:

$$R(z) = \begin{cases} z & \text{if } z > 0 \\ 0 & \text{if } z = 0. \end{cases} \tag{16.3}$$

The ReLU function is non-linear and known to be better performing than the default sigmoid function. In this case study, the MLP uses three hidden layers were used with 15, 10, and 5 neurons each, flanked by the input layer of 12 neurons and the output (prediction) layer of 1 neuron, in line with the MLP structure in Figure 16.3.

In the second model, the input layer is first fed into the autoencoder neural network [85] for feature extraction, and its results are fed into the MLP for prediction. The autoencoder is a feedforward network with similar input and output (target) layers. The hidden layers in the autoencoder are used for the compressed or low-dimensional representation of the data. For the case study here, stack autoencoder (SAE) has been used. SAE has a sequence of autoencoders where the output of the previous autoencoder feeds as input to the next one. With z, w, and b as outcome, weight, and bias vectors, respectively, f as the activation function after preprocessing, and $h^{(l)}$ as the ith hidden layer using information from previous hidden layer ($i-1$), the encoding step is given as:

$$Z^{(l)} = W^{(l)}h^{(l-1)} + b^{(l)}, \text{if } f \text{ is} \begin{cases} relu: & f(z) = \max(0, z) \\ linear: & f(z) = w * z + b'. \end{cases} \tag{16.4}$$

In this case study, four hidden layers are used in the autoencoder neural network with 5, 15, 10, and 5 neurons in the sequence. The fifth layer is the output layer with 1 neuron, representing the MLP. The activation function is applied in all layers after preprocessing. ReLU is used in all layers, except the third hidden layer, where a linear function is used.

Using a 70:30 split for training and testing, respectively, with 200 iterations of training, predictions were made using the MLP models, and their flood susceptibility index is computed. With the natural break classification, the GRB has the five classes (very low, low, moderate, high, and very high) of flood susceptibility covering 26.24%, 29.26%, 22.17%, 16.02%, and 6.31% of the entire study area, respectively, using the MLP model. With the autoencoder-MLP model, the coverage is 10.74%, 21.97%, 35.37%, 23.74%, and 8.18% of the entire area, respectively. The flood susceptibility maps are shown in Figure 16.5.

16.3.2.4 Evaluation and assessment

Similar to the landslide susceptibility assessment, the flood susceptibility map is validated using the area under the ROC. The case study shows 74% and 93% for the training phase for MLP and autoencoder-MLP models, respectively. For the testing phase, it showed 81% and 97%, respectively [13]. The higher values for autoencoder-MLP indicate that it outperforms MLP with respect to accuracy.

One of the limitations of autoencoder-MLP is the randomness in the results for every run, owing to the randomly assigned initial weights. A remedy to address this shortcoming is to implement several runs and select the run with the highest

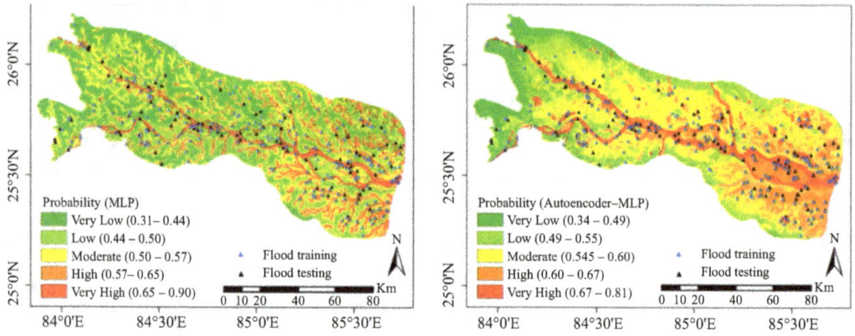

Figure 16.5 Flood susceptibility maps of GRB computed with natural break classification, using (a) MLP model and (b) autoencoder-MLP model (reproduced from work by Ahmadlou et al. [13])

accuracy. A similar remedy is used for the different results owing to a random sampling of data for testing and training in different runs.

16.4 Conclusions

There are many natural disasters that occur on the planet Earth and they continue to wreak havoc, and cause damage to property and casualties worldwide. Some of the recent research efforts are driven to minimize the impact of such disasters. Disaster susceptibility analysis is one of the widely used measures. Such spatial susceptibility analysis plays a vital role in predicting the likelihood of future disaster occurrence in an area. In this chapter, the step-by-step procedure of disaster susceptibility analysis for landslides and floods is presented. Furthermore, specific analysis-related case studies for landslides and floods are discussed. The data-driven methods follow a core procedure, as shown in the case studies, irrespective of the type of natural disaster. Thus, the customizability of this procedure is an advantage in studying several natural disasters.

While the methodology harnesses the state of the art in the machine and deep learning methods, the validation of such work still requires human-intensive data collection. Thus, efforts may be directed toward publishing several benchmark datasets, of which FloodNet [86] is a recent example.

An important aspect of the disaster susceptibility maps is that they are dynamic [45], given the reality of climate change and increased unplanned urbanization. Hence, frequent updation of such maps becomes a requirement. The conventional models may be difficult to scale up to the need owing to the tedious data collection involved. At the same time, machine and deep learning methods are better suited for this requirement. The use of new technology in remote sensing also enables both real-time susceptibility analysis as well as frequent updation of the maps.

The disaster susceptibility maps continue to be beneficial to the land developers, urban planners, policymakers, and related civic bodies and authorities in land-use planning and smarter disaster preparedness of the regions determined to be susceptible. They are needed for deriving insurance premiums for natural hazards [45]. The landslide susceptibility assessment research work at Penang, Malaysia, is part of the collaboration work supported by Ministry of Education, Malaysia, Japan Science and Technology Agency (JST)/Japan International Cooperation Agency (JICA), and Science and Technology Research Partnership for Sustainable Development (SATREPS). A few universities and government departments including Public Work Department Malaysia took part in the research works. The studies are available for the related departments for future planning. In India, the flood susceptibility mapping studies are promoted by the Indian Meteorological Department and the National Disaster Management Division, Ministry of Home Affairs, at the national level, and are conducted in cooperation with concerned departments of the State Government, e.g., Irrigation and Canal Department [87], and other similar water resource management divisions.

Overall, it is important to expand on this research to improve the predictability of such maps.

Acknowledgment

The authors are grateful to B.S. Daya Sagar for enabling this collaboration. The authors thank their respective research groups and institutions for their continued support. This chapter has improved from the comments by anonymous reviewers.

References

[1] Brunsden D. Mass movements. In *Process in Geomorphology*. Arnold, London, 1979, pp. 130–186.

[2] Varnes DJ. *Landslide Hazard Zonation: A Review of Principles and Practice*, vol. 3. UNESCO, Paris, 1984.

[3] Guzzetti F, Mondini AC, Cardinali M, *et al.* Landslide inventory maps: new tools for an old problem. *Earth-Science Reviews*. 2012;112(1–2):42–66.

[4] Nebeokike UC, Igwe O, Egbueri JC, *et al.* Erodibility characteristics and slope stability analysis of geological units prone to erosion in Udi area, southeast Nigeria. *Modeling Earth Systems and Environment*. 2020;6(2): 1061–1074.

[5] Egbueri JC, Igwe O, and Unigwe CO. Gully slope distribution characteristics and stability analysis for soil erosion risk ranking in parts of southeastern Nigeria: a case study. *Environmental Earth Sciences*. 2021;80(7):1–19.

[6] Egbueri JC and Igwe O. The impact of hydrogeomorphological characteristics on gullying processes in erosion-prone geological units in parts of southeast Nigeria. *Geology, Ecology, and Landscapes*. 2021;5(3):227–240.

[7] UNISDR G. Global Assessment Report on Disaster Risk Reduction, Making Development Sustainable: The Future of Disaster Risk Management. United Nations, Geneva, 2015.

[8] Beven KJ, Almeida S, Aspinall WP, *et al.* Epistemic uncertainties and natural hazard risk assessment – Part 1: a review of different natural hazard areas. *Natural Hazards and Earth System Sciences*. 2018;18(10):2741–2768.

[9] Karmakar S, Sherly M, and Mohanty M. Urban flood risk mapping: a state-of-the-art review on quantification, current practices, and future challenges. In: *Advances in Urban Design and Engineering,* New York, NY: Springer, 2022; p. 125–156.

[10] Tehrany MS, Lee MJ, Pradhan B, *et al.* Flood susceptibility mapping using integrated bivariate and multivariate statistical models. *Environmental Earth Sciences*. 2014;72(10):4001–4015.

[11] Swain KC, Singha C, and Nayak L. Flood susceptibility mapping through the GIS-AHP technique using the cloud. *ISPRS International Journal of Geo-Information*. 2020;9(12):720.

[12] Arora A, Pandey M, Siddiqui MA, *et al.* Spatial flood susceptibility prediction in Middle Ganga Plain: comparison of frequency ratio and Shannonâ's entropy models. *Geocarto International*. 2019;36(18):2085–2116.

[13] Ahmadlou M, Al-Fugara A, Al-Shabeeb AR, *et al.* Flood susceptibility mapping and assessment using a novel deep learning model combining multilayer perceptron and autoencoder neural networks. *Journal of Flood Risk Management*. 2021;14(1):e12683.

[14] Gholamnia K, Gudiyangada Nachappa T, Ghorbanzadeh O, *et al.* Comparisons of diverse machine learning approaches for wildfire susceptibility mapping. *Symmetry*. 2020;12(4):604.

[15] Oh JW, Ngarambe J, Duhirwe PN, *et al.* Using deep-learning to forecast the magnitude and characteristics of urban heat island in Seoul Korea. *Scientific Reports*. 2020;10(1):1–13.

[16] Oh HJ, Syifa M, Lee CW, *et al.* Land subsidence susceptibility mapping using Bayesian, functional, and meta-ensemble machine learning models. *Applied Sciences*. 2019;9(6):1248.

[17] Zhang G, Wang M, and Liu K. Forest fire susceptibility modeling using a convolutional neural network for Yunnan province of China. *International Journal of Disaster Risk Science*. 2019;10(3):386–403.

[18] Azarafza M, Azarafza M, Akgün H, *et al.* Deep learning-based landslide susceptibility mapping. *Scientific Reports*. 2021;11(1):1–16.

[19] Islam ARMT, Talukdar S, Mahato S, *et al.* Flood susceptibility modelling using advanced ensemble machine learning models. *Geoscience Frontiers*. 2021;12(3):101075.

[20] Chang KT, Merghadi A, Yunus AP, *et al.* Evaluating scale effects of topographic variables in landslide susceptibility models using GIS-based machine learning techniques. *Scientific Reports*. 2019;9(1):1–21.

[21] Bishop MP, James LA, Shroder Jr JF, *et al.* Geospatial technologies and digital geomorphological mapping: concepts, issues and research. *Geomorphology*. 2012;137(1):5–26.

[22] Beven KJ, Almeida S, Aspinall WP, *et al.* Epistemic uncertainties and natural hazard risk assessment – Part 1: a review of different natural hazard areas. *Natural Hazards and Earth System Sciences.* 2018;18(10):2741–2768.

[23] Gian QA, Tran DT, Nguyen DC, *et al.* Flexible configuration of wireless sensor network for monitoring of rainfall-induced landslide. *Indonesian Journal of Electrical Engineering and Computer Science.* 2018;12(3):1030–1036.

[24] Yaacob N, Tajudin N, and Azize A. Rainfall-landslide early warning system (RLEWS) using TRMM precipitation estimates. *Indonesian Journal of Electrical Engineering and Computer Science.* 2019;13(3):1259–1266.

[25] Lee S and Talib JA. Probabilistic landslide susceptibility and factor effect analysis. *Environmental Geology.* 2005;47(7):982–990.

[26] Pradhan B and Lee S. Delineation of landslide hazard areas on Penang Island, Malaysia, by using frequency ratio, logistic regression, and artificial neural network models. *Environmental Earth Sciences.* 2010;60(5):1037–1054.

[27] Tay LT, Alkhasawneh MS, Lateh H, *et al.* Landslide hazard mapping of Penang Island using Poisson distribution with dominant factors. *Journal of Civil Engineering Research.* 2014;4:72–77.

[28] Bai S, Wang J, Zhang Z, *et al.* Combined landslide susceptibility mapping after Wenchuan earthquake at the Zhouqu segment in the Bailongjiang Basin, China. *Catena.* 2012;99:18–25.

[29] Mandal B and Mandal S. Analytical hierarchy process (AHP) based landslide susceptibility mapping of Lish river basin of eastern Darjeeling Himalaya, India. *Advances in Space Research.* 2018;62(11):3114–3132.

[30] Hong H, Liu J, Bui DT, *et al.* Landslide susceptibility mapping using J48 decision tree with AdaBoost, bagging and rotation forest ensembles in the Guangchang area (China). *Catena.* 2018;163:399–413.

[31] Vieira BC, Fernandes NF, Augusto Filho O, *et al.* Assessing shallow landslide hazards using the TRIGRS and SHALSTAB models, Serra do Mar, Brazil. *Environmental Earth Sciences.* 2018;77(6):1–15.

[32] Youssef AM, Al-Kathery M, and Pradhan B. Landslide susceptibility mapping at Al-Hasher area, Jizan (Saudi Arabia) using GIS-based frequency ratio and index of entropy models. *Geosciences Journal.* 2015;19(1):113–134.

[33] Huqqani IA, Tay LT, and Saleh JM. Analysis of landslide hazard mapping of Penang Island Malaysia using bivariate statistical methods. *Indonesian Journal of Electrical Engineering and Computer Science.* 2019;16(2):781–786.

[34] Basu T and Pal S. Identification of landslide susceptibility zones in Gish River basin, West Bengal, India. *Georisk: Assessment and Management of Risk for Engineered Systems and Geohazards.* 2018;12(1):14–28.

[35] Baeza C and Corominas J. Assessment of shallow landslide susceptibility by means of multivariate statistical techniques. *Earth Surface Processes and Landforms: The Journal of the British Geomorphological Research Group.* 2001;26(12):1251–1263.

[36] Pourghasemi H, Pradhan B, Gokceoglu C, *et al.* A comparative assessment of prediction capabilities of Dempster–Shafer and weights-of-evidence models in landslide susceptibility mapping using GIS. *Geomatics, Natural Hazards and Risk.* 2013;4(2):93–118.

[37] Regmi AD, Devkota KC, Yoshida K, *et al.* Application of frequency ratio, statistical index, and weights-of-evidence models and their comparison in landslide susceptibility mapping in Central Nepal Himalaya. *Arabian Journal of Geosciences.* 2014;7(2):725–742.

[38] Youssef AM, Pourghasemi HR, El-Haddad BA, *et al.* Landslide susceptibility maps using different probabilistic and bivariate statistical models and comparison of their performance at Wadi Itwad Basin, Asir Region, Saudi Arabia. *Bulletin of Engineering Geology and the Environment.* 2016;75(1):63–87.

[39] Alkhasawneh MS, Ngah UK, Tay LT, *et al.* Modeling and testing landslide hazard using decision tree. *Journal of Applied Mathematics.* 2014;2014.

[40] Gorsevski PV, Brown MK, Panter K, *et al.* Landslide detection and susceptibility mapping using LiDAR and an artificial neural network approach: a case study in the Cuyahoga Valley National Park, Ohio. *Landslides.* 2016;13(3):467–484.

[41] Huqqani IA, Tien TL, and Mohamad-Saleh J. Landslide hazard analysis using a multilayered approach based on various input data configurations. *Geosfera Indonesia.* 2021;6(1):20–39.

[42] Huqqani IA, Tay LT, and Mohamad-Saleh J. Landslide hazard mapping of Penang Island Malaysia based on multilayer perceptron approach. In: Guzzetti F, Mihalić Arbanas S, Reichenbach P, *et al.*, (eds.) *Understanding and Reducing Landslide Disaster Risk: Volume 2 From Mapping to Hazard and Risk Zonation.* Cham: Springer International Publishing; 2021. p. 193–199.

[43] Kawabata D and Bandibas J. Landslide susceptibility mapping using geological data, a DEM from ASTER images and an Artificial Neural Network (ANN). *Geomorphology.* 2009;113(1–2):97–109.

[44] Sheela KG and Deepa SN. Review on methods to fix number of hidden neurons in neural networks. *Mathematical Problems in Engineering.* 2013;2013.

[45] Merz B, Thieken A, and Gocht M. Flood risk mapping at the local scale: concepts and challenges. In: *Flood Risk Management in Europe.* Springer; 2007. p. 231–251.

[46] Thangavel R and Sreevalsan-Nair J. CV4FEE: flood extent estimation using consensus voting in ensemble of methods for change detection in Sentinel-1 GRD SAR images. In: *Proceedings of the 7th Asia-Pacific Conference on Synthetic Aperture Radar (APSAR 2021).* IEEE; 2021. p. 1–6.

[47] Bapalu GV and Sinha R. GIS in flood hazard mapping: a case study of Kosi River Basin, India. *GIS Development Weekly.* 2005;1(13):1–3.

[48] Townsend PA and Walsh SJ. Modeling floodplain inundation using an integrated GIS with radar and optical remote sensing. *Geomorphology*. 1998;21(3–4):295–312.

[49] Pham BT, Phong TV, Nguyen HD, *et al.* A comparative study of kernel logistic regression, radial basis function classifier, multinomial naive Bayes, and logistic model tree for flash flood susceptibility mapping. *Water*. 2020;12(1):239.

[50] Fang Z, Wang Y, Peng L, *et al.* Predicting flood susceptibility using LSTM neural networks. *Journal of Hydrology*. 2021;594:125734.

[51] Costache R, Ngo PTT, and Bui DT. Novel ensembles of deep learning neural network and statistical learning for flash-flood susceptibility mapping. *Water*. 2020;12(6):1549.

[52] Janizadeh S, Avand M, Jaafari A, *et al.* Prediction success of machine learning methods for flash flood susceptibility mapping in the Tafresh watershed, Iran. *Sustainability*. 2019;11(19):5426.

[53] Qasim S, Harahap I, and Syed Osman SB. Causal factors of Malaysian landslides: a narrative study. *Research Journal of Applied Sciences, Engineering and Technology*. 2013;5(7):2303–2308.

[54] Yahaya NS, Ramli Z, Thompson A, *et al.* Landslides in Penang Island, Malaysia: insights on emerging issues and the role of geosciences. *Warta Geologi*. 2019;45(4):37–43.

[55] Gao H, Fam PS, Tay LT, *et al.* Three oversampling methods applied in a comparative landslide spatial research in Penang Island, Malaysia. *SN Applied Sciences*. 2020;2(9):1–20.

[56] Gao H, Fam PS, Tay LT, *et al.* Landslide susceptibility analysis using gradient boosting models: a case study in Penang Island, Malaysia. *Disaster Advances*. 2021;14(8):22–37.

[57] Heavy rain causes flash floods and landslides in Penang; 2008. *AsiaOne*, https://www.asiaone.com/print/News/AsiaOne

[58] Kay H. Landslides Keep Penang Islanders in Fear; 2013. *The Malay Mail*, https://www.malaymail.com/news/malaysia/2013/09/17/landslides-keep-penang-islanders-in-fear/526039.

[59] Dermawan A and Basyir M. A regular working day turns tragic for construction workers in Tanjung Bungah; 2017. *New Straits Times*, https://www.nst.com.my/news/nation/2017/10/293498/regular-working-day-turns-tragic-construction-workers-tanjung-bungah.

[60] The Star. Penang landslide: SAR Operations Suspended Due to Heavy Rain at 4.45 am; 2018. *The Star*, https://www.thestar.com.my/news/nation/2018/10/20/penang-landslide-sar-operations-suspended-due-to-heavy-rain-at-445am/.

[61] Xian L. Batu Ferringhi Landslide was a Construction Mishap, says Penang mayor; 2019. *The Star*, https://www.thestar.com.my/news/nation/2019/06/27/batu-ferringhi-landslide-was-a-construction-mishap-instead-says-penang-mayor/.

[62] Khodadad S. Landslide susceptibility mapping in the Penang Island, Malaysia. *Journal of The Korean Geomorphological Association*. 2015;22(1):109–121.

[63] Khai-Wern KL, Tien TL, and Lateh H. Landslide hazard mapping of Penang island using probabilistic methods and logistic regression. In: *2011 IEEE International Conference on Imaging Systems and Techniques*. IEEE; 2011. p. 273–278.

[64] Huqqani IA, Tay LT, and Mohamad-Saleh J. Assessment of landslide susceptibility mapping using artificial bee colony algorithm based on different normalizations and dimension reduction techniques. *Arabian Journal for Science and Engineering*. 2022;47(6):7243–7260.

[65] Varnes DJ. Slope movement types and processes. *Special Report*. 1978;176:11–33. Available from: http://onlinepubs.trb.org/Onlinepubs/sr/sr176/176-002.pdf.

[66] Ali S, Biermanns P, Haider R, *et al.* Landslide susceptibility mapping by using a geographic information system (GIS) along the China–Pakistan Economic Corridor (Karakoram Highway), Pakistan. *Natural Hazards and Earth System Sciences*. 2019;19(5):999–1022.

[67] Arabameri A, Pradhan B, Rezaei K, *et al.* Assessment of landslide susceptibility using statistical-and artificial intelligence-based FR–RF integrated model and multiresolution DEMs. *Remote Sensing*. 2019;11(9):999.

[68] Van Westen C, Rengers N, and Soeters R. Use of geomorphological information in indirect landslide susceptibility assessment. *Natural Hazards*. 2003;30(3):399–419.

[69] Yalcin A, Reis S, Aydinoglu A, *et al.* A GIS-based comparative study of frequency ratio, analytical hierarchy process, bivariate statistics and logistics regression methods for landslide susceptibility mapping in Trabzon, NE Turkey. *Catena*. 2011;85(3):274–287.

[70] Kumar V, Gupta V, and Sundriyal Y. Spatial interrelationship of landslides, litho-tectonics, and climate regime, Satluj valley, Northwest Himalaya. *Geological Journal*. 2019;54(1):537–551.

[71] Wang X, Huang Z, Hong MM, *et al.* A comparison of the effects of natural vegetation regrowth with a plantation scheme on soil structure in a geological hazard-prone region. *European Journal of Soil Science*. 2019;70(3):674–685.

[72] Ding Q, Chen W, and Hong H. Application of frequency ratio, weights of evidence and evidential belief function models in landslide susceptibility mapping. *Geocarto International*. 2017;32(6):619–639.

[73] Yang Z, Cai H, Shao W, *et al.* Clarifying the hydrological mechanisms and thresholds for rainfall-induced landslide: in situ monitoring of big data to unsaturated slope stability analysis. *Bulletin of Engineering Geology and the Environment*;78(4):2139–2150.

[74] McCulloch WS, and Pitts W. A logical calculus of the ideas immanent in nervous activity. *The Bulletin of Mathematical Biophysics*. 1943;5(4):115–133.

[75] Haykin S. *Neural Networks: A Comprehensive Foundation,* 2nd ed. Prentice Hall PTR; 1998.

[76] Garrett JH. Where and why artificial neural networks are applicable in civil engineering. *Journal of Computing in Civil Engineering*. 1994;8(2):129–130.

[77] Chung CJF and Fabbri AG. Probabilistic prediction models for landslide hazard mapping. *Photogrammetric Engineering in Remote Sensing.* 1999;65:1389–1399.

[78] Project Teams of DMSG, RSA, NRSC, India. Flood Hazard Atlas – Bihar – A Geospatial Approach; July 2020. https://bhuvan.nrsc.gov.in/pdf/Flood-Hazard-Atlas-Bihar.pdf.

[79] WRIS, India. Ganga Basin – Version 2.0; 2014. https://indiawris.gov.in/downloads/Ganga

[80] Islam S, Tahir M, and Parveen S. GIS-based flood susceptibility mapping of the lower Bagmati basin in Bihar, using Shannon's entropy model. *Modeling Earth Systems and Environment.* 2022;8(3):3005–3019.

[81] Arora A, Arabameri A, Pandey M, *et al.* Optimization of state-of-the-art fuzzy-metaheuristic ANFIS-based machine learning models for flood susceptibility prediction mapping in the Middle Ganga Plain, India. *Science of the Total Environment.* 2021;750:141565.

[82] Tripathi G, Pandey AC, and Parida BR. Flood hazard and risk zonation in north Bihar using satellite-derived historical flood events and socio-economic data. *Sustainability.* 2022;14(3):1472.

[83] CGIAR CSI. SRTM 90m Digital Elevation Database v4.1. Last accessed on October 6, 2022. https://cgiarcsi.community/data/srtm-90m-digital-elevation-database-v4-1/.

[84] Nair V and Hinton GE. Rectified linear units improve restricted Boltzmann machines. In: *Proceedings of the 27th International Conference on International Conference on Machine Learning*; 2010. p. 807–814.

[85] Baldi P. Autoencoders, unsupervised learning, and deep architectures. In: *Proceedings of ICML Workshop on Unsupervised and Transfer Learning. JMLR Workshop and Conference Proceedings*; 2012. p. 37–49.

[86] Rahnemoonfar M, Chowdhury T, Sarkar A, *et al.* Floodnet: a high resolution aerial imagery dataset for post flood scene understanding. *IEEE Access.* 2021;9:89644–89654.

[87] Saini SS and Kaushik S. Risk and vulnerability assessment of flood hazard in part of Ghaggar Basin: a case study of Guhla block, Kaithal, Haryana, India. *International Journal of Geomatics and Geosciences.* 2012;3(1):42–54.

Chapter 17

Topological assessment of road transportation network

Rahisha Thottolil[1] *and Uttam Kumar*[1]

Abstract

Land areas in urban agglomerations and metropolitan cities have been increasing worldwide, and cities are witnessing rise in both number and their corresponding populations, with an average annual growth rate of 1.7% in 2021. As a consequence, larger cities are facing significant challenges in road transportation system, urban mobility and traffic flow. Road transportation system is one of the most important subsystems in any city because it ensures better accessibility and connectivity between different geographic spaces and improves the travel demands in cities. However, urban mobility is impacted by the topology of transportation networks, therefore, understanding the topological patterns of the road and street networks is essential for enhancing vehicular transportation performance. In this chapter, we discuss various forms of spatial networks that are practiced in the real-world scenarios, and elucidate on the road transportation network structures and their properties. Several measures used for quantifying the transportation network performance such as connectivity and coverage using graph theory are deliberated.

Keywords: Urban mobility; Road transportation network; Connectivity; Coverage; Regression

17.1 Introduction

Scale-free, small-world, and modular networks are examples of complex spatial networks that are defined throughout space and have complicated patterns of connectedness. There are numerous organisational scales present in these networks, which are distinguished by their abundance of nodes and edges. Significant ramifications exist for many disciplines, including geography, urban planning, transportation engineering, ecology, and social sciences in the fast-developing field of inquiry known as

[1]Spatial Computing Laboratory, Center for Data Sciences, International Institute of Information Technology Bangalore (IIITB), India

the study of *complex spatial networks*. So, it is vital to characterise and comprehend the structure and growth of spatial networks in many different fields, ranging from urbanism to epidemiology. Complex geographical networks are also used in social sciences to investigate the composition and dynamics of social interactions, including the transmission of illnesses or information. In case of ecological studies, the interactions between species and their spatial distributions have been represented as complex spatial networks. The theory of these spatial networks attempts to understand the mechanisms that generate their emergent features and highlights the properties that impact the dynamics of systems they represent. Analysis of socio-ecological as well as spatial transportation networks are examples of geospatial applications of these complex networks. Spatial networks are also used in transportation engineering to model traffic flow and improve transportation infrastructures.

Recently, newer methods for modelling and evaluation of complex geographic networks have been developed and together with developments in geospatial technologies like remote sensing, Geographic Information System (GIS) and Global Positioning System (GPS), it is now possible to collect and analyse massive amounts of spatial data [1]. Network science, spatial statistics, and machine learning techniques are now widely used to analyse and model spatial data. In general, the study of complex spatial networks has significant theoretical and practical implications for comprehending the dynamics and structure of geographically embedded systems, as well as for creating efficient management and optimisation strategies for these systems in a variety of domains.

An advanced transportation system should ideally provide adequate access to its infrastructure for all the citizens. Consequently, it will provide a better quality of life for residents without exception. As a matter of fact, expansion of road transportation has greatly improved economic growth, sparked a social transformation and also boosted mobility. Moreover, political discussion on the importance, purposes, and goals of transportation development has always been vital, and transportation planning has long been recognised as an essential instrument for supporting policymaking. Road infrastructure is the backbone of any urban system [2], and limited accessibility of road networks significantly disrupts the traffic flow.

Road and street networks are fundamental components of urban form, and the resilience of cities can be greatly impacted by their design. As such, many researchers have explored resilient street design and street networks. For streets to perform well under normal circumstances and to be able to react to disturbances, resilient street design is of utmost importance. So, in order to do this, researchers have proposed a number of design solutions, such as building streets to accommodate different modes of transportation, lowering the frequency of cul-de-sacs and dead ends, and boosting street connection to increase accessibility and shorten travel times [3]. Therefore, a street network's resilience is significantly influenced by its interconnectedness as well. Earlier research have shown that street networks with more connections can handle traffic disruptions better because traffic can be diverted more effectively. Moreover, connected street networks improve accessibility to essential services during emergencies by facilitating easier movement and increased accessibility.

Network structure measures are often used to characterise a network. These measurements can shed light on the structure and operation of complex network systems, including social, biological, technological, and transportation networks. The structure of road network can be described in terms of mathematical expressions and the level of quality of these networks in a city can also be determined based on quantifiable network structure measures [4]. Some common measures of network structure are degree centrality, betweenness centrality, closeness centrality, clustering coefficient, degree distribution, pan length, and modularity. These measures are commonly used to locate important nodes, subgroups, and patterns in any network and are also frequently used in network analysis to understand the behaviour and structure of complex systems.

Furthermore, uniformity of the flow of traffic in an urban transportation system is referred to as traffic smoothness. It indicates how smoothly traffic flows through the network without unanticipated delays, traffic jams or accidents. It is crucial to model projected traffic smoothness in urban transportation systems as these models can be used to spot possible issues and suggest fixes to enhance traffic flow. Hence, understanding the spatial patterns of the road network is necessary to determine the traffic flow and network disturbances caused by congestion [5]. The expected traffic smoothness can be modelled using a variety of techniques. For instance, simulation models are used to mimic the behaviour of traffic in certain networks. These models are capable of accounting specific elements like traffic density, intersection density and speed limitations. They can also mimic the effects of many variables including weather, accidents and road construction. Recently, machine learning techniques are also being used for modelling traffic smoothness. For example, large datasets can be analysed by these algorithms to find trends and forecast traffic behaviour. They can also use previous traffic data analysis to forecast how traffic would behave at certain periods of the day or week. After the projected traffic smoothness has been modelled, improvements to the traffic flow can be suggested. Changes to the infrastructure, such as building additional roads or putting traffic management systems in place, can be related propositions. Studying the relationship between smoothness and intensity of traffic flow is necessary to optimise the issues of traffic assignment in a road network [5]. Therefore, one crucial step in enhancing urban transportation systems is modelling anticipated traffic smoothness. Following this, a more efficient and sustainable transportation network can be established that are advantageous to all the parties by detecting possible issues and suggesting solutions.

Recently, urban researchers have explored various methods that may be used to analyse the connectedness of the road and street network in terms of dividing the city area. Typically, borders and administrative boundaries are used to divide the Earth's surface. Yet, new study demonstrate that by dividing the city region into smaller portions (basic fields/grids), a thorough investigation of the road network may be carried out [6]. Analysis of the correlation between these field's size and indices of the road network's connectivity offers new understanding to comprehend the intricate urban road network at micro level. This form of study emphasises the value of urban road and street networks and seeks to comprehend the dynamism and complexity of the built environment in a small area of the city. Therefore, studying road transportation network has a vital role in transportation activities in urban planning and development.

Following the above context, the primary objectives of this chapter are to explain the fundamental ideas of graph theory and investigate examples of road and street networks in terms of topological measurements. The subsequent sections of this chapter review earlier research along with graph theory and its related applications in transport research. Methodology for computing transportation metrics with examples is also provided. Finally, a case study is included to illustrate different scenarios in this domain.

17.2 Review of earlier research on spatial transportation network structures

Since the functionality of the entire urban system heavily relies on the spatial connections between the basic variables of the road networks, several researchers and transportation scientists have investigated the relationship between urban system and transportation structures. The field of road networks is gaining more interest and attention under the wider spectrum of urban transportation, notably for assessing urban mobility and the effects of network form on traffic flow and safety.

Sun *et al.* [7] assessed the characteristics of road networks to understand how occurrences affect urban traffic movement. Their study revealed that these characteristics have an effect on how traffic moves through the network, for example, they can be used to identify congested spots and choose different paths/routes for traffic to deviate. The network features such as intersection points in a road network and the accident locations have been used for evaluation of the impacts of incidents on traffic mobility in northern China [7]. Also, resolving challenges with traffic congestion, mobility management and intelligent transportation systems were used to describe the sustainable management of an urban transportation system [5]. The expected smoothness of traffic flow and waiting time in traffic for various intelligent transportation system services were modelled by analysing the complex urban transportation network.

Early work on analysing road transport network structure dates back to 1960s, while some meaningful metrics to describe these spatial network structures were proposed later [4]. Any network's structures can be encapsulated by a mathematical expression, so it is always useful to compare related structures and evaluate how the networks link to one another in a meaningful way using quantitative measures of the structural network. Several indices have been used in the past to quantify network performance, with the major emphasis on estimating a network's connection. These quantitative measurements frequently define a road network's complexity, characteristics and overall effectiveness. Further, the numerical values of metrics can also gain special evidence when comparing two or more sets of network structures or analysing the same road network under different periods (temporal study). As an example, Kansky and Danscoine [4] applied certain fundamental graph indices to the railroad network of Sardinia in Cagliari and proposed 14 indices to represent the characteristics of a transportation network.

Assessing the simplicity of travel using active modes of transportation between various destinations and origins is a key component of measuring network connectivity for bicycle and walking. Analysing the street network and its attributes such as street width, traffic volume, intersection design and the existence of bike lanes or sidewalks are common part of this process. A similar approach evaluated various connectivity measures to increase the walking and cycling in Portland metro area [8]. Overall, evaluating network connectivity for walking and bicycling requires a comprehensive strategy that considers a variety of elements such as physical attributes of the street network, convenience and safety of users of active transportation, and accessibility of destinations. Thereafter, Buhl *et al.* [9] analysed and compared the topological patterns and network efficiency of street network of a few urban settlements in India, Europe, Africa, Central America, and one shantytown using the approach of a complex network theory. They distinguished self-organised urban communities (informal settlements) by their impulsive and unplanned expansion and frequent lack of formal government or regulation. Although these settlements are informal, their street networks often display specific topological characteristics that reflect the underlying socio-economic dynamics of the community. Narrow and winding streets were one of the most prevalent topological characteristics in self-organised urban settlements. These roadways frequently follow the topography of the ground naturally, with homes and other structures arranged in crowded and haphazard ways. This reflects the requirement to make best use of the available space while also fostering social interaction and a sense of community. The study focused in improving the accessibility, mobility and standard of living for those who live in informal settlements by understanding these patterns.

Later, to define the structural characteristics of a complex road network, Xie and Levinson [10] analysed three new measures out of Kansky's metrics in terms of entropy, connection pattern, and continuity. Following this, the influence of network structure on individual travel behaviour was also studied using regression model in the US [11]. The study showed that travel distance can be greatly impacted by the design of a transportation network. The number of available routes, their lengths and the easiness with which they can be traversed have an impact on the trip distance between two places in a network. Network connectivity and the existence of bottlenecks or choke spots in the transportation network are important factors that affect travel time or journey distance. Bottlenecks are areas where traffic movement is constrained, such as congested intersections or constrained bridges, whereas choke points may force cars to take longer routes in order to reach their destinations, thus lengthening travel times. So, shorter travel distances are often the result of more direct routes being available on networks with high levels of connectivity. On the other hand, less connected networks might necessitate more indirect paths, thereby increasing travel times.

In recent years, the fundamental road network measures were also related to socio-economic indicators such as urban mobility, population, and market information in the region of Thessaly in Greece [12]. Travel times, traffic congestion, and accessibility to various parts of the city can all be impacted by the layout and connectivity of the city's roads. Therefore, transportation planners and engineers should design more effective and efficient transportation systems that can better meet the mobility needs of urban residents. This can be accomplished by understanding the topological characteristics of urban road networks such as grid-like networks, hierarchical networks, radial networks, and circulatory networks. These kinds of studies on urban transportation systems provide evidence for considering the network topology measures as acceptable factors for understanding the mobility pattern and therefore, development of a city.

Network studies have also been used to investigate road transportation in developing nations. Also, to the best of knowledge from available literature, none of the published articles provide a detailed study on both coverage and connectivity together to examine the complexity of road and street networks in terms of smaller grids. Micro-level examination of urban road networks in relation to connectivity and coverage measures is therefore crucial for bridging the research gap. Even though existing literature indicate that there are many ways to gauge how complicated a road network is, this chapter discusses seven connection measures and three coverage measures that are extremely helpful and pertinent in this direction. Transport indices create a metrical hierarchy between various roads and characteristics of these networks. As planar graphs form the structure of the road network systems in this case, the network is fully connected for a single gird. Selected indices computation are discussed in detail below.

17.3 Graph theory and its application in modelling transportation networks

Graph theory is a branch of mathematics to analyse the structure of the networks through graphs. Planar graphs are simplified visual representation of a two-dimensional network as a set of points and lines associated with topological relationships. In this context, the points are called vertices or nodes and those connected to the features are called edges or links. A simple graph is a collection of vertices and edges and it is denoted as $G = (v, e)$ where v is a non-empty set of vertices and e (representing adjacency relation among the vertices) is composed of pairs of distinct vertices from v called edges. In other words, each element of v is called a vertex or a node point of a graph G and each element of e is called an edge of a graph G. A simple graph is an undirected graph where each edge connects two different vertices and no two edges connect the same pair of vertices.

Transportation networks have complex structure which can also be analysed using the theoretical concept of graphs. The network structure can be adequately represented through 2D (2-dimensional) or 3D (3-dimensional) graphs and its spatial features can be studied using graph theorem. Consider the example of a road network as a set of nodes and edges (see Figure 17.1), such that each node/point represents an intersection, dead-end, origin or destination on a road network structure and each line

(a) (b)

Figure 17.1 Road and street network. (a) Illustration of 2D road map and (b) converted network dataset with nodes and edges.

corresponds to existing road segment. Robin (1996) presented the fundamental ideas of graph theory and its applications to the systems of road transportation. Later, many researchers explored graph theory to design the topology measures of any transportation networks. The magnitudes of structural measures like connectedness and coverages were calculated using the vertices and edges, which served as a representation of the main network components.

By utilising various GIS technologies, the structure of the transportation network can be mapped as a line feature layer. Digitisation from scanned topographical maps or high-resolution satellite imagery is one of the most popular and practical methods for creating road maps. The road network maps can also be downloaded from a world spatial database called Open Street Map (OSM) (www.geofabrik.de). Representation of the transportation network structure and its network dataset in the form of nodes and edges are shown in Figure 17.1(a) and (b).

Let us take a look at the different network structures to demonstrate how graph metrices can be applied to transportation networks (see Figure 17.2). The degree distribution, which depicts the distribution of the number of edges occurring on each node, is one crucial graph statistic. Nodes in a transportation network indicate places, and edges represent the paths taken to reach those places. The degree distribution of transportation networks is frequently significantly skewed, with the majority of nodes having few edges compared to a small number of nodes with many edges. Betweenness centrality, which gauges a node's significance in the network based on the number of shortest paths that cross through it, is another crucial graph parameter. High betweenness centrality nodes in transportation networks are frequently significant transportation hubs that connect numerous routes and provide effective movement between various places. The third crucial graph parameter is clustering coefficient, which gauges how closely nodes in a network prefer to group together. High clustering coefficients in transportation networks can

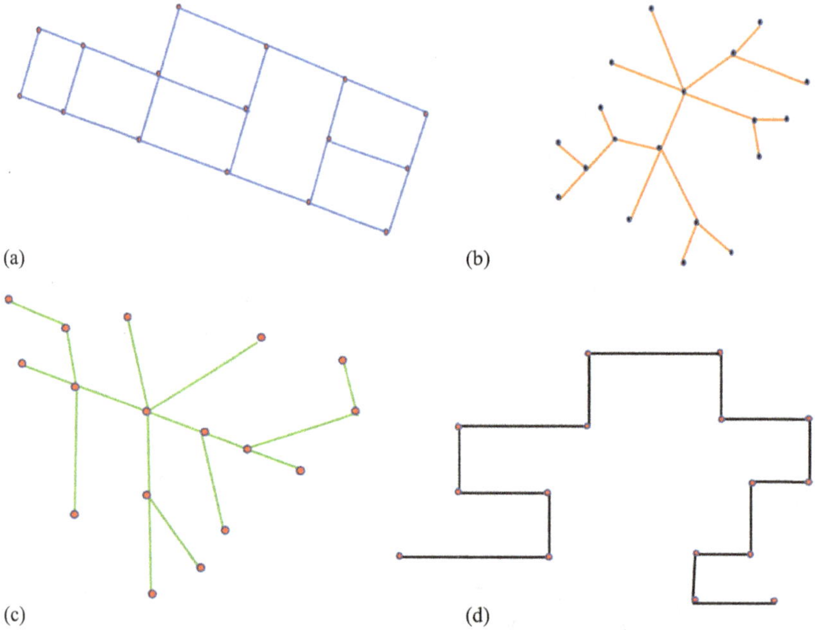

Figure 17.2 Different types of transportation network structure. (a) Mesh/grid, (b) tree, (c) hub and spoke, and (d) linear [13].

indicate that some sites are highly interconnected and might be significant transportation hubs.

Each node in a grid network is connected to its four adjacent nodes. Nodes are placed in a regular grid pattern. This network has a very consistent degree distribution, with majority of the nodes having a moderate degree. While there are numerous equally efficient routes connecting any two points, the betweenness centrality of the grid network may be rather low. Therefore, nearby nodes frequently have strong connections to one another; this network has a high clustering coefficient [13]. A hub-and-spoke network links several densely connected hub locations to numerous less-connected spoke locations. This network's degree distribution is significantly skewed, with a few number of nodes having high degrees and majority of nodes having low degrees. The hub locations have high betweenness centrality, as they are crucial for connecting many distinct routes. Lack of direct connectivity between the hub locations results in a low clustering coefficient for this type of network. A network with a random structure is created when edges between nodes are added with a specific probability. Most nodes in this network have a moderate degree, with the degree distribution generally following a Poisson distribution. While there are numerous, equally efficient routes connecting any two points, the betweenness centrality of this network is rather low. The network's clustering coefficient is also low since nodes are more likely to connect to one another randomly than clusters.

17.4 Assessment of topological measures in transportation network

This section discusses estimation of topological network structure with connectivity and coverage indices for cities and their usefulness in urban transportation planning and management. These structures are better suited for assessing the road transport system and for identifying urban disparities in the city's transportation infrastructure in emerging countries. Moreover, a better city management and planning process depends on the network measurements' relationships with one another.

The key idea of topological structure of a road network denotes the geometry, layout, or spatial pattern of the network system. The topological pattern of a road system can be expressed in a mathematical form of a simple planar graph which is denoted as $G(v, e)$, where G represents a graph, vertices or nodes (v) express the intersection of roads and individual links are edges (e). Such graphical representations and the variables have been a common practice for modelling road networks [4,12,14–16]. The fundamental network measures have been adopted by many urban researchers [6,9,13].

The following section explains selected metrics to estimate the connectivity and coverage measures to depict spatial patterns of a road network. Once the graphical elements of road networks (v, e) are determined from the network dataset, the topological measures are calculated using mathematical equations. These network measures can also be used to compare the characteristics of various transportation networks.

17.4.1 Different measures of the connectivity of network structure

Generally, different topological measures have been used to examine the performance of a complex road network. The common measures for assessing the connectivity of the road and street networks and their computational procedures using various statistical and empirical techniques are discussed here.

Alpha Index is the ratio of the number of *fundamental cycles* in a road network (graph) to the maximum possible number of cycles in the network. The number of cycles or circuits in any network is also called a cyclomatic number and it is denoted as $'f'$,

$$f = (e - v + p) \tag{17.1}$$

where 'e' is the number of edges and 'v' is the number of nodes. The number of subgraphs in a network is denoted by 'p'; the value of p is 1 for a fully connected road network in any planar graph. *Fundamental cycles* (cyclomatic number) are also called Betti number, which is one of the primary indices of graph theory.

The maximum number of independent cycles in a linear graph G is equal to its cyclomatic number and the maximum number of fundamental circuits in a connected graph G is equal to its cyclomatic number. These two characteristics imply the cyclomatic number as a gauge of transportation network structure. Any disconnected

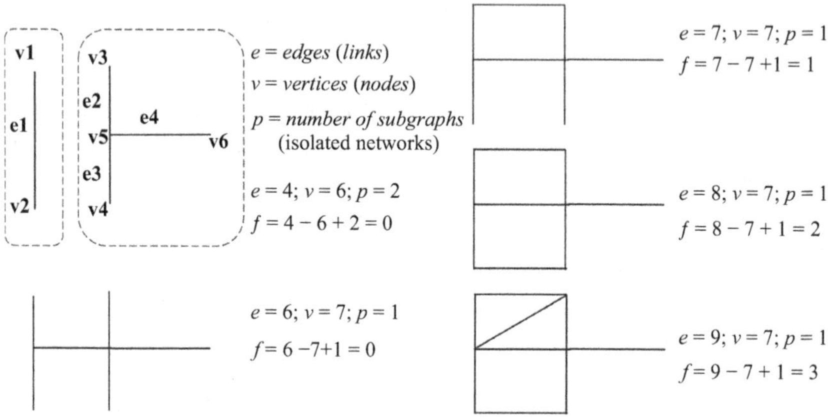

Figure 17.3 Application of cyclomatic number

graph ($p > 1$) and any tree can have a cyclomatic number equal to zero (see Figure 17.3). Higher cyclomatic numbers are seen in graphs with more connections. A cursory study of the transportation systems of various countries would lead one to believe that less developed nations are served by networks of transportation that more closely resemble disconnected graphs or trees. On the other hand, highly developed countries profit from well-connected transportation systems. The cyclomatic number may be a valuable indicator of spatial organisation of the transportation networks considering this commonality. Alpha index is denoted as 'α'

$$\alpha = f/(2(v-5). \tag{17.2}$$

The numerator represents the number of faces associated with any planar graph and the number of edges 'e' of a network with 'v' nodes are bounded by $(3v - 6)$. Therefore, the maximum number of possible cycles is $(2v - 5)$ obtained by substituting into the numerator (f). Alpha index is also called 'meshedness coefficient' [2]. The standard value ranges from 0 to 1, whereas a simple tree network structure has the value 0. 1 represents a highly connected road network like a grid pattern where, every possible road segment is connected by existing nodes (see Figure 17.4).

- *Beta Index* (β) analyses the number of edges with the number of vertices in a network. The value becomes 0 with no edges and is 1 when the network has one circuit, and more than 1 for a complex network with many cycles. The maximum measure of the Beta index in a planar graph could be three and for a nonplanar network, the values are infinite [13]:

$$\beta = \frac{e}{v} \tag{17.3}$$

- *Gamma Index* (γ) compares the existing number of edges with the maximum possible number of edges in the network. The value also ranges from 0 to 1, where 1 indicates a more complex network with innumerable connections:

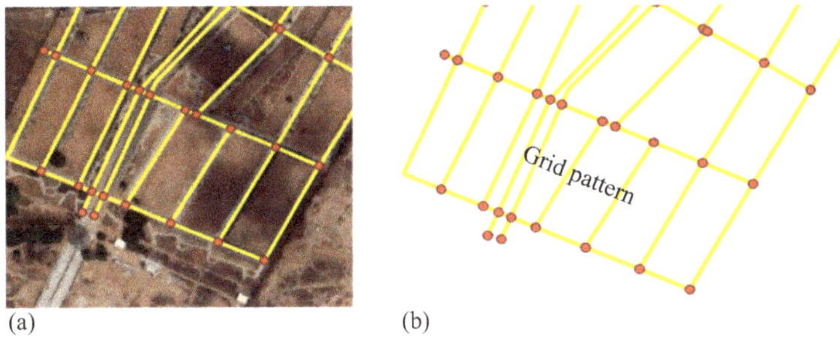

Figure 17.4 *Depiction of a grid pattern of road transportation network structure*

$$\gamma = e/(3v - 2) \tag{17.4}$$

- *Grid Tree Proportion* (GTP) is a measure to diagnose the topology of a network. Similar to Alpha and Gamma indices, GTP index also ranges from 0 for a tree pattern to 1 for a complex grid pattern:

$$GTP = f/\left(\sqrt{v} - 1\right)^2 \tag{17.5}$$

- *Eta Index* (η) represents the average length of a road segment in the network. This index is an indicator of the speed within the network by the assumption that extensive roads are better to ensure maximum speed. Here, the network is considered as a whole and the total length of the road is represented as 'L':

$$\eta = L/e \tag{17.6}$$

- *Three-way and four-way intersection proportions* are a coarse representation of road connectivity. Three-way node denoted as $\overline{v3}$, represents the intersection with three road segments (examples are T and Y junctions) and four-way intersection nodes $(\overline{v4})$ is with four paths [17]. While these measures are more related to connectivity, they do not explain the structural behaviour of the transportation system.

Higher number of three-way and four-way intersection nodes represents higher number of paths between any two locations [6]. The value of three-way and four-way intersection proportions ranges from 0 to 1, where higher proportion corresponds to higher connectivity network. To estimate these measures, a set of transport nodes (v) are considered by determining the number of paths from each transport node for every individual grid:

$$\textit{Three-way intersection proportion} = \overline{v3}/v \tag{17.7}$$

$$\textit{Four-way intersection proportion} = \overline{v4}/v \tag{17.8}$$

17.4.2 Different measures of the coverage of network structure

A road network system's density measurements are interpreted by coverage measures. The calculated data can be used to illustrate how a road network is being developed. Higher the road development, greater is the value of density. Here, network density, edge graph density, and nodal graph density are elaborated for determining coverage measures. *Network length (L)* specifies the total span of the road network and *network density* (A) is measured according to the area occupied by road network.

- *Road network density* is the ratio of the total length of the road network to the area occupied by the road network (which may be a grid). Higher density indicates higher development of road transport system:

$$Network\ Density = L/A \tag{17.9}$$

- *Edge graph density* is the relation between the number of edges and the grid area of the graph computed as in (17.10):

$$Edge\ Graph\ Density = e/A \tag{17.10}$$

- *Nodal graph density* is the ratio of the total number of nodes to the area of the grid:

$$Nodal\ Graph\ Density = v/A \tag{17.11}$$

Road network densities are measures of the development and accessibility of a road transport system. Therefore, average values of the topological measures of the network represent the respective measures for the region of interest.

17.4.3 Partitioning an urban area for micro-scale study

As discussed earlier, urban road networks are extremely complicated systems, but for in-depth investigation, the metropolitan area can be divided into a grid-like pattern [6]. The method of dividing a city area depends on the goal of the study and should be carefully planned. One of the possible approaches is to divide the study space into square grids that are consistently formed. Several instances of dividing an area of interest can be considered to determine the effect of grid size on network topology measurements. For instance, the group of strategies for dividing the area into smaller grids has been denoted as $k = \{1, 2, 3, 4\}$, and the set of grid size examples are denoted as N,

$$N = \{1{,}000\text{m}, 800\text{m}, 600\text{m}, 400\text{m}\}$$

where k is the number of individual cases of partitioning the area, and N is the appropriate size of the grid for partitioning the area.

The selection of grid size can significantly affect the numerical values of connectivity and coverage of the road networks, hence it is important to choose the effective grid size.

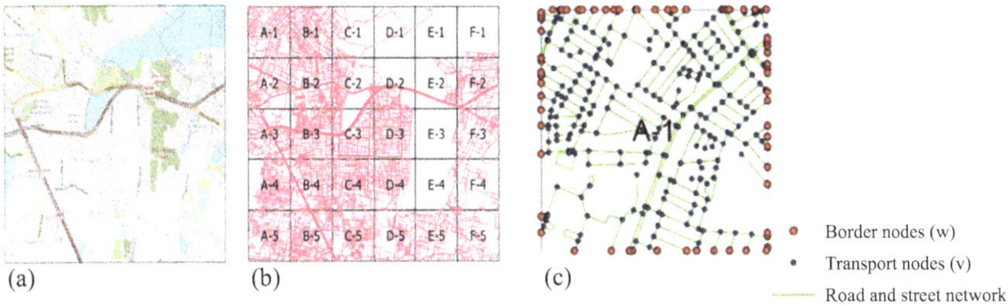

Figure 17.5 Development of graphical model of the road network for a single sampled grid. (a) Examined area, (b) division of the area into grids, road and street network, and (c) elements of the network structure in the sampled grid.

A road and street network in the center business district of Bangalore City, India is as shown in Figure 17.5(a), presuming that grids have a square-like regular shape. Hence, the area of analysis was created by overlaying a square mesh with a side of 1,000 m on a portion of the region of interest.

The position of an individual grid is described by the number of rows and columns of a matrix depicted in Figure 17.5(b) and (c), and each grid has a specific quantifiable value to represent the road network. Initially, road network for the study region are cleaned (i.e., graphic element editing and coordinate thinning are performed) and are extracted for each grid. To analyse the network structure, it is necessary to construct the graphical model that is a simplified representation of a transportation model, and then the network features of the individual grids are identified as nodes and edges. Road and street network structure in a single grid is denoted as $G(w + v, e)$, where v is set of individual transport nodes of the graph G, w is the set of individual border nodes of the graph, and e is the set of individual edges of the graph.

All three elements of a road network structure are the real representation of transportation infrastructures. The transport nodes (v) are intersection points of the road within the grid and border nodes (w) are the intersection points of road segments within a border of a grid. Since square grid patterns are more compact and symmetrical, it is possible to examine grid-wise road and street network. To estimate the topological measures of network structure and to explain the functionality of the road networks, the spatial data must comprise location information as well as network elements like the number of edges, number of nodes, and border and transport nodes.

17.4.4 Grid analysis and statistical modelling

The topological measures of road networks consist of a numerical value of every individual grid which can be compared. The range of values for each topological measure of the network can be categorised into various classes to which the calculated values are attributed. Additionally, the average values for each measure of the whole area is also calculated by simple arithmetic mean operation. This approach aims to compare the numerical values of topology measures for the different ways of partitioning the urban area in terms of various grid sizes. The findings provide insights on how the city division might affect network measurements, and further attempt to explain the relationship between grid size and topological measures that are helpful to city planners.

17.4.4.1 Assessment of variability within grids

The appropriate statistical methods can be used to analyse the variability of individual connectivity and coverage measures in terms of different grid sizes. A systematic way of partitioning the study area, subsequently calculating the topological measures of each grid and later finding the variability among the measures are important phases in transportation planning. Variance as a measure expresses the dispersion around the average values of network for different grid sizes, that is the average value outlining the center of a network measure. Out of the different connectivity and coverage measures, more consistent ones in all the grid sizes can be determined based on the variability measure such as standard deviation and coefficient of variations.

17.4.4.2 Assessment of sensitivity within grids

To study sensitivity among variables, the connectivity and coverage measures can be dependent variables (*Y*) which are the most important measures to describe the complexity of road and street networks. The size of the grid (study region) is independent variable (*X*) and it is an important feature to assess the structure of the road and street network. Prediction of the network measures (connection and coverage) with one input variable as grid size using a linear function aids in predicting the indices accurately as a function of the grid size:

$$Y = \beta_0 + \beta_1 X \tag{17.12}$$

where β_0 and β_1 are regression coefficients and the model can be built for all individual connectivity and coverage measures independently, with the slope of the regression line (β_1) representing the sensitivity of each model. Slope actually represents the predicted change in the response variable per unit change in grid size.

17.5 A case study

For the purpose of examining the spatial pattern of road transportation system, grids and the accompanying road maps were extracted for a small region of interest in Bangalore City as mentioned earlier. Euler's formula was used to calculate the number of cycles or circuits in a network of roads with the assumption that all the road networks are planar graphs ($p=1$). To determine the relevant network features, grid-based approach was executed independently. Current analysis used the grid size of 1 km × 1 km, and numerical value of each network measure were computed. The applicable equations to determine the network measures are explained in Sections 17.4.1 and 17.4.2. Finally, the estimated values of selected network measures were categorised into five classes with a comparable colour code and were composed as maps with legend. Choropleth maps are useful to illustrate the spatial pattern of a complex network structures, which means, greater the connectivity measure, greater is the connectedness of a road network for the given grid.

Geoprocessing tools in GIS software were used for building the network dataset and attribute parameters for road networks including 'e' and 'v'. Examples of sample calculation of the various connection indices are shown in Table 17.1.

Spatial distributions of all the connectivity indices varied among the grids. Each grid's variation (see Figure 17.6) indicates that the quantifiable measures of the connection indices are not uniform. Grid wise connectivity measures were derived using the total length of the road network (L), area of the wards/grids (A), total number of edges (e), and the total number of nodes (v).

In case of connectivity analysis, the results of Alpha, Beta, Gamma and GTP index of a grid size 1,000 m ($k = 1; N = 1{,}000$ m) are shown in Figure 17.6. Here, the values of Alpha index ranges from 0.38 to 0.93, low value of Beta index is 1.64 (red), and maximum value is 2.85 (green) which indicates that most of the grids have complex road networks. Gamma index ranges from 0.60 to 0.96, and GTP ranges from 1.05 to 2.95, where red has been used to signify all low value measures and green represents excellent road network connectivity.

30 grids had 1 km^2 area in the study region, however, only 4 grids had Alpha values greater than 0.81. On the other hand, 14 grids had a medium range of GTP (1.50–1.79). Here, the number of grids is higher in the medium level of connection measures across all four connectivity indices.

Grid-based maps of the other three connectivity measures such as Eta index, three-way intersection proportion and four-way intersection proportion are shown in Figure 17.7.

Network coverage measures have a similar spatial pattern over the region. Analysis of the chosen network density, edge graph density, and nodal graph density revealed that road network density values ranged from 0.85 to 32.22 km/grid, however, there are only 5 grids out of 30 grids that have low network density, indicating the possibility of natural land use pattern or less road development. The

Table 17.1 Grid area (A), number of nodes (v), number of edges (e), total length of the road network (L), and connectivity indices: (α, β, γ and GTP)

Grid	Name	A (sq. km)	v	e	L (km)	Alpha, α	Beta, β	Gamma, γ	GTP
1	A-1	1	299	730	23.068	0.728	2.441	0.819	1.628
2	A-2	1	241	586	22.399	0.725	2.432	0.817	1.640
3	A-3	1	303	728	24.202	0.709	2.403	0.806	1.583
4	A-4	1	337	816	23.372	0.717	2.421	0.812	1.593
5	A-5	1	314	706	23.547	0.631	2.248	0.754	1.406

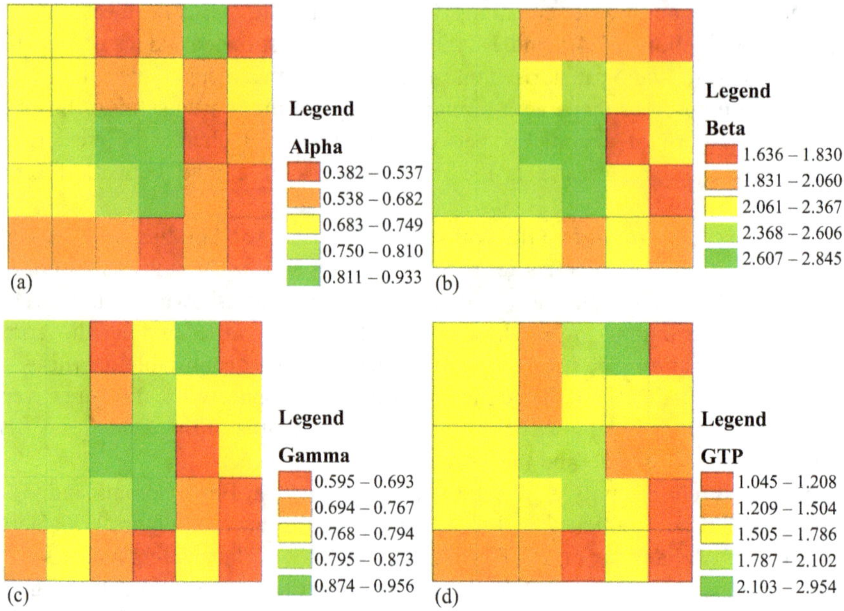

Figure 17.6 The level of connectivity indices: (a) Alpha, (b) Beta, (c) Gamma, and (d) GTP for each grid (where $k = 1$ and $N = 1{,}000$ m)

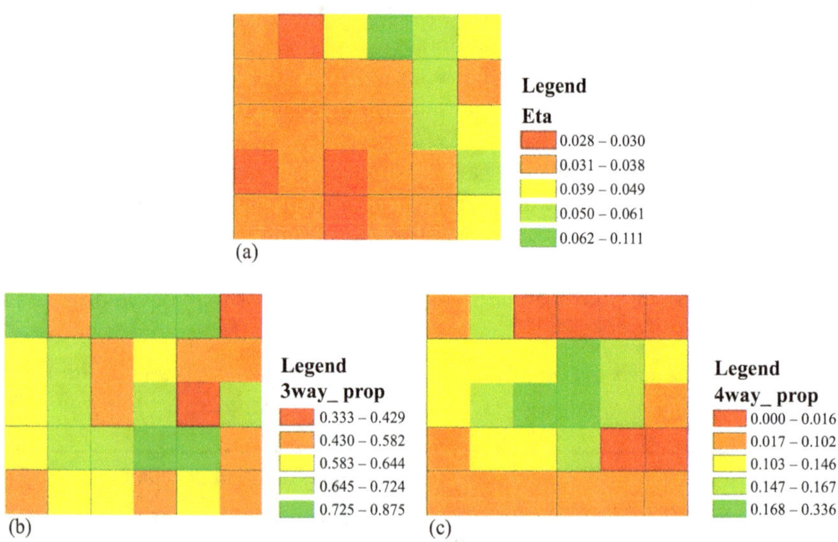

Figure 17.7 The level of connectivity indices. (a) Eta index, (b) three-way intersection proportion, and (c) four-way intersection proportion for each grid (where $k = 1$ and $N = 1{,}000$ m).

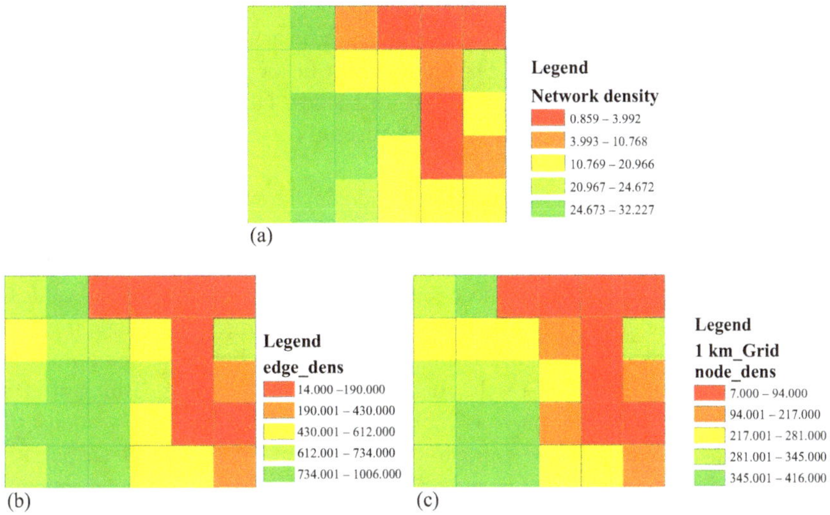

Figure 17.8 The level of coverage measures. (a) Network density, (b) edge graph density and (c) nodal graph density for each grid (where k = 1 and N = 1,000 m).

edge graph density was based on the total number of road segments present in each grid and its value ranged from 734 to 1,008; the highest edge graph density was found in only 8 grids. In contrast, the nodal graph density was built based on the total number of nodes present in a grid. The minimum number of nodes/grid is 7 and it attains a maximum of 416. There were only 5 grids that had a high degree of nodal graph density when the grid size became 1,000 m. Figure 17.8 shows the result of the calculation for coverage indices for $k = 1$ and $N = 1{,}000$ m.

17.6 Discussion and conclusion

The goal of this chapter was to discuss the impact of network structure measures on transportation. Some of the existing methodologies and applications of network measures in spatial analysis were reviewed. The network layout and connectivity of nodes and links, as well as the spatial pattern that exists within the network are all captured by the measurements of network structure. It also describes the computation process for calculating the connectivity and coverage of a transportation structure and mathematical justification for dividing the study region into regular sized grids.

Urban inequalities in transportation infrastructure distributions associated with grids were analysed to compare the performance of the surrounding grids. Results showed that the road network is not evenly spread among the grids. Moreover, the characteristics of a road network have high impact on how well the transportation system operates and how individuals choose to travel on these networks. Similar research are useful for the study of microscale models of the demand and supply for transportation in cities. Decisions regarding which measures to use for a given size of a territorial unit may be made easier as a result of the analysis of the variability and sensitivity of measures with various grids sizes. However, the mathematical descriptions and implementations are not currently available in any specialised software for computation.

Future extension of this analysis will include more network measures to be considered, for example, different modes of transportation network such as railway, bus route, metalled, and non-metalled roadways separately. Besides the selection of grid size and type, there must be more indicators pointing to the different values of measures. Likewise, analysis of the effect of socio-economic parameters, travelling and land-use variables to address urban infrastructure challenges temporally are also relevant here. Depending on the models used, a variety of measures of effectiveness can be considered, such as mobility, congestion, emissions, equity, toll revenue, transit ridership, and infrastructure maintenance needs. Transportation engineers and researchers develop and run models to predict the impact of potential projects or policies. Many other aspects related to the urban road network can be considered for urban transportation planning and the numerical values of the network metrics can be compared with the neighbouring region to evaluate the performance of the road transportation system. Planning helps to ensure that the transportation spending and policies are as effective as possible. Hence, urban sustainability challenges can be addressed more efficiently.

References

[1] Anderson, T. and Dragićević, S. (2020). Complex spatial networks: theory and geospatial applications. *Geography Compass*, *14*(9), e12502.

[2] Zhang, W., Wang, S., Tian, X., Yu, D., and Yang, Z. (2017). The backbone of urban street networks: degree distribution and connectivity characteristics. *Advances in Mechanical Engineering*, *9*(11), 1687814017742570.

[3] Sharifi, A. (2019). Resilient urban forms: a review of literature on streets and street networks. *Building and Environment*, *147*, 171–187.

[4] Kansky, K. and Danscoine, P. (1989). Measures of network structure. *FLUX Cahiers scientifiques internationaux Réseaux et Territoires*, *5*(1), 89–121.

[5] Karoń, G. and Żochowska, R. (2015). Modelling of expected traffic smoothness in urban transportation systems for ITS solutions. *Archives of Transport*, 33, 33–45.

[6] Soczówka, P., Żochowska, R., and Karoń, G. (2020). Method of the analysis of the connectivity of road and street network in terms of division of the city area. *Computation*, *8*(2), 54.

[7] Sun, C., Pei, X., Hao, J., Wang, Y., Zhang, Z., and Wong, S. C. (2018). Role of road network features in the evaluation of incident impacts on urban traffic mobility. *Transportation Research Part B: Methodological*, *117*, 101–116.

[8] Dill, J. (2004, January). Measuring network connectivity for bicycling and walking. In *83rd Annual Meeting of the Transportation Research Board*, Washington, DC (pp. 11–15).

[9] Buhl, J., Gautrais, J., Reeves, N., *et al.* (2006). Topological patterns in street networks of self-organized urban settlements. *The European Physical Journal B—Condensed Matter and Complex Systems*, *49*, 513–522.

[10] Xie, F. and Levinson, D. (2007). Measuring the structure of road networks. *Geographical Analysis*, *39*(3), 336–356.

[11] Parthasarathi, P., Hochmair, H., and Levinson, D. M. (2009). The influence of network structure on travel distance. Available at SSRN 1736326.

[12] Tsiotas, D. and Polyzos, S. (2017). The topology of urban road networks and its role to urban mobility. *Transportation Research Procedia*, *24*, 482–490.

[13] Żochowska, R. and Soczówka, P. (2018). Analysis of selected transportation network structures based on graph measures. *Zeszyty Naukowe. Transport/ Politechnika Śląska.*

[14] Sahitya, K. S. and Prasad, C. S. R. K. (2020). Evaluation of opportunity based urban road network accessibility using GIS. *Spatial Information Research*, *28*(4), 487–493.

[15] Sreelekha, M. G., Krishnamurthy, K., and Anjaneyulu, M. V. L. R. (2016). Assessment of topological pattern of urban road transport system of Calicut city. *Transportation Research Procedia*, *17*, 253–262.

[16] Thottolil, R., and Kumar, U. (2021, December). Assessment of topological pattern of road network: a case study of Bangalore City. In *2021 IEEE International India Geoscience and Remote Sensing Symposium (InGARSS)* (pp. 246–249). IEEE.

[17] Boeing, G. (2019). Urban spatial order: street network orientation, configuration, and entropy. *Applied Network Science*, *4*(1), 1–19.

Chapter 18

Future perspective and research direction in cognitive sensing technologies

G.R. Sinha[1], Bidyadhar Subudhi[2], Chih-Peng Fan[3] and Humaira Nisar[4]

Abstract

Detailed discussion of cognitive sensing technologies, their fundamentals, review and appropriate applications provide lot of insights as what would be future applications of smart and cognitive sensing. In years to come, the sensing technologies are going to dominate various industries such as driver-less cars, environment monitoring, smart transport, and smart city. In all such applications, unlimited automation, control, and monitoring shall be salient features that are expected to be realized in Metaverse, Web 3.0 and similar technologies using digital twins, digital avatars, etc. This chapter presents future direction of smart and cognitive sensing for various sectors and their sustainable growth through digital transformation and Industry 4.0.

Keywords: Future sensing; Metaverse; Research direction; Digital twins; Spectrum sensing; Performance

18.1 Future perspective of cognitive sensing

Cognitive sensor technologies, as we have discussed in previous chapters, are used in numerous applications, and the capabilities of sensitive technologies are expanding due to communication capabilities and 5G communication technologies. In the success of Industry 4.0, where the role of sensor is very important, the smart sensing is helping the industries to better performing, quality production, smart monitoring, and quality assessment. Using digital twins as creating virtual environment, real-time monitoring of industries and other systems is being done more

[1]International Institute of Information Technology Bangalore (IIITB), India
[2]Electrical Engineering, Indian Institute of Technology Goa, India
[3]Electrical Engineering, National Chung Hsing University, Taiwan
[4]Electronic Engineering, Universiti Tunku Abdul Rahman, Malaysia

effectively and efficiently. The sustainability of industry is also depending on the quality, production and the continuous improvement cycle. So, the data which is being captured in various industries and other applications with the help powerful smart sensors and the intelligent interpretation of those sensory data helps in detecting any fault or error in any part of the system. The feedback signal is sent and then the corrective measures are taken, and this is why the improved performance along with sustainability is being provided. In the area of healthcare, especially in telemedicine, medical, Internet of Things, the role of cognitive sensing is very important due to the emergence of 5G technology. The medical modalities from any person from anywhere can be captured and the powerful sensing technology is able to capture very correctly, so the appropriate diagnosis of that medical modality is made by the physicians anywhere in the world. Similarly, in the all forecast and prediction technologies, performance indicators are improving by emergence of Industry 4.0 using industrial Internet of Things (IIoTs). There is a bigger role of smart sensors to play in the industrial applications. Brain computer interface (BCI) is another exciting area that combined with robots can create wonderful impact in terms of assistive technologies for disabled persons, and also to generate the feedback signal to rehabilitate the persons who are suffering with certain brain disorder and disability. The areas which are covered by cognitive sensing are limitless. The quantum computing probably will enrich the capabilities of the data analysis captured through the sensors. The self-driving car is and its operation and performance will always be improved by bringing new sensing technologies, and modern sensors. The 5G has been already rolled out and 6G [1] is on the way and therefore the cognitive sensing has an important role to play, specially using edge devices where at each node has some understanding and control abilities.

Wi-Fi sensing [2] is moving in right direction and hopefully the standards for Wi-Fi sensing will be open by 2024 where the Wi-Fi motion using new technology can bring the revolution in the area of smart sensing by improving the capabilities of sensing technology. The Wi-Fi sensing also brings technology enabled sensing, spectrum sensing, compressed sensing together to bring more immersive experience when it comes to using augmented reality (AR), virtual reality (VR), and mixed reality (MR) for a wide range of applications such as Metaverse and Avatars. The cognitive systems which are developed using different types of Wi-Fi sensors are Wi-Fi sensing, spectrum sensing, and other technologies that would bring a different experience in terms of predictive analytics, improved reliability, and efficient localization of different objects. Home monitoring, wellness monitoring in the area of community wellness, environment monitoring, smart city implementation are going to be most exciting areas of future cognitive sensing methods.

18.1.1 Web 3.0

Web 3.0 [3] provides immersive experience to the customers in a number of applications as compared to its previous versions. Looking into the main features of Web 3.0., 3D graphics, connectivity, ubiquity, blockchain, and edge computing are salient technologies behind the success of the new web technology. In edge computing, 3D graphics, the role of smart sensors and the sensing technologies becomes very important in a number of areas related to computer games, ecommerce, geospatial applications, and so on. IoTs for Web 3.0 require edge computing to implement at different edges, and for which enhanced capabilities of sensing are needed. With the improved connectivity and using semantic web, sensing and understanding in real time needs a new paradigm of sensing methods and modern IoTs with better cognitive abilities. Thus, sensing technology is very crucial for digital transformation [4] as well where sensing technology can be considered as the backbone to understand the different types of data and its real-time assessment in digital transformation of an enterprise.

18.1.2 Industry 4.0

Cognitive manufacturing is synonymous with Industry 4.0 [5] and sometimes we call it as a smart manufacturing which further deploys huge number of IIoTs and therefore sensors and sensing technologies in Industry 4.0 are significant for sustainable production and quality in industries. Data analytics using AI gets improved if sensing method provides robust information using smart sensing technologies and also cognitive processes and operations start performing successfully. Analysis of the process environment becomes efficient and decision-making ability is also enhanced. The optimization of resources and localization of cognitive insight smoothens the process of intelligent operation of manufacturing. The sensing techniques are of huge importance that are integrated with suitable ML technique, and thus cognitive technologies are required to be equipped with IoTs, cognitive automation, artificial intelligence. Cognitive Internet of Things with cognitive abilities utilizing all modern smart sensors, Industry 4.0 can have asset performance management, process and quality improvement, resource optimization, and supply chain optimization. In process and quality improvement as well as in resource optimization and supply chain optimization, we can create some digital twins [6] of different components and the systems which are there in the industry and from some remote place, or virtual space, the optimization can be suggested both for the whole supply chain as well as resources. This particular area is going to very useful and can provide improved productivity, and also predictive technologies can help reduce the error or the faults in various parts of the industry. Industry 4.0 has four 4.0 technologies that include cyber physical systems, smart manufacturing, smart factories, cloud computing, and cognitive computing where the cognitive computing and cyber physical systems require intervention of sensing methods and artificial intelligence. In time to come, with the emergence of 5G/6G technologies, digital twin technologies will find very much relevance that can create virtual versions of real-world installations in industries, processes and their applications. Similarly, in cyber physical systems, the virtual systems can communicate and cooperate with each other in order to get better performance and productivity of the industries. The digital twin-based manufacturing using virtual prototyping and the modern sensors would be the future technologies in the areas of sensing for industry for meeting the requirement of sustainable development goals related to Industry 4.0.

18.1.3 Digital twins and avatars

Digital twins [6] as virtual representations of living and nonliving physical object and also virtual forms of persons as avatars are going to transform in the areas of customer experience. As per different studies, the future of digital twins is going to be extremely exciting and this technology and forecast of digital twin evaluation depends on communication fabric, artificial intelligence, swarm intelligence, etc. Implementing with swarm intelligence using digital twins for digital transformation-based applications shall expand the scope of digital twins and their capabilities, and this will hugely depend on sensing types and characteristics. The Meta digital twins and data that is handled by digital system employing the digital twins are of paramount importance when it comes to proper sensing and acquiring the data. The digital twins are useful in healthcare with the cognitive functionalities and also going to be playing very important role in the society and sustainable world. The future of twins and avatars provide great and immersive experience to the customers and thus areas of digital avatars provide better immersive experience using AR, VR, and MR techniques to the customer and employee experience. The role of sensors, especially the cognitive sensors is of paramount importance.

18.1.4 Metaverse

Paradigm change in sensing technology requirement and capability is demanded in Metaverse and its applications. Compatibility, interoperability, standards and future generation of sensors, sensing technologies would be required when Metaverse comes on ground reality for real-time applications. Metaverse in medicine, assistive technologies and also predictive technologies require lot of 3D imaging data and virtual environment where most significant role shall be played by AI/ML techniques and sensing methods. Interoperability and standardizations issues are pertinent due to the fact that multimodal signals and data will be involved in sampling, interpretation and data classification in Metaverse applications. Cognitive devices and cognitive Internet of Things (CIoT) will be commonly required to realize the anticipated potential of Metaverse by AI community and researchers. This also needs paradigm shift in sensing mechanism, design and development so that sensors can adapt to disruptive technologies.

18.1.5 Driverless cars and smart transport

In the area of smart transport and driverless car, the next-generation sensing technologies would include a number of salient features such as a new dimension of photonics, optical communication, optical wireless power transfer that will also include a new family of imaging and sensing technologies including modern cameras. Data systems with hyperspectral imaging and new series of computational methods, polarization sensing and imaging, Lidar technology are few important components in terms of technological requirement in smart sensing.

All these new cognitive sensing technologies for smart transport and driverless car applications, and also for other applications like environmental and remote sensing operation, sensing for security and safety, sensor-based quality control applications and many more sensor information processing applications are required. Thus, the cognitive sensing is an important area where the autonomous vehicle, new trends of driverless car and smart transport such as robo-taxi will make use of the modern sensors, sensing technologies and the analysis. The autonomous cars and robo-taxi will require a new generation of sensors that further require AI intervention for optimal operation of automotive radar-based systems in the transport. Then, 4D imaging is also one of the important areas which is supported by automotive LIDAR system, and that is also going to play major role in driverless car.

Future cars will deploy a number of MEMS sensors and in the operation of those MEMS sensors, there is limited performance, and this is why future of cognitive sensing lies in quantum sensors and quantum sensing technology. The quantum sensing will be an important area of cognitive sensing that will even create more exciting and immersive experience in the operation of self-driving cars and considered as the future sensing technique.

18.2 Research direction

As mentioned in the previous section that spectrum sensing for cognitive radio network is going to dominate in the sensing technologies for wireless IoT applications in classification and sensor related research problem-solving schemes. This requires lot of conventional spectrum sensing techniques to be updated and modified, and the spectrum sensing and the compressive sensing will also undergo a number of changes while working with cognitive radio networks. To meet this, the mathematical modeling has to be revisited for a spectrum sensing technique and to deal with the challenges involved in the spectrum sensing. There are a variety of applications related to spectrum sensing such as public safety network, smart disaster management system, smart military network system, and so on and in all such challenging applications the spectrum sensing is crucial for expanding scope of cognitive sensing.

The compressive techniques of sensing for cognitive network applications will also require optimal spectrum sensing, management, spectrum sharing, and mobility and therefore sensing technologies need to have strong mathematical background so that energy detection, feature detection and other processes involved in system modeling of cognitive model and the network are taken care of. In the similar manner, sensing technologies need to be in such a manner that can handle different types of attacks, and one such technique is recommended as a cooperative spectrum sensing that does not only handle attacks but also improves the performance of sensing and analysis, in addition to handling with security related attack in the system.

18.2.1 Edge computing, 5G, and 6G

Edge computing and its scope of implementation in a large scale for a variety of sensing and comprehension applications will be expanded more and more with emergence of 5G and 6G technologies. This obviously demands for huge number of sensors working in tandem with smart systems intelligently. The communication capability of 5G technology and then 6G technology will make computation demand faster as well, and this necessitates the smart role of sensors at edge devices in edge computing framework. The edge computing will find wide scope of applications in smart agriculture, smart city, smart environment monitoring, smart parking, smart transport, smart manufacturing, smart healthcare, tele-healthcare, tele-consultancy, and several others. In all such applications cognitive sensing has most important role to play as what kind of and what quality of data being captured by sensors, shall determine the quality of further analysis and classification for various real-time purposes such as robotics in assistive technology. The edge computing demands for novel sensors having tremendous cognitive ability that can help to perform certain amount of computation and nodes of edge network.

Working with edge computing, smart sensors and their architecture in the design of network, implementation of a smart sensing devices, edge computing models, appropriate signal processing for edge computing contribute to the success of sensing applications. The emerging cognitive sensing for e-health, data privacy in e-health, new standardization for smart sensors used for edge computing will be the new areas of sensing technologies. The edge computing has to cover a lot of geographical distribution, therefore to meet that mobility support; and to handle the low latency and heterogeneity issues, suitable changes are to be incorporated while bringing up new range of smart sensors and sensing technologies. The evaluation of sensing technology depends not only on the performance efficiency, reliability but also on privacy and security of the data which is captured through the smart sensing architecture.

18.2.2 Space technology, remote sensing, and navigation

Research in space technology, remote sensing, and navigation is breaking its new records though very exciting applications used for weather forecasting, landslide detection, land-mapping, navigation of satellites, and other UAVs and drones. Thus, in this emerging area of research as well, smart sensors in their most innovative form will be required.

18.3 Roadmap

The future of cognitive sensing and the research direction shall be focusing on the following to make it adaptive, disruptive, and robust.

- **Massive scaling**: Large-scale testbed of 5G or similar communication network shall demand for expanded potential and functionalities of sensors as well as sensing techniques to better acquire the signals for optimal understanding and use.

- **Creating knowledge and big data:** In knowledge creation of huge amount of data poses challenges in meaningful interpretation and classification and thus this thing has to be kept in mind while designing new family of sensors and their technologies.
- **Robustness**: Robustness is all time challenge in almost all signal processing applications of sensors and sensory data analysis. Thus, this is extremely important to keep the robustness in mind while working on new family of sensors and also development of new sensing techniques. In cognitive radio-enabled IoTs, spectrum sensing is recommended with robustness as one of the salient features in the sensing technology for IoT applications of cognitive radio network. Other metrics such as performance indicators, optimization techniques, and computational complexity assessment are equally important working with spectrum sensing for cognitive network. Robustness against various types of attacks, threats, and damages is also a matter of concern for which future research effort is needed in the area of sensors and their technologies.
- **Security and privacy**: Security of data and its privacy is of utmost important in sensory data, irrespective of any applications of sensors and their techniques for acquisition, analysis and classification. Innovative security solutions and protocols are therefore needed in future cognitive sensing technologies. Privacy of physical objects as well as security of virtually created objects needs much attention dealing with new series of sensor technologies for future internet and Metaverse type of applications utilizing digital twins and avatars.
- **Humans in loop:** In AR or MR, human intervention is always there and sensing technique has to come up with innovative solutions related to glitches and challenges that arise due to human in loop for sensor applications. Sensory data is acquired from various types of physical systems and also the machines; so, the sensor modeling becomes imminent while working on multi-agent system. Involving human in loop also poses challenges related to cyber physical system-based threats which again need to be seriously considered in design and development of smart and modern sensors.
- **Digitalization and digital transformation**: For efficient operation and performance, right from converting data into digital; then digitizing the processes and operations; and finally digitalizing the entire system or institution need lot of sensing data analysis where the quality of data matters a lot and thus smart sensing shall play major role in future for digital transformation of various organizations for improved performance, productivity, and learning.

References

[1] Z. Wang, Y. Du, K. Wei, *et al.*, "Vision, application scenarios, and key technology trends for 6G mobile communications," *Sci. China Inf. Sci.*, vol. 65, no. 5, pp. 1–27, 2022, doi: 10.1007/s11432-021-3351-5.

[2] E. Summary, "Future directions in cognitive radio network research NSF workshop report," *Network*, June, pp. 1–40, 2009, http://scholar.google.com/

scholar?hl=en&btnG=Search&q=intitle:Future+Directions+in+Cognitive+Radio+Network+Research#4

[3] S.K. Bhoi, S.K. Panda, K.K. Jena, *et al.*, "IoT-EMS: an Internet of Things based environment monitoring system in volunteer computing environment," *Intell. Autom. Soft Comput.*, vol. 32, no. 3, pp. 1493–1507, 2022, doi: 10.32604/IASC.2022.022833.

[4] O. Vermesan and J. Bacquet, "Internet of things cognitive transformation technology research trends and applications," in *Cognitive Hyperconnected Digital Transformation. Internet of Things Intelligence Evolution*, River Publishers, pp. 17–95, 2017, doi: 10.1201/9781003337584-3.

[5] I. Zolotová, P. Papcun, E. Kajáti, M. Miškuf, and J. Mocnej, "Smart and cognitive solutions for operator 4.0: laboratory H-CPPS case studies," *Comput. Ind. Eng.*, vol. 139, 2020, doi: 10.1016/j.cie.2018.10.032.

[6] P. Angin, M. H. Anisi, F. Göksel, C. Gürsoy, and A. Büyükgülcü, "Agrilora: a digital twin framework for smart agriculture," *J. Wirel. Mob. Networks, Ubiquitous Comput. Dependable Appl.*, vol. 11, no. 4, pp. 77–96, 2020, doi: 10.22667/JOWUA.2020.12.31.077.

Index